Spin in Particle Physics

Motivated by recent dramatic developments in the field, this book provides a thorough introduction to spin and its role in elementary particle physics. Starting with a simple pedagogical introduction to spin and its relativistic generalization, the author successfully avoids the obscurity and impenetrability of traditional treatments of the subject. The book surveys the main theoretical and experimental developments of recent years, as well as discussing exciting plans for the future. Emphasis is placed on the importance of spin-dependent measurements in testing QCD and the Standard Model.

This book will be of value to graduate students and researchers working in all areas of quantum physics and particularly in elementary particle and high energy physics. It is suitable as a supplementary text for graduate courses in theoretical and experimental particle physics.

ELLIOT LEADER is Emeritus Professor in The University of London and Visiting Professor at Imperial College, London. He received his Ph.D. from the University of Cambridge and in 1967 became Professor of Theoretical Physics at Westfield College, London. In 1984 he took up the Chair of Theoretical Physics at Birkbeck College, London, where he worked for 16 years. Professor Leader has done research in universities and laboratories throughout the world, including CERN, Brookhaven, Fermilab, California Institute of Technology and the Lawrence Radiation Laboratory, Berkeley. He has published numerous papers and review articles, and is the joint author of two previous books. *An Introduction to Gauge Theories and the 'New Physics'*, CUP (1982) and *An Introduction to Gauge Theories and Modern Particle Physics*, CUP (1996), both written with Enrico Predazzi.

CAMBRIDGE MONOGRAPHS ON PARTICLE PHYSICS, NUCLEAR PHYSICS AND COSMOLOGY

15

General Editors: T. Ericson, P. V. Landshoff

1. K. Winter (ed.): *Neutrino Physics*
2. J. F. Donoghue, E. Golowich and B. R. Holstein: *Dynamics of the Standard Model*
3. E. Leader and E. Predazzi: *An Introduction to Gauge Theories and Modern Particle Physics, Volume 1: Electroweak Interactions, the 'New Particles' and the Parton Model*
4. E. Leader and E. Predazzi: *An Introduction to Gauge Theories and Modern Particle Physics, Volume 2: CP-Violation, QCD and Hard Processes*
5. C. Grupen: *Particle Detectors*
6. H. Grosse and A. Martin: *Particle Physics and the Schrödinger Equation*
7. B. Andersson: *The Lund Model*
8. R. K. Ellis, W. J. Stirling and B. R. Webber: *QCD and Collider Physics*
9. I. I. Bigi and A. I. Sanda: *CP Violation*
10. A. V. Manohar and M. B. Wise: *Heavy Quark Physics*
11. R. Frühwirth, M. Regler, R. K. Bock, H. Grote and D. Notz: *Data Analysis Techniques for High-Energy Physics*, second edition
12. D. Green: *The Physics of Particle Detectors*
13. V. N. Gribov and J. Nyiri: *Quantum Electrodynamics*
14. K. Winter (ed.): *Neutrino Physics, second edition*
15. E. Leader: *Spin in Particle Physics*

SPIN IN PARTICLE PHYSICS

ELLIOT LEADER
Imperial College, London

PUBLISHED BY THE PRESS SYNDICATE OF THE UNIVERSITY OF CAMBRIDGE
The Pitt Building, Trumpington Street, Cambridge, United Kingdom

CAMBRIDGE UNIVERSITY PRESS
The Edinburgh Building, Cambridge CB2 2RU, UK
40 West 20th Street, New York, NY 10011–4211, USA
10 Stamford Road, Oakleigh, VIC 3166, Australia
Ruiz de Alarcón 13, 28014 Madrid, Spain
Dock House, The Waterfront, Cape Town 8001, South Africa

http://www.cambridge.org

First published 2001

Printed in the United Kingdom at the University Press, Cambridge

Typeface 11/13pt Times *System* TEX [UPH]

A catalogue record for this book is available from the British Library

Library of Congress Cataloguing in Publication data

ISBN 0 521 35281 9 hardback

To Boika

Contents

Preface

Spin is an essential and fascinating complication in the physics of elementary particles. The spin of a particle is a quantum mechanical attribute. Questions about the spin dependence of reactions therefore tend to probe the underlying theoretical structures very deeply.

Spin plays a dramatic Jekyll and Hyde rôle in the theatre of elementary particle physics, acting sometimes as the harbinger of the demise of a current theory, sometimes as a powerful tool in the confirmation and verification of such a theory.

Witness, for example, the parameters of the Standard Model. The world's most precise measurement of the Weinberg angle,

$$\sin^2 \theta_{\mathrm{W}}^{\mathrm{eff}} = 0.23061 \pm 0.00047,$$

comes from the SLD experiment at Stanford, where the use of a polarized electron beam turns out to be equivalent to gaining a factor of 25 in the statistics compared with the unpolarized situation. Or take the LEP collider at CERN. Even though there has never been a serious spin programme there, nonetheless the most precise determination of the beam energy comes from a measurement of the resonant depolarization of the beams. And spin measurements have played a key rôle in elucidating the structure of the weak interactions and in demonstrating the $\mathrm{V}-\mathrm{A}$ form of the weak Lagrangian, and several exquisite and delicate experiments (e.g. the parity-violating optical rotation in bismuth and the longitudinal polarization asymmetry in electron–proton scattering) have had a profound effect upon our fundamental view of the electroweak interaction.

On the 'destructive' side witness the theory of J/Ψ production in hadronic collisions. Measured cross-sections were long ago found to be more than an order of magnitude larger than the predictions of the colour-singlet QCD calculations. So colour-octet enhancement was introduced, thereby apparently providing a successful theory of J/Ψ production. Now it turns out from more refined measurements, wherein the state of polarization of the J/Ψ particles is determined, that there is a serious disagreement between theory and experiment.

On a longer time scale take the case of Regge pole theory. There, an entire and beautiful theoretical structure, highly successful on many fronts, was severely shaken in the face of an accumulating mass of spin-dependent data in contradiction with its predictions.

Spin, because it has no classical correspondence limit to aid our intuition, has tended to be regarded with trepidation and to be seen as surrounded by dangerous pitfalls epitomized by the Thomas precession, which is always mentioned, but rarely explained, in textbooks on quantum mechanics. Indeed there is an unconscious element of witchcraft in the oft found statement that a purely relativistic effect produces a 50% correction to the calculation of the $\mathbf{L} \cdot \mathbf{S}$ coupling in a hydrogenic atom!

Our opening sentence was inspired by a much loved slogan of the 1960s that 'spin is an inessential complication', a view that lent some practical relief in wrestling with the analytic properties of scattering amplitudes and the Mandelstam representation; this was an approach that seemed to offer, for the first time, the possibility of significant results in strong interaction theory. But here too later developments demonstrated clearly that spin could not be ignored and that the high energy behaviour of Feynman diagrams is much influenced by the spin of the virtual particles.

During the 1970s and early 1980s spin physics drifted into a relatively tranquil state of activity, from which it was rudely awakened in 1987 by the extraordinary results of the European Muon Collaboration's experiment, at CERN, on deep inelastic lepton–hadron scattering, using a longitudinally polarized lepton beam on a longitudinally polarized target. Interpreted in simple parton model terms the experiment implied, loosely speaking, that the sum of the spins carried by the quarks in a proton added up to only about one eighth of the proton's spin – a most counter-intuitive result.

The EMC publication became the most-cited experimental paper in the field for the following three years and catalysed an enormous theoretical effort to re-examine, at a more fundamental level, the whole theory of spin effects in deep inelastic scattering. Once again it was found that the explanation of spin-dependent phenomena poses a more profound challenge to a theory than the mere prediction of event rates. The theory of the spin-dependent structure function $g_1(x)$ is much more subtle than expected in the simple parton model and is linked to a deep aspect of field theory, the axial anomaly. And the structure function $g_2(x)$ turns out to have no explanation at all in the simple parton model and requires essential field-theoretic generalizations of the parton model.

The EMC experiment also stimulated massive experimental programmes at SLAC, CERN and DESY, which, in turn, have stimulated the major contemporary experiments, COMPASS at CERN, HERMES at HERA and RHIC, which has just come into operation at Brookhaven.

The information gleaned from decades of unpolarized deep inelastic scattering experiments has played a seminal rôle in our understanding of the internal structure of hadrons and in the testing of certain aspects of quantum chromodynamics. The depth and breadth of this information owes much to the fact that unpolarized deep inelastic scattering can be studied using both charged lepton beams ($e^{\pm}, \mu^{\pm}$) and neutral ones ($\nu_e, \bar{\nu}_e, \nu_\mu, \bar{\nu}_\mu$), the latter requiring gigantic kilotonne targets. The polarized case, by comparison, suffers from the lack of neutrino data – one does not know how to polarize a battleship! But, most extraordinary, it now appears that it may be possible to construct a neutrino factory, based upon a muon storage ring, that produces neutrino fluxes 10^3 or 10^4 times greater than ever before, thus making polarized targets feasible. With this, one can contemplate a new era of polarized deep inelastic scattering, with profound implications for our understanding of the internal spin structure of hadrons.

In purely hadronic physics, too, there are tantalizing questions regarding spin dependence. There exists a whole array of semi-inclusive experiments like $p^{\uparrow}p \to \pi X$, with a transversely polarized proton beam or target, or $pp \to$ hyperon $+ X$, with an unpolarized initial state in which huge hyperon spin asymmetries or polarizations – at the 30%–40% level! – are observed. These experiments are very hard to explain within the framework of QCD. The asymmetries all vanish at the partonic level and one has to invoke soft, non-perturbative mechanisms. All such mechanisms predict that the asymmetries must die out as the momentum transfer increases, yet there is no sign in the present data of such a decrease.

In exclusive reactions like $pp \to pp$ the disagreement between the data on the analysing power at large momentum transfer and the naive QCD asymptotic predictions is even more severe, but here at least there is an escape clause: the theory of exclusive reactions in QCD is horrendously difficult.

On the practical side, the technology of spin measurements has improved dramatically over the past few years. Improvements in polarized sources suggest that proton beams of almost 100% polarization, and with nearly the same intensity as present-day unpolarized beams, will eventually be available. Polarized-target construction is also improving. A highly successful polarized gas cell is in operation in the circulating electron beam at HERA. Experiments using a polarized gas-jet target in a circulating proton beam have been carried out. Polarized electrons and positrons in e^+e^- colliders are commonplace.

Our aim in this book is threefold.

(1) We hope to offer a simple pedagogical treatment of spin in relativistic physics that strips it of its unnecessary mystery. Our approach, based upon the helicity formalism, leads to a unified general treatment for arbitrary exclusive and inclusive reactions at a level that, we hope, should make it of interest to both theorists and experimentalists.
(2) While admitting a lack of expertise in the matter, we have tried, with the help and advice of experimental colleagues, to present and explain some of the absolutely dramatic achievements on the experimental side of spin physics, a continuing endeavour which seems to be part science, part art.
(3) We wish to highlight the importance of spin-dependent measurements in testing QCD and in providing a highly refined probe of the structure of the Standard Model of electroweak interactions. We survey the rich and challenging physics results that have emerged from the major spin-physics experiments of the past few years, EMC and SMC at CERN, E142, E143, E154 and E155 at SLAC, and HERMES at HERA. And we discuss some of the exciting physics that will be explored in the new generation of experiments, COMPASS at CERN and RHIC-SPIN at the RHIC collider at Brookhaven. RHIC will be unique, exploring a formerly undreamed-of regime of spin physics, with its colliding beams of polarized 250 GeV protons.

Looking further ahead, the HERA-$\vec{\mathrm{N}}$ project to polarize the proton beam at HERA would provide a marvellous facility to explore an entirely new regime in polarized deep inelastic lepton–hadron scattering and would, with a fixed polarized nucleon target, offer an experimental set-up beautifully complementary to RHIC in terms of the reactions it could study with high efficiency. We can only hope that a positive decision will be taken to proceed with the project.

In the appendices we have gathered together a large number of useful results, e.g. on the representations of the rotation and Lorentz groups, on Dirac spinors and matrix elements and various representations of the γ-matrices, on the Feynman rules for QCD and on the linearly independent helicity amplitudes and spin-dependent observables for several reactions.

Acknowledgements

I am greatly indebted to a group of colleagues who share my belief in the excitement and importance of spin-dependent measurements in elementary particle physics and from whose advice and expertise I have often benefited: Xavier Artru, Mauro Anselmino, Daniel Boer, Elena Boglione, Claude Bourrely, Gerry Bunce, Nigel Buttimore, Don Crabb,

Tolya Efremov, Thomas Gehrmann, Sergey Goloskokov, Rainer Jakob, Alan Krisch, Peter Kroll, Yousef Makdisi, Gerhard Mallot, Brian Montague, Piet Mulders, Francesco Murgia, Aldo Penzo, Phil Ratcliffe, Oleg Selyugin, Jacques Soffer, Dimiter Stamenov, Oleg Teryaev, Larry Trueman and Werner Vogelsang. In particular the earlier chapters of this book owe much to work done in collaboration with Claude Bourrely and Jacques Soffer.

I am grateful to Elena Boglione for help with diagrams and numerical computations, and to Philip Burrows, Jim Clendenin, Michel Düren, Jim Johnson, Jean-Pierre Koutchouk, Wolfang Lorenzon, Livio Piemontese and Morris Swartz for information about and diagrams of experimental apparatus.

Finally I wish to thank Pasquale Iannelli for his efficient typing of my not always legible manuscript.

Notational conventions

Units

Natural units $\hbar = c = 1$ are used throughout. For the basic unit of charge we use the *magnitude* of the charge of the electron: $e > 0$.

Relativistic conventions

Our notation generally follows that of Bjorken and Drell (1964), in *Relativistic Quantum Mechanics.*

The metric tensor is

$$g_{\mu\nu} = g^{\mu\nu} = \begin{pmatrix} 1 & 0 & 0 & 0 \\ 0 & -1 & 0 & 0 \\ 0 & 0 & -1 & 0 \\ 0 & 0 & 0 & -1 \end{pmatrix}.$$

Space–time points are denoted by the contravariant 4-vector x^μ ($\mu = 0, 1, 2, 3$), where

$$x^\mu = (t, \mathbf{x}) = (t, x, y, z),$$

and the 4-momentum vector for a particle of mass m is

$$p^\mu = (E, \mathbf{p}) = (E, p_x, p_y, p_z),$$

where

$$E = \sqrt{\mathbf{p}^2 + m^2}.$$

Using the equation for the metric tensor, the scalar product of two 4-vectors A, B is defined as

$$A \cdot B = A_\mu B^\mu = g_{\mu\nu} A^\mu B^\mu = A^0 B^0 - \mathbf{A} \cdot \mathbf{B}.$$

γ-matrices
The γ matrices for spin-1/2 particles satisfy

$$\gamma^\mu \gamma^\nu + \gamma^\nu \gamma^\mu = 2g^{\mu\nu}$$

and we use a representation in which

$$\gamma^0 = \begin{pmatrix} I & 0 \\ 0 & -I \end{pmatrix}, \qquad \gamma^j = \begin{pmatrix} 0 & \sigma_j \\ -\sigma_j & 0 \end{pmatrix}, \quad j = 1, 2, 3,$$

where σ_j are the usual Pauli matrices. We define

$$\gamma^5 = \gamma_5 = i\gamma^0\gamma^1\gamma^2\gamma^3 = \begin{pmatrix} 0 & I \\ I & 0 \end{pmatrix}.$$

In this representation one has, for the transpose of the γ-matrices,

$$\gamma^{jT} = \gamma^j \qquad \text{for } j = 0, 2, 5,$$

but

$$\gamma^{jT} = -\gamma^j \qquad \text{for } j = 1, 3.$$

For the hermitian conjugates one has

$$\gamma^{0\dagger} = \gamma^0, \qquad \gamma^{5\dagger} = \gamma^5,$$

but

$$\gamma^{j\dagger} = -\gamma^j \qquad \text{for } j = 1, 2, 3.$$

The combination

$$\sigma^{\mu\nu} \equiv \frac{i}{2}[\gamma^\mu, \gamma^\nu]$$

is often used.

The scalar product of the γ matrices and any 4-vector A is defined as

$$\not{A} \equiv \gamma^\mu A_\mu = \gamma^0 A^0 - \gamma^1 A^1 - \gamma^2 A^2 - \gamma^3 A^3.$$

For further details and properties of the γ-matrices see Appendix A of Bjorken and Drell (1964).

Spinors and normalization
The particle spinors u and the antiparticle spinors v, which satisfy the Dirac equations

$$(\not{p} - m)u(p) = 0$$
$$(\not{p} + m)v(p) = 0$$

respectively, are related by

$$v = i\gamma^2 u^*$$
$$\bar{v} = -iu^T \gamma^0 \gamma^2$$

where $\bar{v} \equiv v^\dagger \gamma^0$; similarly $\bar{u} \equiv u^\dagger \gamma^0$.

Note that our spinor normalization differs from Bjorken and Drell. We utilize

$$u^\dagger u = 2E, \qquad v^\dagger v = 2E,$$

the point being that this normalization can be used equally well for massive fermions and for neutrinos. For a massive fermion or antifermion the above implies

$$\bar{u}u = 2m, \qquad \bar{v}v = -2m.$$

Cross-sections

With our normalization the cross-section formula (B.1) of Appendix B in Bjorken and Drell (1964) holds for both mesons and fermions, massive or massless.

Fields

Often a field such as $\psi_\mu(x)$ for the muon is simply written $\mu(x)$ or just μ if there is no danger of confusion.

In fermion lines in Feynman diagrams the arrow indicates the direction of flow of *fermion number*.

Group symbols and matrices

In dealing with the electroweak interactions and QCD the following symbols often occur.

- n_f is the number of flavours.
- N specifies the gauge group $SU(N)$. Note that $N = 3$ for the colour gauge group QCD.
- The Pauli matrices are written either as σ_j or τ_j ($j = 1, 2, 3$).
- The Gell-Mann $SU(3)$ matrices are denoted by λ^a ($a = 1, \ldots, 8$).
- For a group G with structure constants f_{abc} one defines $C_2(G)$ via

$$\delta_{ab} C_2(G) \equiv f_{acd} f_{bcd}$$

and one writes

$$C_A \equiv C_2[SU(3)] = 3.$$

If there are n_f multiplets of particles, each multiplet transforming according to some representation R under the gauge group, wherein the group generators are represented by matrix $\mathbf{t}^a$, then $T(R)$ is defined by

$$\delta_{ab} T(R) \equiv n_f \; Tr \; (\mathbf{t}^a \mathbf{t}^b).$$

For $SU(3)$ and the triplet (quark) representation one has $\mathbf{t}^a = \lambda^a/2$ and

$$T \equiv T(SU(3);\ \text{triplet}) = \tfrac{1}{2} n_f.$$

For the above representation R one defines $C_2(R)$ analogously to $C_2(G)$ via

$$\delta_{ij}C_2(R) \equiv t^a_{ik}t^a_{kj}.$$

For $SU(3)$ and the triplet representation one has

$$C_F \equiv C_2(SU(3); \text{triplet}) = \tfrac{4}{3}.$$

Colour sums in weak and electromagnetic currents
Since the weak and electromagnetic interactions are 'colour-blind' the colour label on a quark field is almost never shown explicitly when dealing with electroweak interactions. In currents involving quark field operators a *colour sum is always implied.* For example, the electromagnetic current of a quark of flavour f and charge Q_f (in units of e) is written

$$J^\mu_{\text{em}}(x) = Q_f\bar{q}_f(x)\gamma^\mu q_f(x)$$

but if the colour of the quark is labelled j ($j = 1, 2, 3$) then what is implied is

$$J^\mu_{\text{em}}(x) = Q_f \sum_{\substack{\text{colours}\\ j}} \bar{q}_{f_j}(x)\gamma^\mu q_{f_j}(x).$$

Subscripts referring to the laboratory frame (Lab)
Normally a subscript upper-case 'L' is used, e.g. p_{L}. However, sometimes the subscript 'Lab' is used, for further clarification.

1
Spin and helicity

Traditionally, in textbooks on quantum mechanics, spin is introduced via an idealized Stern–Gerlach experiment in which a non-relativistic beam of silver atoms passes through an inhomogeneous magnetic field. Each atom is treated as a single valence electron of charge $-e$ in an s-state. The subsequent splitting of the beam into two indicates the two-valuedness of s_z, which is related to the value $1/2$ for s, and the magnitude of the splitting shows that the magnetic moment $\boldsymbol{\mu}$ is related to $\mathbf{s}$ by

$$\boldsymbol{\mu} = -\frac{e}{mc}\mathbf{s},$$

the proportionality factor (the gyromagnetic ratio) being twice as big as the factor that classically gives the magnetic moment due to the orbital angular momentum of a point charge.

Historically, however, it seems that the early Stern–Gerlach experiments, begun in 1922, had no influence at all upon the discovery of spin, simply because they were too imprecise. Rather, the concept of spin appeared after a long and tedious battle to understand the splitting patterns and separations in line spectra. Several people had for various reasons discussed classical models of rotating charge distributions but Kronig, in 1924, was the first to show that an electron with spin $1/2$ would explain the pattern of what we would today call $\mathbf{L} \cdot \mathbf{S}$ splitting, as well as anomalies in the Zeeman effect. He realized, though, that the gyromagnetic ratio $(-e/mc)$ needed for the latter would give $\mathbf{L} \cdot \mathbf{S}$ splittings twice as big as those observed. It is said that Pauli expressed his negative reaction to Kronig's idea with such vehemence that Kronig never published his work (Mehra and Rechenberg, 1982). Soon thereafter, in 1925, the same idea occurred to Uhlenbeck and Goudsmit (1925), who proceeded to a detailed analysis of the splittings, concluding at first that everything worked beautifully, but then becoming aware, as a consequence of a comment by Heisenberg, of the factor-of-2 inconsistency mentioned above.

Some months later Thomas demonstrated that a careful relativistic treatment produced exactly the factor of one half needed to bring about agreement between the theory of $\mathbf{L}\cdot\mathbf{S}$ splitting and experiment (Thomas, 1926).

In this work appears for the first time the infamous 'Thomas precession', which is mentioned, yet almost never explained, in all textbooks on quantum mechanics. We shall return to it later, but we should like, immediately, to demistify one aspect of it. It is usually said that relativistic effects produce a factor of one half. Now that would indeed be mysterious! What is forgotten is the fact that the $\mathbf{L}\cdot\mathbf{S}$ coupling is itself a relativistic effect. By means of a Lorentz transformation, we can understand that the electron, moving through the Coulomb field of the nucleus, sees a magnetic field in its rest frame. So the Thomas result is simply a correction to an already intrinsically relativistic effect.

1.1 Spin and rotations in non-relativistic quantum mechanics

In non-relativistic quantum mechanics the spin of a particle is introduced as an additional rotational degree of freedom. Analogously to orbital angular momentum one introduces three *spin operators*

$$\hat{\mathbf{s}} \equiv (\hat{s}_x, \hat{s}_y, \hat{s}_z);$$

the *spin states* $|sm\rangle$ are the simultaneous eigenstates of the commuting operators $\hat{\mathbf{s}}^2$ and $\hat{s}_z$, with eigenvalues $s(s+1)$ and m respectively. The *spin* s of the particle can be zero or a positive integer or half integer, while m can take values $-s \le m \le s$ in unit steps. The quantity m is referred to as the 'z-component of the spin'.

The three spin operators $\hat{s}_j$ satisfy the usual angular momentum commutation relations

$$\left[\hat{s}_j, \hat{s}_k\right] = i\epsilon_{jkl}\hat{s}_l. \tag{1.1.1}$$

For a free particle the spin degree of freedom is totally decoupled from the usual kinematic degrees of freedom, and this fact is implemented by writing the state vector in the form of a product, one factor referring to the usual degrees of freedom and the other to the spin degree of freedom. Thus, for a particle of momentum $\mathbf{p}$,

$$|\mathbf{p}; sm\rangle = |\mathbf{p}\rangle \otimes |sm\rangle \tag{1.1.2}$$

or, equivalently, for the wave function,

$$\psi_{\mathbf{p};sm}(\mathbf{x}) = \varphi_{\mathbf{p}}(\mathbf{x})\eta_{(m)} \tag{1.1.3}$$

where $\eta_{(m)}$ is a $(2s+1)$-component *spinor* and $\varphi_{\mathbf{p}}(\mathbf{x})$ is a standard Schrödinger wave function.

Since the labelling of the above spin states uses $m = \hat{s}_z$ and therefore makes reference to 'the z-direction' it is tacitly assumed that we are working in a well-defined, fixed coordinate reference system with origin O.

We wish now to discuss the effect of rotations upon the spin states. To begin with we recall the well-known rules for ordinary vectors. We shall denote by r the *physical operation* of a rotation. Thus, if we say that an object is rotated by e.g. $r_z(\theta)$, where θ is positive, then we mean that we are to physically push that object around the Z-axis through an angle θ in the sense of a right-hand screw advancing along OZ.

If we apply r to a given three-dimensional vector $\mathbf{A}$ we shall call the resultant rotated vector $r\mathbf{A}$ or $\mathbf{A}^r$. The action we have described is often referred to in the literature as the 'active' point of view as distinct from the 'passive' one, in which the axis system is rotated. We think that this is a confusing nomenclature. *All* our rotations *act* as described in the previous paragraph and if we wish to rotate axes we shall simply state that r acts on the coordinate axes.

The components of the rotated vector are related to the components A_i of $\mathbf{A}$ by

$$(r\mathbf{A})_i \equiv \mathbf{A}_i^r = R_{ij}A_j \tag{1.1.4}$$

where the 3×3 matrix R with elements R_{ij} depends, of course, on r. Strictly speaking, we should write it as $R(r)$. Sometimes it is convenient to write the components A_i in the form of a column vector

$$A = \begin{pmatrix} A_x \\ A_y \\ A_z \end{pmatrix}, \tag{1.1.5}$$

in which case (1.1.4) can be written in matrix notation as

$$A^r = RA. \tag{1.1.6}$$

As an example, if $r = r_y(\theta)$ then

$$R\left[r_y(\theta)\right] = \begin{pmatrix} \cos\theta & 0 & \sin\theta \\ 0 & 1 & 0 \\ -\sin\theta & 0 & \cos\theta \end{pmatrix}. \tag{1.1.7}$$

For a tensor T, say of rank 2, the components of the rotated tensor T^r will be given by

$$T_{ij}^r = R_{ik}R_{jm}T_{km} \tag{1.1.8}$$

with obvious generalization to tensors of higher rank. It should be noted that tensors of rank ≥ 2 do not transform *irreducibly* under rotations. (The irreducible representations of the rotation group are discussed briefly in Appendix 1.)

Often one wishes to utilize a set of three orthogonal unit '*basis vectors*' $\mathbf{e}_{(i)}$ along the three coordinate axes. If we rotate one of them, say $\mathbf{e}_{(j)}$, the n components of $\mathbf{e}^r_{(j)}$ will be related to those of $\mathbf{e}_{(j)}$ by (1.1.4). But we can also consider $\mathbf{e}^r_{(j)}$ as a linear superposition of the $\mathbf{e}_{(i)}$, and one easily shows that

$$\mathbf{e}^r_{(j)} = R_{ij}\mathbf{e}_{(i)} = \left(\mathbf{R}^T\right)_{ji}\mathbf{e}_{(i)} \tag{1.1.9}$$

where R^T is the transpose of the matrix R. (Recall that for rotations R is orthogonal i.e. $R^T R = RR^T = I$.)

Note that whereas R appears in (1.1.4) it is R^T that occurs in (1.1.9).

We come now to the physical rôle of rotations. We are interested in the relationship between the descriptions that different observers give to the same physical phenomenon. Let $\mathbf{A}$ be a fixed vector, which observer O in our fundamental reference system S describes as having components A_j. Thus

$$\mathbf{A} = \sum_j A_j \mathbf{e}_{(j)} \tag{1.1.10}$$

Let O^r be an observer using a reference system S^r that has been rotated from S by a rotation r. Using the basis vectors $\mathbf{e}^r_{(l)}$ the observer describes $\mathbf{A}$ as having components $(A_l)_{S^r}$. Thus

$$\mathbf{A} = \sum_l (A_l)_{S^r}\, \mathbf{e}_{(l)} \tag{1.1.11}$$

and via (1.1.9) one finds, using $[R(r)]^{-1} = R(r^{-1})$, that

$$(A_i)_{S^r} = R_{ij}(r^{-1})A_j. \tag{1.1.12}$$

Although slightly misleading it is convenient to abbreviate (1.1.12) in the form

$$(A)_{S^r} = r^{-1}A. \tag{1.1.13}$$

In summary, if the *reference system* is rotated by r then the components of a *fixed vector*, as described in S^r and in S, are related via $R(r^{-1})$, in contradistinction to (1.1.4) in which R is shorthand for $R(r)$.

Spin-s spinors are dealt with in complete analogy to the above. We introduce $2s+1$ unit basis spinors $\eta_{(m)}$, where

$$\eta_{(s)} = \begin{pmatrix} 1 \\ 0 \\ 0 \\ \vdots \\ 0 \\ 0 \end{pmatrix}, \qquad \eta_{(s-1)} = \begin{pmatrix} 0 \\ 1 \\ 0 \\ \vdots \\ 0 \\ 0 \end{pmatrix}, \qquad \ldots, \qquad \eta_{(-s)} = \begin{pmatrix} 0 \\ 0 \\ 0 \\ \vdots \\ 0 \\ 1 \end{pmatrix};$$

the $\eta_{(s)}$ represent eigenstates of $\hat{s}_z$. We write for a general spinor

$$\chi = \sum_m \chi_m \eta_{(m)}; \tag{1.1.14}$$

The numbers χ_m are the 'components' of χ. The components $(\chi_m)_{S^r}$ attributed to the spinor χ in the rotated reference frame S^r are related to χ_m analogously to (1.1.12):

$$(\chi_i)_{S^r} = \mathscr{D}^{(s)}_{ij}(r^{-1})\chi_j \tag{1.1.15}$$

where the matrices $\mathscr{D}^{(s)}(r)$ are the $(2s+1)$-dimensional representation matrices of the rotations r. (See Appendix 1; recall that the $\mathscr{D}$ are unitary matrices, i.e. $\mathscr{D}^\dagger\mathscr{D} = 1$.) By analogy with the inverse of (1.1.9) we have

$$\eta_{(m)} = \mathscr{D}^{(s)}_{m'm}(r^{-1})\eta^r_{(m')}. \tag{1.1.16}$$

The physical interpretation of (1.1.16) is that the state described by observer O in the frame S as $\eta_{(m)}$ is described by the rotated observer O' as a superposition of the states $\eta^r_{(m')}$.

Because of its importance we restate this in more general terms. If an observer O with reference system S sees a spin s particle in a state $|sm\rangle$ then the observer O^r whose reference frame S^r is rotated from S by the rotation r describes the state of the particle as $|sm\rangle_{S^r}$, where

$$|sm\rangle_{S^r} = \mathscr{D}^{(s)}_{m'm}(r^{-1})|sm'\rangle. \tag{1.1.17}$$

It is implicit in (1.1.17) that the states on the right-hand side are the $|sm\rangle$ of O^r.

Although it is not simple to see what we mean by physically rotating a spinor, by analogy with the vector case we shall talk about the active rotation of a state $|sm\rangle$ to $|sm\rangle^r$. Comparing with eqn (1.1.9) for the vector case, we shall interpret $|sm\rangle^r$ as given by

$$|sm\rangle^r = \mathscr{D}^{(s)}_{m'm}(r)|sm'\rangle. \tag{1.1.18}$$

It is very convenient in quantum mechanics to represent the effect of an operation by an *operator* acting directly on the state vectors. Thus we rewrite (1.1.18) in the form

$$|sm\rangle^r = U(r)|sm\rangle \tag{1.1.19}$$

where $U(r)$ is the *operator* representing the rotation r.

From (1.1.18) and (1.1.19) follows the well-known relation

$$\mathscr{D}^{(s)}_{m'm}(r) = \langle sm'|U(r)|sm\rangle. \tag{1.1.20}$$

In this operator notation (1.1.17) becomes

$$|sm\rangle_{S^r} = U(r^{-1})|sm\rangle. \tag{1.1.21}$$

In the case of spin 1/2, the spin operators $\hat{s}_j$ when acting on the two-dimensional spinors $\chi^{1/2}$ are represented by the set $\boldsymbol{\sigma}/2$ of 2×2 hermitian matrices $\sigma_j/2$, the σ_j being the usual Pauli matrices. In the case of arbitrary spin **s** the operators $\hat{s}_j$ when operating on the $(2s+1)$-dimensional spinor χ^s can similarly be represented by a set of three $(2s+1)$-dimensional hermitian matrices S_j, the S_j being the generalization of the Pauli matrices σ_j. There is an important and vital distinction, however, between the σ_j and the S_j, which in a sense makes the spin-1/2 case unique. It is a fact that the most general 2×2 hermitian matrix M can be specified by four independent real parameters and, as a consequence, because the σ_j are hermitian and independent, such a matrix M can always be written as

$$M = \tfrac{1}{2}\,(aI + \mathbf{b}\cdot\boldsymbol{\sigma}) \tag{1.1.22}$$

where the factor 1/2 is for convenience, I is the unit matrix, $\mathbf{b}\cdot\boldsymbol{\sigma}$ is short for $b_j\sigma_j$ and the four numbers a, b_j are all real. The form of (1.1.22) is particularly convenient since it is trivial to solve for a and b_j. One has

$$a = \mathrm{Tr}\ M, \qquad b_j = \mathrm{Tr}\ (\sigma_j M) \tag{1.1.23}$$

where Tr $\equiv$ trace means the sum of the diagonal elements of the matrix.

The Pauli σ_j thus play a dual rôle. On the one hand, they represent the spin operators $\hat{s}_j$; on the other they furnish a basis for expressing any 2×2 hermitian matrix. *It is the confusion of these two rôles that sometimes leads to difficulties in understanding spin effects in relativistic situations.*

In the case of higher spin s the most general hermitian matrix is specified by $(2s+1)^2$ real parameters, so the set of the three S_j matrices is far from adequate as a basis for an expansion analogous to (1.1.22).

The special rôle of spin 1/2 shows itself in yet another way. The most general two-component spinor χ can be specified by four-real parameters, of which one, the overall phase, is totally irrelevant.

If, further, the spinor is normalized to unity, i.e.

$$\chi^\dagger\chi = 1,$$

we are left with two independent real parameters. Thus we can write, without loss of generality,

$$\chi = \begin{pmatrix} \cos\frac{1}{2}\theta\ e^{-i\phi/2} \\ \sin\frac{1}{2}\theta\ e^{i\phi/2} \end{pmatrix}. \tag{1.1.24}$$

If now we compute the *spin-polarization vector* $\mathcal{P}_\chi$ defined by

$$\mathcal{P}_\chi \equiv \langle\boldsymbol{\sigma}\rangle_\chi \equiv \chi^\dagger\boldsymbol{\sigma}\chi \tag{1.1.25}$$

we shall find that

$$\mathcal{P}_\chi = (\sin\theta\cos\phi, \sin\theta\sin\phi, \cos\theta) \tag{1.1.26}$$

with $\mathcal{P}_\chi^2 = 1$. We see trivially that a knowledge of $\mathcal{P}_\chi$ completely specifies the quantum state χ. In the case of higher spin, one can still define a spin-polarization vector for a state χ such that

$$\mathcal{P}_\chi \equiv \langle \hat{\mathbf{s}} \rangle_\chi / s \equiv \mathbf{s}_\chi / s \tag{1.1.27}$$

where $\mathbf{s}$ is the *mean spin vector*, but now the three components of $\mathcal{P}$ are insufficient to fix the $2(2s+1) - 2$ independent parameters of the $(2s+1)$-dimensional spinor χ. Besides the case of spin 1/2 there is no other situation in nature where a knowledge of the spin-polarization vector completely specifies the quantum state. (Of course $\mathcal{P}$ and $\mathbf{s}$ are really pseudovectors. $\mathcal{P}$ is commonly referred to as the *polarization vector* but it is not at all the same thing as the polarization vector $\boldsymbol{\varepsilon}$ used in the description of photons or massive spin-1 particles. For this reason we shall refer to it as the *spin-polarization vector*.)

Finally we note a very important property of the matrices S_i representing the spin operator $\hat{s}_i$ for spin s, namely that they 'transform as vectors under rotation'. More precisely:

$$\mathscr{D}^{(s)}(r) S_i \mathscr{D}^{(s)\dagger}(r) = R_{ij}(r^{-1}) S_j. \tag{1.1.28}$$

This relation is best known in the spin-1/2 case in the simpler looking, but really equivalent, form

$$\langle \sigma_i \rangle_{S^r} = R_{ij}(r^{-1}) \langle \sigma_j \rangle \tag{1.1.29}$$

relating expectation values in S^r to those in S.

1.2 Spin and helicity in a relativistic process

The pioneering work of Dirac (1927) showed that spin emerges automatically in a relativistic theory and that it could no longer be treated as an independent additional degree of freedom. Nevertheless it is not trivial to see precisely how the spin is to be described relativistically, nor how it is to be interpreted physically. We shall give a brief discussion of this question, and then turn to consider the helicity states of Jacob and Wick (1959). Here our emphasis will be upon the physical interpretation and is somewhat complementary to the approach used by other authors.

We assume that the reader has some familiarity with homogeneous and inhomogeneous Lorentz transformations. A clear account can be found in Gasiorowicz (1967).

In a relativistic quantum theory the fundamental operators are the generators of the inhomogeneous Lorentz transformations. There are 10 of these. The three momentum operators $\hat{P}^j$ and the hamiltonian operator $\hat{P}^0$ generate translations in space and time respectively, and the six operators $\hat{M}^{\mu\nu} (= -\hat{M}^{\nu\mu})$ generate the homogeneous Lorentz transformations. It is physically more revealing to work not with the $\hat{M}^{\mu\nu}$ but with the combinations

$$\hat{J}_i \equiv -\tfrac{1}{2}\epsilon_{ijk}\hat{M}^{jk}, \qquad \hat{K}_i \equiv \hat{M}^{i0}, \tag{1.2.1}$$

which can be shown to be the generators of pure rotations and of pure Lorentz transformations ('boosts') respectively. Thus the $\hat{J}_i$ are identified as the total angular momentum operators.

As a consequence of the inherent characteristics of the inhomogeneous Lorentz transformations, one can derive commutation relations that must be satisfied by the generators. In particular, and in accordance with the interpretation of the $\hat{J}_i$ as angular momentum operators, one naturally finds

$$\left[\hat{J}_j, \hat{J}_k\right] = i\epsilon_{jkl}\hat{J}_l. \tag{1.2.2}$$

The operator $\hat{P}_\mu \hat{P}^\mu$ is invariant, i.e. it commutes with all the generators and its eigenvalues can thus be used to label states. Indeed, what we mean when we talk of an elementary particle of mass m is nothing other than matter that is an eigenstate of $\hat{P}_\mu \hat{P}^\mu$ with eigenvalue m^2.

The question that now arises is the following. If the theory already *contains* the spin then which operators are to be identified as the spin operators? Is there a set of operators $\hat{s}_i$, with commutation relations akin to eqn (1.2.2)?

The nearest one can get to a covariant spin operator is the set of Pauli–Lubanski operators $\hat{W}_\mu$, defined as follows:

$$\hat{W}_\sigma = -\tfrac{1}{2}\epsilon_{\mu\nu\rho\sigma}\hat{M}^{\mu\nu}\hat{P}^\rho \tag{1.2.3}$$

(with $\epsilon_{0123} = +1$), whose commutation relations can be shown to be

$$\left[\hat{W}_\lambda, \hat{W}_\mu\right] = i\epsilon_{\lambda\mu\rho\sigma}\hat{W}^\rho \hat{P}^\sigma. \tag{1.2.4}$$

These are not quite what we hoped for, but we notice that *if* we consider the action of these operators on states of momentum $\mathbf{p} = 0$, i.e. on 'rest' states, then for the space parts of the commutation relations (1.2.4) one will have

$$\left[\hat{W}_j, \hat{W}_k\right] = i\epsilon_{jk\rho 0}\hat{W}^\rho m = -im\epsilon_{jkl}\hat{W}_l. \tag{1.2.5}$$

Thus, for the case $m \neq 0$ the three operators

$$\hat{s}_i = \frac{1}{m}\hat{W}^i \tag{1.2.6}$$

have the commutation relations

$$\left[\hat{s}_j, \hat{s}_k\right] = i\epsilon_{jkl}\hat{s}_l \tag{1.2.7}$$

provided they act on the states of particles at rest.

Further, the operator $\hat{W}_\mu \hat{W}^\mu$ is invariant[1] and its eigenvalues, as can be deduced from(1.2.4)–(1.2.7), are of the form $m^2 s(s+1)$ with $s = 0, \frac{1}{2}, 1, \ldots$. It is the number s that is defined as the 'spin' of a particle in a relativistic theory.

In summary, in a relativistic theory a particle is assigned an invariant spin quantum number s. But only when the particle is at rest can one identify a set of spin operators $\hat{s}_i$ and proceed to invoke the usual formalism of non-relativistic quantum mechanics. Indeed from (1.2.3) one sees that when $\hat{W}_\mu$ acts upon a particle at rest it has the form

$$\hat{W}_\sigma = -\frac{m}{2}\epsilon_{\mu\nu 0\sigma}\hat{M}^{\mu\nu}$$

or, from (1.2.1)

$$\hat{W}^i = m\hat{J}_i. \tag{1.2.8}$$

Thus the $\hat{s}_i$ when acting on states at rest are just the $\hat{J}_i$, so that all the rotational properties of non-relativistic spin hold for particles at rest. The possibility that a particle at rest has non-zero total angular momentum has emerged automatically.

For a particle at rest it is convenient to fix a reference frame and then to classify the states of the particle as in the non-relativistic case, i.e. using eigenstates $|ss_z\rangle$ of $\hat{s}^2$ and $\hat{s}_z$. For a particle in motion, however, the labelling of the states is not so clear cut.

The standard approach is to generate states of arbitrary momentum by acting upon the rest states with suitable Lorentz transformations. We shall adopt an equivalent but more physical approach, considering Lorentz transformations in a similar spirit to our discussion of rotations in Section 1.1.

We denote an arbitrary physical Lorentz transformation by l. We continue to denote physical rotations by r, and we denote by l_j, $j = x, y, z$, physical pure Lorentz transformations ('boosts') along the axes. We remind the reader that care must be taken when specifying a *sequence* of operations acting on the reference system. For example, if we first rotate a system S about its Y axis through angle θ (call this frame S') and then boost to a new frame S'' moving with speed v *along the Z-axis of S'*, then we should represent the complete transformation from S to S'' as

$$S \to S'' = l_{z'}(v)r_y(\theta)S;$$

it is essential for clarity to use the primed label z' on l. A *pure* Lorentz transformation or boost in an arbitrary direction is denoted by $l(\mathbf{v})$, where

[1] $\hat{W}_\mu \hat{W}^\mu$ and $\hat{P}_\mu \hat{P}^\mu$ are the only invariant operators of the inhomogeneous Lorentz group.

conventionally

$$l(\mathbf{v}) \equiv \left[r^{-1}(\mathbf{v})\right]'' l_{z'}(v) r(\mathbf{v}). \tag{1.2.9}$$

Here $r(\mathbf{v})$ is the rotation about $\mathbf{e}_{(z)} \times \mathbf{v}$ that rotates the Z-axis into the direction of $\mathbf{v}$ and $(r^{-1}(\mathbf{v}))''$ is its inverse, applied to the boosted frame. We shall refer to (1.2.9) as a *canonical* boost.

The reason for calling (1.2.9) a pure boost is clear from Fig. 1.1, which shows (for the case of $\mathbf{v}$ lying in the XZ plane) that the final reference system S''' has its Z-axis at the same angle θ to $\mathbf{v}$ as did OZ of S.

If a 4-vector A is acted upon by a physical Lorentz transformation l then it is transformed to a new vector, which we shall denote by lA or A_l. Its components are related to those of A by

$$(lA)^\mu \equiv A_l^\mu = \Lambda^\mu{}_\nu(l) A^\nu. \tag{1.2.10}$$

When using matrix notation we shall denote by Λ the 4×4 matrix whose elements are $\Lambda^\mu{}_\nu$, μ referring to the row, ν to the column. The column matrix A is defined to have as components the contravariant components A^ν. Thus (1.2.10) reads as

$$A_l = \Lambda(l) A$$

Explicit forms for $\Lambda^\mu{}_\nu$ for a few cases of special importance follow. If l is simply a rotation r, then we have

$$\Lambda = \begin{pmatrix} 1 & 0 & 0 & 0 \\ 0 & & & \\ 0 & & R & \\ 0 & & & \end{pmatrix} \tag{1.2.11}$$

where R is the matrix defined in (1.1.4).

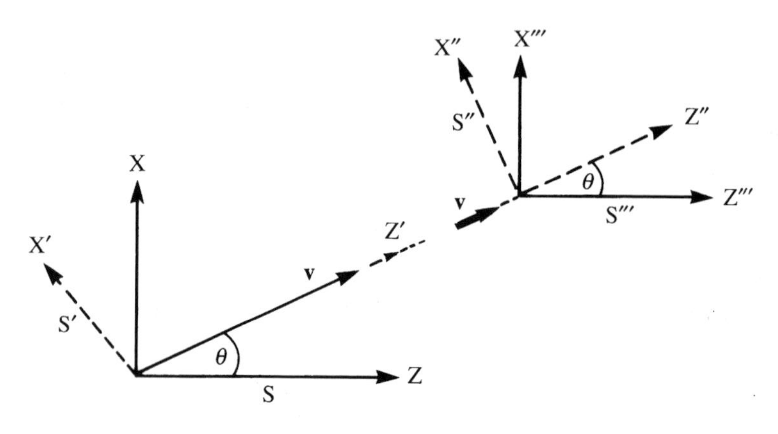

Fig. 1.1. A canonical boost along $\mathbf{v}$ to $S \to S'''$ as shown.

If l is a boost of speed v along the Z axis then

$$\Lambda\,[l_z(v)] = \begin{pmatrix} \gamma & 0 & 0 & \gamma\beta \\ 0 & 1 & 0 & 0 \\ 0 & 0 & 1 & 0 \\ \gamma\beta & 0 & 0 & \gamma \end{pmatrix} \tag{1.2.12}$$

where $\gamma = (1-\beta^2)^{-1/2}$, $\beta = v/c$.

For the canonical boost (1.2.9) one has

$$\Lambda\,[l(\mathbf{v})] = \begin{pmatrix} \gamma & \gamma\beta_x & \gamma\beta_y & \gamma\beta_z \\ \gamma\beta_x & 1+\alpha\beta_x^2 & \alpha\beta_x\beta_y & \alpha\beta_x\beta_z \\ \gamma\beta_y & \alpha\beta_y\beta_x & 1+\alpha\beta_y^2 & \alpha\beta_y\beta_z \\ \gamma\beta_z & \alpha\beta_z\beta_x & \alpha\beta_z\beta_y & 1+\alpha\beta_z^2 \end{pmatrix} \tag{1.2.13}$$

where $\boldsymbol{\beta} = \mathbf{v}/c$, $\gamma = (1-\beta^2)^{-1/2}$ and $\alpha = \gamma^2(\gamma+1)^{-1}$.

If S^l is a frame obtained by applying a Lorentz transformation l to a frame S then analogously to (1.1.12) a fixed 4-vector A in S will appear in S^l to have components

$$(A^\mu)_{S^l} = \Lambda^\mu{}_\nu(l^{-1})A^\nu, \tag{1.2.14}$$

which for convenience will be written, somewhat loosely, as

$$(A)_{S^l} = l^{-1}A. \tag{1.2.15}$$

A brief discussion of the finite-dimensional representations of the Lorentz group is given in Appendix 2.

1.2.1 Particles with non-zero mass

Let us suppose that we are given a definite reference frame S_A in which a particle A of mass m is at rest in state $|s;s_z\rangle$. Let O be an observer moving at velocity $-\mathbf{v}$ with respect to O_A. Choose $\mathbf{v} = \mathbf{p}/\sqrt{\mathbf{p}^2+m^2}$ where $\mathbf{p}$ is some arbitrary momentum. Then observer O looking at A which is at rest in S_A, will see a particle moving with momentum $\mathbf{p}$. Thus in describing this state O will use a label $\mathbf{p}$, i.e. $|\mathbf{p};\cdots\rangle$. However, there are infinitely many reference frames S all attached to O and moving with velocity $-\mathbf{v}$ with respect to O_A in which particle A will still appear to have momentum $\mathbf{p}$: if S_1 is one such frame then so will be any other frame obtained by rotating S_1 bodily about $\mathbf{p}$. Clearly, although all these observers report A as having momentum $\mathbf{p}$ they will each report a different spin state since their reference frames are all rotated from each other. Thus the 'spin' label given to the state of motion of A must depend on a choice as to which reference frame O is using. This choice is a matter of convention.

There are two main choices in the literature.

(1) *The canonical choice.* Here O chooses his/her reference frame S in such a way that it is obtained from S_A by a pure Lorentz transformation $l(-\mathbf{v})$ as in (1.2.9). O then labels the state of motion that he/she sees as $|\mathbf{p}; s_z\rangle$.

(2) *The helicity choice.* Let $\mathbf{p}$ have polar angles θ, ϕ. Then O chooses his frame S as follows. To begin with, O transforms to a frame S' boosted by a speed

$$v = |\mathbf{p}|/\sqrt{\mathbf{p}^2 + m^2}$$

in the direction of the negative Z-axis of S_A.

O then applies a rotation to S' designed to make the momentum of A appear as $\mathbf{p} = (p, \theta, \phi)$. The rotation we use is the simplest one: first through angle $-\theta$ about OY' then through $-\phi$ about the new Z-axis OZ'', i.e. the overall transformation is

$$S_A \to S = r_{z''}(-\phi) r_{y'}(-\theta) l_z^{-1}(v) S_A. \tag{1.2.16}$$

We note that if the usual notation $r(\alpha, \beta, \gamma)$ is used for a rotation through the Euler angles α, β, γ, i.e.

$$r(\alpha, \beta, \gamma) = r_{z''}(\gamma) r_{y'}(\beta) r_z(\alpha) = r_z(\alpha) r_y(\beta) r_z(\gamma) \tag{1.2.17}$$

(the latter equality is explained in Hamermesh (1964)) then

$$S = r^{-1}(\theta, \phi, 0) l_z^{-1}(v) S_A. \tag{1.2.18}$$

If the state A in the rest frame S_A is $|\mathring{p}; s, s_z = \lambda\rangle$, where $\mathring{p} = (m, 0, 0, 0)$, then O using the frame S sees the state $|\mathring{p}; s, s_z = \lambda\rangle_S$, which O labels as $|\mathbf{p}; \lambda\rangle$, i.e. the *helicity state* $|\mathbf{p}; \lambda\rangle$ is defined by

$$|\mathbf{p}; \lambda\rangle \equiv |\mathring{p}; s, s_z = \lambda\rangle_S, \tag{1.2.19}$$

in which S is specified as above. The mathematical relationship between $|\mathbf{p}; \lambda\rangle$ and $|\mathring{p}; s_z = \lambda\rangle$ will be given later.

In what follows we shall seldom use the canonical basis, so that, unless specifically indicated, all our states are helicity states. The formalism is much simplified thereby and the treatment of massive and massless particles is unified.

It should be noted that the rotation defined in (1.2.16) is simpler than the one used in the original paper of Jacob and Wick. Their rotation corresponds to having $r^{-1}(\phi, \theta, -\phi)$ in (1.2.18). However, both Jacob and Wick in later papers adopted the simpler rotation given in (1.2.16).

1.2.2 The physical interpretation of helicity and canonical spin states

Equations (1.2.18) and (1.2.19) are the crucial tools in understanding the physical content of a helicity state. Suppose in a frame S we are told that the particle A is in a state of motion described by $|\mathbf{p};\lambda\rangle$. Then, according to eqns (1.2.18) and (1.2.19), particle A will be found at rest with spin component $s_z = \lambda$ if one observes it in the frame S_A related to S by (1.2.16), i.e. in the frame

$$S_A = l_{z'}(v)r(\phi,\theta,0)S. \qquad (1.2.20)$$

We refer to this particular one of the infinitely many rest frames for A as its '*helicity rest frame*'.

The relation between S_A and S, for the case $\phi = 0$, can be seen in Fig. 1.2. In general, for arbitrary ϕ, Z_A will lie along $\mathbf{p}$ and Y_A along $\mathbf{e}_{(z)} \times \mathbf{p}$. For $\theta = 0$ or π we take Y_A along or opposite to OY respectively. The transformation in (1.2.20) is often given the special symbol $h(\mathbf{p})$, i.e.

$$h(\mathbf{p}) \equiv l_{z'}(v)r(\phi,\theta,0). \qquad (1.2.21)$$

Another way of specifying $h(\mathbf{p})$, which is more common in the literature, is to refer all the operations involved to just one reference frame. In this case it can be shown (Hamermesh, 1964) that

$$h(\mathbf{p}) = r(\phi,\theta,0)l_z(v), \qquad (1.2.22)$$

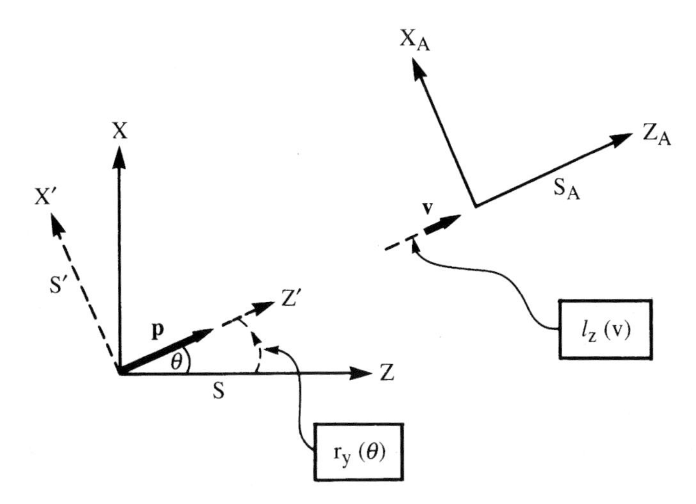

Fig. 1.2 Definition of the helicity rest frame S_A for particle A, which has momentum $\mathbf{p}$ in a reference frame S.

where the absence of primes on the axis labels is crucial. We note that

$$\Lambda[h(\mathbf{p})] = \begin{pmatrix} \gamma & 0 & 0 & \beta\gamma \\ \gamma\beta_x & \cos\theta\cos\varphi & -\sin\varphi & \gamma\beta_x/\beta \\ \gamma\beta_y & \cos\theta\sin\varphi & \cos\varphi & \gamma\beta_y/\beta \\ \gamma\beta_z & -\sin\theta & 0 & \gamma\beta_z/\beta \end{pmatrix}, \tag{1.2.23}$$

where $\boldsymbol{\beta} = \mathbf{p}/E$ has polar angles θ, φ and $\gamma = E/m$. Note that the form for $\boldsymbol{\beta}$ and γ for massless particles will be given in subsection 1.2.3.

We note also, that by its very construction, $h(\mathbf{p})$ operating on the 4-vector $\overset{\circ}{p}$ turns it into p.

We now have

$$S = h^{-1}(\mathbf{p})S_A \tag{1.2.24}$$

and therefore in complete analogy with eqn (1.1.21)

$$|\mathbf{p};\lambda\rangle \equiv |\overset{\circ}{p}; s, s_z = \lambda\rangle_S = U[h(\mathbf{p})]|\overset{\circ}{p}; s, s_z = \lambda\rangle, \tag{1.2.25}$$

where U is the unitary operator corresponding to the Lorentz transformation $h(\mathbf{p})$.

In the usual treatment of helicity states, the $|\mathbf{p};\lambda\rangle$ are simply defined by eqn (1.2.25). The advantage of our treatment is that it clarifies the interpretation of the label λ.

We must at this point add a note of warning to the reader. In building up helicity states for two particles, Jacob and Wick (1959) make a distinction between the states for what they call 'particle 1' and 'particle 2'. For us the definition of a helicity state of any particle is the *same*.

For the moment the crucial point to be drawn from the above is simply that if a helicity state for a particle A is defined in some frame S by

$$|\mathbf{p};\lambda\rangle = U[h_A(\mathbf{p})]|\overset{\circ}{p}; s_z = \lambda\rangle \tag{1.2.26}$$

then λ is the z-component of the spin of the particle A when measured in the helicity rest frame S_A obtained from S_A via

$$S \to S_A = h_A(\mathbf{p})S \tag{1.2.27}$$

and illustrated in Fig. 1.2.

Because of the subtle question of phases, some care must be exercised in talking about the vector $-\mathbf{p}$. If $\mathbf{p} = (p, \theta, \phi)$ then we shall always use $-\mathbf{p} \equiv (p, \pi - \theta, \phi + \pi)$ even when $\theta = \phi = 0$.

The canonical spin states $|\mathbf{p}; s_z\rangle_{\text{can}}$ are introduced in complete analogy with the above, the only difference being that $h_A(\mathbf{p})$ is replaced by the pure boost $l(\mathbf{v})$ where $\mathbf{v} = \mathbf{p}_A/\sqrt{\mathbf{p}_A^2 + m^2}$. Thus instead of (1.2.26) we have for a particle A

$$|\mathbf{p}; s_z\rangle_{\text{can}} = U[l_A(\mathbf{v})]|\overset{\circ}{p}; s_z\rangle \tag{1.2.28}$$

and the physical interpretation is that s_z is the spin component of A as measured in its *canonical rest frame* S_A^0 reached from S by the boost $l_A(\mathbf{v})$. If $\mathbf{v}$ in Fig. 1.1 refers to particle A then the frame S''' is just S_A^0. Comparing Figs. 1.1 and 1.2 we see that the two rest frames differ by a rotation and thus the physical situations described by say $|\mathbf{p}; \lambda = \frac{1}{2}\rangle$ and $|\mathbf{p}; s_z = \frac{1}{2}\rangle_{\text{can}}$ are different. In classical physics what is loosely termed '*the rest frame*' or 'a comoving frame' is what we have called the canonical rest frame.

Finally we note that it is easy to show that the state $|\mathbf{p}; \lambda\rangle$ is an eigenstate of the *helicity operator* $\hat{\mathbf{J}} \cdot \hat{\mathbf{P}}/|\hat{\mathbf{P}}|$, i.e.

$$\frac{\hat{\mathbf{J}} \cdot \hat{\mathbf{P}}}{|\hat{\mathbf{P}}|}|\mathbf{p}; \lambda\rangle = \lambda|\mathbf{p}; \lambda\rangle. \tag{1.2.29}$$

Thus the helicity is the projection of the *total* angular momentum onto the direction of the linear momentum, for a free particle.

1.2.3 Particles with zero mass

In all the above we leant heavily upon the existence of a rest frame for our particle. In fact helicity states can still be defined for massless particles and this is one of the many reasons for preferring them to canonical states. They unify completely the treatment of spin for particles of any spin and mass.

The generalization to mass-zero particles starts from the realization that the helicity states defined by (1.2.25) are eigenstates of the helicity operator $\hat{\mathbf{J}} \cdot \hat{\mathbf{P}}/|\hat{\mathbf{P}}|$.

With this interpretation there is no reference to the mass or a rest frame. We may therefore adopt eqn (1.2.29) as the *definition* of a helicity state for a massless particle. There is to begin with just *one* value of λ, and the spin s is defined by $s = |\lambda|$. If, however, the interactions of the particle are invariant under space reflection then the state obtained by acting with the parity operator $\mathscr{P}$ on $|\mathbf{p}; \lambda\rangle$ must also be a physical state.

Since $\hat{\mathbf{J}} \cdot \hat{\mathbf{P}}/|\hat{\mathbf{P}}|$ is manifestly invariant under rotations and is a pseudo-scalar under space reflection, it is clear that for the state with $\mathbf{p}_z = (0, 0, p)$,

$$\mathscr{Y}|\mathbf{p}_z; \lambda\rangle \equiv e^{-i\pi\hat{J}_y}\mathscr{P}|\mathbf{p}_z; \lambda\rangle \tag{1.2.30}$$

has momentum $\mathbf{p}_z$ and is an eigenstate of $\hat{\mathbf{J}} \cdot \hat{\mathbf{P}}/|\hat{\mathbf{P}}|$ with eigenvalue $-\lambda$. It can thus be written as

$$\mathscr{Y}|\mathbf{p}_z; \lambda\rangle = \eta|\mathbf{p}_z; -\lambda\rangle, \tag{1.2.31}$$

where η is called the *'intrinsic parity factor'* of the massless particle.

In summary, if its interactions conserve parity then a massless particle of spin s has two independent helicity states for a given momentum, namely $|\mathbf{p}; \lambda = s\rangle$ and $|\mathbf{p}; \lambda = -s\rangle$, and they are related by (1.2.31). Thus a photon, whose interactions conserve parity, has two helicity states $\lambda = \pm 1$ whereas a neutrino, whose interactions violate parity, can exist only as a 'left-handed particle', i.e. $\lambda = -\frac{1}{2}$ only.

Although (1.2.29) gives a meaning to λ, it does not specify the state $|\mathbf{p};\lambda\rangle$ uniquely.

In order to specify the relationship between states seen by different observers we can begin in some *standard* reference frame S_{st} in which the particle is moving in the Z-direction with some definite momentum $p_{\text{st}} = (\overline{p}, 0, 0, \overline{p})$. This state is labelled $|\mathbf{p}_{\text{st}};\lambda\rangle$.

In analogy with (1.2.25) the helicity state $|\mathbf{p};\lambda\rangle$ can be defined by

$$|\mathbf{p};\lambda\rangle = U[h(\mathbf{p},\mathbf{p}_{\text{st}})]|\mathbf{p}_{\text{st}};\lambda\rangle \tag{1.2.32}$$

where now $h(\mathbf{p},\mathbf{p}_{\text{st}})$ is the Lorentz transformation of the form

$$r(\phi,\theta,0)l_z(v)$$

such that $h^{-1}(\mathbf{p},\mathbf{p}_{\text{st}})$ changes the frame S_{st} to the frame S in which the momentum is $\mathbf{p}^\mu = (p,\mathbf{p})$ with $\mathbf{p} = (p,\theta,\phi)$. Alternatively, $h(\mathbf{p},\mathbf{p}_{\text{st}})$ acting on p_{st} turns it into p. One can of course show that the $|\mathbf{p};\lambda\rangle$ defined in (1.2.32) do satisfy (1.2.29). But (1.2.32) goes beyond (1.2.29) in that it specifies also the relative phases of the states.

The matrix $\Lambda[h(\mathbf{p},\mathbf{p}_{\text{st}})]$ is still given by (1.2.23) but now

$$\beta = (p^2 - \overline{p}^2)/(p^2 + \overline{p}^2).$$

For a massless particle it is not possible to define the spin s directly from the eigenvalues of $\hat{W}_\mu \hat{W}^\mu$. This can be seen as follows. From (1.2.4) we have that

$$[\hat{W}_\mu, \hat{W}_\nu]|\mathbf{p}_{\text{st}};\lambda\rangle = i\overline{p}[\epsilon_{\mu\nu\rho 0} + \epsilon_{\mu\nu\rho 3}]\hat{W}^\rho|\mathbf{p}_{\text{st}};\lambda\rangle.$$

Therefore one has the following commutation relations for the $\hat{W}^\mu$ *when acting on the* $|\mathbf{p}_{\text{st}};\lambda\rangle$:

$$[\hat{W}^1, \hat{W}^2] = 0 \tag{1.2.33}$$

and

$$\begin{aligned} [\hat{W}^3, \hat{W}^1] &= i\overline{p}\hat{W}^2 \\ [\hat{W}^3, \hat{W}^2] &= -i\overline{p}\hat{W}^1. \end{aligned} \tag{1.2.34}$$

Now consider the fundamental commutation relations of the translation generators with the angular momentum operators. One has

$$[\hat{M}^{\mu\nu}, \hat{P}^\rho] = i\left(g^{\mu\rho}\hat{P}^\nu - g^{\nu\rho}\hat{P}^\mu\right) \tag{1.2.35}$$

and of course

$$[\hat{P}^\mu, \hat{P}^\nu] = 0. \tag{1.2.36}$$

It follows from (1.2.35) and (1.2.1) that

$$\begin{aligned}[\hat{J}_3, \hat{P}^1] &= i\hat{P}^2 \\ [\hat{J}_3, \hat{P}^2] &= -i\hat{P}^1\end{aligned} \tag{1.2.37}$$

and

$$[\hat{P}^1, \hat{P}^2] = 0.$$

On comparing (1.2.37) with (1.2.34) and (1.2.33) we see that, acting on the states $|\mathbf{p}_{\rm st}; \lambda\rangle$ the set of operators $(\hat{W}^3/\overline{p}, \hat{W}^1, \hat{W}^2)$ obeys an algebra isomorphic to that of $(\hat{J}_3, \hat{P}^1, \hat{P}^2)$. Thus the eigenvalues of $\hat{W}^1, \hat{W}^2$ will be, just like momentum eigenvalues, unquantized! It is *postulated* that the *physical* massless particles in nature correspond to eigenvalue zero for $\hat{W}^{1,2}$:

$$\hat{W}^{1,2}|\mathbf{p}_{\rm st}; \lambda\rangle = 0. \tag{1.2.38}$$

It then follows from (1.2.3) that, when acting on these physical states $|\mathbf{p}_{\rm st}; \lambda\rangle$,

$$\hat{W}^\mu|\mathbf{p}_{\rm st}; \lambda\rangle = (\overline{p}J_3, 0, 0, \overline{p}J_3)|\mathbf{p}_{\rm st}; \lambda\rangle \tag{1.2.39}$$

so that

$$\hat{W}^\mu \hat{W}_\mu|\mathbf{p}_{\rm st}; \lambda\rangle = 0 \tag{1.2.40}$$

and

$$\begin{aligned}\frac{\hat{W}^3}{\overline{p}}|\mathbf{p}_{\rm st}; \lambda\rangle = \hat{J}_3|\mathbf{p}_{\rm st}; \lambda\rangle &= \frac{\hat{\mathbf{J}} \cdot \hat{\mathbf{P}}}{|\hat{\mathbf{P}}|}|\mathbf{p}_{\rm st}; \lambda\rangle \\ &= \lambda|\mathbf{p}_{\rm st}; \lambda\rangle.\end{aligned} \tag{1.2.41}$$

Thus the physical helicity states may be thought of as eigenstates of $(\hat{W}^3/\overline{p}, \hat{W}^1, \hat{W}^2)$ with eigenvalues $(\lambda, 0, 0)$.

Since s is not now given by its usual rules, i.e. as the eigenvalue $s(s+1)$ of the square of some spin operator, we have to ask how we determine that s is to have only integer or half-integer values.

The answer can be seen as follows: although the *value* of λ is clearly invariant under rotations (of course for both zero-mass and massive cases) the helicity states do pick up a phase under some rotations. Thus from (1.2.29) it is clear that for a rotation by angle α about the direction of $\mathbf{p}$ one will have

$$|\mathbf{p}; \lambda\rangle \to \exp(-i\alpha\hat{\mathbf{J}} \cdot \mathbf{p}/|\mathbf{p}|)|\mathbf{p}; \lambda\rangle = e^{-i\alpha\lambda}|\mathbf{p}; \lambda\rangle. \tag{1.2.42}$$

For a rotation of 2π we require that this phase be equal to ± 1 in the bosonic and fermionic cases respectively and conclude that λ must be an integer or half integer.

2
The effect of Lorentz and discrete transformations on helicity states, fields and wave functions

In discussing experiments it is often necessary to refer a given physical situation to different reference frames, e.g. to the laboratory or centre-of-mass system. Thus we need to understand how helicity states are affected by Lorentz transformations. The approach is quite similar to the discussion of rotations in Section 1.1 and we seek the analogue of eqn (1.1.17). However, because sequences of Lorentz transformations are more complicated than sequences of rotations the result will look a little less simple. We shall compare and contrast this situation with the transformation properties of fields and wave functions.

2.1 Particles with non-zero mass

Let us suppose that in a given reference system S an observer O sees a particle A in motion with momentum $\mathbf{p}$ and helicity λ, i.e. the observer reports a state of motion specified by $|\mathbf{p};\lambda\rangle$.

Let S^l be a reference frame obtained by carrying out a physical Lorentz transformation l on S. We wish to know how observer O^l describes the motion of A.

By analogy with the rotational case (see eqn (1.1.21)) O^l will describe the state as

$$|\mathbf{p};\lambda\rangle_{S^l} = U(l^{-1})|\mathbf{p};\lambda\rangle \tag{2.1.1}$$

when $U(l)$ is the operator effecting a Lorentz transformation l.

Let us denote by $\mathbf{p}'$ the momentum vector that O^l attributes to A, i.e. $\mathbf{p}' = l^{-1}\mathbf{p}$. Its components p'^{μ} are clearly the components of $\mathbf{p}$ as seen by O^l, i.e. (see eqns (1.2.14), (1.2.15))

$$p'^{\mu} \equiv (p^{\mu})_{S^l} = \Lambda^{\mu}{}_{\nu}(l^{-1})p^{\nu}. \tag{2.1.2}$$

It is obvious that we must expect to find that $|\mathbf{p};\lambda\rangle_{S^l} = |\mathbf{p}';\lambda'\rangle$ with $\mathbf{p}'$ given by (2.1.2). The only question is what values of λ' should appear. To answer this one writes

$$U(l^{-1})|\mathbf{p};\lambda\rangle = U(l^{-1})U[h(\mathbf{p})]|\mathring{p};\lambda\rangle \tag{2.1.3}$$

using the definition of helicity states (1.2.25). One then invokes the brilliant stratagem of multiplying eqn (2.1.3) by unity in the form

$$U[h(\mathbf{p}')]U^{-1}[h(\mathbf{p}')]$$

where $h(\mathbf{p}')$ is the helicity transformation that would be used to define a state $|\mathbf{p}';\lambda\rangle$, i.e. $h(\mathbf{p}')$ is such that

$$|\mathbf{p}';\lambda\rangle = U[h(\mathbf{p}')]|\mathring{p};\lambda\rangle. \tag{2.1.4}$$

One can now write (2.1.3) in the form

$$U(l^{-1})|\mathbf{p};\lambda\rangle = U[h(\mathbf{p}')]\mathscr{R}|\mathring{p};\lambda\rangle \tag{2.1.5}$$

where $\mathscr{R}$ is short for the product $U^{-1}[h(\mathbf{p}')]U(l^{-1})U[h(\mathbf{p})]$. Since the operators U *represent* the various physical operations we can simplify and write

$$\mathscr{R} = U[h^{-1}(\mathbf{p}')l^{-1}h(\mathbf{p})] \equiv U[h^{-1}(l^{-1}\mathbf{p})l^{-1}h(\mathbf{p})]. \tag{2.1.6}$$

The crucial observation is that the sequence of physical operations in U is just a *rotation* no matter what l is. The simplest way to see this is to study the effect of the sequence of operations $h^{-1}(\mathbf{p}')l^{-1}h(\mathbf{p})$ on the 4-vector $\mathring{p} = (m, 0, 0, 0)$. We have

(1) $h(\mathbf{p}) : \mathring{p} \to p$
(2) $l^{-1} : p \to p'$
(3) $h(\mathbf{p}')$ is such that it takes $\mathring{p} \to p'$, thus $h^{-1}(\mathbf{p}') : p' \to \mathring{p}$.

Hence the sequence (1), (2), (3) takes $\mathring{p} \to \mathring{p}$. From the form of $\mathring{p}$ it is clear that only a rotation could have this property. Hence $\mathscr{R}$ represents a rotation no matter what l is. Let us label this physical rotation as $r(l, \mathbf{p})$, i.e.

$$r(l, \mathbf{p}) \equiv h^{-1}(l^{-1}\mathbf{p})l^{-1}h(\mathbf{p}). \tag{2.1.7}$$

We shall refer to this as the *Wick helicity rotation* for the transformation l of *axes* that takes $\mathbf{p}$ to $\mathbf{p}' = l^{-1}\mathbf{p}$. (It is not the same as the Wigner rotation, as will be explained later.)

Once it is recognized that $\mathscr{R}$ corresponds to a rotation the completion of the evaluation of $|\mathbf{p};\lambda\rangle_{S^l}$ becomes simple. From (1.1.18) and (1.1.19) we know what rotations do to particles at rest. Thus

$$\mathscr{R}|\mathring{p};\lambda\rangle = \mathscr{D}^{(s)}_{\lambda'\lambda}[r(l, \mathbf{p})]|\mathring{p};\lambda'\rangle \tag{2.1.8}$$

and since the $\mathscr{D}_{\lambda'\lambda}$ are just numbers, substituting back into (2.1.5) and (2.1.1) and then using (2.1.4) gives

$$|\mathbf{p};\lambda\rangle_{S^l} = \mathscr{D}^{(s)}_{\lambda'\lambda}[r(l,\mathbf{p})]U[h(\mathbf{p}')]|\mathring{p};\lambda'\rangle = \mathscr{D}^{(s)}_{\lambda'\lambda}[r(l,\mathbf{p})]|\mathbf{p}';\lambda'\rangle \qquad (2.1.9)$$

with $\mathbf{p}' = l^{-1}\mathbf{p}$.

This is the desired relationship between the description used in frames S^l and S for the motion of the particle. In the above form it is valid for an arbitrary Lorentz transformation from S to S^l. The reason why $|\mathbf{p};\lambda\rangle_{S^l}$ and $|\mathbf{p}';\lambda'\rangle$ are related by a rotation is that the helicity rest frame of the particle reached from S is not the same as the one reached from S^l. Indeed if we call these helicity rest frames S_A and S'_A respectively, then one can show that

$$S_A = r(l,\mathbf{p})S'_A \qquad (2.1.10)$$

It should be clear that for canonical states we have a result analogous to (2.1.9). The only difference is that $r(l,\mathbf{p})$ is replaced by

$$r_{\mathrm{Wig}}(l,\mathbf{p}) \equiv l^{-1}(\mathbf{v}')l^{-1}l(\mathbf{v}) \qquad (2.1.11)$$

where $l(\mathbf{v})$ and $l(\mathbf{v}')$ are pure boosts corresponding to the momenta $\mathbf{p}$ and $\mathbf{p}' = l^{-1}\mathbf{p}$. The rotation in (2.1.11) is known as the Wigner spin rotation. If S^0 and $S^{0'}$ are the *canonical* rest frames reached from S and S^l respectively, then analogously to (2.1.10) one finds

$$S^0 = r_{\mathrm{Wig}}(l,\mathbf{p})S^{0'}. \qquad (2.1.12)$$

To gain some physical intuition for the rotations involved we shall look at a few cases of practical interest.

2.2 Examples of Wick and Wigner rotations

We here derive explicit expressions for these rotations for several cases of practical interest and we end with a discussion of the Thomas precession.

2.2.1 *Pure rotation of axes*

In frame S let $\mathbf{p}$ lie in the XZ-plane, $\mathbf{p} = (p,\theta,0)$. Apply a rotation through angle β about OY to the frame S such that $l = r_y(\beta)$. Then in S^r we have $l^{-1}\mathbf{p} = (p,\theta-\beta,0)$. One finds trivially $r(l,\mathbf{p}) = 1$, i.e. there is no Wick helicity rotation. Thus, in this case,

$$|\mathbf{p};\lambda\rangle_{S^R} = |\mathbf{p}';\lambda\rangle \qquad \text{with} \qquad \mathbf{p}' = r_y^{-1}\mathbf{p}. \qquad (2.2.1)$$

For a general rotation $r(\alpha,\beta,\gamma)$ of S, with $\mathbf{p} = (p,\theta,\varphi)$ and $r^{-1}\mathbf{p} = (p,\theta',\varphi')$, one finds

$$|\mathbf{p};\lambda\rangle_{S^r} = e^{i\lambda\zeta}|\mathbf{p}';\lambda\rangle \qquad \text{with} \qquad \mathbf{p}' = r^{-1}\mathbf{p} \qquad (2.2.2)$$

where

$$\cos\zeta = \frac{\cos\beta - \cos\theta\cos\theta'}{\sin\theta\sin\theta'} \tag{2.2.3}$$

(In the event that $\cos\zeta$ appears indeterminate it is simpler to use eqn (2.1.7) to determine the rotation involved.)

Both the above results are in accord with the fact that λ is invariant under rotations.

For the canonical spin states for $l = r_y(\beta)$, $\mathbf{p}$ in the XZ-plane, one finds $r_{\text{Wig}}(l, \mathbf{p}) = r_y(-\beta)$. Here the spin transforms just as it would non-relativistically (see eqn (1.1.17)).

2.2.2 *Pure Lorentz boost of axes*

To begin with, take the boost velocity $\boldsymbol{\beta}$ to lie along OZ so that $l = l_z(\beta)$. In the original and boosted frames we have:

$$\begin{aligned} S: &\quad \mathbf{p} = (p, \theta, \varphi), E \quad \text{speed } v \\ S^l: &\quad \mathbf{p}' = l^{-1}\mathbf{p} = (p', \theta', \varphi), E' \quad \text{speed } v' \end{aligned}$$

and from eqn (1.2.22)

$$\begin{aligned} h(\mathbf{p}) &= r(\varphi, \theta, 0) l_z(v) \\ h(\mathbf{p}') &= r(\varphi, \theta', 0) l_z(v'). \end{aligned}$$

The Wick helicity rotation is now

$$r\,[l_z(\beta), \mathbf{p}] = h^{-1}(\mathbf{p}') l_z^{-1}(\beta) h(\mathbf{p}). \tag{2.2.4}$$

It is easy to see that this is just a rotation about the Y-axis: simply examine the effect of the sequence of operations in (2.2.4) on the unit vector in the Y-direction $\mathbf{e}_{(y)} = (0, 0, 1, 0)$. It remains unchanged. Thus

$$r\,[l_z(\beta), \mathbf{p}] = r_y(\theta_{\text{Wick}})$$

so that

$$\mathscr{D}^{(s)}_{\lambda'\lambda}[r_{\text{Wick}}] = d^s_{\lambda'\lambda}(\theta_{\text{Wick}}) \tag{2.2.5}$$

and the angle θ_{Wick} can be found most easily by checking the effect of r_{Wick} upon the unit vector $\mathbf{e}_{(x)} = (0, 1, 0, 0)$. Carrying out the sequence of operations one ends up with

$$\mathbf{e}'_{(x)} = \left\{0,\ \cos\theta\cos\theta' + \gamma\sin\theta\sin\theta',\ 0,\ -\frac{m}{E}(\sin\theta\cos\theta' - \gamma\cos\theta\sin\theta')\right\}$$

where $\gamma = (1 - \beta^2)^{-1/2}$.

Comparing with (1.1.6) and (1.1.7) and using the relation between θ and θ' we end up with

$$\begin{aligned}\cos\theta_{\text{Wick}} &= \frac{\gamma}{p'}(p - \beta E\cos\delta)\\ \sin\theta_{\text{Wick}} &= -\frac{m}{p'}\gamma\beta\sin\delta\end{aligned} \tag{2.2.6}$$

where δ $(0 \leq \delta \leq \pi)$ is the angle between $\boldsymbol{\beta}$ and $\mathbf{p}$. (In *this* case $\delta = \theta$.)

For the general case of a boost $l(\boldsymbol{\beta})$ of the axes, with $\boldsymbol{\beta} = (\beta, \theta_\beta, \varphi_\beta)$, one has

$$|\mathbf{p};\lambda\rangle_{S^{l(\boldsymbol{\beta})}} = e^{i\eta(\lambda-\lambda')}d_{\lambda'\lambda}(\theta_{\text{Wick}})|l^{-1}\mathbf{p};\lambda'\rangle \tag{2.2.7}$$

corresponding to $r_{\text{Wick}} = r(\eta, \theta_{\text{Wick}}, -\eta)$, where η is given by

$$\begin{aligned}\cos\eta &= \frac{\sin\theta\cos\theta_\beta - \cos\theta\sin\theta_\beta\cos(\varphi-\varphi_\beta)}{\sin\delta}\\ \sin\eta &= \frac{\sin\theta_\beta\sin(\varphi-\varphi_\beta)}{\sin\delta}.\end{aligned} \tag{2.2.8}$$

As in (2.2.6), δ is the angle between $\boldsymbol{\beta}$ and $\mathbf{p}$, $0 \leq \delta \leq \pi$.

When both $\mathbf{p}$ and $\boldsymbol{\beta}$ lie in the XZ-plane the general result simplifies to

$$|\mathbf{p};\lambda\rangle_{S^{l(\boldsymbol{\beta})}} = d_{\lambda'\lambda}(\pm\theta_{\text{Wick}})|l^{-1}\mathbf{p};\lambda'\rangle \tag{2.2.9}$$

with θ_{Wick} given by (2.2.6); the $\pm$ correspond to $\boldsymbol{\beta}\times\mathbf{p}$ being along or opposite to OY respectively.

2.2.3 Boost along or opposite to p

It is clear that if S^l is boosted from S in a direction *opposite* to the momentum of $\mathbf{p}$ of the particle then

$$|\mathbf{p};\lambda\rangle_{S^l} = |l^{-1}\mathbf{p};\lambda\rangle. \tag{2.2.10}$$

This holds also for boosts *along* $\mathbf{p}$ provided that the boost speed v satisfies $v < p/E$. For higher boost speeds along $\mathbf{p}$ the particle direction will have reversed in S^l and one finds

$$|\mathbf{p};\lambda\rangle_{S^l} = (-1)^{s+\lambda}|l^{-1}\mathbf{p};-\lambda\rangle. \tag{2.2.11}$$

2.2.4 Transformation from CM to Lab

A case of practical importance is the transformation from centre-of-mass frame (CM) to laboratory frame (Lab). Let the particle, mass m, have momentum $\mathbf{p} = (p, \theta, \varphi)$ in the CM and $l^{-1}\mathbf{p} \equiv \mathbf{p}_{\rm L} = (p_{\rm L}, \theta_{\rm L}, \varphi)$ in the Lab. The boost is along the negative Z-axis with speed $\beta_{\rm Lab}$ (i.e. the speed of the Lab as seen in the CM frame).

In (2.2.8) and (2.2.7) we have $\delta = \pi - \theta$, $\varphi_\beta = \theta_\beta = \pi$ and thus $\eta = \pi$ so that

$$\begin{aligned} |\mathbf{p};\lambda\rangle_{\rm Lab} &= (-1)^{\lambda-\lambda'} d_{\lambda'\lambda}(\theta_{\rm Wick})|\mathbf{p}_{\rm L};\lambda'\rangle \\ &= d_{\lambda'\lambda}(\alpha)|\mathbf{p}_{\rm L};\lambda'\rangle \end{aligned} \tag{2.2.12}$$

where

$$\begin{aligned} \cos\alpha &= \frac{\gamma_{\rm Lab}}{p_{\rm L}}(p + \beta_{\rm Lab} E\cos\theta) \\ \sin\alpha &= \frac{m\gamma_{\rm Lab}\beta_{\rm Lab}}{p_{\rm L}}\sin\theta \end{aligned} \quad . \tag{2.2.13}$$

Another convenient expression for $\sin\alpha$ is

$$\sin\alpha = \frac{m}{E_{\rm L}}(\sin\theta\cos\theta_{\rm L} - \gamma_{\rm Lab}\cos\theta\sin\theta_{\rm L}) \tag{2.2.14}$$

For an elastic reaction $A + B \to A + B$, with B the target in the Lab, one finds for the final state B particle

$$\alpha_B = \theta_{\rm R} \equiv \text{Lab recoil angle.} \tag{2.2.15}$$

For elastic scattering of equal-mass particles, e.g. $pp \to pp$, in addition one finds for the final state A particle, which is scattered through $\theta_{\rm L}$ in the Lab frame,

$$\alpha_A = \theta_{\rm L} \equiv \text{Lab scattering angle.} \tag{2.2.16}$$

2.2.5 Non-relativistic limit of CM to Lab transformation

For a non-relativistic collision we have $\gamma_{\rm Lab} \to 1$, $E_{\rm L} \to m$, and from (2.2.14) we find

$$\alpha = \theta - \theta_{\rm L}, \tag{2.2.17}$$

which is what we would expect non-relativistically given that the helicity is the spin projection along the direction of motion.

2.2.6 Ultra high energy collisions

Consider a very high energy collision in the Lab, which produces particles all of which are highly relativistic in the CM. Then $\beta_{\mathrm{Lab}} \approx 1$, $E \approx p$ and $p_{\mathrm{L}} \approx \gamma_{\mathrm{Lab}} E(1+\cos\theta)$, provided that $\theta \neq 180°$. Then from (2.2.13)

$$\sin\alpha \approx \frac{m}{E}\left(\frac{\sin\theta}{1+\cos\theta}\right) = \frac{m}{E}\tan\left(\frac{\theta}{2}\right). \qquad (2.2.18)$$

For a two-body reaction $A + B \to C + D$ we have $E_C \approx E_D \approx \sqrt{s}/2$ where $\sqrt{s}$ is the total CM energy. Thus

$$\sin\alpha \approx \frac{2m}{\sqrt{s}}\tan\left(\frac{\theta}{2}\right), \qquad (2.2.19)$$

showing that $\alpha \to 0$ as $s \to \infty$ at fixed θ or at fixed momentum transfer to the scattered particle.

Hence follows the important result that a particle for which $m/\sqrt{s} \to 0$ does not undergo a Wick helicity rotation in the transformation CM to Lab.

2.2.7 Massless particles

The transformation of the helicity state for a massless particle can be deduced from the previous results by putting $m = 0$. Thus, under an arbitrary rotation, (2.2.2) continues to hold but under an arbitrary boost $l(\boldsymbol{\beta})$, $\theta_{\mathrm{Wick}} = 0$ and instead of (2.2.7) we have

$$|\mathbf{p};\lambda\rangle_{S^{l(\boldsymbol{\beta})}} = |l^{-1}\mathbf{p};\lambda\rangle, \qquad (2.2.20)$$

so that the helicity label is unaltered by a boost.

2.2.8 The Thomas precession

We shall give what we hope is an intelligible derivation of this famous effect, which so baffled physicists at the time of the discovery of intrinsic spin.

Let $\mathbf{s}$ be the expectation or mean value of the spin operator $\hat{\mathbf{s}}$ for an electron of charge $-e$. The electron's intrinsic magnetic moment $\boldsymbol{\mu}$ is given in Gaussian units by

$$\boldsymbol{\mu} = -\frac{ge}{2mc}\mathbf{s}, \qquad (2.2.21)$$

where g is the gyromagnetic factor, which is very nearly equal to 2. For non-relativistic motion we expect $\mathbf{s}$ to obey a classical equation of motion. In particular, for a magnetic field in the rest frame of the particle, $\mathring{\mathbf{B}}$, we expect to have

$$\frac{d\mathbf{s}}{dt} = \boldsymbol{\mu} \times \mathring{\mathbf{B}} = -\frac{ge}{2mc}\mathbf{s} \times \mathring{\mathbf{B}}. \qquad (2.2.22)$$

Consider an electron that at time t has velocity $\mathbf{v}$ in some fixed reference frame, in which there is an electric field $\mathbf{E}$. If we Lorentz-transform to the electron's comoving canonical rest frame S_t^0 at that instant we shall find a magnetic field $\overset{\circ}{\mathbf{B}}$ that, to order v/c, is given by

$$\overset{\circ}{\mathbf{B}} = -\frac{\mathbf{v}}{c} \times \mathbf{E}. \tag{2.2.23}$$

It was originally supposed that a correct description of the motion of $\mathbf{s}$ was thus given by

$$\frac{d\mathbf{s}}{dt} = -\frac{\mu}{c}\mathbf{s} \times (\mathbf{v} \times \mathbf{E}) = \frac{ge}{2mc^2}\mathbf{s} \times (\mathbf{v} \times \mathbf{E}), \tag{2.2.24}$$

but this leads, in hydrogenic-type atoms, to a spin–orbit interaction that is too large by a factor of 2.

To see that (2.2.24) is incorrect, imagine a situation in which there is *no* torque acting on $\boldsymbol{\mu}$ or $\mathbf{s}$ in the canonical rest frame. We shall use the canonical definition of the spin, so that $\mathbf{s}(t)$ is the non-relativistic spin vector in the canonical rest frame S_t^0 reached from our reference frame, the Lab S_L, say, at time t when the electron has velocity $\mathbf{v}$. Thus $\mathbf{s}(t)$ is the spin vector in

$$S_t^0 = l(\mathbf{v})S_\mathrm{L}. \tag{2.2.25}$$

In the following we ignore time dilatations since they turn out to be irrelevant to our accuracy.

As viewed from the canonical rest frame S_t^0, the electron is at rest at time t but has accelerated to some infinitesimal velocity $d\overset{\circ}{\mathbf{v}}$ at time $t + dt$. The motion is wholly non-relativistic and there is no physical torque, so the mean spin vector in S_t^0 at time $t + dt$ should still be $\mathbf{s}(t)$. But this is equivalent to saying that $\mathbf{s}(t)$ is the mean spin vector in the canonical rest frame $S_{t+dt}^{0'}$ reached from S_t^0 by the infinitesimal boost $l(d\overset{\circ}{\mathbf{v}})$ (see Fig. 2.1),

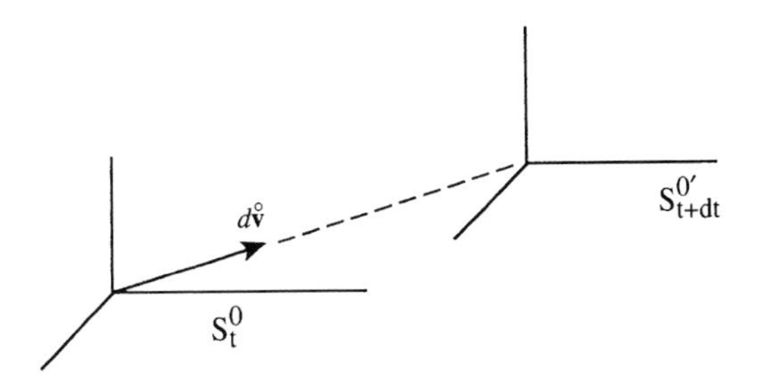

Fig. 2.1. Boost from S_t^0 to $S_{t+dt}^{0'}$.

i.e.

$$\left(\mathbf{s}(t+dt)\right)_{S^{0'}_{t+dt}} = \mathbf{s}(t) \tag{2.2.26}$$

We now have the following situation at time $t+dt$. The mean spin vector is $\mathbf{s}(t+dt)$ in the canonical rest frame S^0_{t+dt} reached from S_L; it is $\mathbf{s}(t)$ in the canonical rest frame $S^{0'}_{t+dt}$ reached from the Lorentz-transformed frame $S^0_t = l(\mathbf{v})S_L$.

From our earlier discussion we know that S^0_{t+dt} and $S^{0'}_{t+dt}$ are not generally the same rest frame and are related by a Wigner rotation. From (2.1.12)

$$S^0_{t+dt} = r_{\text{Wig}}[l(\mathbf{v}), \mathbf{v}+d\mathbf{v}]S^{0'}_{t+dt}. \tag{2.2.27}$$

It follows that

$$\mathbf{s}(t+dt) = r^{-1}_{\text{Wig}}\left(\mathbf{s}(t+dt)\right)_{S^{0'}_{t+dt}} = r^{-1}_{\text{Wig}}\mathbf{s}(t). \tag{2.2.28}$$

Thus, even in the absence of a physical torque, $\mathbf{s}(t+dt) \neq \mathbf{s}(t)$. To find the intrinsic rate of change of $\mathbf{s}$ we study the Wigner rotation, taking into account that $d\mathbf{v}$ is infinitesimal.

We have from (2.1.11), since $d\mathring{\mathbf{v}} = \left[l^{-1}(\mathbf{v})\right](\mathbf{v}+d\mathbf{v})$,

$$r_{\text{Wig}}[l(\mathbf{v}), \mathbf{v}+d\mathbf{v}] = l^{-1}(d\mathring{\mathbf{v}})l^{-1}(\mathbf{v})l(\mathbf{v}+d\mathbf{v}). \tag{2.2.29}$$

To identify the rotation involved we evaluate the matrix $\Lambda(r_{\text{Wig}})$, using (1.2.13) and working to first order in $d\mathbf{v}$. Note that to this order

$$d\mathring{\mathbf{v}} = \gamma^2 d\mathbf{v}_{\parallel} + \gamma d\mathbf{v}_{\perp} \tag{2.2.30}$$

where $\parallel$ and $\perp$ are relative to the direction of $\mathbf{v}$ and $\gamma = (1 - v^2/c^2)^{-1/2}$.

We find eventually

$$\mathbf{s}(t+dt) = \left[r^{-1}(d\boldsymbol{\vartheta})\right]\mathbf{s}(t) \tag{2.2.31}$$

where

$$d\boldsymbol{\vartheta} = \frac{\gamma^2}{1+\gamma}\left(\frac{\mathbf{v}\times d\mathbf{v}}{c^2}\right). \tag{2.2.32}$$

From this follows

$$\frac{d\mathbf{s}}{dt} = \boldsymbol{\omega}_T \times \mathbf{s}, \tag{2.2.33}$$

where the Thomas angular velocity is

$$\boldsymbol{\omega}_{\text{T}} = \frac{\gamma^2}{1+\gamma}\left(\frac{\mathbf{a}\times\mathbf{v}}{c^2}\right) \approx \frac{1}{2}\left(\frac{\mathbf{a}\times\mathbf{v}}{c^2}\right), \tag{2.2.34}$$

$\mathbf{a} = d\mathbf{v}/dt$ being the electron's acceleration at time t.

Thus owing to the interpretation of $\mathbf{s}(t)$ as a vector in the canonical rest frame we find that $\mathbf{s}(t)$ rotates even when no physical torque acts on it in the rest frame. Clearly, then, in the presence of a magnetic torque (2.2.24) should be modified to

$$\frac{d\mathbf{s}}{dt} = \frac{ge}{2mc^2}\mathbf{s} \times (\mathbf{v} \times \mathbf{E}) + \boldsymbol{\omega}_{\mathrm{T}} \times \mathbf{s}. \tag{2.2.35}$$

For a one-electron Coulombic atom, with potential $V(r)$,

$$(-e)\mathbf{E} = -\frac{1}{r}\frac{dV}{dr}\mathbf{r}$$

and

$$\mathbf{a} = -\frac{e\mathbf{E}}{m},$$

leading, via (2.2.34), to

$$\frac{d\mathbf{s}}{dt} = \frac{g-1}{2m^2c^2}\left(\frac{1}{r}\frac{dV}{dr}\right)\mathbf{L} \times \mathbf{s}. \tag{2.2.36}$$

We see that for $g = 2$ the Thomas term just halves the strength of the spin–orbit interaction.

In Section 3.4 we shall introduce a *covariant* mean spin 4-vector and in subsection 6.3.1 derive relativistically covariant equations for its motion. They will offer a more direct derivation of the above results.

2.3 The discrete transformations

We now consider how helicity states transform under space inversion and time reversal. These results are crucial to an understanding of the physical consequences of these symmetries in specific reactions. We also briefly discuss charge conjugation.

2.3.1 Parity

Under space inversion, $S \to S^{\mathscr{P}} = l_{\mathscr{P}}S$ such that $x \to x' = (t, -\mathbf{x})$. The Hilbert space operator $U(l_{\mathscr{P}}^{-1})$ is usually written as $\mathscr{P}$ and has the following effect on the Lorentz generators $\hat{\mathbf{J}} = \left\{\hat{J}_i\right\}, \hat{\mathbf{K}} = \left\{\hat{K}_i\right\}$, see (1.2.1):

$$\mathscr{P}^{-1}\hat{\mathbf{J}}\mathscr{P} = \hat{\mathbf{J}} \tag{2.3.1}$$

$$\mathscr{P}^{-1}\hat{\mathbf{K}}\mathscr{P} = -\hat{\mathbf{K}}. \tag{2.3.2}$$

The operator $\mathscr{P}$ is unitary and taken to satisfy $\mathscr{P}^2 = 1$. Under $S \to S^{\mathscr{P}}$ we have, as in (2.1.1),

$$|\mathbf{p}; \lambda\rangle \to U(l_{\mathscr{P}}^{-1})|\mathbf{p}; \lambda\rangle \equiv \mathscr{P}|\mathbf{p}; \lambda\rangle \tag{2.3.3}$$

Consider the action of $\mathscr{P}$ on the helicity state of a massive particle with spin s

$$|\mathbf{p};\lambda\rangle \equiv |p,\theta,\varphi;\lambda\rangle = U[h(\mathbf{p})]|\overset{\circ}{p};\lambda\rangle$$
$$= U[r(\varphi,\theta,0)l_z(v)]|\overset{\circ}{p};\lambda\rangle; \tag{2.3.4}$$

we have

$$\mathscr{P}|p,\theta,\varphi;\lambda\rangle = U[r(\varphi,\theta,0)l_z(-v)]\mathscr{P}|\overset{\circ}{p};\lambda\rangle. \tag{2.3.5}$$

The intrinsic parity $\eta_{\mathscr{P}}$ is defined by

$$\mathscr{P}|\overset{\circ}{p};\lambda\rangle = \eta_{\mathscr{P}}|\overset{\circ}{p};\lambda\rangle \tag{2.3.6}$$

with $\eta_{\mathscr{P}}^2 = 1$. After some manipulation, using

$$l_z(-v) = r_y(-\pi)l_z(v)r_y(\pi)$$

we find

$$\mathscr{P}|p,\theta,\varphi;\lambda\rangle = \eta_{\mathscr{P}}e^{-i\pi s}|p,\pi-\theta,\varphi+\pi;-\lambda\rangle.^1 \tag{2.3.7}$$

For massless particles we have already defined the intrinsic parity in (1.2.31). For the operator for reflections in the XZ-plane, $\mathscr{Y} = r_y(\pi)\mathscr{P}$, we have

$$\mathscr{Y}|p,\theta,\varphi;\lambda\rangle = \eta_{\mathscr{P}}(-1)^{s-\lambda}|p,\theta,-\varphi;-\lambda\rangle \tag{2.3.8}$$

which is consistent with (1.2.31) since there $\lambda = s$ and $\varphi = 0$.

2.3.2 Time reversal

The time-reversal operator $\mathscr{T}$ is an anti-unitary operator (i.e. $\mathscr{T}$ is anti-linear with $\mathscr{T}^{-1} = \mathscr{T}^\dagger$), which has the following action on the Lorentz generators:

$$\begin{aligned} \mathscr{T}^{-1}\hat{\mathbf{J}}\mathscr{T} &= -\hat{\mathbf{J}} \\ \mathscr{T}^{-1}\hat{\mathbf{K}}\mathscr{T} &= \hat{\mathbf{K}}. \end{aligned} \tag{2.3.9}$$

Because of the anti-linearity these imply

$$\begin{aligned} \mathscr{T}^{-1}r\mathscr{T} &= r \\ \mathscr{T}^{-1}l\mathscr{T} &= l^{-1} \end{aligned} \tag{2.3.10}$$

for any rotation r and pure boost l.

---

[1] Of course the vector $(p,\pi-\theta,\varphi+\pi)$ is just $-\mathbf{p}$, but we are loth to use that notation since e.g. $|-(-\mathbf{p});\lambda\rangle \neq |\mathbf{p};\lambda\rangle$. Indeed, with $-\mathbf{p} = (p,\pi-\theta,\varphi+\pi)$

$$|-(-\mathbf{p});\lambda\rangle = |p,\theta,\varphi+2\pi;\lambda\rangle = (-1)^{2s}|p,\theta,\varphi;\lambda\rangle$$
$$= \pm|\mathbf{p};\lambda\rangle,$$

the plus sign corresponding to bosons and the minus sign to fermions.

Because of its anti-linearity care must be exercised when using $\mathscr{T}$ inside matrix elements, and it is safer to revert to a Hilbert-space notation for these rather than the Dirac notation. We recall that for any operator O

$$\langle\beta|O|\alpha\rangle \equiv (\beta, O\alpha). \tag{2.3.11}$$

For a linear operator $\hat{L}$ the hermitian conjugate $\hat{L}^\dagger$ is defined by

$$(\beta, \hat{L}\alpha) = (\hat{L}^\dagger\beta, \alpha) = (\alpha, L^\dagger\beta)^*, \tag{2.3.12}$$

so that, as usual,

$$\langle\beta|\hat{L}|\alpha\rangle = \langle\alpha|\hat{L}^\dagger|\beta\rangle^*. \tag{2.3.13}$$

For the anti-linear operator $\mathscr{T}$ the hermitian conjugate $\mathscr{T}^\dagger$ has to be defined by

$$(\beta, \mathscr{T}\alpha) = (\mathscr{T}^\dagger\beta, \alpha)^* = (\alpha, \mathscr{T}^\dagger\beta). \tag{2.3.14}$$

It is therefore safer to use the notation $|\mathscr{T}\alpha\rangle$ rather than $\mathscr{T}|\alpha\rangle$ for the time-reversed state of $|\alpha\rangle$. Thus, under $S \to S^{\mathscr{T}} = l_{\mathscr{T}}S$ such that $x' = l_{\mathscr{T}}^{-1}x = (-t, \mathbf{x})$,

$$|\mathbf{p};\lambda\rangle \to |\mathscr{T}(\mathbf{p},\lambda)\rangle \tag{2.3.15}$$

We follow the convention used by Jacob and Wick (1959) and take, for a particle at rest,

$$|\mathscr{T}(\mathring{p},\lambda)\rangle = (-1)^{s-\lambda}|\mathring{p};-\lambda\rangle. \tag{2.3.16}$$

Note that with this convention $\mathscr{T}^2 = (-1)^{2s}$.

It follows from (2.3.16) and (2.3.10) that

$$|\mathscr{T}(p,\theta,\varphi;\lambda)\rangle = e^{-i\pi\lambda}|p,\pi-\theta,\varphi+\pi;\lambda\rangle \tag{2.3.17}$$

and the same result holds for massless particles.

Note that for any linear operator $\hat{L}$ one has

$$\begin{aligned}\langle\mathscr{T}\alpha|\hat{L}|\mathscr{T}\beta\rangle &= (\mathscr{T}\alpha, \hat{L}\mathscr{T}\beta) = (\alpha, \mathscr{T}^\dagger\hat{L}\mathscr{T}\beta)^* \\ &= \langle\alpha|\mathscr{T}^\dagger\hat{L}\mathscr{T}|\beta\rangle^* \\ &= \langle\beta|\mathscr{T}^\dagger\hat{L}^\dagger\mathscr{T}|\alpha\rangle,\end{aligned} \tag{2.3.18}$$

the last step following since $\mathscr{T}^\dagger\hat{L}\mathscr{T}$ is a linear operator.

Time-reversal invariance is usually taken to mean that, for transition amplitudes or S-matrix elements,

$$\langle\mathscr{T}\alpha|S|\mathscr{T}\beta\rangle = \langle\beta|S|\alpha\rangle. \tag{2.3.19}$$

From (2.3.18) we see that time-reversal invariance implies

$$\mathscr{T}^{-1}S\mathscr{T} = S^\dagger \tag{2.3.20}$$

in contrast to all linear invariances, where there would be no dagger symbol on the right-hand side.

2.3.3 Charge conjugation

The charge conjugation operator $\mathscr{C}$ ($\mathscr{C}^2 = 1$) changes particles into antiparticles and vice versa. For a particle A at rest

$$\mathscr{C}|A;\mathring{p},\lambda\rangle = \eta_{\mathscr{C}}|\overline{A};\mathring{p},\lambda\rangle \tag{2.3.21}$$

where $\eta_{\mathscr{C}} = \pm 1$ is the *charge parity* of the particle. Since $\mathscr{C}$ has no effect on the kinematic variables, we have also

$$\mathscr{C}|A;\mathbf{p},\lambda\rangle = \eta_{\mathscr{C}}|\overline{A};\mathbf{p},\lambda\rangle. \tag{2.3.22}$$

Note that $\eta_{\mathscr{C}} = +1$ for pions and nucleons, -1 for photons.

We remind the reader that some care must be exercised when dealing with multiplets of an internal symmetry. For example, if protons and neutrons are regarded as forming an isotopic spin doublet of the nucleon N, so that

$$|N;I_z = 1/2\rangle = |p\rangle\,, \qquad |N;I_z = -1/2\rangle = |n\rangle\,, \tag{2.3.23}$$

then the antinucleon multiplet that transforms like an isospin doublet is

$$|\overline{N};I_z = 1/2\rangle = -\,|\overline{n}\rangle\,, \qquad |\overline{N};I_z = -1/2\rangle = |\overline{p}\rangle\,. \tag{2.3.24}$$

This is explained in subsection 2.4.2.

2.4 Fields and wave functions

On the one had we saw in Section 2.1 that under Lorentz transformations the state vector in a relativistic theory transforms in a complicated way, the transformation matrix depending upon the Wick helicity rotation or the Wigner rotation.

On the other hand, in setting up a field theory it is customary to use fields that transform simply under Lorentz transformations. Thus if a Lorentz transformation l acting on the reference frame S takes it to S^l,

$$S \xrightarrow{l} S^l,$$

so that $x \to x' = l^{-1}x$, then the fields $\phi_n(x), n = 1,\ldots,N$, are taken to undergo the transformation $\phi_n(x) \to \phi_n'(x')$ where

$$\phi_n'(x) = U(l)\phi_n(x)U(l^{-1}) = D_{nm}(l^{-1})\phi_m(lx). \tag{2.4.1}$$

here D_{nm} is an N-dimensional representation of the homogeneous Lorentz group (see Appendix 2). Note that the matrices depend *only* on l.

We consider here some aspects of the relationship between the two approaches.

The fields $\phi_n(x)$ are generally not irreducible, in the sense that they have more components (N) than are needed to describe quanta of some given spin s, i.e. $N > 2s + 1$. As a consequence the representation $D^{(N)}$ may be reducible under pure rotations, as, for example, when massive spin-1 quanta are described by a Lorentz 4-vector, or they may even be reducible under all homogeneous Lorentz transformations, as in the case when spin-1/2 quanta are described by a four-component Dirac field. (In the latter case the representation becomes irreducible if the operation of space inversion is included.)

In order to construct Lorentz-invariant lagrangians etc. it is useful to deal with conjugate fields $\overline{\phi}_n(x)$. These may be just the hermitian conjugate fields $\phi_n^\dagger(x)$ or some fixed linear combination of these (e.g. $\overline{\Psi}(x) = \Psi^\dagger(x)\beta$ in the Dirac theory) so designed that $\overline{\phi}$ transforms contra-grediently to ϕ, i.e. under $S \xrightarrow{l} S^l$, $\overline{\phi}_n(x) \to \overline{\phi}'_n(x')$ where

$$\overline{\phi}'_n(x) = U(l)\overline{\phi}_n(x)U(l^{-1}) = \overline{\phi}_m(lx)D_{mn}(l). \tag{2.4.2}$$

Thus in matrix notation, regarding ϕ as a column vector and $\overline{\phi}$ as a row vector:

$$\begin{aligned} \phi'(x') &= D^{-1}(l)\phi(x) \\ \overline{\phi}'(x') &= \overline{\phi}(x)D(l), \end{aligned} \tag{2.4.3}$$

so that $\overline{\phi}\phi$ is a scalar, i.e.

$$\overline{\phi}'(x')\phi'(x') = \overline{\phi}(x)\phi(x). \tag{2.4.4}$$

The use of $\overline{\phi}$ and ϕ makes it quite simple to construct quantities with definite transformation properties under Lorentz transformations. But some price has to be paid for the redundant components; this price is the existence of field equations that must be satisfied even by non-interacting fields. These equations are nothing more than invariant conditions of constraint upon the unwanted components. In a series of elegant papers Weinberg (1964a, 1964b) showed how one may construct irreducible fields ϕ_λ with only $2s + 1$ components. These satisfy no field equations (other than the Klein–Gordon equation, which just imposes the correct relation between energy and momentum) but they do not transform simply under Lorentz transformations. They shed an interesting light upon the whole question of fields and field equations and we therefore give a brief discussion of this approach in Appendix 3. Here we continue to deal with the usual fields $\phi_n(x)$.

The fields $\phi_n(x)$, $\overline{\phi}_n(x)$ are Fourier expanded in terms of creation and annihilation operators ($a^\dagger$, a for particles and $b^\dagger$, b for antiparticles), which create and annihilate quanta of spin s with definite momenta and helicity.

Thus one writes

$$\phi_n(x) = \sum_\lambda \int \frac{d^3\mathbf{p}}{(2\pi)^{3/2}2p^0} \left[u_n(\mathbf{p},\lambda)a(\mathbf{p},\lambda)e^{-ip\cdot x} + v_n(\mathbf{p},\lambda)b^\dagger(\mathbf{p},\lambda)e^{ip\cdot x}\right] \quad (2.4.5)$$

$$\overline{\phi}(x) = \sum_\lambda \int \frac{d^3\mathbf{p}}{(2\pi)^{3/2}2p^0} \left[\overline{u}_n(\mathbf{p},\lambda)a^\dagger(\mathbf{p},\lambda)e^{ip\cdot x} + \overline{v}_n(\mathbf{p},\lambda)b(\mathbf{p},\lambda)e^{-ip\cdot x}\right] \quad (2.4.6)$$

where the u and v are 'wave functions' for the quanta (in the Dirac case they just correspond to the Dirac 4-spinors u, v).

Since $a^\dagger(\mathbf{p},\lambda)$ creates the state $|\mathbf{p};\lambda\rangle$ from the Lorentz invariant vacuum, it follows from eqn (2.1.9) and the unitarity of the representations of the rotation group that

$$U(l)a(\mathbf{p},\lambda)U(l^{-1}) = \mathscr{D}^{(s)}_{\lambda\lambda'}(r)a(l\mathbf{p},\lambda'), \quad (2.4.7)$$

where $r = r(l,\mathbf{p})$ is the Wick rotation defined in eqn (2.1.7).

For free fields or fields in the interaction representation and with particle states such that

$$|\mathbf{p};\lambda\rangle = a^\dagger(\mathbf{p},\lambda)|0\rangle, \quad (2.4.8)$$

where the operators satisfy commutation or anticommutation relations

$$\left[a(\mathbf{p},\lambda), a^\dagger(\mathbf{p}',\lambda')\right]_\pm = 2p^0\delta(\mathbf{p}'-\mathbf{p})\delta_{\lambda'\lambda}, \quad (2.4.9)$$

one has

$$\langle 0|\phi_n(x)|\mathbf{p};\lambda\rangle = \frac{u_n(\mathbf{p},\lambda)e^{-ip\cdot x}}{(2\pi)^{3/2}} \quad (2.4.10)$$

and for antiparticles

$$\langle 0|\overline{\phi}_n(x)|\overline{p};\lambda\rangle = \frac{\overline{v}_n(\mathbf{p},\lambda)e^{-ip\cdot x}}{(2\pi)^{3/2}}. \quad (2.4.11)$$

The set of wave functions $u_n(\mathbf{p},\lambda)$ will be said to correspond to the state $|\mathbf{p};\lambda\rangle$:

$$|\mathbf{p};\lambda\rangle \longleftrightarrow u_n(\mathbf{p},\lambda). \quad (2.4.12)$$

Clearly the $u_n(\mathbf{p},\lambda)e^{-ip\cdot x}$ satisfy the same free-field equations as do the $\phi_n(x)$. Thus the u_n are usually obtained by solving those equations, but care must be exercised in order to have consistent phase conventions. Thus if

$$|\mathbf{p};\lambda\rangle \longleftrightarrow u_n(\mathbf{p},\lambda)$$

and

$$|\mathring{p};s_z=\lambda\rangle \longleftrightarrow u_n(\mathring{p},\lambda)$$

then from (1.2.25) and (2.4.1), using the Lorentz invariance of the vacuum,

$$\begin{aligned}\langle 0|\phi_n(x)|\mathbf{p},\lambda\rangle &= \langle 0|\phi_n(x)U[h(\mathbf{p})]|\overset{\circ}{p};\lambda\rangle \\ &= \langle 0|U^{-1}[h(\mathbf{p})]\phi_n(x)U[h(\mathbf{p})]|\overset{\circ}{p};\lambda\rangle \\ &= D_{nm}[h(\mathbf{p})]\langle 0|\phi_m(h^{-1}x)|\overset{\circ}{p};\lambda\rangle,\end{aligned} \tag{2.4.13}$$

which leads, via (2.4.10), to the requirement that

$$u_n(\mathbf{p},\lambda) = D_{nm}[h(\mathbf{p})]u_m(\overset{\circ}{p},\lambda). \tag{2.4.14}$$

A similar argument, for antiparticles, leads to

$$\bar{v}_n(\mathbf{p},\lambda) = \bar{v}_m(\overset{\circ}{p},\lambda)D_{mn}[h^{-1}(\mathbf{p})]. \tag{2.4.15}$$

Consider now the effect of an arbitrary Lorentz transformation $S \xrightarrow{l} S^l$. Using eqns (2.1.3), (2.1.9) and (2.4.1) in (2.4.10), we have the correspondence

$$|\mathbf{p};\lambda\rangle \longleftrightarrow u_n(\mathbf{p},\lambda)$$

and

$$\begin{aligned}U(l^{-1})|\mathbf{p};\lambda\rangle \longleftrightarrow\ & D_{nm}(l^{-1})u_m(\mathbf{p},\lambda) \\ &= u_n(l^{-1}\mathbf{p},\lambda')\mathscr{D}^{(s)}_{\lambda'\lambda}(r)\end{aligned} \tag{2.4.16}$$

where $r = r(l,\mathbf{p})$.

In a similar way one finds for antiparticles

$$|\bar{\mathbf{p}};\lambda\rangle \longleftrightarrow \bar{v}_n(\mathbf{p},\lambda)$$

and

$$\begin{aligned}U(l^{-1})|\bar{\mathbf{p}};\lambda\rangle \longleftrightarrow\ & \bar{v}_m(\mathbf{p},\lambda)D_{mn}(l) \\ &= \bar{v}_n(l^{-1}\mathbf{p},\lambda')\mathscr{D}^{(s)}_{\lambda'\lambda}(r)\end{aligned} \tag{2.4.17}$$

and, in addition,

$$\begin{aligned}\langle\mathbf{p},\lambda| \longleftrightarrow\ & \bar{u}_n(\mathbf{p},\lambda) \\ \langle\mathbf{p},\lambda|\,U(l) \longleftrightarrow\ & \bar{u}_m(\mathbf{p},\lambda)D_{mn}(l) \\ &= \bar{u}_n(l^{-1}\mathbf{p},\lambda')\mathscr{D}^{(s)}_{\lambda\lambda'}(r^{-1})\end{aligned} \tag{2.4.18}$$

and for antiparticles

$$\begin{aligned}\langle\bar{\mathbf{p}},\lambda| \longleftrightarrow\ & v_n(\mathbf{p},\lambda) \\ \langle\bar{\mathbf{p}},\lambda|\,U(l) \longleftrightarrow\ & D_{nm}(l^{-1})v_m(\mathbf{p},\lambda) \\ &= v_n(l^{-1}\mathbf{p},\lambda')\mathscr{D}^{(s)}_{\lambda\lambda'}(r^{-1}).\end{aligned} \tag{2.4.19}$$

2.4.1 The discrete transformations of the fields

Consider the discrete transformations. Under space inversion

$$S \xrightarrow{l_{\mathscr{P}}} S^{\mathscr{P}} = l_{\mathscr{P}} S$$

with $x \to x' = l_{\mathscr{P}} x = (t, -\mathbf{x})$, one takes $\phi_n(x) \to \phi_n^{\mathscr{P}}(x')$ with (see Section 2.3)

$$\phi_n^{\mathscr{P}}(x) = \mathscr{P}^{-1}\phi_n(x)\mathscr{P} = P_{nm}\phi_m(t, -\mathbf{x}) \tag{2.4.20}$$

where P is an $N \times N$ matrix ($P^2 = I$) chosen so that $\phi_n^{\mathscr{P}}(x')$ satisfies the space-inverted field equations. This does not fix the absolute phase of P. However, using eqn (2.3.7) we have for a particle of spin s

$$\begin{aligned}\langle 0|\phi_n(x)\mathscr{P}|p, \theta, \varphi; \lambda\rangle &= \eta_{\mathscr{P}} e^{-i\pi s}\langle 0|\phi_n(x)|p, \pi - \theta, \varphi + \pi; -\lambda\rangle. \\ &= \langle 0|\mathscr{P}^{-1}\phi_n(x)\mathscr{P}|p, \theta, \varphi; \lambda\rangle \\ &= P_{nm}\langle 0|\phi_m(t, -\mathbf{x})|p, \theta, \varphi; \lambda\rangle \end{aligned} \tag{2.4.21}$$

from which, via (2.4.10), we have that P must be chosen such that

$$P_{nm}u_m(p, \theta, \varphi; \lambda) = \eta_{\mathscr{P}} e^{-i\pi s}u_n(p, \pi - \theta, \varphi + \pi; -\lambda). \tag{2.4.22}$$

For antiparticles one has, since $P^2 = I$, i.e. $P^{-1} = P$,

$$\overline{v}_m(p, \theta, \varphi; \lambda)P_{mn} = \overline{\eta}_{\mathscr{P}} e^{-i\pi s}\overline{v}_n(p, \pi - \theta, \varphi + \pi; -\lambda) \tag{2.4.23}$$

where $\overline{\eta}_{\mathscr{P}}$ is the intrinsic parity of the antiparticle.

We also have the following correspondence between states and wave functions:

$$\mathscr{P}|\mathbf{p}; \lambda\rangle \longleftrightarrow P_{nm}u_m(\mathbf{p}, \lambda) \tag{2.4.24}$$

$$\mathscr{P}|\overline{\mathbf{p}}; \lambda\rangle \longleftrightarrow \overline{v}_m(\mathbf{p}, \lambda)P_{mn}. \tag{2.4.25}$$

As an example, in the Dirac case it is conventional to choose $P = \gamma^0$. For the particle at rest, the use of (2.4.24) and (2.4.25) in (2.3.6) and its analogue for antiparticles shows that we must then choose $\eta_{\mathscr{P}} = 1$ and $\overline{\eta}_{\mathscr{P}} = -1$.

Consider now the anti-unitary time-reversal operation

$$S \xrightarrow{l_{\mathscr{T}}} S^{\mathscr{T}} = l_{\mathscr{T}} S$$

(see subsection 2.3.2) with $x \to x' = l_{\mathscr{T}}^{-1} x = (-t, \mathbf{x})$. One takes $\phi_n^{\mathscr{T}}(x) \to \phi_n^{\mathscr{T}}(x')$ with

$$\phi_n^{\mathscr{T}}(x) = \mathscr{T}^{-1}\phi_n(x)\mathscr{T} = T_{nm}\phi_m(-t, \mathbf{x}), \tag{2.4.26}$$

where T is an $N \times N$ matrix with $T^*T = (-1)^{2s}I$, chosen such that $\phi_n^{\mathscr{T}}(x)$ satisfies the time-reversed equations. Its phase is fixed as follows. Using eqns (2.3.16), (2.3.18) and (2.4.26) we find

$$\begin{aligned}\langle 0|\phi_n(x)|\mathscr{T}(p,\theta,\varphi;\lambda)\rangle &= e^{-i\pi\lambda}\langle 0|\phi_n(x)|p,\pi-\theta,\varphi+\pi;\lambda\rangle \\ &= \langle 0|\mathscr{T}^{-1}\phi_n(x)\mathscr{T}|p,\theta,\varphi;\lambda\rangle^* \\ &= T^*_{nm}\langle 0|\phi_m(-t,\mathbf{x})|p,\theta,\varphi;\lambda\rangle^* \end{aligned} \tag{2.4.27}$$

from which we have the requirement

$$T^*_{nm}u^*_m(p,\theta,\varphi;\lambda) = e^{-i\pi\lambda}u_n(p,\pi-\theta,\varphi+\pi;\lambda) \tag{2.4.28}$$

or

$$T_{nm}u_m(p,\theta,\varphi;\lambda) = e^{i\pi\lambda}u^*_n(p,\pi-\theta,\varphi+\pi;\lambda). \tag{2.4.29}$$

Similarly, for antiparticles

$$\bar{v}_m(p,\theta,\varphi;\lambda)T_{mn} = e^{i\pi\lambda}\bar{v}^*_n(p,\pi-\theta,\varphi+\pi;\lambda). \tag{2.4.30}$$

Note that one has the correspondence between states and wave functions

$$|\mathscr{T}(p,\theta,\varphi;\lambda)\rangle \longleftrightarrow e^{-i\pi\lambda}u_n(p,\pi-\theta,\varphi+\pi;\lambda) \tag{2.4.31}$$

and for antiparticles

$$|\mathscr{T}(\overline{p,\theta,\varphi;\lambda})\rangle \longleftrightarrow e^{-i\pi\lambda}\bar{v}_n(p,\pi-\theta,\varphi+\pi;\lambda). \tag{2.4.32}$$

With the conventions (1.2.22), for the Dirac case one has $T=\gamma^3\gamma^1$ if we use the standard representation of the γ-matrices, given for example in Bjorken and Drell (1964), in which γ^3 and γ^1 are real.

Finally, under charge conjugation (see subection 2.3.3) we have from eqns (2.3.22) and (2.4.10)

$$\begin{aligned}\frac{u_n(\mathbf{p},\lambda)}{(2\pi)^{3/2}} &= \langle 0|\phi_n(0)|\mathbf{p};\lambda\rangle \\ &= \eta_{\mathscr{C}}\langle 0|\phi_n(0)\mathscr{C}|\overline{\mathbf{p}};\lambda\rangle \\ &= \eta_{\mathscr{C}}\langle 0|\mathscr{C}^{-1}\phi_n(0)\mathscr{C}|\overline{\mathbf{p}};\lambda\rangle, \end{aligned} \tag{2.4.33}$$

which is only possible, via (2.4.11), if

$$\mathscr{C}^{-1}\phi_n(x)\mathscr{C} = \eta_{\mathscr{C}}C_{nm}\overline{\phi}_m(x), \tag{2.4.34}$$

where $\mathscr{C}^2=I$.

Substituted into (2.4.33) this implies that

$$u_n(\mathbf{p},\lambda) = C_{nm}\bar{v}_m(\mathbf{p},\lambda). \tag{2.4.35}$$

For the Dirac case, in the standard representation of the γ-matrices one has $C=i\gamma^2\gamma^0$, with $C^2=-I$.

2.4.2 Isospin multiplets for antiparticles

We mentioned in subsection 2.3.3 that if protons and neutrons are regarded as forming a doublet under isotopic spin rotations,

$$|N; I_z = \tfrac{1}{2}\rangle = |p\rangle \qquad |N; I_z = -\tfrac{1}{2}\rangle = |n, \tag{2.4.36}$$

then the antiparticle doublet that transforms as an isodoublet is

$$|\overline{N}; I_z = \tfrac{1}{2}\rangle = -|\overline{n}\rangle \qquad |\overline{N}; I_z = -\tfrac{1}{2}\rangle = |\overline{p}\rangle\rangle \tag{2.4.37}$$

The source of the minus sign or, for a general isospin multiplet, of certain phase factors can be understood as follows.

Let $|A; I_z\rangle$ be an isospin multiplet of *particles* of type A. Under an isospin rotation r, in complete analogy to ordinary rotations (see (1.1.18) and (1.1.19)) one will have

$$U(r)|A; I_z\rangle = \mathscr{D}^{(I)}_{I'_z I_z}(r)|A; I'_z\rangle \tag{2.4.38}$$

where $U(r)$ is the unitary operator that represents the isotopic spin rotation acting on the state vectors and the $\mathscr{D}^{(I)}$ are the $SU(2)$ representation matrices, whose properties are discussed in Appendix 1.

If the creation operators for the particles are labelled $a^\dagger_{I_z}$ then (2.4.38) is tantamount to having

$$U(r)a^\dagger_{I_z}U^{-1}(r) = \mathscr{D}^{(I)}_{I'_z I_z}(r)a^\dagger_{I'_z} \tag{2.4.39}$$

where we do not display arguments such as momentum, helicity etc. that are irrelevant to the discussion.

Consider now the set of usual fields $\Phi_{I_z}(x)$ corresponding to the set of particles of type A and isospin I. They ought to transform analogously to (2.4.1), except that there is here obviously no effect on the space–time coordinates. So we wish to have

$$U(r)\Phi_{I_z}(x)U^{-1}(r) = \mathscr{D}^{(I)}_{I_z I'_z}(r^{-1})\Phi_{I'_z}(x). \tag{2.4.40}$$

Now the field $\Phi_{I_z}(x)$ contains the *annihilation* operator a_{I_z} as in (2.4.5), so we have to check that (2.4.39) and (2.4.40) are compatible. Indeed they are, since taking the hermitian conjugate of (2.4.39) yields

$$\begin{aligned} U(r)a_{I_z}U^{-1}(r) &= \mathscr{D}^{(I)\,*}_{I'_z I_z}(r)a_{I'_z} \\ &= \left[\mathscr{D}^{(I)\dagger}(r)\right]_{I_z I'_z} a_{I'_z} \end{aligned}$$

which, using the unitarity of the matrices $\mathscr{D}^{(I)}$, gives

$$U(r)a_{I_z}U^{-1}(r) = \mathscr{D}^{(I)}_{I_z I'_z}(r^{-1})a_{I'_z}, \tag{2.4.41}$$

as required for (2.4.40).

However, the field $\Phi_{I_z}(x)$ also contains the *creation* operators $b^\dagger_{I_z}$, which create the states $|\overline{A, I_z}\rangle$ corresponding to the antiparticles of the particles A_{I_z}. For consistency with (2.4.40) they will have to transform as follows:

$$U(r) b^\dagger_{I_z} U^{-1}(r) = \mathscr{D}^{(I)}_{I_z I'_z}(r^{-1}) b^\dagger_{I'_z}$$

which, as before, via the unitarity nature of $\mathscr{D}^{(I)}$ gives

$$\begin{aligned} U(r) b^\dagger_{I_z} U^{-1}(r) &= \mathscr{D}^{(I)\dagger}_{I_z I'_z}(r) b^\dagger_{I'_z} \\ &= \mathscr{D}^{(I)*}_{I'_z I_z}(r) b^\dagger_{I'_z}. \end{aligned} \tag{2.4.42}$$

Comparing with (2.4.39) and (2.4.38) we have, for the isospin multiplet made up of particles,

$$U(r)|A; I_z\rangle = \mathscr{D}^{(I)}_{I'_z I_z}(r)|A; I'_z\rangle \tag{2.4.43}$$

and, for their antiparticles,

$$U(r)|\overline{A; I_z}\rangle = \mathscr{D}^{(I)*}_{I'_z I_z}(r)|\overline{A; I'_z}\rangle. \tag{2.4.44}$$

In other words the set of antiparticles states $|\overline{A; I_z}\rangle$ does not transform as a standard isospin multiplet.

However, for the group of isospin rotations $SU(2)$ the representations $\mathscr{D}^{(I)}$ and $\mathscr{D}^{(I)^*}$ are *equivalent*, i.e. there exists a unitary matrix $C^{(I)}$, independent of r, such that

$$\mathscr{D}^{(I)^*}(r) = C^{(I)} \mathscr{D}^{(I)}(r) C^{(I)-1} \tag{2.4.45}$$

for all r.

Then the antiparticle multiplet $|\bar{A}; I_z\rangle$ that transforms as a standard isospin multiplet is clearly

$$|\bar{A}; I_z\rangle \equiv C^{(I)}_{I'_z I_z} |\overline{A; I'_z}\rangle, \tag{2.4.46}$$

i.e.

$$U(r)|\bar{A}; I_z\rangle = \mathscr{D}^{(I)}_{I'_z I_z}(r)|\bar{A}; I'_z\rangle. \tag{2.4.47}$$

In fact the matrix $C^{(I)}$ is very simple. It can be taken, conventionally, as

$$C^{(I)}_{ij} = (-1)^{I-i} \delta_{i,-j}. \tag{2.4.48}$$

As an example of (2.4.46) and (2.4.48), for the nucleon isodoublet one finds just the results (2.4.36) and (2.4.37). (Of course the *overall* sign in (2.4.37) is irrelevant and sometimes the opposite convention is used.)

3
The spin density matrix

The state of an ensemble of particles is specified by a density matrix. In any reaction one starts with a knowledge of the density matrix of the initial system (known from its mode of preparation) and one attempts to measure the density matrix of the final system.

The properties of a density matrix are of three kinds.

(1) Firstly, there are properties of a very general nature that follow from the very definition of a density matrix and from the basic postulates of quantum mechanics. To check that a measured density matrix conforms to those requirements is best thought of as a test of the reliability of the experimental measurements.

We shall refer to these as *basic properties*.

(2) Secondly, there are properties of a kinematical-dynamical origin, which reflect the general properties of an interaction, for example its symmetries, but which do not depend on a detailed knowledge of the dynamics. It is important to check that the density matrix measured in a particular reaction does satisfy these properties.

We shall refer to these as *general kinematical-dynamical properties*.

(3) Thirdly, there are properties which depend upon the specific dynamical mechanism in a reaction and which can therefore be used either to learn about these mechanisms or to test dynamical models.

We shall refer to these as *model-dependent properties*.

In this chapter we shall discuss only the *basic properties* of the density matrix. It will turn out that *all* the basic properties of the non-relativistic case hold also for the helicity density matrix provided care is taken with the physical interpretation of the latter.

We discuss, amongst other things, the expression of the density matrix in terms of multipole parameters or statistical tensors, the concept of 'degree-of-rank-L polarization' and the transformation properties of the density matrix under rotations and Lorentz transformations.

We also give a detailed discussion of the density matrix for spin-1 particles, bearing in mind that polarized deuteron beams are already in use and will become more commonly available in the near future.

The treatment of the *general kinematical-dynamical properties* will follow in Chapter 4 after the discussion of scattering amplitudes. Some *model-dependent properties* will be found in the discussion of specific dynamical models.

In all experiments involving the use of polarized targets or polarized beams we are dealing with a system of quantum mechanical particles that is not in a definite, pure quantum state. Rather we have an incoherent mixture or statistical ensemble of particles about which our knowledge is limited to the average of certain dynamical variables, an average, that is, for the whole ensemble. Strictly speaking this ought to apply also to variables such as momentum, but the averaging processes involved therein are usually quite uninteresting for hadron physics and therefore we shall ignore them, adopting the fiction that each particle in the beam emerging from an accelerator has precisely the same momentum. Our sole concern will be with the spin properties of these ensembles.

We review the main properties of the density matrix in the next section. A more general exposition can be found in the review article of Fano (Fano, 1957).

3.1 The non-relativistic density matrix

3.1.1 *Definition*

For a particle of spin s, a *pure* quantum mechanical spin state $|\psi\rangle$ is defined and identified by the coefficients c_m involved in its expansion into a sum of basic states $|s;m\rangle$; these are usually taken as eigenstates of $\hat{s}_z$, the z-component of the spin operator, i.e. one has

$$|\psi\rangle = \sum_{m=-s}^{s} c_m|sm\rangle. \tag{3.1.1}$$

For an arbitrary operator $\hat{O}$ with matrix elements

$$O_{mm'} = \langle sm|\hat{O}|sm'\rangle \tag{3.1.2}$$

the mean value in the state $|\psi\rangle$, normalized to unity, is given by

$$\left\langle \hat{O} \right\rangle_{\psi} \equiv \langle\psi|\hat{O}|\psi\rangle = \sum_{m,m'} c^*_{m'} O_{m'm} c_m. \tag{3.1.3}$$

For a non-pure state we might have an incoherent mixture of a number of pure states $|\psi^{(i)}\rangle$, each occurring in an ensemble with probability or statistical weight $p^{(i)}$ with $\sum_i p^{(i)} = 1$. For *each state* the operator $\hat{O}$ will have a mean value

$$\left\langle \hat{O} \right\rangle_{\psi^{(i)}} = \sum_{m,m'} c_{m'}^{(i)*} O_{m'm} c_m^{(i)}$$

and therefore its mean value over the *whole ensemble* will be

$$\left\langle \hat{O} \right\rangle = \sum_i p^{(i)} \left\langle \hat{O} \right\rangle_{\psi^{(i)}} = \sum_{m,m'} O_{m'm} \sum_i p^{(i)} c_{m'}^{(i)*} c_m^{(i)}. \tag{3.1.4}$$

The spin density matrix in the basis $|s; s_z = m\rangle$ is now defined by

$$\rho_{mm'} = \sum_i p^{(i)} c_m^{(i)} c_{m'}^{(i)*} \tag{3.1.5}$$

so that equation (3.1.4) becomes

$$\left\langle \hat{O} \right\rangle = \sum_{m,m'} O_{m'm} \rho_{mm'} = \text{Tr}\,(O\rho) \tag{3.1.6}$$

where O and ρ are the *matrices* whose elements are $O_{m'm}$ and $\rho_{mm'}$.

Equation (3.1.6) allows us to calculate the mean value for the ensemble of every physical operator once we know the density matrix ρ.

Conversely, and of most interest in hadron physics, a knowledge of the mean values for the ensemble of a sufficiently large number of physical observables will enable the inversion of eqn (3.1.6) and thus determination of the density matrix.

3.1.2 Some general properties of $\rho_{mm'}$

The $\rho_{mm'}$ are the elements of the density matrix referred to a particular choice of basis states $|s; s_z = m\rangle$. We can also give the density matrix in any other basis unitarily related to $|s; s_z = m\rangle$. If T is any unitary $(2s+1) \times (2s+1)$ matrix we can take as basis states

$$|n\rangle' = \sum_m T_{mn} |s; s_z = m\rangle$$

and if we label the density elements in the new basis as $\rho'_{nn'}$ we will have

$$\rho'_{nn'} = \sum_{m,m'} T_{nm}^{-1} \rho_{mm'} T_{m'n'} \tag{3.1.7}$$

or in matrix notation

$$\rho' = T^{-1} \rho T. \tag{3.1.8}$$

We note the important property that the trace is invariant under change of basis:

$$\text{Tr}\,\rho' = \text{Tr}\,\rho. \tag{3.1.9}$$

From the definition (3.1.5) and the condition $\sum_i p^{(i)} = 1$ the following properties can easily be read off.

(1) The trace of ρ is unity, i.e.

$$\mathrm{Tr}\ \rho = 1. \tag{3.1.10}$$

(2) ρ is a hermitian matrix, i.e.

$$\rho^*_{mm'} = \rho_{m'm}. \tag{3.1.11}$$

(3) For each m, the diagonal elements are positive semi-definite, i.e.

$$\rho_{mm} \geq 0 \tag{3.1.12}$$

and this holds in any unitarily related basis.

(4) The hermitian properties of ρ guarantee the existence of a unitary matrix U that will diagonalize ρ, i.e. we have

$$U^{-1}\rho U = \rho^{\mathrm{D}} \tag{3.1.13}$$

where ρ^{D} is the diagonal matrix

$$\left(\rho^{\mathrm{D}}\right)_{mn} = \lambda_m \delta_{mn}, \tag{3.1.14}$$

with $\lambda_m \geq 0$.

(5) From (3.1.9), (3.1.12), (3.1.13), (3.1.14)

$$\mathrm{Tr}\ \rho^2 = \mathrm{Tr}\ \left(\rho^{\mathrm{D}}\right)^2 = \sum_m \lambda_m^2 \leq \left(\sum_m \lambda_m\right)^2 = (\ \mathrm{Tr}\ \rho)^2 = 1.$$

Thus

$$\mathrm{Tr}\ \rho^2 = \sum_{m,m'} |\rho_{mm'}|^2 \leq 1. \tag{3.1.15}$$

(6) If it happens that all members of an ensemble are in a single pure quantum state, then all except one $p^{(i)}$ will be zero, and the non-zero one $p^{(j)}$ say, will be equal to unity. In this case ρ will be a rank-1 matrix[1] and it will have one eigenvalue equal to unity and all the rest equal to zero. It can then be written in a factorized form, e.g.

$$\rho_{mm'} = c_m c^*_{m'}. \tag{3.1.16}$$

For this case the equality holds in (3.1.15).

[1] The rank of ρ is the dimension of the largest non-zero determinant that can be formed from the rows and columns of ρ. Equivalently, the rank of an $n \times n$ matrix ρ is $r = n - k$, where k is the dimension of the eigenspace corresponding to the eigenvalue zero of ρ.

3.1.3 Combined systems of several particle types

If the overall system is a mixture of several systems of different particles then it can be described by a joint density matrix. For example for two types of particles A, B one would have $\rho(A, B)$ with matrix elements $\rho(A, B)_{mn;m'n'}$, the labels m, n referring to the eigenstates

$$|s^A, s_z^A = m; s^B, s_z^B = n\rangle$$

of the system of two particles A and B.

The mean value for the whole system of an arbitrary operator $\hat{O}$ is again given by an equation like (3.1.6):

$$\left\langle \hat{O} \right\rangle = \text{Tr } [O\rho(A, B)] \tag{3.1.17}$$

where now the trace is used in the generalized sense

$$\text{Tr } [O\rho(A, B)] = \sum_{m,n} [O\rho(A, B)]_{mn;mn}$$

with

$$[O\rho(A, B)]_{mn;mn} = \sum_{m',n'} O_{mn;m'n'} \rho(A, B)_{m'n';mn}.$$

If on the one hand we wish to calculate the joint expectation values of an observable $\hat{O}^{(A)}$ of the particle A and an observable $\hat{O}^{(B)}$ of the particle B then we must take the expectation value of the operator product $\hat{O}^{(A)} \otimes \hat{O}^{(B)}$ defined in such a way that

$$\left(\hat{O}^{(A)} \otimes \hat{O}^{(B)}\right)_{mn;m'n'} \equiv O^{(A)}_{mm'} O^{(B)}_{nn'}. \tag{3.1.18}$$

If on the other hand we wish to calculate the expectation value for the measurement of a physical observable of just *one* type, A, then if $\hat{O}^{(A)}$ is the operator corresponding to this observable we get the mean value of $\hat{O}^{(A)}$ by calculating the mean value of $\hat{O}^{(A)} \otimes \hat{1}^{(B)}$, where $\hat{1}^{(B)}$ is the unit operator in the space of the labels referring to particle B.

Thus

$$\begin{aligned}
\left\langle \hat{O}^{(A)} \right\rangle &= \text{Tr } \left[\hat{O}^{(A)} \otimes \hat{1}^{(B)} \rho(A, B)\right] \\
&= \sum_{\substack{m,m' \\ n,n'}} O^{(A)}_{mm'} \delta_{nn'} \rho(A, B)_{m'n';mn} \\
&= \sum_{m,m'} O^{(A)}_{mm'} \sum_{n} \rho(A, B)_{m'n;mn} \\
&= \text{Tr}_A \left[O^{(A)} \rho(A)\right]
\end{aligned} \tag{3.1.19}$$

where $\rho(A)$ is the $(2s_A + 1) \times (2s_A + 1)$ *effective density matrix* for type-A particles, defined by

$$\rho(A)_{mm'} \equiv \sum_n \rho(A, B)_{mn;m'n}. \tag{3.1.20}$$

We note that if the rank of $\rho(A, B)$ is r, then for the rank of $\rho(A)$ one has

$$\text{rank } \rho(A) \leq (2s_B + 1)r.$$

Of course a similar result holds for an observable of the particles of type B.

If the state of the combined system is *uncorrelated* then the mean values of all measurements carried out on particles of type A and B must factorize into mean values over the separate ensembles of A and B, i.e.

$$\left\langle \hat{O}^{(A)} \otimes \hat{O}^{(B)} \right\rangle = \left\langle \hat{O}^{(A)} \right\rangle \left\langle \hat{O}^{(B)} \right\rangle$$

must hold for every observable $\hat{O}^{(A)}$, $\hat{O}^{(B)}$. This is only possible if the joint density matrix itself factorizes. We thus have the important result that

$$\rho(A, B)_{mn;m'n'} = \rho(A)_{mm'} \rho(B)_{nn'} \tag{3.1.21}$$

if and only if the ensemble of particles A and particles B is uncorrelated. An example of such ensembles is the incoming beam and the target in a scattering experiment *prior* to interaction.

In general, if several spinning particles C, D, E,... are produced in a reaction then the full density matrix for the final state is a joint matrix $\rho(C, D, E, \ldots)$ with matrix elements

$$\rho(C, D, E, \ldots)_{c,d,e,\ldots;c',d',e',\ldots}.$$

If, as often happens in practice, the properties of only *one* of the particles are measured, say those of type C, then the mean values are to be calculated using the *effective density matrix* $\rho(C)$ where

$$\rho(C)_{c;c'} = \sum_{d,e,\ldots} \rho(C, D, E, \ldots)_{c,d,e,\ldots;c',d,e,\ldots}. \tag{3.1.22}$$

Usually one refers to this simply as 'the density matrix for C'.

3.1.4 The independent parameters specifying ρ

We saw in Chapter 1 that a *pure state* for a particle of spin s can be specified by $2(2s+1)-2 = 4s$ real parameters. For an incoherent mixture made up of particles of spin s the ensemble is completely characterized by the $(2s + 1) \times (2s + 1)$ hermitian matrix ρ. Taking into account the normalization condition (3.1.10) one requires $(2s+1)^2-1$ independent real parameters to specify ρ fully. The direct listing of the individual elements of ρ could then be limited to $2s$ of the $2s+1$ (real) diagonal elements, and the real and imaginary parts of the elements above the diagonal. This is not always the most convenient set of numbers to deal with, from the point of view of either experiment or theory.

Various 'representations' of ρ can be introduced, expressing ρ as a sum over certain standard matrices, the properties of a particular ρ being then specified by the coefficients in the expansion.

The best known of these is the density matrix for spin-1/2 particles. Since ρ is now a 2×2 matrix, it can always be written, see (1.1.22), as

$$\rho = \tfrac{1}{2}(I + \boldsymbol{\mathcal{P}} \cdot \boldsymbol{\sigma}) \tag{3.1.23}$$

where $\boldsymbol{\mathcal{P}}$ is now the spin-polarization vector for the ensemble,

$$\boldsymbol{\mathcal{P}} = \langle \boldsymbol{\sigma} \rangle = \text{Tr}\, \rho \boldsymbol{\sigma}. \tag{3.1.24}$$

Thus the three real numbers $\boldsymbol{\mathcal{P}}$ can be used to specify ρ. We note that whereas for a *pure* state $\boldsymbol{\mathcal{P}}^2 = 1$, in general for an ensemble we have

$$\boldsymbol{\mathcal{P}}^2 \leq 1 \tag{3.1.25}$$

as follows from (3.1.15).

What is the generalization of (3.1.23) for spins $s > 1/2$? Clearly it is not sufficient to replace $\boldsymbol{\sigma}$ by the set of three hermitian matrices $\mathbf{S} = \{S_j\}$ that represents the spin operator $\hat{\mathbf{s}}$. We need to construct many more basis matrices and this can be done in principle by using products of the S_j. (It must of course be remembered that results like $\sigma_1\sigma_2 = i\sigma_3$ are specific to spin 1/2; higher-spin products, such as S_1S_2, are independent and cannot be expressed in terms of S_3.)

3.1.5 The multipole parameters

A very useful and convenient set of basis matrices can be obtained by forming sets of products of the spin operators that transform very simply under rotations. These so-called *spherical tensor operators* $\hat{T}^L_M$, $0 \leq L \leq 2s$, $-L \leq M \leq L$, and the matrices T^L_M that represent them can be chosen in such a way that the elements of these matrices are given by *vector-addition coefficients* (Edmonds, 1957 and Appel, 1968). Thus

$$\left(T^L_M\right)_{mm'} \equiv \langle sm|\hat{T}^L_M|sm'\rangle \equiv \langle sm|sm'; LM\rangle\,; \tag{3.1.26}$$

L is called the *rank* of the tensor operator. Some examples are as follows

Scalar:

$$\hat{T}^0_0 = 1.$$

Vector or rank 1 tensor:

$$\hat{T}_0^1 = \frac{1}{\sqrt{s(s+1)}}\hat{s}_z$$
$$\hat{T}_1^1 = -\frac{1}{\sqrt{2s(s+1)}}(\hat{s}_x + i\hat{s}_y). \qquad (3.1.27)$$
$$\hat{T}_{-1}^1 = \frac{1}{\sqrt{2s(s+1)}}(\hat{s}_x - i\hat{s}_y)$$

Further examples may be found in the review article by Jackson (Jackson, 1965).

For our purposes it is not necessary to know the precise form of the operator $\hat{T}_M^L$. The crucial information is contained in equation (3.1.26). We note, incidentally, that there are $(2s+1)^2$ different T_M^L but from the properties of the vector-addition coefficients it can be shown that

$$T_{-M}^L = (-1)^M T_M^{L\dagger}. \qquad (3.1.28)$$

We now proceed to derive the expansion of the density matrix ρ in terms of the matrices T_M^L.

Let us define the set of complex parameters t_M^L $(0 \le L \le 2s)$ by

$$t_M^{L\,*} = \sum_{m,m'} \langle sm|sm'; LM\rangle\, \rho_{mm'}. \qquad (3.1.29)$$

The inverse of this is

$$\rho_{mm'} = \frac{1}{2s+1}\sum_{L,M}(2L+1)\,\langle sm|sm'; LM\rangle\, t_M^{L\,*} \qquad (3.1.30)$$

and, using (3.1.26),

$$\rho_{mm'} = \frac{1}{2s+1}\sum_{L,M}(2L+1)t_M^{L\,*}(T_M^L)_{mm'}. \qquad (3.1.31)$$

Thus the matrix ρ is expanded in terms of the matrix set T_M^L as

$$\rho = \frac{1}{2s+1}\sum_{L,M}(2L+1)t_M^{L\,*}\,T_M^L. \qquad (3.1.32)$$

This is the desired generalization of eqn (3.1.23). Now from the definition (3.1.26) it follows that

$$\mathrm{Tr}\,\left(T_{M'}^{L'} T_M^{L\dagger}\right) = \frac{2s+1}{2L+1}\delta_{LL'}\delta_{MM'} \qquad (3.1.33)$$

and hence that

$$\mathrm{Tr}\,\left(\rho T_M^{L\dagger}\right) = t_M^{L\,*}$$

or, since ρ is hermitian,

$$t_M^L = \text{Tr}\,(\rho T_M^L). \tag{3.1.34}$$

Thus the t_M^L, which are called either *multipole parameters* (of rank L) or *statistical tensors*, are a generalization of the spin-polarization vector. Indeed for the lowest-rank multipole parameters one has

$$\begin{aligned} t_0^0 &= \text{Tr}\,\rho = 1 \\ t_0^1 &= \mathscr{P}_z\sqrt{\frac{s}{s+1}} \\ t_1^1 &= -(\mathscr{P}_x + i\mathscr{P}_y)\sqrt{\frac{s}{2(s+1)}} \\ t_{-1}^1 &= (\mathscr{P}_x - i\mathscr{P}_y)\sqrt{\frac{s}{2(s+1)}} \end{aligned} \tag{3.1.35}$$

where the spin-polarization vector $\boldsymbol{\mathcal{P}}$ is defined in eqn (1.1.27).

We note that, from eqn (3.1.29) and the properties of the vector-addition coefficients, one has

$$t_{-M}^L = (-1)^M t_M^{L\,*}. \tag{3.1.36}$$

In particular the t_0^L are real.

Thus the set of t_M^L is actually specified by $(2s+1)^2$ real numbers. Bearing in mind that $t_0^0 = 1$ we see that the $(2s+1)^2 - 1$ remaining real parameters are just the right number to specify ρ completely.

The condition (3.1.15) leads to the inequality

$$\frac{1}{2s+1}\sum_{L,M}(2L+1)|t_M^L|^2 \leq 1. \tag{3.1.37}$$

We stress the fact that whether we choose to specify the set of numbers $\rho_{mm'}$ or the set of numbers t_M^L is merely a question of convenience. They are directly related by (3.1.29) or (3.1.30).

3.1.6 Multipole parameters for combined systems of particles

In the case of a combined system of different particles A, B, ..., in analogy with the discussion in subsection 3.1.3 the joint density matrix $\rho(A, B, \ldots)$ will be expanded in terms of the direct product of matrices $T(A)_M^L \otimes T(B)_{M'}^{L'} \otimes \cdots$ with coefficients $t_{MM'\ldots}^{LL'\ldots}(A, B, \ldots)$, the *joint multipole parameters*.

(Note that $T_M^L(A)$ is a matrix of dimension $2s_A + 1$. We will usually leave out the particle label on the T_M^L.)

If the different types of particles are uncorrelated, we will have

$$t_{MM'\ldots}^{LL'\ldots} = t_M^L t_{M'}^{L'} \cdots \tag{3.1.38}$$

The *effective multipole parameters* $t^L_M(A)$ for particles of type A, say, when no spin measurement is carried out on the other particles, will be

$$t^L_M(A) = t^{L00...0}_{M00...0}. \tag{3.1.39}$$

3.1.7 Even and odd polarization

It sometimes happens that *only* the even-rank multipoles or *only* the odd-rank multipoles are non-zero. We refer to such states of polarizations as 'even' or 'odd' (Doncel *et al.*, 1970). When this happens the density matrix has a special symmetry, namely

$$\rho_{-\mu-\lambda} = \pm(-1)^{\lambda-\mu}\rho_{\lambda\mu}, \tag{3.1.40}$$

the $(\pm)$ corresponding to an $\binom{\text{even}}{\text{odd}}$ state of polarization. In fact, it is sometimes convenient to break ρ up into its even and odd parts for a general state of polarization. Thus we write

$$\rho = \rho_+ + \rho_- \tag{3.1.41}$$

and

$$\begin{aligned} \rho_+ &= \frac{1}{2s+1}\sum_{\substack{L\text{ even}\\ M}} (2L+1)t^{L\,*}_M T^L_M \\ \rho_- &= \frac{1}{2s+1}\sum_{\substack{L\text{ odd}\\ M}} (2L+1)t^{L\,*}_M T^L_M. \end{aligned} \tag{3.1.42}$$

Equivalently

$$(\rho_\pm)_{\lambda\mu} = \tfrac{1}{2}\left[\rho_{\lambda\mu} \pm (-1)^{\lambda-\mu}\rho^*_{-\lambda-\mu}\right]. \tag{3.1.43}$$

Thus, in general,

$$(\rho_\pm)_{-\mu-\lambda} = \pm(-1)^{\lambda-\mu}\rho_{\lambda\mu}. \tag{3.1.44}$$

It will be seen later that it is usually easier to measure the elements of ρ_+ than those of ρ_-.

We note that if rank $\rho = r$ then

$$\text{rank } \rho_\pm \leq 2r. \tag{3.1.45}$$

3.1.8 The effect of rotations on the density matrix

Since the density matrix elements $\rho_{mm'}$ are given in a basis specified by the spin states $|s; s_z = m\rangle$, they are implicitly dependent on the choice of axis system.

We denote by $\rho^S_{mm'}$ the elements of the density matrix defined using as basis states the $|sm\rangle$ appropriate to the reference frame S and by $\rho^{S^r}_{mm'}$ the elements of the density matrix defined using as basis states the rotated states $|sm\rangle^r = U(r)|sm\rangle$, see (1.1.18), appropriate to S^r. Then, similarly to eqn (3.1.8), we have

$$\rho^{S^r} = \mathscr{D}^{(s)\dagger}(r)\rho^S\mathscr{D}^{(s)}(r) \tag{3.1.46}$$

or

$$\rho^{S^r}_{mm'} = \mathscr{D}^{(s)}_{nm}{}^*(r)\rho^S_{nn'}\mathscr{D}^{(s)}_{n'm'}(r)$$

It is clear that ρ^{S^r} thus defined is the correct density matrix to use when evaluating expectation values as seen in reference frame S^r. For this reason we shall refer to ρ^{S^r} as the *density matrix in the frame* S^r.

The relationship between $\rho^{S^r}_{mm'}$ and $\rho^S_{mm'}$ is rather complicated. The formula can be simplified a little using the rules for the reduction of products of rotation matrices. One finds

$$\rho^{S^r}_{mm'} = \sum_{J=0}^{2s}\sum_{n,n'=-s}^{s}(-1)^{m'-n'}\langle s,m;s,-m'|J,m-m'\rangle \\ \times\langle s,n;s,-n'|J,M\rangle\,\mathscr{D}^{(J)}_{m-m',M}(r^{-1})\rho^S_{nn'}. \tag{3.1.47}$$

The multipole parameters t^L_M transform very simply, however. If $(t^L_M)_S$ and $(t^L_M)_{S^r}$ denote the components of the statistical tensors in the frames S and S^r then, from (3.1.46) and (3.1.29), one finds the simple result

$$(t^L_M)_{S^r} = \sum_{M'}\mathscr{D}^{(L)}_{M'M}(r)(t^L_{M'})_S, \tag{3.1.48}$$

which is the usual rule relating the components of a spherical tensor in different reference frames.

3.1.9 Diagonalization of ρ. The quantization axis

Although it is always possible to diagonalize ρ it is *not* always possible to do so by means of an actual physical rotation of axes. If, however, the ensemble consists of a mixture of magnetic substates, i.e. eigenstates $|sm\rangle$ where m is the projection of $\hat{s}$ along the quantization direction, then in a frame that has OZ along the quantization direction clearly ρ will be diagonal, and all multipole parameters t^L_M with $M \neq 0$ will be zero. We shall refer to the direction OZ that makes ρ diagonal as the *quantization axis*. For spin-1/2 particles the quantization axis coincides with the polarization vector $\boldsymbol{\mathcal{P}}$, but the quantization axis is a somewhat more general concept since for higher spins one can easily have the vector polarization zero yet still have some 'alignment' along the quantization axis (see subsection 3.1.12 below).

3.1.10 Other choices of basis matrices

The case we know best, namely spin 1/2, is misleadingly simple. Here we succeed in expanding the hermitian ρ in terms of the *hermitian* Pauli matrices σ_j with real coefficients of direct observable relevance, and at the same time we enjoy the very simple properties of the σ_i under rotations.

The T^L_M used for the general case have simple rotation properties but are not hermitian. As a result the t^L_M are complex and are not so closely related to what is actually measured.

In fact, it is very easy for arbitrary spin s to introduce a set of *hermitian* basis matrices (Doncel *et al.*, 1970) Q^L_M, defined as follows:

$$\begin{aligned}
M &\geq 1 \qquad & Q^L_M &= \frac{(-1)^M}{2}\sqrt{\frac{2L+1}{s}}\left\{T^L_M + {T^L_M}^\dagger\right\} \\
M &= 0 \qquad & Q^L_0 &= \sqrt{\frac{2L+1}{2s}}\,T^L_0 \\
M &\leq -1 \qquad & Q^L_M &= \frac{(-1)^M}{2i}\sqrt{\frac{2L+1}{s}}\left\{T^L_{-M} + {T^L_{-M}}^\dagger\right\}
\end{aligned} \tag{3.1.49}$$

with corresponding *real* multipole parameters r^L_M given by

$$\begin{aligned}
M &\geq 1 \qquad & r^L_M &= (-1)^M\sqrt{\frac{2L+1}{s}}\ \mathrm{Re}\ t^L_M \\
M &= 0 \qquad & r^L_0 &= \sqrt{\frac{2L+1}{2s}}\,t^L_0 \\
M &\leq -1 \qquad & r^L_M &= (-1)^M\sqrt{\frac{2L+1}{s}}\ \mathrm{Im}\ t^L_{-M}
\end{aligned} \tag{3.1.50}$$

and the density matrix expansion

$$\rho = \frac{1}{2s+1}\left(I + 2s\sum_{L=1}^{2s}\sum_{M=-L}^{L} r^L_M Q^L_M\right). \tag{3.1.51}$$

This approach is especially useful for discussing the 'domain' of the density matrix, i.e. the range of permitted values for the parameters specifying ρ. However, the price one pays is that the rotational properties of the Q^L_M and hence of the r^L_M are more complicated.

For this reason, we have chosen to develop our general treatment of reactions in terms of the usual T^L_M.

3.1.11 Invariant characterization of the state of polarization of an ensemble

Full information about the state of an ensemble requires a knowledge of the whole density matrix. It is useful, however, to have a simple, invariant, albeit cruder, characterization of the ensemble. Thus for spin 1/2 we talk of an *unpolarized* ensemble or a polarized ensemble with *degree of polarization* $\mathscr{P} = \sqrt{\boldsymbol{\mathcal{P}}^2}$. We wish to generalize these concepts to arbitrary spin.

An unpolarized or *isotropic* ensemble of spin-s particles has equal probabilities $p^{(i)} = 1/(2s+1)$ of being in any pure state $|\psi^{(i)}\rangle$ and is therefore given by the density matrix

$$\rho_{\text{iso}} = \frac{1}{2s+1} I \tag{3.1.52}$$

in any basis.

Therefore the matrix

$$\rho - \rho_{\text{iso}} = \frac{1}{2s+1} \sum_{\substack{L \geq 1 \\ M}} (2L+1) t^{L\,*}_M T^L_M \tag{3.1.53}$$

measures the *departure from isotropy* (Doncel *et al.*, 1972).

To characterize this difference in a rotationally invariant fashion we have to introduce some measure of the 'difference' between two matrices, or, as it is often described, the 'distance' between them.

A suitable invariant measure is

$$\text{Tr}\,(\rho - \rho_{\text{iso}})^2 = \text{Tr}\,\rho^2 - \text{Tr}\,\rho^2_{\text{iso}} = \text{Tr}\,\rho^2 - \frac{1}{2s+1} \tag{3.1.54}$$

by (3.1.10) and (3.1.52). In fact, the ratio $(\text{Tr}\,\rho^2 - \text{Tr}\,\rho^2_{\text{iso}})/\,\text{Tr}\,\rho^2_{\text{iso}}$ takes the value zero for an unpolarized ensemble and the value $2s$ for a pure state. We thus define the *overall degree of polarization*

$$d \equiv \frac{1}{\sqrt{2s}} \left[(2s+1)\ \text{Tr}\,\rho^2 - 1\right]^{1/2} \tag{3.1.55}$$

so that

$$0 \leq d \leq 1.$$

For spin 1/2, as expected,

$$d = \mathscr{P} = \sqrt{\boldsymbol{\mathcal{P}}^2} \tag{3.1.56}$$

but this case is misleadingly simple. For higher spin we can have vector polarization, rank-2 tensor polarization etc. and the magnitude of the vector polarization is no longer the overall degree of polarization.

The representation in terms of Cartesian spin matrices gets very clumsy for higher spin so we restrict ourselves to the multipole parameter expansion (3.1.32). One can define a measure of rank-L polarization ($L \geq 1$) by

$$d_L = \sqrt{\frac{2L+1}{2s}}\left(\sum_M |t^L_M|^2\right)^{1/2} \tag{3.1.57}$$

and the overall degree of polarization is then

$$d = \left\{\sum_{L\geq 1} d_L^2\right\}^{1/2}. \tag{3.1.58}$$

However, the d_L can be a little misleading since the individual d_L cannot usually attain the value 1, although d itself can. (For example, for spin-1 particles $(d_1)_{\max} = \sqrt{3}/2$.)

3.1.12 Spin-1 particles and photons

(i) Massive particles

With the production and general use of polarized deuteron beams this case has become of great interest and we therefore treat it in some detail.

The density matrix can either be written in the standard form (3.1.32) involving multipole parameters or it can be given in a Cartesian form as follows:

$$\rho = \tfrac{1}{3}\left[1 + \tfrac{3}{2}\boldsymbol{\mathcal{P}}\cdot\mathbf{S} + \sqrt{\tfrac{3}{2}}\,T_{ij}(S_iS_j + S_jS_i)\right] \tag{3.1.59}$$

with T_{ij} real and symmetric, and traceless: $\sum_i T_{ii} = 0$. Here $\mathbf{S}$ stands for the 3×3 traceless matrices S_j representing the spin operators $\hat{s}_j$ for spin 1:

$$S_x = \frac{1}{\sqrt{2}}\begin{pmatrix}0&1&0\\1&0&1\\0&1&0\end{pmatrix} \qquad S_y = \frac{i}{\sqrt{2}}\begin{pmatrix}0&-1&0\\1&0&-1\\0&1&0\end{pmatrix}$$
$$S_z = \begin{pmatrix}1&0&0\\0&0&0\\0&0&-1\end{pmatrix}. \tag{3.1.60}$$

The three real parameters $\mathcal{P}_j$ and the five independent T_{ij} are all independent of each other.

Other definitions of $\boldsymbol{\mathcal{P}}$ and T_{ij} are sometimes given in the literature (Werle, 1966) but ours are designed to have the simplest physical interpretation. One finds that $\boldsymbol{\mathcal{P}}$ is the spin-polarization vector

$$\boldsymbol{\mathcal{P}} = \langle\hat{\mathbf{s}}\rangle. \tag{3.1.61}$$

in agreement with (1.1.27) for the case $s = 1$ and T_{ij} measures the rank-2 spin tensor

$$T_{ij} = \frac{1}{2}\sqrt{\frac{3}{2}}\left(\langle \hat{s}_i\hat{s}_j + \hat{s}_j\hat{s}_i\rangle - \frac{4}{3}\delta_{ij}\right). \tag{3.1.62}$$

The degrees of vector polarization $\mathscr{P}$ and of tensor polarization T are

$$\mathscr{P} = \sqrt{\boldsymbol{\mathcal{P}}^2} \qquad 0 \leq \mathscr{P} \leq 1 \tag{3.1.63}$$

and

$$T = \sqrt{\sum_{ij} (T_{ij})^2} \qquad 0 \leq T \leq 1. \tag{3.1.64}$$

The overall degree of polarization is

$$d = \left(\tfrac{3}{4}\mathscr{P}^2 + T^2\right)^{1/2}. \tag{3.1.65}$$

The *multipole parameters* are related to $\mathscr{P}_j$ and T_{ij} via

$$t_0^1 = \tfrac{1}{\sqrt{2}}\mathscr{P}_z \qquad t_{\pm 1}^1 = \mp\tfrac{1}{2}\left(\mathscr{P}_x \pm i\mathscr{P}_y\right) \tag{3.1.66}$$

and

$$\begin{aligned} t_0^2 = \sqrt{\tfrac{3}{5}}T_{zz} \qquad t_{\pm 1}^2 &= \mp\sqrt{\tfrac{2}{5}}\left(T_{xz} \pm iT_{yz}\right) \\ t_{\pm 2}^2 &= \sqrt{\tfrac{1}{10}}\left(T_{xx} - T_{yy} \pm 2iT_{xy}\right). \end{aligned} \tag{3.1.67}$$

They are related to the elements of the density matrix itself via

$$t_1^1 = -\tfrac{1}{\sqrt{2}}(\rho_{10} + \rho_{0-1})^* \qquad t_0^1 = \tfrac{1}{\sqrt{2}}(\rho_{11} - \rho_{-1-1})^* \tag{3.1.68}$$

and

$$\begin{aligned} t_2^2 = \sqrt{\tfrac{3}{5}}\rho_{1-1}^* \qquad t_1^2 &= -\sqrt{\tfrac{3}{10}}(\rho_{10} - \rho_{0-1})^* \\ t_0^2 &= \sqrt{\tfrac{1}{10}}(\rho_{11} + \rho_{-1-1} - 2\rho_{00})^*. \end{aligned} \tag{3.1.69}$$

Often the ensemble is made up of particles whose spin is quantized along the Z-axis. Let p_+, p_0 and p_- be the probabilities of finding a particle with spin projection 1, 0, -1 respectively along the quantization axis. Then from (3.1.61) and (3.1.62)

$$\begin{aligned} &\mathscr{P}_x = \mathscr{P}_y = 0 \qquad \mathscr{P}_z = (p_+ - p_-) \\ T_{ij} = 0 \text{ if } i \neq j \qquad &T_{xx} = T_{yy} = -\tfrac{1}{2}T_{zz} \qquad T_{zz} = \tfrac{1}{\sqrt{6}}(1 - 3p_0) \end{aligned} \tag{3.1.70}$$

The degree of vector and tensor polarization are then

$$\mathscr{P} = |p_+ - p_-| \qquad T = \tfrac{1}{2}|1 - 3p_0| \tag{3.1.71}$$

and the density matrix is

$$\rho = \frac{1}{3}\begin{pmatrix} 1+\frac{3}{2}\mathscr{P}_z+\sqrt{\frac{3}{2}}T_{zz} & 0 & 0 \\ 0 & 1-\sqrt{6}T_{zz} & 0 \\ 0 & 0 & 1-\frac{3}{2}\mathscr{P}_z+\sqrt{\frac{3}{2}}T_{zz} \end{pmatrix} \tag{3.1.72}$$

In this case, in the frame with OZ along the quantization axis, the multipole parameters take the simple form

$$\begin{gathered} t_0^1 = \sqrt{\tfrac{1}{2}}\mathscr{P}_z \qquad t_0^2 = \sqrt{\tfrac{3}{5}}T_{zz} = \sqrt{\tfrac{1}{10}}\mathscr{A} \\ t_{\pm 1}^1 = t_{\pm 1}^2 = t_{\pm 2}^2 = 0 \end{gathered} \tag{3.1.73}$$

where

$$\mathscr{A} = 1 - 3p_0 \tag{3.1.74}$$

is referred to as the *alignment* (Steenberg, 1953).

It should be noted that ensembles of the above type are by no means the most general ones for spin-1 particles. To discuss the general case, consider the orthonormal basis states

$$\begin{aligned} |\mathbf{e}_{(x)}\rangle &\equiv \frac{1}{\sqrt{2}}\left(|\lambda=-1\rangle - |\lambda=+1\rangle\right) \\ |\mathbf{e}_{(y)}\rangle &\equiv \frac{i}{\sqrt{2}}\left(|\lambda=-1\rangle + |\lambda=+1\rangle\right) \\ |\mathbf{e}_{(z)}\rangle &\equiv |\lambda=0\rangle \end{aligned} \tag{3.1.75}$$

The most general normalized pure spin state for a spin-1 particle is then

$$|\boldsymbol{\varepsilon}\rangle = \epsilon_x|\mathbf{e}_{(x)}\rangle + \epsilon_y|\mathbf{e}_{(y)}\rangle + \epsilon_z|\mathbf{e}_{(z)}\rangle \tag{3.1.76}$$

where $\boldsymbol{\varepsilon} = (\epsilon_x, \epsilon_y, \epsilon_z)$, the polarization vector, is a complex vector with

$$\boldsymbol{\varepsilon}^* \cdot \boldsymbol{\varepsilon} = 1. \tag{3.1.77}$$

It is $\boldsymbol{\varepsilon}$ that is the analogue of the polarization vector in classical electrodynamics.

For a *pure* state one finds that the spin polarization vector $\boldsymbol{\mathcal{P}}$ is related to the polarization vector $\boldsymbol{\varepsilon}$ via

$$\boldsymbol{\mathcal{P}} = \mathrm{Im}\,(\boldsymbol{\varepsilon}^* \times \boldsymbol{\varepsilon}); \tag{3.1.78}$$

this will be given a covariant form in Section 3.4. Thus $\boldsymbol{\mathcal{P}} = 0$ for any pure state with real $\boldsymbol{\varepsilon}$.

For the tensor T_{ij} one finds

$$T_{ij} = \sqrt{\tfrac{3}{2}}\left[\tfrac{1}{3}\delta_{ij} - \mathrm{Re}\,(\epsilon_i^*\epsilon_j)\right]. \tag{3.1.79}$$

Note that there is no pure state for which $T_{ij} = 0$ for *all* i and j.

For example, for states with $s_z = \pm 1, 0$ we have

$$\begin{aligned} \boldsymbol{\varepsilon}^{(\pm)} &= \tfrac{1}{\sqrt{2}}(\mp 1, -i, 0) \\ \boldsymbol{\varepsilon}^{(0)} &= (0, 0, 1) \end{aligned} \tag{3.1.80}$$

so that

$$\begin{aligned} \boldsymbol{\mathcal{P}}^{(+)} &= (0, 0, 1) \\ \boldsymbol{\mathcal{P}}^{(-)} &= (0, 0, -1)\,. \\ \boldsymbol{\mathcal{P}}^{(0)} &= (0, 0, 0) \end{aligned} \tag{3.1.81}$$

(ii) Photons

Although intrinsically relativistic we may treat photons as above provided, as will be justified in Section 3.2, we interpret the states $|s_z = \pm 1\rangle$ as helicity states for the photons moving along the direction OZ. Of course the states $|s_z = 0\rangle$ are now absent. As a consequence, an ensemble of photons can never be isotropic. Indeed, from (3.1.73) and (3.1.74) we see that for all ensembles of photons $T_{zz} = 1/\sqrt{6}$ and therefore $t_0^1 = \sqrt{1/10}$.

(α) Circular polarization. A photon with helicity ± 1 is said to be circularly polarized. For a mixture of such states (3.1.72) becomes, since now $T_{zz} = 1/\sqrt{6}$,

$$\rho_\gamma^{\text{circ}} = \frac{1}{2}\begin{pmatrix} 1 + \mathscr{P}_{\text{circ}} & 0 & 0 \\ 0 & 0 & 0 \\ 0 & 0 & 1 - \mathscr{P}_{\text{circ}} \end{pmatrix}. \tag{3.1.82}$$

$\mathscr{P}_{\text{circ}}$ is conventionally referred to as the *circular polarization* of the photons. From (3.1.70) $\mathscr{P}_{\text{circ}}$ is given in terms of the probabilities for finding helicity $+1$ and helicity -1 polarized photons as

$$\mathscr{P}_{\text{circ}} = p_+ - p_- \tag{3.1.83}$$

as expected.

Note that $\mathscr{P}_{\text{circ}} = +1$ corresponds to photons with *positive* helicity. In terms of the electric field vector of a classical electromagnetic wave propagating along OZ, the case $\mathscr{P}_{\text{circ}} = +1$ corresponds to the case when the electric field vector is seen to rotate *anticlockwise* when looking into the wave. In optics this is referred to as *left-circularly polarized* light.

In the case of circular polarization the spin-polarization vector and the multipole parameters are given by

$$\boldsymbol{\mathcal{P}}_{\text{circ}} = (0, 0, \mathscr{P}_{\text{circ}}) \tag{3.1.84}$$

$$t_0^1 = \tfrac{1}{\sqrt{2}}\mathscr{P}_{\text{circ}} \qquad t_0^2 = \tfrac{1}{\sqrt{10}} \qquad t_{\pm 1}^1 = t_{\pm 2}^2 = t_{\pm 1}^2 = 0$$

Note that $\mathscr{P}_{\text{circ}}$ is, in magnitude, a measure of the degree of *vector* polarization.

Because of the absence of the $|m = 0\rangle$ states it is sometimes convenient to write (3.1.82) in the form

$$\rho_\gamma^{\text{circ}} = \tfrac{1}{2}\left(I + \mathscr{P}_{\text{circ}}\sigma_z\right). \tag{3.1.85}$$

(β) Linear polarization. A photon is said to be linearly polarized along OX or OY if its state is $|\mathbf{e}_{(x)}\rangle$ or $|\mathbf{e}_{(y)}\rangle$ respectively as defined in (3.1.75).

Consider a mixture of photons linearly polarized along the directions OX', OY' in the XY-plane, where OX' and OY' make an angle γ with OX and OY respectively. The *linear polarization along* OX' is defined by

$$\mathscr{P}_{\text{lin}} = p_{X'} - p_{Y'}$$

where $p_{X'}$, $p_{Y'}$ are the probabilities for finding photons linearly polarized along OX' and OY' respectively.

Using the fundamental definition (3.1.5) of the density matrix, and eqn (3.1.46), we get the density matrix for photons linearly polarized in the XY-plane at angle γ to the X-axis:

$$\rho_\gamma^{\text{lin}} = \frac{1}{2}\begin{pmatrix} 1 & 0 & -\mathscr{P}_{\text{lin}}e^{-2i\gamma} \\ 0 & 0 & 0 \\ -\mathscr{P}_{\text{lin}}e^{2i\gamma} & 0 & 1 \end{pmatrix}. \tag{3.1.86}$$

In this case the spin-polarization vector and multipole parameters are given by

$$\begin{gathered}\boldsymbol{\mathcal{P}}_{\text{lin}} = (0,0,0) \\ t_2^2 = -\tfrac{1}{2}\sqrt{\tfrac{3}{5}}\mathscr{P}_{\text{lin}}e^{2i\gamma} \qquad t_0^2 = \tfrac{1}{\sqrt{10}} \qquad t_m^1 = t_{\pm 1}^2 = 0\end{gathered} \tag{3.1.87}$$

and $\mathscr{P}_{\text{lin}}$ contributes only to the tensor polarization, as follows from (3.1.67).

Again, it is sometimes useful to abbreviate (3.1.86) in the form

$$\rho_\gamma^{\text{lin}} = \tfrac{1}{2}\left[I - \mathscr{P}_{\text{lin}}(\cos 2\gamma\ \boldsymbol{\sigma}_x + \sin 2\gamma\ \boldsymbol{\sigma}_y)\right]. \tag{3.1.88}$$

The physical interpretation of (3.1.82) and (3.1.86) when the photon has momentum $\mathbf{p} = (p, \theta, \varphi)$ will be explained in subsection 3.2.1.

(γ) Mixed polarization. Although light sources are usually either linearly or circularly polarized, it is in principle possible to have a mixture of both.

Let f be the fraction of circularly polarized photons and $1 - f$ the fraction linearly polarized. Then the density matrix for the mixture is just

$$\rho_\gamma = f\rho_\gamma^{\text{circ}} + (1-f)\rho_\gamma^{\text{lin}}.$$

3.1.13 Positivity of the density matrix

The density matrix, being hermitian, can always be diagonalized. In a basis in which it is diagonal it is clear that its elements $\rho_{mm} = \lambda_m$ simply measure the probability p_m of finding the state $|m\rangle$ in the ensemble. Thus the eigenvalues of ρ are either positive or zero.

A hermitian matrix whose diagonal elements have this property is called a *positive semi-definite* matrix. When a density matrix is measured experimentally it is essential to check that the matrix so obtained is indeed positive semi-definite. If it is not, this is a sure indicator of experimental error. Unfortunately it is a non-trivial task to get enough information experimentally to allow the calculation of the eigenvalues of ρ; it requires a knowledge of the whole matrix ρ.

Often, however, ρ is only partially known and it is important to be able to test whether this partial knowledge is compatible with the ultimate positive semi-definiteness of ρ. *Thus we require criteria for the positivity of ρ that do not involve a knowledge of its eigenvalues.*

The most useful result is the following. Let ρ_{ij} be the elements of ρ in *any basis.* Then *every principal minor* of the matrix must be positive semi-definite, i.e. if in some basis

$$\rho = \begin{pmatrix} \rho_{11} & \rho_{12} & \cdots & \rho_{1n} \\ \rho_{21} & \rho_{22} & \cdots & \rho_{2n} \\ \vdots & & & \vdots \\ \rho_{n1} & \rho_{n2} & \cdots & \rho_{nn} \end{pmatrix} \tag{3.1.89}$$

then one must have

(1) $\rho_{jj} \geq 0$ for every j

(2) $\begin{vmatrix} \rho_{jj} & \rho_{jk} \\ \rho_{kj} & \rho_{kk} \end{vmatrix} \geq 0$ for every j and all $k > j$

(3) $\begin{vmatrix} \rho_{jj} & \rho_{jk} & \rho_{jl} \\ \rho_{kj} & \rho_{kk} & \rho_{kl} \\ \rho_{lj} & \rho_{lk} & \rho_{ll} \end{vmatrix} \geq 0$ for every j and all $l > k > j$

$\vdots$

(n) $\begin{vmatrix} \rho_{11} & \rho_{12} & \cdots & \rho_{1n} \\ \rho_{21} & \rho_{22} & \cdots & \rho_{2n} \\ \vdots & & & \vdots \\ \rho_{n1} & \rho_{n2} & \cdots & \rho_{nn} \end{vmatrix} \geq 0.$

Failure of any one of these conditions will imply that ρ is not positive semi-definite.

Thus even partial measurements of ρ can be tested for compliance.

When ρ is diagonal it is trivial to see the consequences of positivity. For example in the case of spin 1, from (3.1.72) one has clearly

$$T_{zz} \leq 1/\sqrt{6} \tag{3.1.90}$$

and

$$-\tfrac{2}{3}\left(1+\sqrt{\tfrac{3}{2}}T_{zz}\right) \leq \mathscr{P}_z \leq \tfrac{2}{3}\left(1+\sqrt{\tfrac{3}{2}}T_{zz}\right)$$

which, combined with (3.1.90), gives

$$-1 \leq \mathscr{P}_z \leq 1. \tag{3.1.91}$$

For a more detailed analysis of the positivity conditions and an introduction to the concept of the *polarization domain* the reader should consult the review of Bourrely, Leader and Soffer (1980).

3.2 The relativistic case

We turn now to the relativistic case and introduce the helicity density matrix. All the properties discussed in Section 3.1 remain valid provided that care is exercised in the physical interpretation.

3.2.1 Definition of the helicity density matrix

Using as basis the helicity states discussed in Section 1.2 we can formally define the density matrix ρ in a given reference frame in which the particle is moving with momentum $\mathbf{p}$, in exact analogy with the non-relativistic case. If we have an ensemble of particles, all with momentum $\mathbf{p}$ but distributed with probability $p^{(i)}$ over various states $|\psi^{(i)};\mathbf{p}\rangle$, where

$$|\psi^{(i)};\mathbf{p}\rangle = \sum_{\lambda=-s}^{s} c_\lambda^{(i)}|\mathbf{p};\lambda\rangle, \tag{3.2.1}$$

then we define ρ by

$$\rho_{\lambda\lambda'} = \sum_i p^{(i)} c_\lambda^{(i)} c_{\lambda'}^{(i)*}. \tag{3.2.2}$$

The only question is: what is the physical meaning and use of this matrix?

In Section 1.2 we discussed the physical interpretation of helicity states. From this it is clear that $\rho_{\lambda\lambda'}$ for a given particle A is the ordinary, non-relativistic spin density matrix for particle A if we observe *A in the helicity rest frame of A*.

Thus for any observable $\hat{O}$ connected with particle A, Tr (ρO) *is the expectation value of $\hat{O}$ for the ensemble, in the helicity rest frame of A.*

If there is a mixture of several particle types, as in the initial or final state of a reaction, then one can, as in the non-relativistic case, define a joint density matrix using helicity states as a basis. For example, for two types of particles A, B one will have $\rho(A,B)$ with matrix elements $\rho(A,B)_{\lambda,\mu;\lambda',\mu'}$ defined in terms of simple direct products of the helicity states of A and B. This density matrix thus describes the spin distributions in the respective helicity rest frames of A and B. If $\hat{O}$ is an observable connected with both particles A and B then Tr $[\rho(A,B)O]$ gives the ensemble expectation value of $\hat{O}$ for measurements on A carried out in the helicity rest frame of A and measurements on B carried out in the helicity rest frame of B. As in the non-relativistic case, if we measure an observable belonging only to one of the particles, say A, then we require the $(2s_A+1)\times(2s_A+1)$ effective density matrix for A, $\rho(A)$, where

$$\rho(A)_{\lambda\lambda'} = \sum_{\mu} \rho(A,B)_{\lambda,\mu;\lambda',\mu}; \tag{3.2.3}$$

then

$$\left\langle \hat{O}^{(A)} \right\rangle = \text{Tr}\ [\rho(A)O^{(A)}]. \tag{3.2.4}$$

An identical result holds for B.

In a similar fashion, for massless particles ρ gives the density matrix of a particle in the standard frame where its momentum is $p^{\mu} = (p,0,0,p)$.

3.2.2 Definition of helicity multipole parameters

Because of the simple, i.e. non-relativistic, meaning of the helicity density matrix in the respective helicity rest frames, it is clear that multipole parameters defined in terms of ρ, as in the non-relativistic case, will also enjoy the same simple rotational properties.

Thus for any particle A we define

$$t^{L}_{M}{}^{*}(A) = \sum_{\lambda,\lambda'} \langle s\lambda|s\lambda';LM\rangle\, \rho_{\lambda\lambda'}(A) \tag{3.2.5}$$

as the helicity-basis *multipole parameters* for A.

In A's helicity rest frame S_A, the $t^L_M(A)$ are just the non-relativistic multipole parameters corresponding to an axis system coinciding with S_A.

Joint helicity multipole parameters are defined in terms of joint helicity density matrices, exactly as in Section 3.1, and all the properties derived there hold equally well.

3.2.3 *The effect of Lorentz transformations on the helicity density matrix*

(*i*) *Rotations of rest frame.* Let $\rho(A)$ be the helicity density matrix of A. As discussed above $\rho(A)$ *is* the density matrix for A in its helicity rest frame. The density matrix for A in any other rest frame S_A^r is simply obtained from $\rho(A)$ by a rotation. If $S_A^r = rS_A$ then by (3.1.46)

$$\rho^{S_A^r}(A) = \mathscr{D}^{(s)\dagger}(r)\rho(A)\mathscr{D}^{(s)}(r). \tag{3.2.6}$$

(*ii*) *Lorentz transformations.* The density matrix in a Lorentz-transformed frame is obtained as follows.

Let $\rho^S(\mathbf{p})$ be the helicity density matrix in frame S where the particle has momentum $\mathbf{p}$. Let $\rho^{S^l}(\mathbf{p}')$ be the helicity density matrix in the frame $S^l = lS$ obtained from S by an arbitrary Lorentz transformation l and in which the particle has momentum $\mathbf{p}' = l^{-1}\mathbf{p}$.

If we think of $\rho^{S^l}_{\mu\mu'}(\mathbf{p}')$ as the matrix of an operator $\hat{\rho}'$,

$$\rho^{S^l}_{\mu\mu'}(\mathbf{p}') = \langle \mathbf{p}'; \mu|\hat{\rho}'|\mathbf{p}'; \mu'\rangle,$$

then it is clear, from the meaning of $\rho_{\lambda\lambda'}$ as a probability correlation, that we must have the numerical relation

$$\rho^{S}_{\lambda\lambda'}(\mathbf{p}) = {}_{S^l}\langle \mathbf{p}; \lambda|\hat{\rho}'|\mathbf{p}; \lambda'\rangle_{S^l}$$

from which, using (2.1.9), we eventually obtain

$$\rho^{S^l}(\mathbf{p}') = \mathscr{D}[r(l,\mathbf{p})]\rho^S(\mathbf{p})\mathscr{D}^\dagger[r(l,\mathbf{p})] \tag{3.2.7}$$

where $r(l,\mathbf{p})$ is the Wick helicity rotation defined in Section 2.1.

Note that if the particle is *not at rest* in S, then for any frame S^r obtained from S by a rotation r, eqn (3.2.7) reduces to

$$\rho^{S^r}_{\lambda\lambda'}(\mathbf{p}') = \rho^{S}_{\lambda\lambda'}(\mathbf{p})e^{i\zeta(\lambda-\lambda')} \tag{3.2.8}$$

where ζ is given in (2.2.3). If the particle *is at rest* in S then (3.2.6) holds.

3.2.4 *Transformation law for multipole parameters*

It was stressed earlier that the multipole parameters t^L_M transform more simply under rotations than does ρ. Because Lorentz transformations are effected ultimately by just the Wick helicity rotation, also in this case the helicity-basis multipole parameters defined in eqn (3.2.5) will enjoy simpler transformation properties. Thus the analogue of (3.2.7) is

$$(t^L_M)_{S^l} = \mathscr{D}^{(L)\dagger}_{M'M}(r(l,\mathbf{p}))[t^L_{M'}]_S. \tag{3.2.9}$$

Note that for a spin-1/2 particle, if we write

$$\rho^{S^l} = \tfrac{1}{2}(I + \boldsymbol{\mathcal{P}}' \cdot \boldsymbol{\sigma}) \qquad \rho = \tfrac{1}{2}(I + \boldsymbol{\mathcal{P}} \cdot \boldsymbol{\sigma})$$

then from (3.2.7) and (1.1.28)

$$\mathscr{P}'_i = R_{ij}(r)\mathscr{P}_j \tag{3.2.10}$$

where r is short for $r(l, \mathbf{p})$.

Analogous results will hold for $\boldsymbol{\mathcal{P}}$ and T_{ij} for spin-1 particles, etc.

In the case of a pure rotation for which ζ in (3.2.8) happens to be zero, so that $\rho^{S^r} = \rho$, putting

$$\rho^{S^r} = \tfrac{1}{2}\left(I + \boldsymbol{\mathcal{P}}' \cdot \boldsymbol{\sigma}\right)$$

implies the perhaps surprising result $\boldsymbol{\mathcal{P}}' = \boldsymbol{\mathcal{P}}$. It must not be forgotten that $\mathscr{P}'_i$ and $\mathscr{P}_i$ are the components of the spin-polarization vector in the helicity rest frames reached from S^r and S respectively and that these rest frames coincide in this particular case.

3.3 Choices of reference frame for a reaction

We consider a general reaction

$$A + B \to C + D + E + \cdots$$

taking place either in the Lab frame, corresponding to a fixed-target experiment, where B is at rest, or in a frame, corresponding to collider physics, where A and B collide head-on. In the latter case the frame may or may not be the actual CM frame.

The actual choice of axes is partly a matter of convention, partly a matter of convenience in the context of the particular experiment.

Quite generally the collision axis is taken as the Z-axis. In the Lab frame OZ is taken to lie along the incoming beam.

The choice of Y-axis depends on the kind of experiment. Much of the early work on spin-dependent reactions utilized fixed spectrometers, which therefore defined the reaction plane; the spin-polarization vectors of beam and target particles, which could be varied in the experiment, were referred to this reaction plane.

For the $2 \to 2$ reaction $A + B \to C + D$, according to the so-called Basel convention OY is defined to lie along the *normal to the reaction plane*, defined as the direction of $\mathbf{p}_A \times \mathbf{p}_C$.

In more modern experiments, where the collision axis is surrounded by detectors, there is no obviously preferred fixed reaction plane and OY is then chosen arbitrarily, according to convenience. This is particularly important when a reaction is being used as an analysing reaction, i.e. to measure the direction and magnitude of the spin-polarization vectors of a beam and/or target; it may then be necessary to perform weighted integrals over the ϕ-dependent angular distributions in some fixed reference frame in order to determine the components of the spin-polarization vectors in that frame.

In doing calculations it is generally simplest to work with the CM helicity amplitudes, which are directly related to the CM helicity density matrix (see Section 5.3). Thus it is helpful to specify the states of the particles and to carry out analysing measurements on the particles, in frames that are related to the CM simply by a Lorentz transformation.

3.3.1 Density matrix for the initial particles

It is simplest to give the density matrices or multipole parameters of A and B in their helicity rest frames S_A, S_B, reached from the CM of the reactions as shown in Fig. 3.1. (Note that Y_A and Y_B are in opposite directions.) These are then the correct parameters to use in specifying the initial state in the reaction CM.

The laboratory (Lab) frame will always be taken to have the same orientation of axes as the CM frame and is to be thought of as reached in the limit as we boost along the negative Z-axis until B is just barely at rest. The helicity rest frames for the initial particles A, B reached from the Lab frame will then coincide with those shown in Fig. 3.1.

Note that for the target the axes of S_B do not point in the same direction as the Lab axes. One has

$$S_B = r_{y'}(\pi) r_z(\pi) S_{\mathrm{L}}$$

and spin information about the target, if specified in the Lab frame, must always be transformed into S_B.

Often, however, for magnetically prepared beam and target, the polarization information is given in a rest frame whose Z-axis ($\hat{Z}$) lies along the quantization axis. Let this axis have polar angles $\theta = \beta, \phi = \gamma$ relative to the axes of the CM frame (or of the Lab frame) as shown in Fig. 3.2.

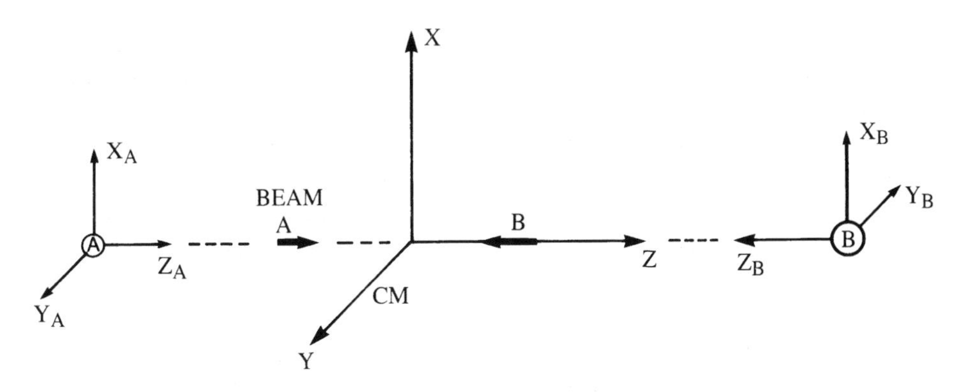

Fig. 3.1. Helicity rest frames for beam and target reached from the CM.

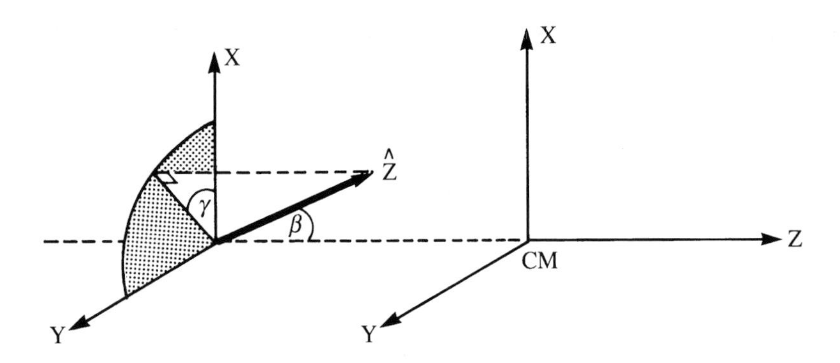

Fig. 3.2. Rest frame with $O\hat{Z}$ along quantization axis.

In this rest frame ρ_{ij} is diagonal, its elements being just the probability of the various magnetic substates. Equivalently, the multipole parameters are such that $t^l_m = 0$ for $m \neq 0$. Let us label the non-zero multipole parameters in this frame by $\hat{t}^l_0$.

Then for particle A, coming in along OZ, the CM helicity multipole parameters are

$$t^l_m(A) = e^{i\gamma_A m} d^l_{m0}(\beta_A)\hat{t}^l_0(A) \tag{3.3.1}$$

as follows from eqn (3.1.48).

For particle B, moving in the negative OZ direction, the CM helicity multipole parameters are

$$t^l_m(B) = e^{-i\gamma_B m} d^l_{m0}(\pi - \beta_B)\hat{t}^l_0(B) \tag{3.3.2}$$

wherein account has been taken of the fact that Y_B in Fig. 3.1 is opposite in direction to Y_{CM}.

As an example consider an electron or nucleon with spin-polarization vector $(0, 0, \mathscr{P}_{\hat{z}})$ along or opposite to $O\hat{Z}$. From (3.1.35)

$$\hat{t}^1_0 = \sqrt{\tfrac{1}{3}}\mathscr{P}_{\hat{z}}. \tag{3.3.3}$$

For the case of *longitudinal* polarization, $O\hat{Z}$ lies along OZ or opposite to it. For particle A, (3.3.3) then gives, for degree of polarization $\mathscr{P}_A$,

$$t^1_0(A) = \pm\sqrt{\tfrac{1}{3}}\mathscr{P}_A \qquad t^1_{\pm 1} = 0 \tag{3.3.4}$$

for longitudinal polarization along or opposite to A's motion.

Similarly

$$t^1_0(B) = \mp\sqrt{\tfrac{1}{3}}\mathscr{P}_B \qquad t^1_{\pm 1} = 0 \tag{3.3.5}$$

for longitudinal polarization along or opposite to B's motion.

For the case of *transverse* polarization, say perpendicular to the plane XZ, we get

$$t_0^1(A) = 0 \qquad t_1^1(A) = t_{-1}^1(A) = \mp\sqrt{\tfrac{1}{6}}\,i\mathscr{P}_A \tag{3.3.6}$$

where $\mathscr{P}_A$ is the degree of polarization along or opposite to OY_{CM}.

For particle B one has

$$t_0^1(B) = 0 \qquad t_1^1(B) = t_{-1}^1(B) = \pm\sqrt{\tfrac{1}{6}}\,i\mathscr{P}_B \tag{3.3.7}$$

where $\mathscr{P}_B$ is the degree of polarization along or opposite to OY_{CM}.

As a second example consider a beam of photons incident along OZ_{CM}. Here we may directly use the results (3.1.82), (3.1.83) and (3.1.86) for the density matrix. For the case of circular polarization one has from (3.1.82), (3.1.68) and (3.1.69)

$$t_0^1 = \frac{1}{\sqrt{2}}\mathscr{P}_{\mathrm{circ}} \qquad t_0^2 = \frac{1}{\sqrt{10}} \tag{3.3.8}$$

with all other $t_m^l = 0$ for $l \neq 0$.

For the case of linear polarization in the XY-plane at angle γ to the X-axis, from (3.1.86), (3.1.68) and (3.1.69),

$$t_2^2 = -\tfrac{1}{2}\sqrt{\tfrac{3}{5}}\,\mathscr{P}_{\mathrm{lin}}e^{2i\gamma} \qquad t_0^2 = \frac{1}{\sqrt{10}} \tag{3.3.9}$$

with all other $t_m^l = 0$ for $l \neq 0$.

If the linear polarization of the photon is specified with respect to the CM X- and Y-axes and if the photon is incident in the *negative* Z-direction, then, bearing in mind Fig. 3.1, one must take $\gamma \to -\gamma$ in (3.3.9).

3.3.2 Density matrix of final state particles

The density matrix of a produced particle may be obtained experimentally from studying the decay of the particle or by letting it undergo a secondary analysing reaction.

Unfortunately many conventions exist and many different frames have been used in the past for this analysis. A comprehensive discussion of the Adair, Gottfried–Jackson, and transversity frames can be found in Bourrely, Leader and Soffer (1980).

The frame in which one wishes to know the density matrix is dictated by the kind of reaction under study. There are basically two situations:

(i) reactions in which a resonance is produced and its decay studied;
(ii) reactions in which a stable final state particle undergoes a secondary, analysing reaction.

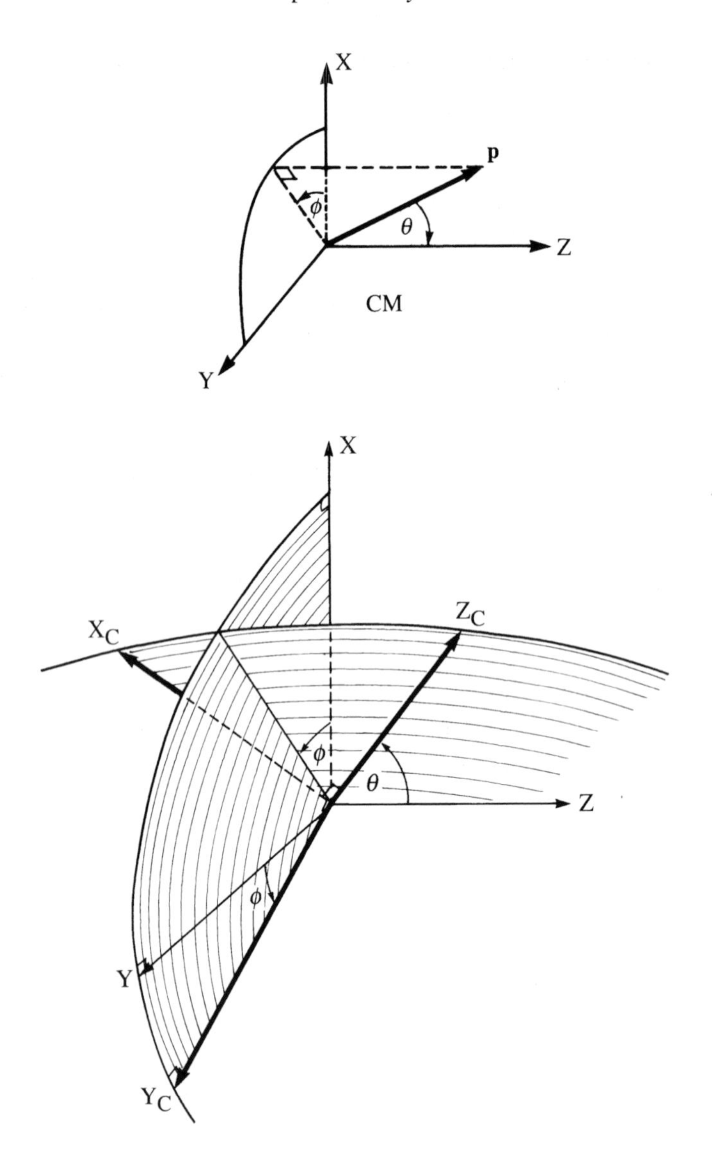

Fig. 3.3. Helicity rest frame for particle produced with angles θ, ϕ in CM.

(i) Resonance production

If one or several final state particles are unstable we will be interested in their decay distributions, which yield information about the production mechanism.

It is simplest to analyse the decay of some particle or resonance C in its own *helicity rest frame* (see subsection 1.2.2) reached from the CM frame of the production reaction $A + B \to C + D + E + \dots$, since in that case the initial helicity density matrix of C before it decays is just equal to the helicity density matrix of C in the CM frame of the production reaction, i.e. it is given directly in terms of the CM helicity amplitudes for the production process.

Let C have momentum $\mathbf{p} = (p, \theta, \phi)$ in the CM frame. Then as explained in subsection 1.2.2 its helicity rest frame S_C has its Z-axis, Z_C, along $\mathbf{p}$ and its Y-axis along $\mathbf{e}_{(z)} \times \mathbf{p}$, where $\mathbf{e}_{(z)}$ is a unit vector along the Z-axis of the CM frame. This is illustrated in Fig. 3.3. Note that for $\theta = 0$ or π we take Y_C along or opposite to the CM Y-axis.

The case of a reaction taking place in the XZ-plane is easier to visualize. The relative orientation of the helicity rest frames reached from the CM frame is shown in Fig. 3.4.

The relationship between the decay characteristics and the density matrix of C is discussed in Section 8.2. For an analysis done in S_C the relevant density matrix is then just the density matrix of C in the CM of the production reaction – no transformation is needed.

(ii) Secondary scattering

Consider a stable particle K produced in the reaction CM with momentum (p_K, θ_K, ϕ_K). We shall define the *natural analysing frame* for K, S_{LK}, to be a frame reached from the laboratory frame S_{L} by mean of a pure rotation such as to give particle K polar angles equal to zero. In other words

$$\mathbf{p}_K^{S_{LK}} = (p_K^{\mathrm{L}}, 0, 0). \tag{3.3.10}$$

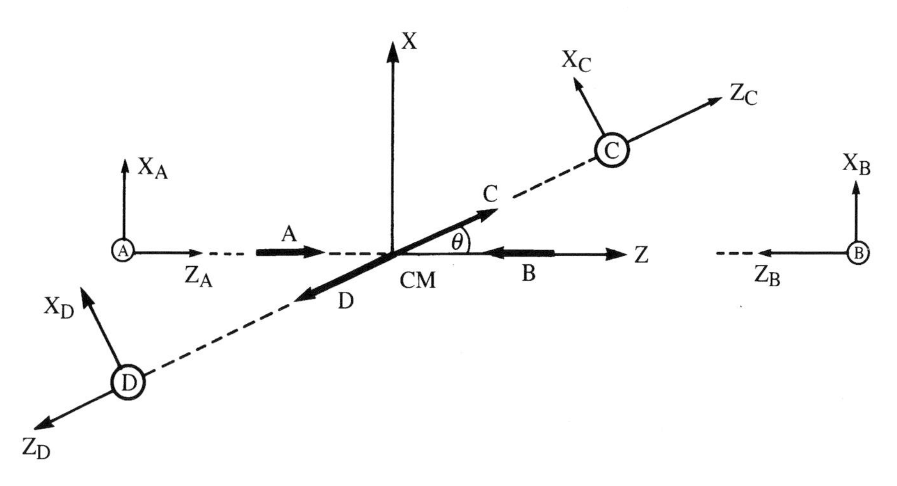

Fig. 3.4 Helicity rest frames reached from the CM of the reaction $A + B \to C + D$. Note that the Y-axes for B and D are opposite to those for A and C.

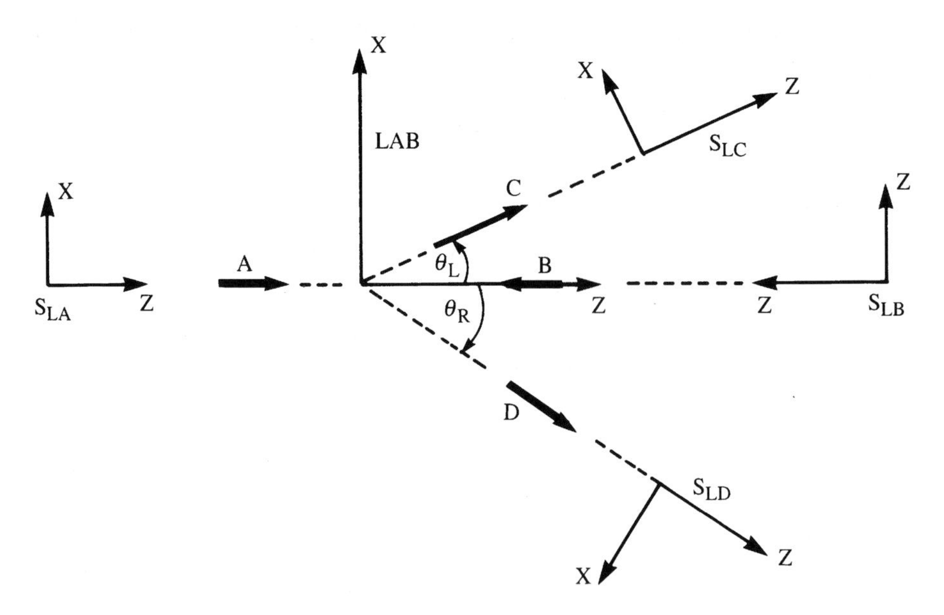

Fig. 3.5 The laboratory 'analysing' frames S_{LA}, S_{LB}, S_{LC} and S_{LD} for the reaction $A + B \to C + D$.

The natural analysing frames for the $2 \to 2$ reaction $AB \to CD$ are illustrated in Fig. 3.5. We include frames for the initial particles.

The Lab frame S_{LK} is the *simplest* and most natural frame in which to study the analysing reaction for K for the following reasons.

(α) Because S_{LK} is reached from S_L by the rotation

$$r_{y'}(\theta_K^{\mathrm{L}})r_z(\phi_K) = r(\phi_K, \theta_K^{\mathrm{L}}, 0) \tag{3.3.11}$$

it is easy to see from (2.1.7) that the helicity density matrix for K is the same in S_{LK} as in S_{L}.

(β) Because the CM frame for the *analysing reaction* is reached from S_{LK} by boosting along the positive Z-axis of S_{LK}, the helicity density matrix for K in the CM of the analysing reaction is the same as it is in S_{LK}. By (α) it is then the same as in the main Lab frame.

Thus we have the result

$$[\rho(K)]_{\substack{\text{CM of}\\ \text{analysing}\\ \text{reaction}}} = \rho^{S_{LK}}(K) = \rho^{S_{\mathrm{L}}}(K) \tag{3.3.12}$$

and the *initial* helicity density matrix of K needed for the analysing reaction is simply given by $\rho^{S_{\mathrm{L}}}(K)$.

Finally, then, in terms of the helicity density matrix in the CM of the production reaction we have

$$[\rho(K)]_{\substack{\text{CM of} \\ \text{analysing} \\ \text{reaction}}} = d(\alpha_K)\rho(K)d^\dagger(\alpha_K) \tag{3.3.13}$$

with α_K given by (2.2.13), or, equivalently,

$$[t^l_m(K)]_{\substack{\text{CM of} \\ \text{analysing} \\ \text{reaction}}} = d^l_{mm'}(\alpha_K)t^l_{m'}(K). \tag{3.3.14}$$

We remind the reader that α_K takes on special values when the production reaction is an elastic reaction; see (2.2.15) and (2.2.16).

3.4 Covariant spin vectors

In Section 1.1 the spin-polarization vector for a non relativistic state $|\chi\rangle$ was defined as

$$\boldsymbol{\mathcal{P}}_\chi \equiv \frac{1}{s}\langle\chi|\hat{\mathbf{s}}|\chi\rangle \equiv \frac{\mathbf{s}_\chi}{s}, \tag{3.4.1}$$

where $\mathbf{s}_\chi$ is the mean spin vector, and this was generalized in subsection 3.1.5 to an ensemble or mixture of pure states. As stressed in Section 3.2 these non-relativistic quantities and, more generally, the multipole parameters can continue to be used in the relativistic case for a massive particle, provided that any physical statements about the spin are understood to hold in the helicity rest frame of the particle.

It is nonetheless advantageous sometimes to deal with relativistic, covariant generalizations of these quantities. We showed in Section 1.2 that the natural covariant generalization of the non-relativistic spin operators $\hat{s}_j$ is given by the space components of the Pauli–Lubanski operators $\hat{W}^\mu$; namely, when acting on the state of a massive particle at rest,

$$\hat{W}^j|\mathring{p}; s_z = \lambda\rangle = m\hat{s}^j|\mathring{p}; s_z = \lambda\rangle. \tag{3.4.2}$$

Note, in addition, that

$$\hat{W}^0|\mathring{p}; s_z = \lambda\rangle = 0. \tag{3.4.3}$$

Now for any rest state $|\mathring{p};\chi\rangle$ that is a linear superposition of states of spin s with different values of s_z, we can define

$$\mathring{\mathscr{S}}^\mu_\chi \equiv \frac{1}{s}\langle\mathring{p};\chi|\hat{W}^\mu|\mathring{p};\chi\rangle \tag{3.4.4}$$

and we see from (3.4.2), (3.4.3) and (3.4.1) that

$$\mathring{\mathscr{S}}^\mu_\chi = \frac{m}{s}(0, \mathbf{s}_\chi) = m(0, \boldsymbol{\mathcal{P}}_\chi). \tag{3.4.5}$$

(The reason for our convention of including the factor m in this definition will become clear later.)

Moreover, because $\overset{\circ}{p} = (m, 0, 0, 0)$, we have

$$\overset{\circ}{\mathscr{S}}\chi \cdot \overset{\circ}{p} = 0 \tag{3.4.6}$$

$$\overset{\circ}{\mathscr{S}}\chi \cdot \overset{\circ}{\mathscr{S}}\chi = -\frac{m^2}{s^2}\mathbf{s}^2_\chi.$$

Now consider the expectation value of $\hat{W}^\mu$ for a relativistic helicity state $|\mathbf{p};\lambda\rangle$. We have from (1.2.25)

$$\begin{aligned}\langle \mathbf{p};\lambda|\hat{W}^\mu|\mathbf{p};\lambda\rangle &= \langle \overset{\circ}{p};\lambda|U^{-1}[h(\mathbf{p})]\,\hat{W}^\mu U[h(\mathbf{p})]|\overset{\circ}{p};\lambda\rangle \\ &= \Lambda^\mu{}_\nu \langle \overset{\circ}{p};\lambda|\hat{W}^\nu|\overset{\circ}{p};\lambda\rangle \end{aligned}\tag{3.4.7}$$

since the $\hat{W}^\mu$ transform covariantly as a 4-vector. Hence we can define the *covariant helicity mean spin vector*

$$\mathscr{S}^\mu(\mathbf{p},\lambda) \equiv \frac{1}{s}\langle \mathbf{p};\lambda|\hat{W}^\mu|\mathbf{p};\lambda\rangle \tag{3.4.8}$$

and we then have, in full detail,

$$\mathscr{S}^\mu(\mathbf{p},\lambda) = (\Lambda[h(\mathbf{p})])^\mu{}_\nu\, \overset{\circ}{\mathscr{S}}{}^\nu_\lambda \tag{3.4.9}$$

where

$$p^\mu = (E,\mathbf{p}) = (\Lambda[h(\mathbf{p})])^\mu{}_\nu\, \overset{\circ}{p}{}^\nu.$$

For a state χ that is a linear combination of states with different values of s_z we clearly have

$$\mathscr{S}^\mu(\mathbf{p},\chi) = (\Lambda[h(\mathbf{p})])^\mu{}_\nu\, \overset{\circ}{\mathscr{S}}{}^\nu\chi. \tag{3.4.10}$$

If one uses the canonical spin states (see subsection 1.2.1) then one defines

$$\mathscr{S}^\mu_{\text{can}}(\mathbf{p},s_z) \equiv \frac{1}{s}\,{}_{\text{can}}\langle \mathbf{p};s_z|\hat{W}^\mu|\mathbf{p};s_z\rangle_{\text{can}} \tag{3.4.11}$$

and so, in contrast to (3.4.10),

$$\mathscr{S}^\mu_{\text{can}}(\mathbf{p},\chi) = (\Lambda[l(\mathbf{v})])^\mu{}_\nu\, \overset{\circ}{\mathscr{S}}{}^\nu\chi \tag{3.4.12}$$

where $l(\mathbf{v})$ is the pure boost that takes $\overset{\circ}{p}$ to p^μ.

Note that (3.4.6) generalizes to

$$\mathscr{S}(\mathbf{p},\chi)\cdot p = 0 \tag{3.4.13}$$

and one has

$$\mathscr{S}^2(\mathbf{p},\chi) = -\frac{m^2}{s^2}\mathbf{s}^2_\chi. \tag{3.4.14}$$

We shall see in subsection 6.3.1 that $\mathscr{S}^\mu$ provides a convenient approach to the relativistic motion of the mean spin vector.

As an example let us compute the covariant helicity mean spin vector for a spin-1/2 particle in a definite helicity state $\lambda = \pm 1/2$. In this case from (3.4.1)

$$\mathbf{s}_\chi \equiv \mathbf{s}(\pm 1/2) = (0, 0, \pm 1/2)$$

and using eqn (1.2.23) we arrive at

$$\mathscr{S}^\mu(\mathbf{p}, \lambda) = 2\lambda(p, E\hat{\mathbf{p}}) \tag{3.4.15}$$

where $\hat{\mathbf{p}}$ is a unit vector along $\mathbf{p}$. Note the important result that, for spin 1/2,

$$\lim_{m \to 0} \mathscr{S}^\mu(\mathbf{p}, \lambda) = 2\lambda p^\mu. \tag{3.4.16}$$

Had we not included the factor m in the definition of $\overset{\circ}{\mathscr{S}}_\mu$ the limit $m \to 0$ of (3.4.15) could not have been taken. Thus, with our convention, $\mathscr{S}^\mu$ applies equally well for massless particles.

In the general case of massless particles we replace (3.4.4), (3.4.5) by

$$\mathscr{S}^\mu(\mathbf{p}_{\text{st}}, \lambda) \equiv \frac{1}{s}\langle \mathbf{p}_{\text{st}}; \lambda | \hat{W}^\mu | \mathbf{p}_{\text{st}}; \lambda \rangle \tag{3.4.17}$$

using the 'standard' states, defined in subsection 1.2.3, where $p_{\text{st}} = (\overline{p}, 0, 0, \overline{p})$. It follows from (1.2.39) and (1.2.41) that, for a massless helicity state with arbitrary 4-momentum p^μ,

$$\mathscr{S}^\mu(\mathbf{p}, \lambda) = \frac{\lambda}{s} p^\mu, \tag{3.4.18}$$

which is perfectly consistent with (3.4.15)when $s = 1/2$.

The tensor operators $\hat{T}_{ij}$, introduced in subsection 3.1.12 for *massive* spin-1 particles, are all expressed as products of the basic spin operators $\hat{s}_j$. A relativistic generalization of (3.1.62) is then

$$\hat{T}^{\mu\nu} = \frac{1}{2}\sqrt{\frac{3}{2}}\left[\frac{1}{m^2}\left(\hat{W}^\mu \hat{W}^\nu + \hat{W}^\nu \hat{W}^\mu\right) + \frac{4}{3}\left(g^{\mu\nu} - \frac{\hat{P}^\mu \hat{P}^\nu}{m^2}\right)\right] \tag{3.4.19}$$

with expectation values

$$\mathscr{T}^{\mu\nu}(\mathbf{p}, \chi) = \langle \mathbf{p}; \chi | \hat{T}^{\mu\nu} | \mathbf{p}; \chi \rangle \tag{3.4.20}$$

such that

$$\mathscr{T}^{\mu\nu}(\mathbf{p}, \chi) = \Lambda^\mu{}_\alpha \Lambda^\nu{}_\beta \overset{\circ}{\mathscr{T}}{}^{\alpha\beta}_\chi \tag{3.4.21}$$

where $\Lambda = \Lambda[h(\mathbf{p})]$. The relation to the non-relativistic expectation values T_{ij} is then

$$\overset{\circ}{\mathscr{T}}_{ij} = T_{ij} \qquad \overset{\circ}{\mathscr{T}}_{0j} = \overset{\circ}{\mathscr{T}}_{j0} = \overset{\circ}{\mathscr{T}}_{00} = 0. \tag{3.4.22}$$

We note that

$$\begin{aligned} \sum_\mu \mathscr{T}^\mu{}_\mu &= 0 \\ p_\mu \mathscr{T}^{\mu\nu}(\mathbf{p}, \chi) = \mathscr{T}^{\mu\nu}(\mathbf{p}, \chi) p_\nu &= 0. \end{aligned} \tag{3.4.23}$$

To specify the most general state for a massive spin-1 particle at rest we introduced in eqn (3.1.76) a polarization vector $\boldsymbol{\varepsilon}$ (in general complex). Copying the above procedure, we can define

$$\overset{\circ}{\epsilon}{}^\mu \equiv (0, \boldsymbol{\varepsilon}) \tag{3.4.24}$$

and take

$$\epsilon^\mu(\mathbf{p}) = (\Lambda[h(\mathbf{p})])^\mu{}_\nu \overset{\circ}{\epsilon}{}^\nu. \tag{3.4.25}$$

We have

$$p_\mu \epsilon^\mu(\mathbf{p}) = 0 \tag{3.4.26}$$

and from (3.1.77)

$$\epsilon^*(\mathbf{p}) \cdot \epsilon(\mathbf{p}) = -1. \tag{3.4.27}$$

The relations between $\mathscr{S}^\mu(\mathbf{p}, \boldsymbol{\varepsilon})$, $\mathscr{T}^{\mu\nu}(\mathbf{p}, \boldsymbol{\varepsilon})$ and $\epsilon^\mu(\mathbf{p})$, which generalize (3.1.78) and (3.1.79), are

$$\mathscr{S}_\mu(\mathbf{p}, \boldsymbol{\varepsilon}) = -\epsilon_{\mu\alpha\beta\gamma} p^\alpha \ \mathrm{Im}\left(\epsilon^{*\beta} \epsilon^\gamma\right) \tag{3.4.28}$$

and

$$\mathscr{T}_{\mu\nu}(\mathbf{p}, \boldsymbol{\varepsilon}) = -\sqrt{\frac{3}{2}} \left[\mathrm{Re}\,(\epsilon^*_\mu \epsilon_\nu) + \frac{1}{3}\left(g_{\mu\nu} - \frac{p_\mu p_\nu}{m^2} \right) \right]. \tag{3.4.29}$$

For states of definite helicity, $\boldsymbol{\varepsilon}$ is given by eqn (3.1.80). Equation (3.4.28) illustrates very clearly that the spin-polarization vector for a spin-1 particle is quite different from the polarization vector ϵ^μ. Indeed, the complex polarization vector contains all the information needed to specify the state of the particle whereas this is not true for the spin-polarization vector.

It is instructive to link the above discussion of the polarization vector ϵ^μ with the more familiar use of polarization vectors in field theory when describing a field of spin-1 quanta by means of a 4-vector field $A_\mu(x)$. In that case the analogue of eqn (2.4.5) is usually written

$$A_\mu(x) = \sum_\lambda \int \frac{d^3\mathbf{p}}{(2\pi)^{3/2} 2p^o} \left[\epsilon_\mu(\mathbf{p}, \lambda) a(\mathbf{p}, \lambda) e^{-ip\cdot x} + \epsilon^*_\mu(\mathbf{p}, \lambda) b^\dagger(\mathbf{p}, \lambda) e^{ip\cdot x} \right]. \tag{3.4.30}$$

To ensure that there are no spin-0 quanta present one imposes the invariant condition

$$\partial^\mu A_\mu(x) = 0 \tag{3.4.31}$$

from which we have the requirement

$$p^\mu \epsilon_\mu(\mathbf{p}, \lambda) = 0 \tag{3.4.32}$$

as in (3.4.26).

Moreover, using (3.4.25) and (3.1.80), if one takes the simple case $\mathbf{p} = (0, 0, p)$ then, via eqn (1.2.23),

$$\epsilon^\mu(p_z, \lambda = \pm 1) = \frac{1}{\sqrt{2}}(0, \mp 1, -i, 0) \tag{3.4.33}$$

$$\epsilon^\mu(p_z, \lambda = 0) = \frac{1}{m}(p, 0, 0, E) \tag{3.4.34}$$

and we check that

$$\sum_{\lambda=0,\pm 1} \epsilon^{*\mu}(\mathbf{p}, \lambda)\epsilon^\nu(\mathbf{p}, \lambda) = \frac{p^\mu p^\nu}{m^2} - g^{\mu\nu}. \tag{3.4.35}$$

But (3.4.33), (3.4.34) and (3.4.35) are just the usual properties of the polarization vectors in (3.4.30). (See, for example, Gasiorowicz, 1967.) Thus the polarization vectors introduced in (3.4.25) coincide exactly with those used in conventional field theory.

Finally, let us note from (2.4.10) that $\epsilon_\mu(\mathbf{p}, \lambda)$ plays the rôle of the wave function for the single-particle state $|\mathbf{p}; \lambda\rangle$ annihilated by the field $A_\mu(x)$.

For photons, our standard state $|\mathbf{p}_{\text{st}}; \lambda\rangle$ consists of the photon moving along OZ with momentum $p^\mu_{\text{st}} = (\overline{p}, 0, 0, \overline{p})$ and helicity ± 1 and we may take $\epsilon^\mu(\mathbf{p}_{\text{st}}; \lambda = \pm 1)$ as given by (3.4.33).

For a photon in the state $|\mathbf{p}; \lambda\rangle = U[h(\mathbf{p}, \mathbf{p}_{\text{st}})]|\mathbf{p}_{\text{st}}; \lambda\rangle$ the polarization vector is then

$$\epsilon^\mu(\mathbf{p}, \lambda) = (\Lambda[h(\mathbf{p}, \mathbf{p}_{\text{st}})])^\mu{}_\nu\, \epsilon^\nu(\mathbf{p}_{\text{st}}; \lambda). \tag{3.4.36}$$

Explicitly, one finds, when $\mathbf{p} = (p, \theta, \varphi)$, that

$$\epsilon^\mu(\mathbf{p}, \pm 1) = \tfrac{1}{\sqrt{2}}(0,\ \mp\cos\theta\cos\varphi + i\sin\varphi, \mp\cos\theta\sin\varphi - i\cos\varphi,\ \pm\sin\theta). \tag{3.4.37}$$

Using this, one can check that the connection between $\mathscr{S}_\mu(\mathbf{p}, \boldsymbol{\varepsilon})$ and ϵ^μ given for massive spin-1 particles in (3.4.28) continues to hold for photons, and correctly gives (3.4.18).

It is simple to check that, as expected, for the spatial part of the vectors

$$\boldsymbol{\varepsilon}(\mathbf{p}) \cdot \mathbf{p} = 0. \tag{3.4.38}$$

To define a covariant spin tensor $\mathscr{T}^{\mu\nu}$ for photons is a little clumsy. The rôle of $\overset{\circ}{p}{}^{\mu}/m = (1, 0, 0, 0)$ must here be taken by a unit time-like vector n^{μ}_{st} defined to have components

$$n^{\mu}_{\text{st}} = (1, 0, 0, 0) \tag{3.4.39}$$

in the standard frame. Then

$$\mathscr{T}^{\mu\nu}(\mathbf{p}, \boldsymbol{\varepsilon}) = \sqrt{\tfrac{3}{2}} \left[\tfrac{1}{3} \left(n^{\mu} n^{\nu} - g^{\mu\nu} \right) - \operatorname{Re} \left(\epsilon^{\mu *} \epsilon^{\nu} \right) \right] \tag{3.4.40}$$

where ϵ^{μ} is given by (3.4.37) and

$$\begin{aligned} n^{\mu} \equiv n^{\mu}(\mathbf{p}) &= (\Lambda[h(\mathbf{p}, \mathbf{p}_{\text{st}})])^{\mu}{}_{\nu} n^{\nu}_{\text{st}} \\ &= \frac{1}{2p\bar{p}} \left[p^2 + \bar{p}^2, (p^2 - \bar{p}^2) \hat{\mathbf{p}} \right]. \end{aligned} \tag{3.4.41}$$

This $\mathscr{T}^{\mu\nu}$ satisfies eqns (3.4.21) and (3.4.22) with $\overset{\circ}{\mathscr{T}}$ replaced by $\mathscr{T}_{\text{st}}$.

4
Transition amplitudes

Ultimately our fundamental goal in particle physics is to understand the dynamics, i.e. to have a theory from which we can actually calculate transition amplitudes. Tests of the theory will involve, at the crudest level, measurements of differential cross-sections or decay rates but, at a more sophisticated and more probing level, measurements of all kinds of spin-dependent phenomena. On the one hand, given a dynamical theory it is probably simplest to calculate the helicity transition amplitudes and from them the formulae for the spin-dependent observables that can be tested against experimental data. On the other hand, in the absence of a theory it would seem best to try to obtain information on the behaviour of the transition amplitudes from a sufficiently large number of different independent measurements. In this way one would hope to be led to deduce the nature of the underlying dynamics.

In both these situations it is important to bear in mind that certain properties are intrinsic to transition amplitudes, i.e. they do not depend upon detailed dynamical theory but rather follow from very general conservation laws, principally from the conservation of angular momentum.

The study of reactions thus divides into two phases:

(1) the general properties of transition amplitudes and the connection between their behaviour and the underlying dynamics; and
(2) the relationship between transition amplitudes and observables.

In this chapter we concentrate upon the former. The latter will be discussed in Chapter 5.

4.1 Helicity amplitudes for elastic and pseudoelastic reactions

Many kinds of transition amplitude can be found in the literature, but it seems to us that helicity amplitudes are generally the simplest and most useful amplitudes, and we shall therefore concentrate almost exclusively on them. (However, in some circumstances other types of transition amplitude can be valuable, in particular transversity amplitudes, so we include a brief discussion of these in Appendix 4.)

We consider reactions of the type

$$A + B \to C + D$$

where C and D may be stable or unstable particles. The particles have arbitrary spins s_A, s_B, s_C, s_D.

In defining the scattering amplitudes we shall utilize the simple *helicity states* discussed in Section 1.2, which differ slightly from those of the original Jacob–Wick paper (Jacob and Wick, 1959). We do not adopt the convention that deals asymmetrically with the particles and distinguishes 'particle 2' in the reaction.[1] Nevertheless our *helicity amplitudes* will be almost identical to the Jacob–Wick amplitudes at $\phi = 0$, the difference being an irrelevant constant factor. Our amplitudes will have a simpler ϕ-dependence and this will lead to simpler properties of the final state density matrices.

As in Section 1.2 we define single-particle helicity states $|\mathbf{p};\lambda\rangle \equiv |p,\theta,\phi;\lambda\rangle$ normalized as follows:

$$\langle \mathbf{p}';\lambda'|\mathbf{p};\lambda\rangle = (2\pi)^3 \times 2E\delta^3(\mathbf{p}' - \mathbf{p}). \tag{4.1.1}$$

A two-particle state, or indeed an N-particle state, is defined as a direct product of one-particle states. Thus our two-particle CM helicity state with relative momentum $\mathbf{p}' = (p',\theta,\phi)$ is

$$|\mathbf{p}';\lambda_C\lambda_D\rangle = |p',\theta,\phi;\lambda_C\rangle \otimes |p',\pi-\theta,\phi+\pi;\lambda_D\rangle. \tag{4.1.2}$$

For consistency, the initial state with A along OZ and relative momentum $\mathbf{p} = (p,0,0)$ is then

$$|\mathbf{p};\lambda_A\lambda_B\rangle = |p,0,0;\lambda_A\rangle \otimes |p,\pi,\pi;\lambda_B\rangle. \tag{4.1.3}$$

The transition amplitudes are essentially the matrix elements of the $\hat{S}$-operator taken between initial and final CM helicity states. We shall write these as $H_{\lambda_C\lambda_D;\lambda_A\lambda_B}(\theta,\phi)$ and they will be normalized in such a way that for an *unpolarized initial state* the invariant differential cross-section is given by

$$\frac{d\sigma}{dt} = \frac{1}{(2s_A+1)(2s_B+1)} \sum_{\text{all } \lambda} |H_{\lambda_C\lambda_D;\lambda_A\lambda_B}(\theta)|^2 \tag{4.1.4}$$

---

[1] So long as one works *only* in the CM the Jacob–Wick convention is sensible, but the moment one wishes to transform to other systems, e.g. to the Lab, the asymmetric treatment of the particle becomes a nuisance. Indeed Wick himself discarded the convention in later papers. (Wick, 1962).

where t is the invariant square of the 4-momentum transfer,

$$t \equiv (p_C - p_A)^2. \tag{4.1.5}$$

For photons, the factor $2s+1$ is replaced by 2 in (4.1.4). Here, as throughout, $H(\theta)$ means $H(\theta, \phi = 0)$.

Our amplitudes are then related to those of Jacob–Wick as follows. For $\phi = 0$,

$$H_{\lambda_C\lambda_D;\lambda_A\lambda_B}(\theta) = \exp\left[i\pi(s_B - s_D)\right]\sqrt{\frac{\pi}{pp'}}\, f_{\lambda_C\lambda_D;\lambda_A\lambda_B}(\theta) \tag{4.1.6}$$

in which the *constant* phase factor is basically irrelevant.

However, the ϕ-dependence of our amplitudes is simpler than in Jacob–Wick. We have

$$H_{\lambda_C\lambda_D;\lambda_A\lambda_B}(\theta, \phi) = \exp\left\{i\phi(\lambda_A - \lambda_B)\right\} H_{\lambda_C\lambda_D;\lambda_A\lambda_B}(\theta). \tag{4.1.7}$$

With our normalization and conventions the partial-wave expansion is

$$\begin{aligned} H_{\lambda_C\lambda_D;\lambda_A\lambda_B}(\theta, \phi) &= e^{i\pi(s_B - s_D)}\sqrt{\frac{\pi}{pp'}}\frac{e^{i\phi\lambda}}{p} \\ &\quad\times \sum_j \left(J + \tfrac{1}{2}\right) \langle\lambda_C\lambda_D|\hat{T}^j(E)|\lambda_A\lambda_B\rangle d^J_{\lambda\mu}(\theta) \end{aligned} \tag{4.1.8}$$

where

$$\lambda = \lambda_A - \lambda_B \qquad \mu = \lambda_C - \lambda_D \qquad \hat{S} = 1 + i\hat{T} \tag{4.1.9}$$

and the *partial-wave amplitudes* are identical to those of Jacob–Wick.

4.2 Symmetry properties of helicity amplitudes

We now list the symmetry properties of the $H_{\{\lambda\}}$ when the reaction possesses certain invariant properties.

4.2.1 Parity

Let η_j be the intrinsic parity of particle j and suppose that invariance under space inversion holds. Then, using also rotational invariance, one finds

$$H_{-\lambda_C-\lambda_D;-\lambda_A-\lambda_B}(\theta, \phi) = \eta e^{-i\pi\mu} H_{\lambda_C\lambda_D;\lambda_A\lambda_B}(\theta, \pi - \phi) \tag{4.2.1}$$

where μ is defined in (4.1.9) and

$$\eta = \frac{\eta_C\eta_D}{\eta_A\eta_B}(-1)^{s_A+s_B-s_C-s_D}. \tag{4.2.2}$$

Taking $\phi = 0$ and using (4.1.7) yields a condition on the $\phi = 0$ amplitudes:

$$H_{-\lambda_C-\lambda_D;-\lambda_A-\lambda_B}(\theta) = \eta(-1)^{\lambda-\mu} H_{\lambda_C\lambda_D;\lambda_A\lambda_B}(\theta). \tag{4.2.3}$$

4.2.2 Time reversal

If time-reversal invariance holds then the sets of helicity amplitudes $H_{\lambda_C\lambda_D;\lambda_A\lambda_B}(\theta)$ for the process $A + B \to C + D$ and $H'_{\lambda_A\lambda_B;\lambda_C\lambda_D}(\theta)$ for the process $C + D \to A + B$ are related by

$$H'_{\lambda_A\lambda_B;\lambda_C\lambda_D}(\theta) = (-1)^{2(s_B - s_D)}(-1)^{\lambda-\mu} H_{\lambda_C\lambda_D;\lambda_A\lambda_B}(\theta). \tag{4.2.4}$$

If the reaction is an *elastic*, one $A + B \to A + B$, then (4.2.4) constitutes a set of relations amongst the amplitudes for the reaction:

$$H_{\lambda_A\lambda_B;\lambda'_A\lambda'_B}(\theta) = (-1)^{\lambda-\mu} H_{\lambda'_A\lambda'_B;\lambda_A\lambda_B}(\theta). \tag{4.2.5}$$

4.2.3 Identical particles

If C and D are identical particles with $s_C = s_D = s'$ then correctly symmetrized final states

$$\begin{aligned}\frac{1}{\sqrt{2}}\Big(&|C;\theta,\phi;\lambda_C\rangle \otimes |D;\pi-\theta,\phi+\pi;\lambda_D\rangle \\ &+ (-1)^{2s}|C;\pi-\theta,\phi+\pi;\lambda_D\rangle \otimes |D;\theta,\phi;\lambda_C\rangle\Big)\end{aligned} \tag{4.2.6}$$

must be used instead of (4.1.2); similarly for the initial states if $A = B$.

Let $\mathscr{P}_{12}$ be the operator that exchanges the space and spin quantum numbers of the first and second particles in the state. Under this exchange for particles C and D one finds, using the definition of the helicity amplitudes, that for $\phi = 0$

$$H_{\lambda_C\lambda_D;\lambda_A\lambda_B}(\theta) \to (-1)^{2s'} \exp\left[i\pi(\lambda_A - \lambda_B)\right] H_{\lambda_D\lambda_C;\lambda_A\lambda_B}(\pi - \theta) \tag{4.2.7}$$

and a similar result for $A \leftrightarrow B$.

The correctly symmetrized amplitudes for processes involving identical particles, either fermions or bosons, are then as follows (we label the helicities a, b, c, d for simplicity):

For $A + A \to C + D$

$$H^{\mathscr{S}}_{cd,aa'}(\theta) = \frac{1}{\sqrt{2}}\left[H_{cd,aa'}(\theta) + (-1)^{c-d}H_{cd,a'a}(\pi-\theta)\right]. \tag{4.2.8}$$

For $A + B \to C + C$

$$H^{\mathscr{S}}_{cc',ab}(\theta) = \frac{1}{\sqrt{2}}\left[H_{cc',ab}(\theta) + (-1)^{a-b}H_{c'c,ab}(\pi-\theta)\right] \tag{4.2.9}$$

and for $A + A \to C + C$

$$\begin{aligned}H^{\mathscr{S}}_{cc',aa'}(\theta) = \tfrac{1}{2}\Big[&H_{cc',aa'}(\theta) + (-1)^{c-c'+a-a'}H_{c'c,a'a}(\theta) \\ &+ (-1)^{a-a'}H_{c'c,aa'}(\pi-\theta) + (-1)^{c-c'}H_{cc',a'a}(\pi-\theta)\Big]\end{aligned} \tag{4.2.10}$$

The correctly symmetrized amplitudes have the following properties:

For $A + A \to C + D$

$$H^{\mathscr{S}}_{cd,aa'}(\theta) = (-1)^{c-d} H^{\mathscr{S}}_{cd,a'a}(\pi - \theta) \tag{4.2.11a}$$

For $A + B \to C + C$

$$H^{\mathscr{S}}_{cc',ab}(\theta) = (-1)^{a-b} H^{\mathscr{S}}_{c'c,ab}(\pi - \theta) \tag{4.2.11b}$$

For $A + A \to C + C$, both the above apply and, in addition,

$$H^{\mathscr{S}}_{cc',aa'}(\theta) = (-1)^{c-c'+a-a'} H^{\mathscr{S}}_{c'c,a'a}(\theta). \tag{4.2.12}$$

Note that if the particles belong to a multiplet of some internal symmetry group, so that we are dealing with an internal state vector (or wave function) that has a definite symmetry under interchange of the internal quantum numbers of the particles, then this symmetry factor (± 1) must be inserted on the right-hand side of (4.2.11a,b). For example, for a state of definite isospin I a factor $(-1)^{I+1}$ should be inserted. The symmetry (4.2.11a,b) forces certain amplitudes to vanish at 90° in the CM as follows:

For $A + A \to C + D$

$$H^{\mathscr{S}}_{cd,aa'}(\pi/2) = 0 \qquad \text{if } a = a' \quad \text{and} \quad (-1)^{c-d} = -1. \tag{4.2.13}$$

For $A + B \to C + C$

$$H^{\mathscr{S}}_{cc',ab}(\pi/2) = 0 \qquad \text{if } c = c' \quad \text{and} \quad (-1)^{a-b} = -1 \tag{4.2.14}$$

and, as before, both apply to $A + A \to C + C$.

Again, if the state has a definite symmetry under interchange of internal quantum numbers then the symmetry factor must be included in (4.2.13) and (4.2.14). Thus, for definite isospin (4.2.13) becomes $(-1)^{c-d+I+1} = -1$, etc.

There exist powerful phenomenological consequences of the symmetry conditions. We give some classical examples.

(*i*) *Elastic proton–proton scattering*. In the conventional notation

$$\phi_1(\theta) \equiv H_{++;++}(\theta) \qquad \phi_2(\theta) \equiv H_{++;--}(\theta) \qquad \phi_3(\theta) \equiv H_{+-;+-}(\theta)$$
$$\phi_4(\theta) \equiv H_{+-;-+}(\theta) \qquad \phi_5(\theta) \equiv H_{++;+-}(\theta), \tag{4.2.15}$$

we find

$$\begin{aligned} \phi^{\mathscr{S}}_{1,2} &= \phi_{1,2}(\theta) + \phi_{1,2}(\pi - \theta) \qquad & \phi^{\mathscr{S}}_5 &= \phi_5(\theta) - \phi_5(\pi - \theta) \\ \phi^{\mathscr{S}}_3 &= \phi_3(\theta) - \phi_4(\pi - \theta) \qquad & \phi^{\mathscr{S}}_4 &= \phi_4(\theta) - \phi_3(\pi - \theta), \end{aligned} \tag{4.2.16}$$

an immediate consequence of which (see subsection 5.4.1(ii)) is that the polarizing power which is proportional to ϕ_5, vanishes at $\theta = 90°$.

Also note that we have

$$\phi^{\mathscr{S}}_3(\pi - \theta) = -\phi^{\mathscr{S}}_4(\theta). \tag{4.2.17}$$

(*ii*) *Resonance decaying into two identical particles.* As explained in subsection 8.2.1 the decay amplitude for a resonance of spin J into two particles is obtained by just keeping the term with the relevant J in the partial-wave expansion (4.1.8). In addition the partial-wave amplitude is then independent of the helicity of the resonance. Aside from a normalization constant, one has for a spin-J resonance $E \to C + D$, with helicities e, c, d,

$$H_{cd,e}(\theta) = M_E(c,d) d^J_{e,c-d}(\theta) \tag{4.2.18}$$

where the $M_E(c,d)$ are dynamics-dependent parameters that depend only on the helicities of C and D.

For the correctly symmetrized amplitudes for

$$E \to C + C$$

one finds from (4.2.11a,b) and (4.2.18), upon using

$$d^J_{\lambda\mu}(\pi - \theta) = (-1)^{J+\lambda} d^J_{\lambda-\mu}(\theta), \tag{4.2.19}$$

that

$$M^{\mathscr{S}}_E(c,c') = (-1)^J M^{\mathscr{S}}_E(c',c) \tag{4.2.20}$$

from which we see that

$$M^{\mathscr{S}}_E(\lambda,\lambda) = 0 \qquad \text{if } J \text{ is odd.} \tag{4.2.21}$$

A classical example is the decay of a massive spin-1 particle into two photons. To conserve the z-component of angular momentum in the rest frame of the particle we must have $|J_z| \le 1$, so the photons can only have the same helicity, as shown in Fig. 4.1.

Thus, by (4.2.21), a massive spin-1 particle cannot decay into two photons, a result originally due to Landau (1948) and Yang (1950). The result (4.2.21) is thus a generalization of the Landau–Yang theorem.

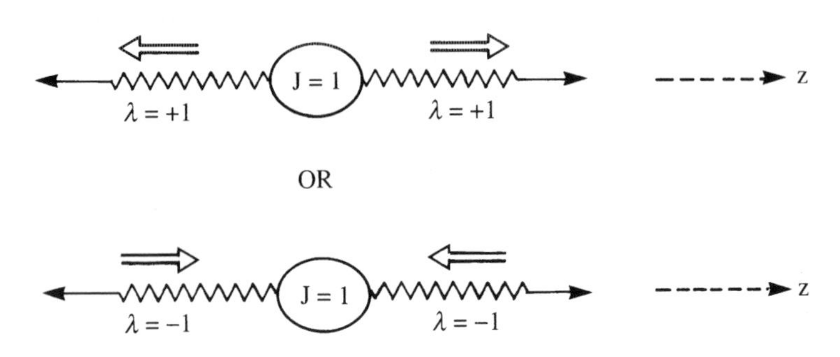

Fig. 4.1. Possible helicities for $J = 1$ decay into two photons.

4.2.4 Charge conjugation

For interactions that are invariant under charge conjugation $\mathscr{C}$, the most interesting cases, as regards helicity dependence, are reactions of the type

$$A + \overline{A} \to D + \overline{D}$$

or the decay of a resonance with definite charge parity of the type

$$E \to D + \overline{D}.$$

Since charge conjugation on a state of the type $|A\overline{A}\rangle$ is equivalent to exchanging the space and spin quantum numbers of the particles together with interchanging their order in the state, we have that

$$\mathscr{C}|A\overline{A}\ldots\rangle = (-1)^{2s_A}\mathscr{P}_{12}|A\overline{A}\ldots\rangle \tag{4.2.22}$$

and analogously to (4.2.12) we find for $A + \overline{A} \to D + \overline{D}$

$$H_{d\bar{d};a\bar{a}}(\theta) = (-1)^{\lambda-\mu}H_{\bar{d}d;\bar{a}a}(\theta) \tag{4.2.23}$$

where $\lambda = a - \bar{a}$ and $\mu = d - \bar{d}$.

In the case that A is its own antiparticle, i.e.

$$A + \overline{A} \to D + \overline{D}$$

with $\overline{A} = A$, one has also

$$H_{d\bar{d};a\bar{a}}(\theta) = (-1)^{a-\bar{a}}H_{\bar{d}d;a\bar{a}}(\pi - \theta). \tag{4.2.24}$$

For a resonance E of spin J that is an eigenstate of $\mathscr{C}$ with charge parity η_C one finds for $E \to D + \overline{D}$

$$M_E(d, \bar{d}) = \eta_C(-1)^J M_E(\bar{d}, d) \tag{4.2.25}$$

so that $M_E(d, \bar{d} = d) = 0$ if $\eta_C(-1)^J$ is odd.

4.3 Some analytic properties of the helicity amplitudes

An important consequence of the analytic structure of the $H_{\{\lambda\}}$ is that some amplitudes must vanish in the forward or backward direction. This is summarized by writing (Wang, 1966)

$$H_{\lambda_C\lambda_D;\lambda_A\lambda_B}(\theta) = (\sin\theta/2)^{|\lambda-\mu|}(\cos\theta/2)^{|\lambda+\mu|}\tilde{H}_{\lambda_C\lambda_D;\lambda_A\lambda_B}(\theta) \tag{4.3.1}$$

where $\tilde{H}_{\{\lambda\}}$ is, in general, finite and non-zero at $\theta = 0$ and $\theta = \pi$.

In particular dynamical models the helicity amplitudes may vanish *more rapidly* as $\theta \to 0$ or π (Leader, 1968). Equation (4.3.1) gives the minimum requirement on this vanishing in the forward and backward direction.

If however there are *dynamical* singularities, e.g. at $t = 0$ in the Coulomb scattering of two charged particles, then $\tilde{H}$ may be singular at $\theta = 0$ or π. In that case the *relative* vanishing of different helicity amplitudes must be at least as fast as given by the $\sin\theta/2$, $\cos\theta/2$ factors in (4.3.1). For example, the electromagnetic (one-photon-exchange) contribution to proton–proton scattering at small t gives

$$\begin{aligned} \phi_1^{\text{em}} &= H_{1/2\ 1/2;1/2\ 1/2} \propto \frac{1}{t} \\ \phi_5^{\text{em}} &= H_{1/2\ 1/2;1/2-1/2} \propto \frac{1}{m\sqrt{-t}} \end{aligned} \tag{4.3.2}$$

the ratio being in accordance with (4.3.1).

There are other kinematic points at which analyticity imposes some particular behaviour, namely the thresholds $s = (m_A + m_B)^2$, $s = (m_C + m_D)^2$, the pseudothresholds $s = (m_A - m_B)^2$, $s = (m_C - m_D)^2$ and the origin $s = 0$. The detailed discussion of Cohen-Tannoudji *et al.* (1968a, b) showed that the behaviour of the helicity amplitudes in the neighbourhood of thresholds and pseudothresholds is complicated and involves constraint equations tying together the behaviour of several different amplitudes. (In Appendix 4 we shall see that, on the contrary, the behaviour of transversity amplitudes is simple at these points while it is complicated at $\theta = 0$ or π.)

At high energies, the behaviour of $H_{\{\lambda\}}$ at thresholds and pseudothresholds is unimportant. If however we construct models of the *t-channel* helicity amplitudes (see below) then care must be taken, because, for them, the singularities occur at points $t = (m_A \pm m_C)^2$ and $t = (m_B \pm m_D)^2$, some of which may be close to the physical scattering region. Care too must be taken to satisfy the constraints at $\theta = 0$ or π. Observed effects originating from the kinematic singularities must not be attributed to the dynamics, and models should be constructed so as to satisfy the constraints automatically.

4.4 Crossing for helicity amplitudes

The amplitudes for the three reactions

$$\begin{aligned} &A + B \to C + D &&s\text{-channel} \\ &\overline{D} + B \to C + \overline{A} &&t\text{-channel} \\ &\overline{C} + B \to \overline{A} + D &&u\text{-channel} \end{aligned} \tag{4.4.1}$$

all depend upon the Mandelstam variables

$$\begin{aligned} s &= (p_A + p_C)^2 \\ t &= (p_A - p_C)^2 \\ u &= (p_A - p_D)^2 \end{aligned} \tag{4.4.2}$$

with

$$s + t + u = m_A^2 + m_B^2 + m_C^2 + m_D^2 \tag{4.4.3}$$

and are described by just one set of analytic functions evaluated in different regions of the variables s, t, u. The reaction amplitudes for any one reaction channel are obtained by analytic continuation from the amplitudes of any other channel. The set of relations amongst the amplitudes constitute the 'crossing relations' (Trueman and Wick, 1964).

For the t- and u-channel reactions, the variables t and u respectively play the rôle of the square of the CM energy, just as s does for the s-channel reaction. Let $H_{\lambda_C\lambda_D;\lambda_A\lambda_B}$ denote the helicity amplitudes for the s-channel reaction and let us denote by $H^{(t)}_{\lambda_C\lambda_{\bar{A}};\lambda_{\bar{D}}\lambda_B}$, $H^{(u)}_{\lambda_{\bar{A}}\lambda_D;\lambda_{\bar{C}}\lambda_B}$ the helicity amplitudes for the t-channel and u-channel reactions, all with $\phi = 0$. Then the $t \to s$ crossing relation states that

$$\begin{aligned} H_{\lambda_C\lambda_D;\lambda_A\lambda_B} = {} & d^{s_C}_{\mu_C\lambda_C}(\psi_C) d^{s_D}_{\mu_{\bar{D}}\lambda_D}(\psi_D) \\ & \times d^{s_A}_{\mu_{\bar{A}}\lambda_A}(\psi_A) d^{s_B}_{\mu_B\lambda_B}(\psi_B) H^{(t)}_{\mu_C\mu_{\bar{A}};\mu_{\bar{D}}\mu_B} \end{aligned} \tag{4.4.4}$$

where the $t \to s$ crossing angles ψ_i are given by

$$\begin{aligned} \cos\psi_A &= -\frac{(s+m_A^2-m_B^2)(t+m_A^2-m_C^2)+2m_A^2\Delta}{\mathscr{S}_{AB}\mathscr{T}_{AC}} \\ \sin\psi_A &= \frac{m_A\mathscr{S}_{CD}}{\sqrt{s}\mathscr{T}_{AC}}\sin\theta \\ \cos\psi_B &= \frac{(s+m_B^2-m_A^2)(t+m_B^2-m_D^2)+2m_B^2\Delta}{\mathscr{S}_{AB}\mathscr{T}_{BD}} \\ \sin\psi_B &= \frac{m_B\mathscr{S}_{CD}}{\sqrt{s}\mathscr{T}_{BD}}\sin\theta \\ \cos\psi_C &= \frac{(s+m_C^2-m_D^2)(t+m_C^2-m_A^2)-2m_C^2\Delta}{\mathscr{S}_{CD}\mathscr{T}_{AC}} \\ \sin\psi_C &= \frac{m_C\mathscr{S}_{AB}}{\sqrt{s}\mathscr{T}_{AC}}\sin\theta \\ \cos\psi_D &= -\frac{(s+m_D^2-m_C^2)(t+m_D^2-m_B^2)+2m_D^2\Delta}{\mathscr{S}_{CD}\mathscr{T}_{BD}} \\ \sin\psi_D &= \frac{m_D\mathscr{S}_{AB}}{\sqrt{s}\mathscr{T}_{BD}}\sin\theta. \end{aligned} \tag{4.4.5}$$

Here

$$\begin{aligned} \mathscr{S}_{ij}^2 &\equiv [s-(m_i-m_j)^2][s-(m_i+m_j)^2] \\ \mathscr{T}_{ij}^2 &\equiv [t-(m_i-m_j)^2][t-(m_i+m_j)^2] \\ \Delta &\equiv m_B^2+m_C^2-m_A^2-m_D^2 \end{aligned} \tag{4.4.6}$$

and θ is the s-channel CM scattering angle.

For the crossing from $u \to s$ we have

$$H_{\lambda_C\lambda_D;\lambda_A\lambda_B} = d^{s_C}_{\mu_{\bar{C}}\lambda_C}(\chi_C) d^{s_D}_{\mu_D\lambda_D}(\chi_D) \times d^{s_A}_{\mu_{\bar{A}}\lambda_A}(\chi_A) d^{s_B}_{\mu_B\lambda_B}(\chi_B) H^{(u)}_{\mu_{\bar{A}}\mu_D;\mu_{\bar{C}}\mu_B} \tag{4.4.7}$$

where each χ_i is obtained from the ψ_i of eqn (4.4.5) by the substitutions

$$t \to u \qquad m_C \to m_D \qquad m_D \to m_C.$$

Note that for a massless particle the crossing rules simplify greatly. If under crossing the particle remains a particle its crossing matrix is simply $d^s_{\mu\lambda}(0) = \delta_{\mu\lambda}$. If an antiparticle crosses into a particle then the crossing matrix is $d^s_{\bar{\mu}\lambda}(\pi) = (-1)^{s+\bar{\mu}}\delta_{\bar{\mu},-\lambda}$.

4.5 Transition amplitudes in field theory

Consider now the calculation of the matrix elements of some operator in quantum field theory. All operators are expressed in terms of products of fields and the particle states are reduced to the vacuum state by the action of the field operators, as shown in eqn (2.4.10), for example. One sees that each particle or antiparticle in a matrix element will give rise to one or other wave-function factor. Thus a general transition amplitude involving particles A, B, ... and antiparticles $\overline{C}$, $\overline{D}$, ... will always be of the form

$$\langle B, \ldots, \overline{D}, \ldots | S | A, \ldots, \overline{C}, \ldots \rangle = \bar{u}_\alpha(B) \cdots \bar{v}_\beta(C) M_{\alpha\ldots\beta\ldots,\gamma\ldots\delta\ldots} u_\gamma(A) \cdots v_\delta(D). \tag{4.5.1}$$

4.6 Structure of matrix elements

The matrix M, which is a function only of the momenta of the particles, will be shown to have simple Lorentz transformation properties. It is therefore possible, in any given case, to write down its most general structure consistent with these properties (and with the requirements of invariance under the discrete transformation). The M's are referred to as M-functions in the literature (Stapp, 1962). We shall not give a general discussion of the theory of M-functions but will illustrate their use in some cases of particular importance.

4.6.1 Matrix elements of a vector current

As a prototype example we shall examine the matrix elements of a 4-vector current $j^\mu(x)$ taken between states of a spin-1/2 Dirac particle. This is germane to the study of the electromagnetic form factors of a nucleon. The method used works equally well for any 'current' that has a well-defined law of transformation under Lorentz transformations, e.g. a scalar, spinor, vector etc.

Under the Lorentz transformation $S \xrightarrow{l} S^l$ let

$$x^\mu \to x'^\mu = (l^{-1}x)^\mu \equiv \Lambda^\mu{}_\nu x^\nu \tag{4.6.1}$$

(with our conventions $\Lambda^\mu{}_\nu = \Lambda^\mu{}_\nu(l^{-1})$; see eqns (1.2.10), (1.2.14)) and, analogously to (2.4.1),

$$j^\mu(x) \to j'^\mu(x')$$

where

$$j'^\mu(x) = U(l)j^\mu(x)U(l^{-1}) = \Lambda^\mu{}_\nu j^\nu(lx). \tag{4.6.2}$$

Consider the 'vertex'

$$\begin{aligned}\Gamma^\mu(\mathbf{p}_2\lambda_2;\mathbf{p}_1\lambda_1) &\equiv \langle \mathbf{p}_2;\lambda_2|j^\mu(0)|\mathbf{p}_1;\lambda_1\rangle & (4.6.3)\\ &= \bar{u}_\alpha(\mathbf{p}_2,\lambda_2)M^\mu_{\alpha\beta}(\mathbf{p}_2,\mathbf{p}_1)u_\beta(\mathbf{p}_1,\lambda_1)\\ &= \bar{u}(\mathbf{p}_2,\lambda_2)M^\mu(\mathbf{p}_2,\mathbf{p}_1)u(\mathbf{p}_1,\lambda_1) & (4.6.4)\end{aligned}$$

where u, $\bar{u}$ are Dirac four-component spinors for particles of definite helicity. Our aim is to study the structure of the 4×4 matrices M^μ.

The transformation matrix $D_{nm}(l^{-1})$ that appears in (2.4.16) is customarily denoted by the 4×4 matrix S_{nm} for the case of Dirac particles. Thus, from (2.4.16) and (2.4.18) we have

$$\left.\begin{aligned}u(l^{-1}\mathbf{p},\lambda')\mathscr{D}^{(1/2)}_{\lambda'\lambda}(r) &= Su(\mathbf{p},\lambda)\\ \mathscr{D}^{(1/2)}_{\lambda\lambda'}(r^{-1})\bar{u}(l^{-1}\mathbf{p},\lambda') &= \bar{u}(\mathbf{p},\lambda)S^{-1}.\end{aligned}\right\} \tag{4.6.5}$$

Now using (4.6.2) we insert

$$\Lambda^\mu{}_\nu j^\nu(0) = U(l)j^\mu(0)U(l^{-1})$$

into (4.6.3) and obtain, using (4.6.4),(2.1.1) and (2.1.9)

$$\begin{aligned}&\bar{u}(\mathbf{p}_2,\lambda_2)\Lambda^\mu{}_\nu M^\nu(\mathbf{p}_2,\mathbf{p}_1)u(\mathbf{p}_1,\lambda_1)\\ &\quad= \langle \mathbf{p}_2;\lambda_2|U(l)j^\mu(0)U(l^{-1})|\mathbf{p}_1;\lambda_1\rangle\\ &\quad= \mathscr{D}^{(1/2)}_{\lambda_2\lambda_2'}(r^{-1})\langle l^{-1}\mathbf{p}_2;\lambda_2'|j^\mu(0)|l^{-1}\mathbf{p}_1;\lambda_1'\rangle\mathscr{D}^{(1/2)}_{\lambda_1'\lambda_1}(r)\\ &\quad= \mathscr{D}^{(1/2)}_{\lambda_2\lambda_2'}(r^{-1})\bar{u}(l^{-1}\mathbf{p}_2,\lambda_2')\\ &\qquad\times M^\mu(l^{-1}\mathbf{p}_2,l^{-1}\mathbf{p}_1)u(l^{-1}\mathbf{p}_1,\lambda_1')\mathscr{D}^{(1/2)}_{\lambda_1'\lambda_1}(r)\\ &\quad= \bar{u}(\mathbf{p}_2,\lambda_2)S^{-1}M^\mu(l^{-1}\mathbf{p}_2,l^{-1}\mathbf{p}_1)Su(\mathbf{p}_1,\lambda_1).\end{aligned} \tag{4.6.6}$$

Thus we end up with the requirement on M^μ

$$\Lambda^\mu{}_\nu M^\nu(\mathbf{p}_2,\mathbf{p}_1) = S^{-1}M^\mu(l^{-1}\mathbf{p}_2,l^{-1}\mathbf{p}_1)S. \tag{4.6.7}$$

The next step is to note that M, being a 4×4 matrix, can be written as a superposition of the complete set of 16 Dirac matrices, which comprises:

the scalar I; the vector γ^μ; the tensor $\sigma^{\mu\nu} = \frac{i}{2}[\gamma^\mu, \gamma^\nu]$; the axial vector $\gamma^\mu\gamma_5$; and the pseudoscalar γ_5. They have the transformation properties

$$\begin{aligned} &S^{-1}IS = I \qquad S^{-1}\gamma_5 S = |\Lambda|\gamma_5 \qquad S^{-1}\gamma^\mu S = \Lambda^\mu{}_\nu\gamma^\nu \\ &S^{-1}\gamma^\mu\gamma_5 S = |\Lambda|\Lambda^\mu{}_\nu\gamma^\nu\gamma_5 \qquad S^{-1}\sigma^{\mu\nu}S = \Lambda^\mu{}_\alpha\Lambda^\nu{}_\beta\sigma^{\alpha\beta} \end{aligned} \tag{4.6.8}$$

where $|\Lambda| = \det(\Lambda^\mu{}_\nu)$.

When l corresponds to the operation of space inversion, $j^\mu(x)$ transforms as a true vector under

$$\begin{aligned} x \to x' = l_{\mathscr{P}}^{-1}x = (t, -\mathbf{x}) = (g^{\mu\mu})x^\mu& \\ (\text{no sum on } \mu)& \\ \mathscr{P}^{-1}j^\mu(x)\mathscr{P} = (g^{\mu\mu})j^\mu(t - \mathbf{x}).& \end{aligned} \tag{4.6.9}$$

Using (2.3.7) and the fact that $S = \gamma^0$ for space inversion, one finds that

$$(g^{\mu\mu})M^\mu(\mathbf{p}_2, \mathbf{p}_1) = \gamma^0 M^\mu(-\mathbf{p}_2, -\mathbf{p}_1)\gamma^0 \tag{4.6.10}$$

must be satisfied.

It is simple to check that the following all satisfy (4.6.7) and (4.6.10); here we write $q^\mu \equiv p_2^\mu - p_1^\mu$:

$$\begin{array}{ccc} Iq^\mu & \gamma^\mu & \sigma^{\mu\nu}q_\nu \\ I(p_1 + p_2)^\mu & \sigma^{\mu\nu}(p_1 + p_2)_\nu & \epsilon^\mu{}_{\nu\rho\sigma}p_1^\nu p_2^\rho\gamma_5\gamma^\sigma. \end{array}$$

However, since M^μ is sandwiched between Dirac spinors, use of the Dirac equation enables the latter three forms to be expressed in terms of the first three.

In addition the current is conserved, i.e. $\partial_\mu j^\mu(x) = 0$, so that, upon using the fact that translations are generated by the momentum operator $[\hat{P}_\alpha, f(x)] = -i\partial_\alpha f(x)$, we find

$$q_\mu\langle\mathbf{p}_2; \lambda_2|j^\mu(0)|\mathbf{p}_1; \lambda_1\rangle = 0, \tag{4.6.11}$$

which is incompatible with a term of the form Iq^μ. Thus we are left with γ^μ and $\sigma^{\mu\nu}q_\nu$.

Finally, under time reversal $x \to x' = l_{\mathscr{T}}^{-1}x = (-t, \mathbf{x})$

$$\mathscr{T}^{-1}j^\mu(x)\mathscr{T} = (g^{\mu\mu})j^\mu(-t, \mathbf{x}). \tag{4.6.12}$$

Using (2.3.17), and remembering that $\mathscr{T}$ is an anti-linear operator (see the discussion in subsection 2.3.2) we have

$$\begin{aligned} &\bar{u}(\mathbf{p}_2, \lambda_2)\ g^{\mu\mu}\ M^\mu(\mathbf{p}_2, \mathbf{p}_1)u(\mathbf{p}_1, \lambda_1) \\ &\quad = \langle\mathbf{p}_2; \lambda_2|\mathscr{T}^{-1}j^\mu(0)\mathscr{T}|\mathbf{p}_1; \lambda_1\rangle \\ &\quad = \langle\mathscr{T}(\mathbf{p}_2; \lambda_2)|j^\mu(0)|\mathscr{T}(\mathbf{p}_1; \lambda_1)\rangle^* \\ &\quad = e^{i\pi(\lambda_1-\lambda_2)}\langle p_2, \pi - \theta_2, \phi_2 + \pi; \lambda_2|j^\mu(0)|p_1, \pi - \theta_1, \phi_1 + \pi; \lambda_1\rangle^* \\ &\quad = \bar{u}(\mathbf{p}_2, \lambda_2)(\gamma^3\gamma^1)^\dagger M^{\mu*}(-\mathbf{p}_2, -\mathbf{p}_1)\gamma^3\gamma^1 u(\mathbf{p}_1, \lambda_1), \end{aligned} \tag{4.6.13}$$

where we have used (2.4.29) with $T = \gamma^3\gamma^1$. Thus we need

$$(\gamma^3\gamma^1)^\dagger M^{\mu *}(-\mathbf{p}_2, -\mathbf{p}_1)\gamma^3\gamma^1 = (g^{\mu\mu})M^\mu(\mathbf{p}_2, \mathbf{p}_1). \qquad (4.6.14)$$

It is easy to check that

$$(\gamma^3\gamma^1)^\dagger \gamma^{\mu *}\gamma^3\gamma^1 = (g^{\mu\mu})\gamma^\mu \qquad (4.6.15)$$

and

$$(\gamma^3\gamma^1)^\dagger \sigma^{\mu\nu *}(q^0, -\mathbf{q})_\nu \gamma^3\gamma^1 = -(g^{\mu\mu})\sigma^{\mu\nu}q_\nu \qquad (4.6.16)$$

so that (4.6.14) is satisfied by the forms γ^μ or $i\sigma^{\mu\nu}q_\nu$, times any real scalar function. Conventionally one writes

$$\begin{aligned} &\langle \mathbf{p}_2; \lambda_2 | j^\mu(0) | \mathbf{p}_1; \lambda_1 \rangle \\ &\qquad = \bar{u}(\mathbf{p}_2, \lambda_2)\left[F_1(q^2)\gamma^\mu + \frac{\kappa}{2m}F_2(q^2)i\sigma^{\mu\nu}q_\nu\right]u(\mathbf{p}_1, \lambda_1) \qquad (4.6.17) \end{aligned}$$

where κ is the anomalous magnetic moment of the fermion of mass m, and $F_{1,2}$ are the Dirac form factors.

The approach used in this section can be applied to the analysis of the matrix elements of any operator that has a well-defined behaviour under Lorentz transformations. If parity and/or time-reversal invariance are broken one simply does not impose the restrictions (4.6.10) and/or (4.6.14).

The analysis that utilizes Lorentz invariance etc. to expose the essential structure of the matrix elements in (4.6.17) is akin to the familiar use of the Wigner–Eckhardt theorem to express a set of matrix elements in terms of just the reduced matrix elements. Thus these 16 matrix elements ($\mu = 0, 1, 2, 3; \lambda_1 = \pm 1/2, \lambda_2 = \pm 1/2$) are expressed in terms of just two independent functions $F_{1,2}$. The *dynamics*, therefore, is entirely contained in these functions.

4.6.2 Vector and axial-vector coupling

The two most fundamental theories at the present time are the electroweak theory of Glashow, Salam and Weinberg and quantum chromodynamics, and some aspects of these will be discussed in detail in Chapters 9 and 10. For a general introduction the reader is referred to Leader and Predazzi (1996). Here we note that these theories contain only vector and axial-vector couplings of the various gauge bosons to the spin-1/2 fermions. It is thus important to have a detailed understanding of the properties and the structure of these vertices.

Firstly we consider the relationship between the expressions for the Feynman diagram vertices shown below involving incoming and outgoing spin-1/2 fermions A, B or antifermions $\bar{A}$, $\bar{B}$.

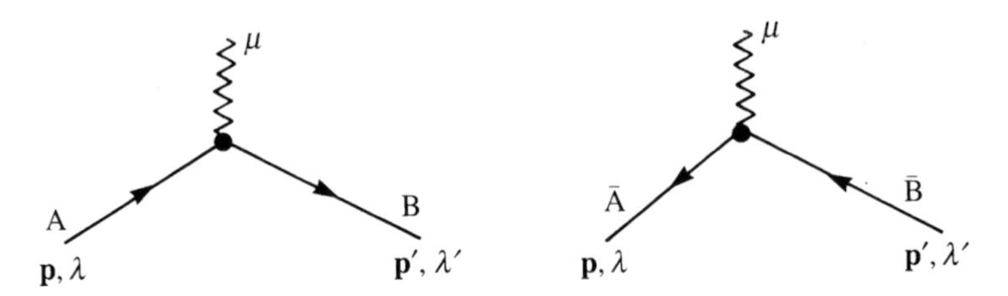

Here the vertex is either γ^μ or $\gamma^\mu\gamma_5$. The transition amplitudes $A \to B$ or $\overline{A} \to \overline{B}$ will involve, see eqn (4.5.1),

$$\Gamma_{B\leftarrow A}(\mathbf{p}', \mathbf{p}) = \bar{u}_B(\mathbf{p}', \lambda') \{\gamma^\mu \text{ or } \gamma^\mu\gamma_5\} u_A(\mathbf{p}, \lambda) \tag{4.6.18}$$

and

$$\Gamma_{\bar{B}\leftarrow\bar{A}}(\mathbf{p}', \mathbf{p}) = \bar{v}_A(\mathbf{p}, \lambda) \{\gamma^\mu \text{ or } \gamma^\mu\gamma_5\} v_B(\mathbf{p}', \lambda'). \tag{4.6.19}$$

Using the charge conjugation result (2.4.35), we have

$$u(\mathbf{p}, \lambda) = C\bar{v}(\mathbf{p}, \lambda) \tag{4.6.20}$$

with

$$C = i\gamma^2\gamma^0. \tag{4.6.21}$$

Adding the fact that

$$C\gamma^\mu C^{-1} = -\gamma^{\mu T} \tag{4.6.22}$$

where $\gamma^{\mu T}$ is the transpose of γ^μ, one arrives at

$$\Gamma_{\bar{B}\leftarrow\bar{A}}(\mathbf{p}', \mathbf{p}) = \bar{u}_B(\mathbf{p}', \lambda') \{\gamma^\mu \text{ or } -\gamma^\mu\gamma_5\} u_A(\mathbf{p}, \lambda). \tag{4.6.23}$$

Comparing with (4.6.18) we see that the amplitudes for $A \to B$ and $\overline{A} \to \overline{B}$ are equal for the vector coupling and opposite in sign for the axial-vector coupling. This will be helpful in comparing, for example,

$$\nu_e + n \to e^- + p$$

with

$$\bar{\nu}_e + p \to e^+ + n.$$

Next we consider the detailed helicity dependence of the vector and axial-vector vertices.

The four-component Dirac spinors which are constructed in accordance with eqns (2.4.14) and (2.4.15) and which respect eqns (4.6.20), (4.6.21) can be written

$$u(\mathbf{p}, \lambda) = \frac{1}{\sqrt{E+m}} \begin{pmatrix} E+m \\ 2p\lambda \end{pmatrix} \chi_\lambda(\hat{\mathbf{p}}) \tag{4.6.24}$$

$$v(\mathbf{p}, \lambda) = \frac{1}{\sqrt{E+m}} \begin{pmatrix} -2p\lambda \\ E+m \end{pmatrix} \chi_{-\lambda}(\hat{\mathbf{p}}) \tag{4.6.25}$$

where $\hat{\mathbf{p}} = (\sin\theta\cos\phi,\ \sin\theta\sin\phi,\ \cos\theta)$, $\lambda = \pm 1/2$ and $\chi_\lambda(\hat{\mathbf{p}})$ is a two-component spinor. In (4.6.24) and (4.6.25) both $E + m$ and $2p\lambda$ are of course understood to be multiplied by $\chi(\hat{\mathbf{p}})$ to yield a four-component spinor. One has

$$\chi_\lambda(\hat{\mathbf{p}}) = e^{-i\phi\sigma_z/2} e^{-i\theta\sigma_y/2} \chi_\lambda. \tag{4.6.26}$$

where λ is $+$ or $-$ and

$$\chi_+ = \begin{pmatrix} 1 \\ 0 \end{pmatrix} \qquad \chi_- = \begin{pmatrix} 0 \\ 1 \end{pmatrix}. \tag{4.6.27}$$

Explicitly,

$$\chi_+(\hat{\mathbf{p}}) = \begin{pmatrix} e^{-i\phi/2}\cos\theta/2 \\ e^{i\phi/2}\sin\theta/2 \end{pmatrix} \qquad \chi_-(\hat{\mathbf{p}}) = \begin{pmatrix} -e^{-i\phi/2}\sin\theta/2 \\ e^{i\phi/2}\cos\theta/2 \end{pmatrix}. \tag{4.6.28}$$

Let us for brevity put

$$\begin{aligned} V^\mu_{\lambda'\lambda} &\equiv N\bar{u}(\mathbf{p}', \lambda')\gamma^\mu u(\mathbf{p}, \lambda) \\ A^\mu_{\lambda'\lambda} &\equiv N\bar{u}(\mathbf{p}', \lambda')\gamma^\mu\gamma_5 u(\mathbf{p}, \lambda), \end{aligned} \tag{4.6.29}$$

with $N = [(E'+m')(E+m)]^{-1/2}$ included to make the result dimensionless, and let us define the angular function

$$h^\mu_{\lambda\lambda'} \equiv \chi^\dagger_{\lambda'}(\hat{\mathbf{p}}')\sigma^\mu\chi_\lambda(\hat{\mathbf{p}}) \tag{4.6.30}$$

where

$$\sigma^\mu \equiv (I, \boldsymbol{\sigma}). \tag{4.6.31}$$

One finds

$$V^0_{\lambda'\lambda} = N^2\left[(E'+m')(E+m) + 4pp'\lambda\lambda'\right] h^0_{\lambda'\lambda} \tag{4.6.32}$$
$$V^j_{\lambda'\lambda} = 2N^2\left[(E'+m')p\lambda + (E+m)p'\lambda'\right] h^j_{\lambda'\lambda} \tag{4.6.33}$$
$$A^0_{\lambda'\lambda} = 2N^2\left[(E'+m')p\lambda + (E+m)p'\lambda'\right] h^0_{\lambda'\lambda} \tag{4.6.34}$$
$$A^j_{\lambda'\lambda} = N^2\left[(E'+m')(E+m) + 4pp'\lambda\lambda'\right] h^j_{\lambda'\lambda}. \tag{4.6.35}$$

We see that only two different energy-dependent factors occur. So we may write

$$V^0_{\lambda'\lambda} = E_{\lambda'\lambda}h^0_{\lambda'\lambda}, \qquad A^0_{\lambda'\lambda} = F_{\lambda'\lambda}h^0_{\lambda'\lambda} \tag{4.6.36}$$
$$V^j_{\lambda'\lambda} = F_{\lambda'\lambda}h^j_{\lambda'\lambda}, \qquad A^j_{\lambda'\lambda} = E_{\lambda'\lambda}h^j_{\lambda'\lambda} \tag{4.6.37}$$

with

$$E_{\lambda'\lambda} \equiv \frac{1}{(E'+m')(E+m)}\left[(E'+m')(E+m) + 4pp'\lambda\lambda'\right] \tag{4.6.38}$$

and

$$F_{\lambda'\lambda} \equiv \frac{2}{(E'+m')(E+m)}\left[(E'+m')p\lambda + (E+m)p'\lambda'\right]. \tag{4.6.39}$$

In dealing with QCD and the parton model we shall be particularly interested in situations in which $E \gg m$ and $E' \gg m'$, corresponding to the partons being essentially massless. In this limit

$$E_{\lambda'\lambda} = 1 + 4\lambda\lambda' + O(m/E) \tag{4.6.40}$$

$$F_{\lambda'\lambda} = 2\,(\lambda + \lambda') + O(m/E). \tag{4.6.41}$$

We have then the remarkable result that for $\lambda' = -\lambda\,(=\pm 1/2)$

$$V^{\mu}_{-\lambda,\lambda} = A^{\mu}_{-\lambda,\lambda} = 0 + O(m/E). \tag{4.6.42}$$

Thus the vector and axial-vector couplings approximately conserve helicity for a fast-moving particle. The helicity is exactly conserved for a massless fermion, e.g. for a neutrino. The impact of this in the parton model is dramatic, since in that model one is supposed to view the collision from an 'infinite momentum frame', i.e. from a frame in which all particles are moving at 'infinite' (i.e. very high) speeds.

For the helicity non-flip matrix elements one has the simple results

$$V^{\mu}_{\lambda,\lambda} = 2\left(h^{0}_{\lambda,\lambda}, 2\lambda h^{j}_{\lambda,\lambda}\right) + O(m/E) \tag{4.6.43}$$

$$A^{\mu}_{\lambda,\lambda} = 2\lambda V^{\mu}_{\lambda,\lambda} + O(m/E). \tag{4.6.44}$$

An analogous simplification arises if we consider the creation or annihilation of a fermion and antifermion via vector or axial-vector coupling in the limit $E \gg m$.

Consider the creation process

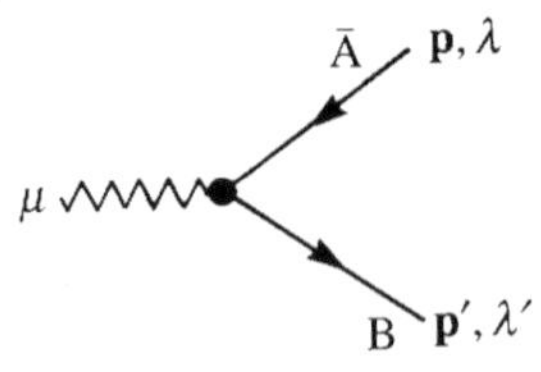

If we define

$$\overline{V}^{\mu}_{\lambda',\lambda} \equiv N\bar{u}_B(\mathbf{p}'\lambda')\gamma^{\mu}v_A(\mathbf{p},\lambda) \tag{4.6.45}$$

$$\overline{A}^{\mu}_{\lambda',\lambda} \equiv N\bar{u}_B(\mathbf{p}'\lambda')\gamma^{\mu}\gamma_5 v_A(\mathbf{p},\lambda) \tag{4.6.46}$$

then we find

$$\overline{V}^{0}_{\lambda'\lambda} = F_{\lambda',-\lambda}h^{0}_{\lambda',-\lambda} \qquad \overline{A}^{0}_{\lambda'\lambda} = E_{\lambda',-\lambda}h^{0}_{\lambda',-\lambda} \tag{4.6.47}$$

$$\overline{V}^{j}_{\lambda'\lambda} = E_{\lambda',-\lambda}h^{j}_{\lambda',-\lambda} \qquad \overline{A}^{j}_{\lambda'\lambda} = F_{\lambda',-\lambda}h^{j}_{\lambda',-\lambda}. \tag{4.6.48}$$

It follows from (4.6.40) and (4.6.41) that in the limit $E \gg m$

$$\overline{V}^{\mu}_{\lambda,\lambda} = \overline{A}^{\mu}_{\lambda,\lambda} = 0 + O(m/E). \tag{4.6.49}$$

Hence the amplitude for producing the fermion and the antifermion with equal helicity is of order m/E. For opposite helicities the result takes the simple form, analogous to (4.6.43) and (4.6.44),

$$\overline{V}^{\mu}_{\lambda,-\lambda} = 2\left(2\lambda h^{0}_{\lambda,\lambda}, h^{j}_{\lambda,\lambda}\right) + O(m/E) \tag{4.6.50}$$

$$\overline{A}^{\mu}_{\lambda,-\lambda} = 2\lambda \overline{V}^{\mu}_{\lambda,-\lambda} + O(m/E). \tag{4.6.51}$$

In a similar way, in the annihilation of a fermion–antifermion pair the matrix element of the form $\bar{v}(\gamma^{\mu}$ or $\gamma^{\mu}\gamma_5)u$ will vanish for $E \gg m$ unless the fermion and the antifermion have opposite helicities.

4.6.3 Chirality

Let us consider now the connection between these results and the concept of *chirality*. A Dirac spinor is said to be either right-handed (R) or left-handed (L) if it is an eigenvector of γ_5. By convention

$$\begin{aligned} \gamma_5 u_R &= u_R \qquad & \gamma_5 u_L &= -u_L \\ \gamma_5 v_R &= -v_R \qquad & \gamma_5 v_L &= v_L. \end{aligned} \tag{4.6.52}$$

An arbitrary spinor can always be split up into right-handed and left-handed pieces by noting that

$$\gamma_5(1 \pm \gamma_5) = \pm(1 \pm \gamma_5),$$

so that

$$u_{R,L} \equiv \tfrac{1}{2}(1 \pm \gamma_5)u \qquad v_{R,L} \equiv \tfrac{1}{2}(1 \mp \gamma_5)v \tag{4.6.53}$$

satisfy (4.6.52), and then

$$u = u_R + u_L \qquad v = v_R + v_L. \tag{4.6.54}$$

It is clear from (4.6.24), (4.6.25) that $u(\mathbf{p}, \lambda)$, $v(\mathbf{p}, \lambda)$ are not eigenvectors of

$$\gamma_5 = \begin{pmatrix} 0 & I \\ I & 0 \end{pmatrix}.$$

However, when $m = 0$ they do become chiral states and we have

$$\left.\begin{aligned} u_R(\mathbf{p}) &= u(\mathbf{p}, 1/2) \qquad & u_L(\mathbf{p}) &= u(\mathbf{p}, -1/2) \\ v_R(\mathbf{p}) &= v(\mathbf{p}, 1/2) \qquad & v_L(\mathbf{p}) &= v(\mathbf{p}, -1/2) \end{aligned}\right\} \quad (m = 0). \tag{4.6.55}$$

Clearly we should expect (4.6.55) to hold also for massive particles in the limit $m/E \to 0$. Upon splitting $u(\mathbf{p}, \lambda)$, $v(\mathbf{p}, \lambda)$ into their right- and left-handed pieces, as in (4.6.53), (4.6.54), we find

$$u(\mathbf{p}, 1/2) = \frac{E+m+p}{2[E(E+m)]^{1/2}} \times \left[u_{\mathrm{R}}(\mathbf{p}, 1/2) + \frac{E+m-p}{E+m+p} u_{\mathrm{L}}(\mathbf{p}, 1/2)\right] \tag{4.6.56}$$

and

$$u(\mathbf{p}, -1/2) = \frac{E+m+p}{2[E(E+m)]^{1/2}} \times \left[u_{\mathrm{L}}(\mathbf{p}, -1/2) + \frac{E+m-p}{E+m+p} u_{\mathrm{R}}(\mathbf{p}, -1/2)\right] \tag{4.6.57}$$

Thus as $m/E \to 0$ we get

$$\begin{aligned} u(\mathbf{p}, 1/2) &= u_{\mathrm{R}}(\mathbf{p}) \left[1 + O(m/E)\right] \\ u(\mathbf{p}, -1/2) &= u_{\mathrm{L}}(\mathbf{p}) \left[1 + O(m/E)\right] \end{aligned} \tag{4.6.58}$$

with analogous results for $v(\mathbf{p})$.

The result (4.6.42) can now be understood from a different point of view. Let us denote chirality eigenstates by $u_\eta(\mathbf{p})$, with $\eta = +1/-1$ corresponding to R/L, so that (4.6.52) reads

$$\begin{aligned} u_\eta(\mathbf{p}) &= \eta\gamma_5 u_\eta(\mathbf{p}) \\ v_\eta(\mathbf{p}) &= -\eta\gamma_5 v_\eta(\mathbf{p}), \end{aligned} \tag{4.6.59}$$

and let us consider the vector and axial-vector matrix elements for states of definite chirality. One has for example

$$\begin{aligned} \bar{u}_{\eta'}(\mathbf{p}')\gamma^\mu u_\eta(\mathbf{p}) &= \eta\bar{u}_{\eta'}(\mathbf{p}')\gamma^\mu\gamma_5 u_\eta(\mathbf{p}) \\ &= -\eta\bar{u}_{\eta'}(\mathbf{p}')\gamma_5\gamma^\mu u_\eta(\mathbf{p}). \end{aligned}$$

From (4.6.59) we have $\bar{u}_{\eta'}\gamma_5 = -\eta'\bar{u}_{\eta'}$, so the right-hand side is

$$\eta\eta'\bar{u}_{\eta'}(\mathbf{p}')\gamma^\mu u_\eta(\mathbf{p}),$$

which is our initial expression multiplied by $\eta\eta'$, so that we must have $\eta\eta' = +1$ for a non-zero matrix element, i.e. $\eta' = \eta$. The same result holds for $\gamma^\mu\gamma_5$.

Thus for *massless* fermions, γ^μ and $\gamma^\mu\gamma_5$ *exactly* conserve chirality. The conservation of the helicity in the limit $m/E \to 0$ follows because of the identification of helicity and chirality in this limit, as shown in (4.6.58).

For fermion-antifermion annihilation or creation, i.e. matrix elements of the type $\bar{v}(\gamma^\mu$ or $\gamma^\mu\gamma_5)u$ or $\bar{u}(\gamma^\mu$ or $\gamma^\mu\gamma_5)v$ one finds that in the massless case the fermion and the antifermion must have opposite chirality, which coincides with our results (4.6.49)–(4.6.51) that the amplitude for annihilation or creation with equal helicities is $O(m/E)$ compared with the opposite helicity case.

These results will play a seminal rôle in our study of the electroweak theory and QCD, where the couplings are just γ^μ and $\gamma^\mu\gamma_5$. (A more general version of these results is given in Section 10.4.) Note, for comparison, that the other couplings (I, γ_5, $\sigma^{\mu\nu}$) can be shown to flip helicity in the limit $m/E \to 0$, in contrast to (4.6.42).

5
The observables of a reaction

Interesting spin effects are seen in many hadronic reactions, such as $pp \to pp$, $np \to np$, $pp \to n\Delta$, $\Lambda p \to \Lambda p$, $pp \to \pi X$ etc. And recently more complete measurements have been made on $\Lambda p \to \Lambda p$ and the related reaction $\bar{p}p \to \bar{\Lambda}\Lambda$, especially at LEAR at CERN. In addition experiments using polarized deuteron beams and targets are becoming relatively commonplace.

Given the interest and variety of reactions that are or will be studied it seems worthwhile to set up a general description for an arbitrary $2 \to 2$ reaction with particles of any spin. Indeed we shall set up a general scheme which is, surprisingly, simpler to work with than the usual one for $NN \to NN$ and from which the relevant information for a specific reaction can be easily read off.

Our emphasis here will be upon those quantities, the *observables* that can be measured and upon how they are related to the helicity amplitudes.

We begin with total cross-section measurements, which yield information about the forward amplitudes, and then consider more general observables. For the latter we work first in the CM and then relate the CM observables to the Lab frames where the measurements are actually made.

A comprehensive list of linearly independent measurable reaction parameters and their relation to the helicity amplitudes, for various reactions, is given in Appendix 10.

5.1 The generalized optical theorem

For spinless particles, in our normalization, the usual optical theorem (see e.g. Messiah, 1958) relates the imaginary part of the forward helicity amplitude H to the total cross-section as follows:

$$\text{Im } H(\theta = 0) = \frac{1}{4\sqrt{\pi}} \sigma_{\text{tot}}. \tag{5.1.1}$$

For particles with spin, the direct generalization of (5.1.1) is

$$\text{Im } H_{\lambda_A\lambda_B;\lambda_A\lambda_B}(\theta = 0) = \frac{1}{4\sqrt{\pi}}\sigma_{\text{tot}}(\lambda_A, \lambda_B) \tag{5.1.2}$$

where $\sigma_{\text{tot}}(\lambda_A, \lambda_B)$ means the total cross-section measured with the initial particles A and B in the unique helicity states λ_A, λ_B respectively, a situation that can sometimes be realized using a polarized beam and target.

The *unpolarized* total cross-section is defined as

$$\sigma_{\text{tot}} = \frac{1}{(2s_A + 1)(2s_B + 1)} \sum_{\lambda_A, \lambda_B} \sigma_{\text{tot}}(\lambda_A, \lambda_B) \tag{5.1.3}$$

so that from (5.1.2)

$$\sum_{\lambda_A, \lambda_B} \text{Im } H_{\lambda_A\lambda_B;\lambda_A\lambda_B} = \frac{(2s_A + 1)(2s_B + 1)}{4\sqrt{\pi}}\sigma_{\text{tot}}, \tag{5.1.4}$$

where $H_{\lambda_A\lambda_B;\lambda_A\lambda_B}$ is evaluated at $\theta = 0$. For photons the factor $2s + 1$ is replaced by 2 in (5.1.3) and (5.1.4).

Relations (5.1.4) and (5.1.2) are very valuable. Equation (5.1.4), which is easy to use in practice, allows a determination of the imaginary part of the forward 'spin-averaged' amplitude whereas (5.1.2), which may be difficult in practice, gives the imaginary parts of the individual amplitudes $H_{\lambda_A\lambda_B;\lambda_A\lambda_B}$ at $\theta = 0$.

However, (5.1.2) is not the most general form of the optical theorem. There are other amplitudes, not of the form $\lambda_A\lambda_B \to \lambda_A\lambda_B$, which need not vanish in the forward direction (see Section 4.3), namely those of the form $\lambda_A\lambda_B \to \lambda'_A\lambda'_B$ where $\lambda'_A - \lambda'_B = \lambda_A - \lambda_B$; *all* these can be measured by suitably preparing the initial states of beam and target.

Let $\rho_{\text{i}}(A, B)$ be the joint helicity density matrix for the initial particles. Then (Bialkowski, 1970) the generalization of (5.1.2) is

$$\sum_{\lambda'_A - \lambda'_B = \lambda_A - \lambda_B} \rho_{\text{i}\lambda'_A\lambda'_B;\lambda_A\lambda_B}(A, B) \,\text{Im } H_{\lambda'_A\lambda'_B;\lambda_A\lambda_B}(\theta = 0) = \frac{1}{4\sqrt{\pi}}\sigma_{\text{tot}}(\rho_{\text{i}}) \tag{5.1.5}$$

where $\sigma_{\text{tot}}(\rho_{\text{i}})$ is the total cross-section measured with the beam and target described by ρ_{i}.

Usually the beam and target are uncorrelated, so that

$$\rho_{\text{i}}(A, B) = \rho_{\text{i}}(A) \otimes \rho_{\text{i}}(B). \tag{5.1.6}$$

We shall illustrate the use of (5.1.5) in nucleon–nucleon scattering and then consider a more general reaction.

5.1.1 Nucleon–nucleon scattering

Let the spin-polarization vectors for the beam ($\boldsymbol{\mathcal{P}}^A$) and for the target ($\boldsymbol{\mathcal{P}}^B$) *both* be specified in the Lab frame, as is commonly done in experiments. Then the CM helicity density matrices for particles A and B will be

$$\begin{aligned} \rho(A) &= \tfrac{1}{2}\left(I + \boldsymbol{\mathcal{P}}^A \cdot \boldsymbol{\sigma}\right) \\ \rho(B) &= \tfrac{1}{2}\left(I + \tilde{\boldsymbol{\mathcal{P}}}^B \cdot \boldsymbol{\sigma}\right) \end{aligned} \tag{5.1.7}$$

where, because the Lab frame is rotated from the helicity rest frame for B (see Fig. 3.1 and discussion thereafter)

$$\tilde{\boldsymbol{\mathcal{P}}}^B \equiv (\mathscr{P}_x^B, -\mathscr{P}_y^B, -\mathscr{P}_z^B). \tag{5.1.8}$$

Substituting in (5.1.5) yields

$$\begin{aligned} &\text{Im } H_{++;++}(1 - \mathscr{P}_z^A \mathscr{P}_z^B) \\ &+ \text{Im } H_{+-;+-}(1 + \mathscr{P}_z^A \mathscr{P}_z^B) \\ &+ \text{Im } H_{++;--}(\mathscr{P}_x^A \mathscr{P}_x^B + \mathscr{P}_y^A \mathscr{P}_y^B) = \frac{1}{2\sqrt{\pi}} \sigma_{\text{tot}}(\boldsymbol{\mathcal{P}}^A, \boldsymbol{\mathcal{P}}^B) \end{aligned} \tag{5.1.9}$$

where $(\pm)$ is short for $(\pm 1/2)$.

The connection between our helicity amplitudes and the notation commonly used in nucleon–nucleon (NN) physics (Goldberger *et al.*, 1960) is, aside from normalization,

$$\begin{aligned} H_{++;++} = \phi_1 \qquad H_{++;--} &= \phi_2 \qquad H_{+-;+-} = \phi_3 \\ H_{+-;-+} = \phi_4 \qquad H_{++;+-} &= \phi_5 \end{aligned} \tag{5.1.10}$$

If $\overset{\leftarrow}{\rightarrow}$ indicates complete polarization along or opposed to the incoming beam direction and $\uparrow\downarrow$ indicates polarizations transverse to the beam then (5.1.9) gives the now familiar results

$$\Delta\sigma_L \equiv \sigma_{\text{tot}\underset{\leftarrow}{\rightarrow}} - \sigma_{\text{tot}\underset{\rightarrow}{\rightarrow}} = 4\sqrt{\pi}\ \text{Im}\ \left(H_{++;++} - H_{+-;+-}\right), \tag{5.1.11}$$

where the top arrow refers to the beam polarization and the bottom arrow to the target polarization, and

$$\Delta\sigma_T \equiv \sigma_{\text{tot}\uparrow\downarrow} - \sigma_{\text{tot}\uparrow\uparrow} = -4\sqrt{\pi}\ \text{Im}\ H_{++;--}, \tag{5.1.12}$$

where the first arrow refers to the beam polarization and the second arrow to the target polarization.

Measurements of $\Delta\sigma_L$ and $\Delta\sigma_T$ have produced rather interesting results, as will be discussed in Chapter 14.

5.1.2 Particles of arbitrary spin

It is now simplest to specify the initial CM helicity density matrix in terms of the multipole parameters $t^l_m(A)$ and $t^L_M(B)$ (see eqn (3.1.32)). Then (5.1.5) becomes

$$\sum_m \sum_{l,L} t^l_m(A) t^L_m(B) \ \mathrm{Im}\ h_{lL}(m) = \frac{1}{4\sqrt{\pi}} \sigma_{\mathrm{tot}}(\rho_\mathrm{i}),$$

where $h_{lL}(m)$ is a linear combination of forward amplitudes:

$$h_{lL}(m) \equiv \frac{(2l+1)(2L+1)}{(2s_A+1)(2s_B+1)} \ \mathrm{Tr}\ \left[T^l_m T^L_m H(\theta = 0)\right]. \tag{5.1.13}$$

Parity invariance gives

$$h_{lL}(-m) = (-1)^{l+L} h_{lL}(m) \tag{5.1.14}$$

and time-reversal invariance yields

$$h_{lL}(m) = (-1)^{l+L} h_{lL}(m). \tag{5.1.15}$$

Thus only even values of $l + L$ can occur and we end up with the result

$$\frac{1}{4\sqrt{\pi}} \sigma_{\mathrm{tot}}(\rho_\mathrm{i}) = \sum_{\substack{l,L \\ l+L \text{ even}}} \sum_{m \geq 0} (2 - \delta_{m0}) \ \mathrm{Im}\ h_{lL}(m) \ \mathrm{Re}\ \left[t^l_m(A) t^L_m(B)\right]. \tag{5.1.16}$$

Notice that there is no interference between even and odd ranks of polarization.

For identical particles one also has

$$h_{lL}(m) = h_{Ll}(m). \tag{5.1.17}$$

By suitably choosing the $t^l_m(A)$ and $t^L_m(B)$ one can measure the linear combinations of forward amplitudes $h_{lL}(m)$.

Note that since $\left(T^l_m\right)_{ij} = 0$ unless $i = j + m$, all amplitudes in the sum (5.1.16) are of the form

$$H_{\lambda_A+m, \lambda_B+m; \lambda_A \lambda_B}$$

with, of course $|m| \leq \min\{2s_A, 2s_B\}$. Thus we have an important result: *The determination of the imaginary part of a forward amplitude of the form* $H_{\lambda_A+m, \lambda_B+m; \lambda_A \lambda_B}$ *requires polarization of rank* $l \geq |m|$ *in* both *beam and target.*

Once the $h_{lL}(m)$ are determined, the individual helicity amplitudes can be obtained via

$$H_{\lambda_A+m,\lambda_B+m;\lambda_A\lambda_B} = (-1)^{s_A+s_B-\lambda_A-\lambda_B}\,[(2s_A+1)(2s_B+1)]^{1/2}$$
$$\times \sum_{lL} \frac{h_{lL}(m)}{[(2l+1)(2L+1)]^{1/2}} \langle l,m|s_A,\lambda_A+m;s_A,-\lambda_A\rangle$$
$$\times \langle L,m|s_B,\lambda_B+m;s_B,-\lambda_B\rangle\,. \tag{5.1.18}$$

5.1.3 *Application to deuteron–nucleon and deuteron–deuteron scattering*

Consider a magnetically prepared beam and target with axes of quantization in the Lab frame specified by polar angles $\theta = \beta_A, \phi = \gamma_A$ and $\theta = \beta_B, \phi = 0$ respectively, as shown in Fig. 5.1.

Let $\hat{t}^l_0$ and $\hat{t}^L_0$ be the multipole parameters of beam and target when referred to the frames in which their quantization axes are along OZ. Then the CM multipole parameters needed are found from (3.3.1) and (3.3.2), and (5.1.16) becomes

$$\frac{1}{4\sqrt{\pi}}\sigma_{\text{tot}}(\rho_{\text{i}}) = \sum_{\substack{l,L \\ l+L\ \text{even}}} \hat{t}^l_0\hat{t}^L_0 \sum_{m\geq 0} \cos(\gamma_A m)\, d^l_{m0}(\beta_A)$$
$$\times\, d^L_{m0}(\pi-\beta_B)(2-\delta_{m0})\ \text{Im}\ h_{lL}(m). \tag{5.1.19}$$

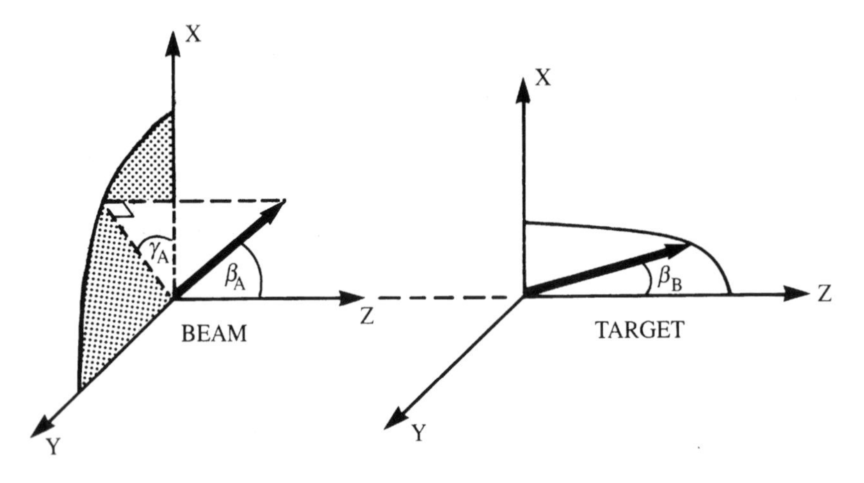

Fig. 5.1. Angles specifying quantization axes of beam and target.

In terms of the polarization and the alignment[1] of beam and target, (5.1.19) gives the following.

For $d + N \to d + N$

$$\begin{aligned}\frac{1}{4\sqrt{\pi}}\sigma_{\text{tot}}(\boldsymbol{\mathcal{P}}^d, \mathscr{A}; \boldsymbol{\mathcal{P}}^N) = {} & \text{Im}\, h_{00}(0) \\ & + \tfrac{1}{\sqrt{6}}\left[\mathscr{P}_x^d\mathscr{P}_x^N\ \text{Im}\, h_{11}(1) - \mathscr{P}_z^d\mathscr{P}_z^N\ \text{Im}\, h_{11}(0)\right] \\ & + \tfrac{3}{\sqrt{10}}\mathscr{A}\left(3\cos^2\beta - 1\right)\ \text{Im}\, h_{20}(0) \qquad (5.1.20)\end{aligned}$$

where

$$\begin{aligned} h_{00}(0) &= \tfrac{1}{3}\left(H_{11/2;11/2} + H_{01/2;01/2} + H_{-11/2;-11/2}\right) \\ h_{11}(1) &= \sqrt{3}H_{11/2;0-1/2} \\ h_{11}(0) &= \sqrt{\tfrac{3}{2}}\left(H_{11/2;11/2} - H_{-11/2;-11/2}\right) \\ h_{20}(0) &= \tfrac{1}{3}\sqrt{\tfrac{5}{2}}\left(H_{11/2;11/2} + H_{-11/2;-11/2} - 2H_{01/2;01/2}\right)\end{aligned} \qquad (5.1.21)$$

and where β is the angle between $\boldsymbol{\mathcal{P}}^d$ and the beam direction. Note that four measurements are needed to find all the amplitudes, and only polarizations along and transverse to the beam are required.

For $d + d \to d + d$, labelling the beam and target deuterons by A and B respectively, one gets[2]

$$\begin{aligned}\frac{1}{4\sqrt{\pi}}&\sigma_{\text{tot}}(\boldsymbol{\mathcal{P}}^B, \mathscr{A}^B; \boldsymbol{\mathcal{P}}^A, \mathscr{A}^A) \\ = {} & \text{Im}\, h_{00}(0) + \tfrac{1}{2}\left[\mathscr{P}_x^A\mathscr{P}_x^B\ \text{Im}\, h_{11}(1) - \mathscr{P}_z^A\mathscr{P}_z^B\ \text{Im}\, h_{11}(0)\right] \\ & + \tfrac{1}{2\sqrt{10}}\left[\mathscr{A}^A\left(3\cos^2\beta_A - 1\right) + \mathscr{A}^B\left(3\cos^2\beta_B - 1\right)\right]\ \text{Im}\, h_{20}(0) \\ & + \tfrac{1}{40}\mathscr{A}^A\mathscr{A}^B\left[\left(3\cos^2\beta_A - 1\right)\left(3\cos^2\beta_B - 1\right)\ \text{Im}\, h_{22}(0)\right. \\ & \qquad - 12\cos\gamma_A\sin\beta_A\cos\beta_A\sin\beta_B\cos\beta_B\ \text{Im}\, h_{22}(1) \\ & \qquad \left. + 3\cos 2\gamma_A\sin^2\beta_A\sin^2\beta_B\ \text{Im}\, h_{22}(2)\right]. \qquad (5.1.22)\end{aligned}$$

[1] See subsection 3.1.12.

[2] This can be written in simpler form using the T_{ij} of eqn (3.1.59). We have not done so because experimentally it is easier to think in terms of the alignment.

Finally, the amplitude combinations measured by seven experiments are:

$$
\begin{aligned}
h_{00}(0) &= \tfrac{2}{9}\left(H_{11;11} + 2H_{10;01} + H_{1-1;1-1} + \tfrac{1}{2}H_{00;00}\right) \\
h_{11}(1) &= H_{11;00} + H_{10;0-1} \\
h_{11}(0) &= H_{11;11} - H_{1-1;1-1} \\
h_{20}(0) &= \tfrac{\sqrt{10}}{9}\left(H_{11;11} - H_{10;10} + H_{1-1;1-1} - H_{00;00}\right) \\
h_{22}(0) &= \tfrac{5}{9}\left(H_{11;11} - 4H_{10;10} + H_{1-1;1-1} + 2H_{00;00}\right) \\
h_{22}(1) &= \tfrac{5}{3}\left(H_{11;00} - H_{10;0-1}\right) \\
h_{22}(2) &= \tfrac{5}{3}H_{11;-1-1}.
\end{aligned}
\tag{5.1.23}
$$

Note that now the polarizations of both beam and target have to be set at some angle other than along or transverse to the beam for at least one measurement. For example, one could choose $\gamma_A = 0$, $\beta_A = \beta_B = 45°$.

5.2 The final state helicity density matrix

We consider now the definition, and some important properties, of the helicity density matrix of the final particles produced in a reaction. Initially we deal with $2 \to 2$ reactions, but this will be generalized in Section 5.8.

5.2.1 Definition

We consider an arbitrary reaction $A + B \to C + D$. For given initial helicities a, b, the helicity amplitudes $H_{cd;ab}$ are a measure of the probability amplitude for finding the final helicities c, d. Thus, in analogy with eqn (3.2.2) the joint CM helicity density matrix for the final state is

$$\rho'_{cd;c'd'}(C,D) = \sum_{\substack{a,b \\ a',b'}} H_{cd;ab}\ \rho_{\text{i}\,ab;a'b'}(A,B)\ H^*_{c'd';a'b'} \tag{5.2.1}$$

where $\rho_\text{i}(A,B)$ is the initial state helicity density matrix. To avoid the profusion of indices we write (5.2.1) in matrix form:

$$\rho'(C,D) = \mathbf{H}\rho_\text{i}(A,B)\mathbf{H}^\dagger. \tag{5.2.2}$$

If $\rho_\text{i}(A,B)$ is correctly normalized, so that $\text{Tr}\ \rho_\text{i}(A,B) = 1$, it will be found that $\rho'(C,D)$ is not normalized to trace 1, so for computing expectation values of observables in the final state we must always use

$$\rho(C,D) \equiv \frac{\rho'(C,D)}{\text{Tr}\ \rho'(C,D)}.$$

With our normalization for $H_{\{\lambda\}}$, eqn (4.1.4),

$$\mathrm{Tr}\ \rho'(C,D) = 2\pi \frac{d^2\sigma}{dtd\phi}(\rho_{\mathrm{i}}), \tag{5.2.3}$$

where $(d^2\sigma/dtd\phi)(\rho_{\mathrm{i}})$ is the differential cross-section into the momentum transfer range $t \to t + dt$ and the azimuthal range $\phi \to \phi + d\phi$ for an initial state specified by ρ_{i}.

5.2.2 *Rank conditions*

Since the rank (see subsection (3.1.2)) of a product of matrices must be less than or equal to the rank of any matrix in the product, (5.2.2) implies that

$$r_f \leq r_{\mathrm{i}} \tag{5.2.4}$$

where r_f and r_{i} are the ranks of the final and initial state density matrices.

This condition can be a very stringent one. For example, in

$$\pi + N \to \pi + N^*(J)$$

where $N^*(J)$ is a high-spin resonance, r_{i} cannot be greater than 2 (ρ_{i} is a 2×2 matrix) and therefore ρ_f, which is $2J \times 2J$ and could thus be a huge matrix, must have rank ≤ 2.

If it happens that only the even part ρ_+ of the final state density matrix can be measured (see subsection 3.1.7), then the weaker rank condition

$$\mathrm{rank}\ \rho_+ \leq 2r_{\mathrm{i}}$$

holds. In our $N^*(J)$ example above, if $J = 3/2$ we end up with rank $\rho_+ \leq 4$ which is no restriction at all, bearing in mind that ρ_+ is a 4×4 matrix! If both C and D have non-zero spin and we consider the effective density matrix of, say, C, then its rank must satisfy a much weaker bound than (5.2.4), namely,

$$\mathrm{rank}\ \rho(C) \leq (2s_D + 1)\ \mathrm{rank}\ \rho_{\mathrm{i}} \tag{5.2.5}$$

with analogous constraint for D.

Generally a large number of relations may exist amongst the elements of ρ_f and they must be taken into account experimentally.

5.2.3 *Angular momentum constraints near* $\theta = 0, \pi$

The behaviour of the $H_{\{\lambda\}}$ near $\theta = 0$ and π (Section 4.3) imposes constraints on $\rho(C,D)$ near the forward and backward regions. These depend upon $\rho_{\mathrm{i}}(A,B)$.

The strongest conditions apply when the initial state is *unpolarized.* Then at $\theta = 0$ or π

$$\rho_{cd;c'd'}(C,D) = 0$$

unless both

$$c - d = c' - d' \quad \text{and} \quad |c - d| \leq s_A + s_B.$$

Near these points, the behaviour is

$$\rho_{cd;c'd'} \propto (\sin\theta/2)^{s_1}(\cos\theta/2)^{c_1} \tag{5.2.6}$$

where

$$s_1 \equiv |c - d - c' + d'| + (1 + \epsilon)M$$
$$c_1 \equiv |c - d + c' - d'| + (1 - \epsilon)M$$

with

$$\epsilon \equiv \text{sign}\,\{(c - d)(c' - d')\}$$

and

$$M = \begin{cases} 0 & \text{when either } |c - d| \text{ or } |c' - d'| \leq s_A + s_B \\ \min\left\{\left||c - d| - s_A - s_B\right|; \left||c' - d'| - s_A - s_B\right|\right\} & \text{otherwise.} \end{cases}$$

For the effective single-particle density matrix, say of particle C, we have

$$\rho_{c'c}(C) = 0, \tag{5.2.7}$$

unless both

$$c = c' \quad \text{and} \quad |c| \leq s_A + s_B + s_D,$$

and

$$\rho_{c'c}(C) \propto (\sin\theta/2)^{\bar{s}_1}(\cos\theta/2)^{\bar{c}_1} \tag{5.2.8}$$

where

$$\bar{s}_1 \equiv |c - c'| + (1 + \bar{\epsilon})\overline{M}$$
$$\bar{c}_1 \equiv |c + c'| + (1 - \bar{\epsilon})\overline{M}$$

with

$$\bar{\epsilon} \equiv \text{sign}\,\{cc'\}$$

and

$$\overline{M} = \begin{cases} 0 & \text{when either } |c| \text{ or } |c'| \leq s_A + s_B + s_D \\ \min\left\{\left||c| - s_A - s_B - s_D\right|; \left||c'| - s_A - s_B - s_D\right|\right\} & \text{otherwise.} \end{cases}$$

The above constraints must be respected in any data analysis. It will be seen in Section 5.4 that the multipole parameters have a much simpler behaviour than ρ at $\theta \simeq 0, \pi$.

5.3 The CM observables and the dynamical reaction parameters

Several discussions of the variables valid for relativistic scattering have been given in the literature for nucleon–nucleon scattering. Detailed references are given in Bourrely, Leader and Soffer, (1980).

Our treatment is more general, applying to any reaction, and is actually simpler. We expand the initial and correctly normalized final density matrices in terms of joint multipole parameters $t^{lL}_{mM}(A,B)$, $t^{l'L'}_{m'M'}(C,D)$, according to eqn (3.1.31) as generalized to combined systems of particles, and substitute in (5.2.2). There results a relation between the initial and final multipole parameters of the reaction:

$$
\begin{aligned}
t^{l'L'}_{m'M'}(C,D)\frac{d^2\sigma}{dtd\phi}(\rho_{\rm i})
&= \left(\frac{2}{3}\right)^{n_\gamma}\frac{1}{2\pi}\frac{d\sigma}{dt}\sum_{\substack{lL\\mM}}(2l+1)(2L+1)\\
&\quad\times (l,m;L,M|l',m';L',M')_\phi\, t^l_m(A)t^L_M(B)
\end{aligned}
\tag{5.3.1}
$$

where n_γ is the number of photons in the initial state. We have assumed that the beam and target are uncorrelated. Equation (5.3.1) gives the value of $t^{l'L'}_{m'M'}(C,D)$ when C's direction is at polar angles θ, ϕ in the CM. The outcome of the experiment is controlled by the fundamental *CM dynamical reaction parameters* (we shall simply call them 'reaction parameters'),

$$
\begin{aligned}
&(l,m;L,M|l',m';L',M')_\phi\\
&\quad\equiv \left(\frac{d\sigma}{dt}\right)^{-1}\frac{1}{(2s_A+1)(2s_B+1)}\\
&\qquad\times\ \mathrm{Tr}\ [HT^{l\dagger}_m(s_A)T^{L\dagger}_M(s_B)H^\dagger T^{l'}_{m'}(s_C)T^{L'}_{M'}(s_D)],
\end{aligned}
\tag{5.3.2}
$$

where H is the matrix whose elements are $H_{cd;ab}(\theta,\phi)$. The use of matrix notation is compact and efficient, but to avoid any confusion we write out the trace in (5.3.2) in full detail:

$$
\begin{aligned}
&\mathrm{Tr}\ [HT^{l\dagger}_m(s_A)T^{L\dagger}_M(s_B)H^\dagger T^{l'}_{m'}(s_C)T^{L'}_{M'}(s_D)]\\
&\qquad = H_{cd;ab}\left(T^{l\dagger}_m(s_A)\right)_{aa'}\left(T^{L\dagger}_M(s_B)\right)_{bb'}\\
&\qquad\quad\times H^*_{c'd';a'b'}\left(T^{l'}_{m'}(s_C)\right)_{c'c}\left(T^{L'}_{M'}(s_D)\right)_{d'd}.
\end{aligned}
$$

The reaction parameters (5.3.2) are a direct generalization of the Wolfenstein parameters. All the dynamics is contained in these parameters, which can be evaluated in terms of the helicity amplitudes. They depend on both θ (or t) and ϕ, but the ϕ-dependence is trivial:

$$
(l,m;L,M|l',m';L',M')_\phi = e^{i\phi(M-m)}\,(l,m;L,M|l',m';L',M'), \tag{5.3.3}
$$

where the right-hand side parameters are at $\phi = 0$. When no ϕ-label is shown we shall always mean $\phi = 0$ in the reaction parameters.

It should be noted that the order of symbols is

$$(\text{beam; target}|\text{scattered; recoil})$$

and the normalization is such that

$$(0,0;0,0|0,0;0,0) = 1. \tag{5.3.4}$$

Note that for some colliding-beam experiments the spin measurements are carried out in the CM, so that (5.3.1) will apply *directly* to the measured quantities.

In using (5.3.1) and various special cases to be derived from it, it must be remembered that for all *photons*, whether polarized or not, because of the absence of states with helicity $\lambda = 0$ one has $t_0^2 = 1/\sqrt{10}$, as is explained in subsection 3.1.12. Also of use in this case is the result

$$(2,0;0,0|0,0;0,0) = 1/\sqrt{10},$$

which follows from (5.3.2) and the properties of T_0^2 as given in (3.1.26).

5.3.1 Properties of the CM reaction parameters

The reaction parameters are not all independent as a consequence of the symmetry properties of the helicity amplitudes and of the T_m^l matrices.

(*i*) *Reality*. From $T_m^{l\dagger} = (-1)^m T_{-m}^l$ follows

$$(l,m;L,M|l',m';L',M')^* \\ = (-1)^{m+M+m'+M'} \times (l,-m;L,-M|l',-m';L',-M'). \tag{5.3.5}$$

(*ii*) *Parity*. Using $\left(T_m^l\right)_{-m_1-m_2} = (-1)^l \left(T_{-m}^l\right)_{m_1 m_2}$ and the space inversion properties eqn (4.2.1) in both the $H_{\{\lambda\}}$ in (5.3.2) yields

$$(l,m;L,M|l',m';L',M') \\ = (-1)^{m+M+m'+M'}(-1)^{l+L+l'+L'} \\ \times (l,-m;L,-M|l',-m';L',-M'). \tag{5.3.6}$$

Thus

$$(l,0;L,0|l',0;L',0) = 0 \tag{5.3.7}$$

if $l + L + l' + L'$ is odd.

When this is combined with (5.3.5) we have the important result

$$(l,m;L,M|l',m';L',M') \quad \text{is} \quad \begin{Bmatrix} \text{real} \\ \text{imaginary} \end{Bmatrix} \\ \text{as } l + L + l' + L' \quad \text{is} \quad \begin{Bmatrix} \text{even} \\ \text{odd} \end{Bmatrix}. \tag{5.3.8}$$

(*iii*) *Time-reversal.* Using eqn (4.2.4) in both the $H_{\{\lambda\}}$ in (5.3.2) and also the fact that the T^l_m are real gives

$$\begin{aligned}(l,m;L,M|l',m';L',M')^{AB\to CD}\\ = (l',m';L',M'|l,m;L,M)^{CD\to AB}_{\phi=\pi}.\end{aligned} \tag{5.3.9}$$

For *elastic* reactions eqn (4.2.5) yields

$$\begin{aligned}(l,m;L,M|l',m';L',M')\\ = (-1)^{m+M+m'+M'}\,(l',m';L',M'|l,m;L,M).\end{aligned} \tag{5.3.10}$$

(*iv*) *Identical particles.* Using (4.2.11a,b) we find the following.
If $A = B$,

$$\begin{aligned}(l,m;L,M|l',m';L',M')^{\theta}\\ = (-1)^{m'+M'}(L,M;l,m|l',m';L',M')^{\pi-\theta}.\end{aligned} \tag{5.3.11}$$

Thus at $\theta = \pi/2$

$$(l,m;l,m|l',m';L',M') = 0 \qquad \text{if } m' + M' \text{ is odd.} \tag{5.3.12}$$

If $C = D$,

$$(l,m;L,M|L',M';l',m')^{\theta} = (-1)^{m+M}(l,m;L,M|l',m';L',M')^{\pi-\theta}. \tag{5.3.13}$$

Thus at $\theta = \pi/2$

$$(l,m;L,M|l',m';l',m') = 0 \qquad \text{if } m + M \text{ is odd.} \tag{5.3.14}$$

Equations (5.3.13), (5.3.14) also hold for reactions of the type

$$A + A \to D + \overline{D}$$

provided the reaction is invariant under charge conjugation.

Finally, if $A = B$ and $C = D$ then

$$(L,M;l,m|L',M';l',m') = (-1)^{m+M+m'+M'}(l,m;L,M|l',m';L',M'). \tag{5.3.15}$$

This also holds for reactions of the type

$$A + \overline{A} \to D + \overline{D}$$

if charge conjugation is a good symmetry.

(*v*) *Additional parity and time-reversal constraints.* The application of the above symmetry results will not, in general, reduce the number of independent reaction parameters to the expected N^2 in the case where there are N independent helicity amplitudes. The additional relations can be obtained by applying the symmetry concerned to just *one* $H_{\{\lambda\}}$ in (5.3.2). The results are as follows.

Parity:

$$\begin{aligned}(l, m; L, M|l', m'; L', M') \\ = \eta \sum_{\substack{\text{repeated}\\ \text{indices}}} \mathscr{A}_{l_1 m_1}(lm)\mathscr{A}_{L_1 M_1}(LM)\mathscr{A}_{l'_1 m'_1}(l'm')\mathscr{A}_{L'_1 M'_1}(L'M') \\ \times (l_1, m_1; L_1, M_1|l'_1, m'_1; L'_1, M'_1) \qquad (5.3.16)\end{aligned}$$

where η is defined in eqn (4.2.1). The $\mathscr{A}(lm)$ are given in terms of vector addition coefficients and are tabulated for $s = 1/2, 1$ and $3/2$ in Appendix 6.

Time reversal: (for elastic reactions):

$$\begin{aligned}(l, m; L, M|l', m'; L', M') = \sum_{\substack{\text{repeated}\\ \text{indices}}} \mathscr{C}^{lm;l'm'}_{l_1 m_1;l'_1 m'_1} \mathscr{C}^{LM;L'M'}_{L_1 M_1;L'_1 M'_1} \\ \times (l_1, m_1; L_1, M_1|l'_1, m'_1; L'_1, M'_1) \quad (5.3.17)\end{aligned}$$

The coefficients $\mathscr{C}$ are explained in Appendix 7 and are tabulated for $s = 1/2$. In Appendix 10 we give a comprehensive list of linearly independent reaction parameters for various reactions and their relation to the helicity amplitudes.

(*vi*) *Behaviour near* $\theta = 0$ *or* π. In the forward and backward scattering regions we find

$$(l, m; L, M|l', m'; L', M')^{\theta \to 0} \propto (\sin\theta/2)^{|m-M-m'+M'|} \qquad (5.3.18a)$$

and

$$(l, m; L, M|l', m'; L', M')^{\theta \to \pi} \propto (\cos\theta/2)^{|m-M+m'-M'|} \qquad (5.3.18b)$$

The phenomenological consequences of these properties will emerge in the following sections.

5.4 Experimental determination of the CM reaction parameters

In this section we assume that we are given the CM multipole parameters for an arbitrarily prepared initial state and that we are able to measure the joint CM multipole parameters of the final state. The connection with measurements carried out in the Lab and the question of *how* one measures the multipole parameters will be dealt with in Sections 5.5 and 5.6.

From (5.3.1), exhibiting explicitly the ϕ-dependence, the outcome of an experiment is controlled by

$$
\begin{aligned}
t^{l'L'}_{m'M'}(C,D)\frac{d^2\sigma}{dtd\phi} = \left(\frac{2}{3}\right)^{n_\gamma}\frac{1}{2\pi}\frac{d\sigma}{dt}&\sum_{lL}(2l+1)(2L+1)\\
&\times\sum_{mM}e^{i\phi(M-m)}t^l_m(A)t^L_M(B)\\
&\times(l,m;L,M|l',m',L',M')
\end{aligned}
\qquad (5.4.1)
$$

where, of course, the left-hand side is measured for an initial state specified by $t^l_m(A)$ and $t^L_M(B)$. (For photons we recall the discussion after eqn (5.3.4).)

There are, in general, two ways to utilize (5.4.1) experimentally in order to learn about the reaction parameters. The first way takes advantage of the simple ϕ-dependence to study asymmetries such as 'up–down' or 'left–right'. The most sophisticated example would involve measuring over the whole range of ϕ at fixed θ and then taking experimental averages of $e^{i\mu\phi}$ over the data at fixed θ, the μ being integers.

The second way looks at the changes induced in a measured observable when the density matrix of the initial state is altered, e.g. by reversal of the ordinary (rank-1) polarization of beam or target. For spin $> 1/2$ the method is less efficacious than for $s = 1/2$, where one can maximize the effect by fully reversing the sign of the polarization. It is not generally possible to reverse the sign of an arbitrary t^l_m when $l \geq 2$. We shall discuss an example where the t^l_m are altered by the passage through a magnetic field.

5.4.1 *Unpolarized initial state*

(i) Measurements of the generalized polarizing power and the final state polarization correlation parameters

Since all $t^l_m(A)$ and $t^L_M(B)$ are zero except $t^0_0(A) = t^0_0(B) = 1$, there is no ϕ-dependence left in (5.4.1) and, remembering that by definition

$$\int_0^{2\pi} d\phi\frac{d^2\sigma}{dtd\phi}(\text{unpol. initial state}) = \frac{d\sigma}{dt},$$

one obtains

$$t^{l'L'}_{m'M'}(C,D;\text{unpol. initial state}) = (0,0;0,0|l',m';L',M'). \qquad (5.4.2)$$

For an elastic reaction time reversal, eqn (5.3.10), then gives the parameters:

$$(l', m'; L', M'|0, 0; 0, 0) = (-1)^{m'+M'}(0, 0; 0, 0|l', m'; L', M').$$

Note that the parameters $(0, 0; 0, 0|l', m'; 0, 0)$ and $(0, 0; 0, 0|0, 0; L', M')$ are analogous to the usual polarizing power of the reaction but are here generalized to specify the rank of the polarization produced. We shall refer to them as the 'lm polarizing power'. The parameters with both l' and L' non-zero are generalizations of the final state polarization correlation parameters C_{ij} used in nucleon–nucleon scattering. All these parameters can be determined, in principle, using an unpolarized initial state.

(ii) Properties of the final state multipole parameters

From (5.4.2) and the properties (5.3.5)–(5.3.17) of the reaction parameters we learn the following properties of the final state CM joint (or effective) helicity-basis multipole parameters for a *parity-conserving* reaction with *unpolarized* initial state:

(a) $t^{lL}_{mM}(C, D)$ is independent of ϕ.

(b) As always $t^{lL}_{-m-M}(C, D) = (-1)^{m+M} t^{lL}_{mM}(C, D)^*$.

(c)
$$t^{lL}_{mM} \text{ is } \begin{Bmatrix} \text{real} \\ \text{or} \\ \text{imaginary} \end{Bmatrix} \quad \text{for } l + L \begin{Bmatrix} \text{even} \\ \text{or} \\ \text{odd} \end{Bmatrix}. \tag{5.4.3}$$

(d) Hence, $t^{lL}_{-m-M}(C, D) = (-1)^{l+L+m+M} t^{lL}_{mM}(C, D)$ and $t^{lL}_{00} = 0 \quad \text{if } l + L$ is odd.

As an example, consider the famous result that the spin-polarization vector $\mathcal{P}$ of the final particles in a parity-conserving two-body reaction with unpolarized initial state must be perpendicular to the reaction plane. The properties (c) and (d) imply that $t^1_0 = 0$ and $t^1_{\pm 1}$ is pure imaginary respectively. The result then follows from eqn (3.1.35).

In reverse, we note that a non-zero value of, say, the longitudinal component of $\mathcal{P}$ (i.e. the component along the particle's momentum) signals a parity violation. Some of the most beautiful electroweak experiments play upon just this feature.

(e) If particles C and D are identical then

$$t^{lL}_{mM}(\theta) = t^{Ll}_{Mm}(\pi - \theta). \tag{5.4.4}$$

(f) If, in addition, particles A and B are identical then

$$t^{lL}_{mM}(\theta) = (-1)^{m+M} t^{Ll}_{Mm}(\theta) \tag{5.4.5}$$

and it follows that

$$t^{lL}_{mM}(\pi/2) = 0 \qquad \text{if } m + M \text{ is odd.} \tag{5.4.6}$$

As an example, in $pp \to pp$ we have $t^1_{\pm 1} = 0$ at $\theta = \pi/2$. Thus the spin-polarization vector $\mathcal{P}$ has magnitude 0 at $\theta = \pi/2$. Equivalently one can say that the polarizing power vanishes at $\theta = \pi/2$.

(g)

$$\begin{aligned} &\text{As } \theta \to 0 \qquad t^{lL}_{mM} \propto (\sin\theta/2)^{|m-M|} \\ &\text{As } \theta \to \pi \qquad t^{lL}_{mM} \propto (\cos\theta/2)^{|m-M|} \end{aligned} \tag{5.4.7}$$

Most of the above properties have obvious consequences for the helicity density matrix itself. The most interesting result follows from (d), namely

$$\rho_{c'd';cd} = (-1)^{c-c'+d-d'} \rho_{-c'-d';-c-d}. \tag{5.4.8}$$

Note that from eqn (3.1.43) the even- and odd-polarization parts of the final state density matrix are, in this case, simply the real and imaginary parts of ρ, i.e.

$$\begin{aligned} \rho_+ &= \text{Re}\ \rho \\ \rho_- &= i\ \text{Im}\ \rho. \end{aligned} \tag{5.4.9}$$

In the transversity basis the analogue for the effective density matrix of either of the final particles is

$$\rho^{\mathrm{T}}_{c'c} = 0 \quad \text{if} \quad c' - c \text{ is odd,} \tag{5.4.10}$$

thus giving ρ^{T} a 'chequerboard' pattern and forcing $[t^l_m]_{S^{\mathrm{T}}} = 0$ if m is odd.

5.4.2 *Polarized beam, unpolarized target*

We consider the measurement of the cross-section and the final state multipole parameters for an arbitrarily polarized beam. We also give some results for specific types of initial polarization.

(i) Measurement of cross-section asymmetries – the generalized analysing power

From (5.4.1) we have, in general (for photons we recall the discussion after eqn (5.3.4))

$$\frac{d^2\sigma}{dtd\phi} = \left(\frac{2}{3}\right)^{n_\gamma} \frac{1}{2\pi}\frac{d\sigma}{dt} \sum_{l,m} (2l+1) t^l_m(A)(l, m; 0, 0|0, 0; 0, 0) e^{-im\phi}. \tag{5.4.11}$$

The parameters $(l, m; 0, 0|0, 0; 0, 0)$ play the rôle of the 'analysing power' of the reaction for lm-type initial polarization, since they govern the magnitude of the asymmetry or ϕ-dependence in $d^2\sigma/dtd\phi$. From eqn

(5.3.10) we see that for an elastic reaction the magnitudes of the lm polarizing power and lm analysing power are equal.[1]

In a typical polarized-beam experiment let the quantization axis for the beam have polar angles $\theta = \beta$, $\phi = \gamma$ in the rest frame of the beam (see Fig. 5.1). Let $\hat{t}^l_0$ be the (known) helicity multipole parameters in the frame whose Z-axis is along the quantization axis. Then the CM multipole parameters $t^l_m(A)$ needed for (5.4.11) are, from (3.3.1),

$$t^l_m(A) = e^{i\gamma m} d^l_{m0}(\beta)\hat{t}^l_0. \tag{5.4.12}$$

We refer to the plane $\phi = \gamma$, i.e. the plane containing the beam and the quantization axis, as the *quantization plane*. For this discussion there is no loss of generality in choosing $\gamma = 0$, so that the quantization plane is the XZ-plane. In detail (5.4.11) now becomes (recall that $(l, m; 0, 0|0, 0; 0, 0)$ is pure imaginary when l is odd)

$$\begin{aligned} \frac{d^2\sigma}{dtd\phi} = \left(\frac{2}{3}\right)^{n_\gamma} \frac{1}{2\pi}\frac{d\sigma}{dt} \Bigg(1 + \tfrac{1}{2}\sum_{l\geq 1}(2l+1)\hat{t}^l_0 \sum_{m\geq 0}(2-\delta_{m0})d^l_{m0}(\beta) \\ \times \left\{[1+(-1)^l]\cos m\phi - i[1-(-1)^l]\sin m\phi\right\} \\ \times (l, m; 0, 0|0, 0; 0, 0)\Bigg) \end{aligned} \tag{5.4.13}$$

where ϕ is the azimuthal angle measured from the quantization plane.

The asymmetries with respect to the quantization plane, or the detailed ϕ-dependence itself, can be used to isolate the combinations such as

$$\mathscr{C}_m = \sum_{l\geq m}(2l+1)\hat{t}^l_0 d^l_{m0}(\beta)(l, m; 0, 0|0, 0; 0, 0) \tag{5.4.14}$$

for each $m \geq 0$.

To measure the individual $(l, m; 0, 0|0, 0; 0, 0)$ one must be able to vary the $t^l_m(A)$ of the beam for each m. One way to do this is to deflect the polarized beam in a magnetic field, between the production reaction and the main reaction. We shall discuss one simple example.

---

[1] Because of this and eqn (5.4.2), an analogue of (5.4.11) appears in the non-relativistic literature with $(l, m; 0, 0|0, 0; 0, 0)$ replaced by $(-1)^m t^l_m(\theta)$, the latter being the CM final state multipole parameters for A when produced from an unpolarized initial state. We avoid this in practice since it confuses properties of the beams in special situations with properties of the reaction. Moreover in *relativistic double-scattering experiments* the $t^l_m(A)$ to be inserted into (5.4.11) are NOT the final state CM multipole parameters of the first reaction but, rather, are the $[t^l_m(\theta)]_{S_{LC}}$ discussed in subsection 3.3.2 (see eqn (3.3.14)).

(ii) Use of a magnetic field to vary the initial state density matrix

Let a polarized beam having multipole parameters t'^{l}_{m} emerge in the XZ-plane and pass through a uniform magnetic field **B** oriented along OY. The particles are deflected around OY through an angle θ_{cyc} (the *cyclotron angle*) as measured in the Lab. (θ_{cyc} is zero for neutral particles.)

The helicity density matrix of the beam, considered, as usual, to be arriving along the Z-direction of the main reaction, will then be described by the CM helicity multipole parameters t^{l}_{m} given by

$$t^{l}_{m}(\delta) = \sum_{m'} d^{l}_{mm'}(\delta) t'^{l}_{m'} \tag{5.4.15}$$

where δ is the *angle of precession* of the spin vector of the particle during the passage through the magnetic field.

For a particle of mass m, charge Q, arbitrary spin s and total magnetic moment μ, the g-factor is defined by

$$\boldsymbol{\mu} = g \frac{Q}{2mc} \mathbf{s}. \tag{5.4.16}$$

Then the precession angle δ is given, in terms of the cyclotron angle, by

$$\delta = \left(\frac{g}{2} - 1\right) \frac{E_{\text{L}}}{mc^2} \theta_{\text{cyc}}. \tag{5.4.17}$$

where E_{L} is the Lab energy of the beam particles. For protons and deuterons one has

$$\left(\frac{g}{2} - 1\right)_{\text{proton}} = 1.79 \qquad \left(\frac{g}{2} - 1\right)_{\text{deuteron}} = -0.14.$$

It is thus difficult to cause a sizeable alteration of the t'^{l}_{m} for deuterons. Nevertheless a successful experiment of this type, using 410 MeV deuterons, was carried out by Button and Mermod (1960), and the idea seems to stem from Lakin (1955).

For neutral particles

$$\delta = -2\mu \left(\frac{E_{\text{L}}}{p_{\text{L}}c}\right) \frac{e_p}{2m_p c} Bd \tag{5.4.18}$$

where μ is the magnetic moment in units of the proton magneton, e_p and m_p are the charge and mass of the proton and d is the distance through the magnetic field traversed by the particle.

If B is measured in gauss and d in metres then

$$\delta \approx -3.2 \times 10^{-5} \mu \left(\frac{E_{\text{L}}}{p_{\text{L}}c}\right) Bd. \tag{5.4.19}$$

We have, for example,

$$\mu_{\text{neutron}} = -1.91 \qquad \mu_{\Lambda} = -0.61.$$

Clearly one must utilize as many different values δ_i of δ as there are l-values appearing in the sum (5.4.14) and measure $C_m(\delta_i)$ for each. The individual $(l,m;0,0|0,0;0,0)$ are then obtained by solving a set of simultaneous equations.

(iii) Measurement of the generalized depolarization and polarization-transfer parameters

Consider the case where we measure the effective multipole parameters of particle C. From (5.4.1) and (5.4.2) we have

$$\left[t^{l'}_{m'}(C)\frac{d^2\sigma}{dtd\phi}\right]_{\text{pol.bm.}} - \left[t^{l'}_{m'}(C)\frac{d^2\sigma}{dtd\phi}\right]_{\text{unpol.}} = \left(\frac{2}{3}\right)^{n_\gamma}\frac{1}{2\pi}\frac{d\sigma}{dt}\sum_{\substack{l\geq 1\\ m}}(2l+1)t^l_m(A) \times (l,m;0,0|l',m';0,0)e^{-im\phi} \tag{5.4.20}$$

where $d\sigma/dt$ is, of course, the unpolarized cross-section.

Equation (5.4.20) indicates the significance of the generalized *depolarization parameters* $(l,m;0,0|l',m';0,0)$ which can be measured by studying the asymmetry in ϕ of the left-hand side for several values of l' and m', bearing in mind the ϕ-independence of the second term on the left-hand side. As in (ii) above, the isolation of individual parameters will be possible only if the initial $t^l_m(A)$ can be varied.

If it is the density matrix of D that is measured, completely analogous equations hold and one determines thereby the generalized $A \to D$ *polarization-transfer parameters* $(l,m;0,0|0,0;L',M')$.

If the joint multipole parameters for C and D can be measured, one learns analogously about the 'three-spin' parameters $(l,m;0,0|l',m';L',M')$.

(iv) Properties of the final state

From (5.4.13), (5.4.20) and the properties (5.3.5)–(5.3.17) of the reaction parameters we find that the special properties of $d^2\sigma/dtd\phi$ and $t^{lL}_{mM}(C,D)$ for our main reaction, as listed below, hold for any of the following situations.

- (s_1) The magnetically prepared beam has $\beta = \pi/2$, i.e. the quantization axis is perpendicular to the beam.
- (s_2) The beam is a secondary beam emerging from a previous parity conserving reaction $R_1 : E + F \to A + G$, with unpolarized initial state, and our Y-axis is along $\mathbf{p}_E \times \mathbf{p}_A$.
- (s_3) As in (s_2), but R_1 can have a polarized beam E, a polarized target F or both, provided that the quantization axes are normal to the scattering plane of R_1.

The properties are:

(a) $d^2\sigma/dtd\phi$ is symmetric under reflection in the beam-containing plane that is perpendicular to the quantization plane, i.e. under $\phi \to -\phi$. (Exceptionally, if particles A have spin 1/2, this holds also for any angle β of the quantization axis; furthermore, Re t^{lL}_{mM} for $l+L$ even and Im t^{lL}_{mM} for $l+L$ odd are symmetric under $\phi \to -\phi$.)

(b) $t^{lL}_{mM}(C,D)$ now depends on ϕ in general.

(c) As always

$$t^{lL}_{-m-M} = (-1)^{m+M} t^{lL*}_{mM}. \tag{5.4.21}$$

(d)

$$t^{lL*}_{mM}(\phi) = (-1)^{l+L} t^{lL}_{-m-M}(-\phi). \tag{5.4.22}$$

Thus at $\phi = 0$

$$t^{lL}_{mM} \text{ is } \left\{ \begin{matrix} \text{real} \\ \text{or} \\ \text{imaginary} \end{matrix} \right\} \text{ as } l+L \text{ is } \left\{ \begin{matrix} \text{even} \\ \text{or} \\ \text{odd} \end{matrix} \right\}. \tag{5.4.23}$$

(e) Hence $t^{lL}_{-m-M}(\phi) = (-1)^{m+M+l+L} t^{lL}_{mM}(-\phi)$, and

$$t^{lL}_{00}(\phi = 0) = 0 \qquad \text{if } l+L \text{ is odd.} \tag{5.4.24}$$

As an example, an incoming beam with its spin-polarization vector $\mathcal{P}^A$ perpendicular to the scattering plane satisfies the condition (s_1). Then use of (5.4.22) together with (3.1.35) tells us that the spin-polarization vectors $\mathcal{P}^C$ and $\mathcal{P}^D$ must also be perpendicular to the scattering plane.

(f) For an arbitrary initial polarization, if $C = D$ one has

$$t^{lL}_{mM}(\theta, \phi) = t^{Ll}_{Mm}(\pi - \theta, \phi + \pi). \tag{5.4.25}$$

(g)

$$\begin{aligned} &\text{As } \theta \to 0 \qquad t^{lL}_{mM} \propto (\sin\theta/2)^{\Lambda} \\ &\text{As } \theta \to \pi \qquad t^{lL}_{mM} \propto (\cos\theta/2)^{\Lambda} \end{aligned} \tag{5.4.26}$$

where $\Lambda = \max\{0, \; |m - M| - \overline{m}'\}$ and $\overline{m}'$ is the largest value of $|m'|$ that occurs in the $t^{l'}_{m'}(A)$ of the polarized beam.

For the density matrix itself, the results given in subsection 5.4.1 hold at $\phi = 0$. For $\phi \neq 0$ one has

$$\rho_{c'd';cd}(\phi) = (-1)^{c-c'+d-d'} \rho_{-c'-d';-c-d}(-\phi). \tag{5.4.27}$$

In particular ρ satisfies (5.4.8) at $\phi = 0$ under the experimental conditions (s_1)–(s_3).

There are other results that hold for rather special circumstances. For example, if the quantization axis lies *in* the scattering plane and if the beam possesses *only* even-rank or *only* odd-rank polarization, then

$$\left(\mathrm{Re}\, t^{lL}_{mM}\right)_{\text{pol.bm.}} \quad \text{for} \left\{ \begin{array}{c} \text{even-rank} \\ \text{odd-rank} \end{array} \right\} \text{polarization}$$

$$= \left(\mathrm{Re}\, t^{lL}_{mM}\right)_{\text{unpol.}} \quad \text{for } l+L \left\{ \begin{array}{c} \text{odd} \\ \text{even} \end{array} \right\} \tag{5.4.28}$$

and

$$\left(\mathrm{Im}\, t^{lL}_{mM}\right)_{\text{pol.bm.}} \quad \text{for} \left\{ \begin{array}{c} \text{even-rank} \\ \text{odd-rank} \end{array} \right\} \text{polarization}$$

$$= \left(\mathrm{Im}\, t^{lL}_{mM}\right)_{\text{unpol.}} \quad \text{for } l+L \left\{ \begin{array}{c} \text{even} \\ \text{odd} \end{array} \right\}. \tag{5.4.29}$$

These are particularly powerful when the beam consists of spin-1/2 particles, since in this case only rank-1 polarization is possible. As an example, if the spin-polarization vector of the beam, $\mathcal{P}^A$, lies *in* the scattering plane then the spin-polarization vector $\mathcal{P}^C$ can have components both in $\left(\mathcal{P}^C_{\parallel}\right)$ and perpendicular to $(\mathcal{P}^C_{\perp})$ to the scattering plane. Equations (5.4.28), (5.4.29) together with (3.1.35) tell us that $\mathcal{P}^C_{\perp}$ is independent of the vector $\mathcal{P}^A$, i.e. it is the same as it would have been if the beam were unpolarized.

5.4.3 Polarized target, unpolarized beam

The transcription of the results of subsection 5.4.2 to the situation where the target is polarized and the beam is unpolarized is absolutely straightforward. Only one point requires mention.

If the experiment involves a stationary target in the laboratory and if the target quantization axis is specified by polar angles $\theta = \beta'$, $\phi = \gamma'$ in the Lab frame, then in place of (5.4.12) one must have (see eqn (3.3.2))

$$t^L_M(B) = e^{-i\gamma' M} d^L_{M0}(\pi - \beta')\hat{t}^L_0. \tag{5.4.30}$$

If, however, the experiment involves colliding beams and if β', γ' refer to the quantization axis for B in its helicity rest frame S_B (see Fig. 3.1) then (5.4.12) should be used to calculate $t^L_M(B)$.

5.4.4 Polarized beam and target

For either the differential cross-section or the final state multipole parameters, the general result when the beam and target are both polarized is, from (5.4.1), of the form (for photons recall the discussion after eqn

(5.3.4))

$$\left[t^{l'L'}_{m'M'}(C,D)\frac{d^2\sigma}{dtd\phi}\right]_{\substack{\text{pol.bm.}\\ \text{pol.targ.}}} - \left[t^{l'L'}_{m'M'}\frac{d^2\sigma}{dtd\phi}\right]_{\text{pol.bm.}}$$

$$- \left[t^{l'L'}_{m'M'}\frac{d^2\sigma}{dtd\phi}\right]_{\text{pol.targ.}} + \left[t^{l'L'}_{m'M'}\frac{d^2\sigma}{dtd\phi}\right]_{\text{unpol.}}$$

$$= \left(\frac{2}{3}\right)^{n_\gamma}\frac{1}{2\pi}\frac{d\sigma}{dt}\sum_{\substack{l,L\geq 1\\ m,M}}(2l+1)(2L+1)t^l_m(A)t^L_M(B)$$

$$\times (l,m;L,M|l',m';L',M')e^{-i(M-m)\phi}. \tag{5.4.31}$$

Here, obviously, the state of polarization of beam (or target) must, where labelled, be the same on both sides of the equation.

The generalized *initial state polarization correlation parameters,*

$$(l,m;L,M|0,0;0,0),$$

which are the analogues of A_{ij} in nucleon–nucleon scattering, can be studied from the ϕ-dependence of the differential cross-section. Other three- and four-spin tensors require measurements of the final state multipole parameters.

For arbitrarily polarized beam and target the final state parameters t^{lL}_{mM} do not possess any special symmetry properties. If, however, the following experimental condition holds,

(s_4) the quantization axes of beam and target are parallel

then the properties (a)–(f) listed in subsection 5.4.2(iv) continue to hold *in the situations* (s_1)–(s_3).

The behaviour near $\theta = 0, \pi$ is now as follows.

$$\text{For } \theta \to 0 \qquad t^{lL}_{mM} \propto (\sin\theta/2)^{\Lambda'}$$
$$\text{For } \theta \to \pi \qquad t^{lL}_{mM} \propto (\cos\theta/2)^{\Lambda'} \tag{5.4.32}$$

where $\Lambda' = \max\{0, \ |m-M| - \bar{\mu}\}$ and $\bar{\mu}$ is the largest value of $|m'-M'|$ that occurs in the $t^{l'}_{m'}(A)$ and $t^{L'}_{M'}(B)$ of the polarized beam and target.

5.5 The laboratory reaction parameters

For some colliding-beam experiments the measurements are carried out in the CM so that the multipole parameters that appear in (5.4.1) are the ones measured. For fixed targets in the laboratory what one actually measures are the multipole parameters in the Lab natural analysing frames (see subsection 3.3.2, especially Fig. 3.5). It is straightforward to translate these measurements into statements about the CM multipole parameters so that (5.4.1) again applies. However, for psychological reasons, experimentalists prefer to utilize the analogue of (5.4.1), which connects directly what goes into the experiment with what comes out in the Lab.

With all quantities measured respectively in the Lab analysing frames S_{LA}, S_{LB}, S_{LC}, S_{LD} one has

$$\left[t^{l'L'}_{m'M'}(C,D)\,\frac{d^2\sigma}{dtd\phi}\right]_{S_{LC},S_{LD}} = \left(\frac{2}{3}\right)^{n_\gamma}\frac{1}{2\pi}\frac{d\sigma}{dt}\sum_{\substack{lL\\mM}}(2l+1)(2L+1) \times \left[t^l_m(A)\right]_{S_{LA}}\left[t^L_M(B)\right]_{S_{LB}} \times (l,m;L,M|l',m';L',M')_{\text{Lab}}\,e^{i(M-m)\phi} \tag{5.5.1}$$

where, from (3.3.14), the Lab *reaction parameters* are

$$(l,m;L,M|l',m';L',M')_{\text{Lab}} = \sum_{m'',M''}(l,m;L,M|l',m'';L',M'') \times d^{l'}_{m''m'}(-\alpha_C)d^{L'}_{M''M'}(-\alpha_D) \tag{5.5.2}$$

with the angles α_C, α_D being given by (2.2.13).

Note that the Lab reaction parameters enjoy the same reality property (5.3.8) as do the CM ones.

Clearly the entire analysis of measurements in the CM can be taken over unchanged to discuss the extraction of the Lab reaction parameters from the Lab experimental data.

The symmetry properties that relate many of the CM parameters to each other will give rise, via (5.5.2), to similar, though more complicated-looking, relations amongst the Lab parameters.

Only the parity result looks simple:

$$(l,-m;L,-M|l',-m';L',-M')_{\text{Lab}} = (-1)^{m+M+m'+M'}(-1)^{l+L+l'+L'} \times (l,m;L,M|l',m';L',M')_{\text{Lab}} \tag{5.5.3}$$

from which one gets

$$(l,0;L,0|l',0;L',0)_{\text{Lab}} = 0 \qquad \text{if } l+L+l'+L' \text{ is odd.} \tag{5.5.4}$$

For the other symmetries there is no point in writing down the general results. In a specific reaction it is best to write them out explicitly for the CM parameters and then to substitute the inverse of (5.5.2) to get the relations amongst the Lab reaction parameters.

The properties of the CM final state multipole parameters $t^{lL}_{mM}(C, D)$ listed in subsections 5.4.1(ii) and 5.4.2(iv) hold also for the Lab multipole parameters provided that they are measured in the respective Lab analysing frames S_{LC} and S_{LD}.

5.6 Applications: Cartesian formalism for initial particles with spin 1/2

For particles of spin ≥ 1 the above formalism is the simplest and most compact. For spin-1/2 particles, however, one is accustomed to working with the Cartesian components of the spin-polarization vectors $\mathcal{P} = \langle \boldsymbol{\sigma} \rangle$. Moreover only the values $l = 0, 1$ occur in the reaction parameters so that a simpler notation is possible.

The transformation between multipole parameters and components of the spin-polarization vector for spin 1/2 is

$$\sqrt{2l+1}\, t^l_m = \sum_\mu U_{lm;\mu} \mathscr{P}^\mu \tag{5.6.1}$$

where lm $(= 00, 11, 10, 1-1)$ labels the rows, and μ $(= 0, X, Y, Z)$ labels the columns, with $\mathscr{P}^0 \equiv 1$. The matrix U is given by

$$U = \begin{pmatrix} 1 & 0 & 0 & 0 \\ 0 & -1/\sqrt{2} & -i/\sqrt{2} & 0 \\ 0 & 0 & 0 & 1 \\ 0 & 1/\sqrt{2} & -i/\sqrt{2} & 0 \end{pmatrix} \tag{5.6.2}$$

with $U^\dagger U = 1$.

5.6.1 *The reaction* spin 1/2 + spin 1/2 → spin 1/2 + spin 1/2

The Cartesian analogue of the CM relation (5.4.1) is then

$$\langle \sigma_{\alpha'}(C) \sigma_{\beta'}(D) \rangle \frac{d^2\sigma}{dt d\phi} = \frac{1}{2\pi} \frac{d\sigma}{dt} \sum_{\alpha,\beta} \langle \sigma_\alpha(A) \rangle \langle \sigma_\beta(B) \rangle \times (\alpha\beta|\alpha'\beta')_\phi, \tag{5.6.3}$$

where $\alpha, \beta, \alpha', \beta'$ take on the values $0, X, Y, Z$, corresponding to the usual three Pauli matrices $\boldsymbol{\sigma}$ supplemented by a fourth matrix $\sigma_0 \equiv I$, the unit 2×2 matrix.[1]

Equation (5.6.3) relates the final state spin expectation values to those of the initial state in the CM. We shall make much use of this result when studying electroweak and QCD reactions.

It must be remembered that the directions X, Y, Z refer to the CM frame but that the physical interpretation of each $\langle \boldsymbol{\sigma}(K) \rangle$ is that it is the

[1] According to convention we use upper-case X, Y, Z here.

mean spin vector for particle K in its helicity rest frame S_K reached from the CM.

The CM *Cartesian reaction parameters* are given by

$$(\alpha\beta|\alpha'\beta')_\phi = \sum_{\substack{\text{repeated}\\ \text{indices}}} \sqrt{2l+1}\sqrt{2L+1}U_{lm;\alpha}U_{LM;\beta}$$
$$\times (l,m;L,M|l',m';L',M')_\phi$$
$$\times U^*_{l'm';\alpha'}U^*_{L'M';\beta'}\sqrt{2l'+1}\sqrt{2L'+1}$$
$$= \frac{1}{4}\left(\frac{d\sigma}{dt}\right)^{-1} \text{Tr}\,(\sigma_\alpha\sigma_\beta H^\dagger\sigma_{\alpha'}\sigma_{\beta'}H) \qquad (5.6.4)$$

in complete analogy to (5.3.2).

The explicit ϕ-dependence of $(\alpha\beta|\alpha'\beta')_\phi$ can be found from (5.6.4) and (5.3.3). One gets

$$(\alpha\beta|\alpha'\beta')_\phi = \mathscr{R}^z_{\alpha\alpha''}(\phi)\mathscr{R}^z_{\beta\beta''}(-\phi)(\alpha''\beta''|\alpha'\beta') \qquad (5.6.5)$$

with

$$\mathscr{R}^z(\phi) = \begin{pmatrix} 1 & 0 \quad 0 \quad 0 \\ 0 & \\ 0 & R[r_z(\phi)] \\ 0 & \end{pmatrix} = \begin{pmatrix} 1 & 0 & 0 & 0 \\ 0 & \cos\phi & -\sin\phi & 0 \\ 0 & \sin\phi & \cos\phi & 0 \\ 0 & 0 & 0 & 1 \end{pmatrix}. \qquad (5.6.6)$$

In (5.6.5), as usual, absence of a ϕ-label implies $\phi = 0$.

In a similar fashion the Lab spin expectation values in the final state are related to those of the initial state, provided each is measured in the natural analysing frame S_{LK}, by

$$\left[\langle\sigma_{\alpha'}(C)\sigma_{\beta'}(D)\rangle \frac{d^2\sigma}{dt d\phi}\right]_{S_{LC},S_{LD}}$$
$$= \frac{1}{2\pi}\frac{d\sigma}{dt}\sum_{\alpha,\beta}\left[\langle\sigma_\alpha(A)\rangle\langle\sigma_\beta(B)\rangle\right]_{S_{LA},S_{LB}}(\alpha\beta|\alpha'\beta')^{\phi}_{\text{Lab}} \qquad (5.6.7)$$

where now the directions X, Y, Z for, say, particle K refer to the spin projections along the X-, Y-, Z-axes of the Lab frame S_{LK}.

From (5.6.4) and (5.5.2) we find

$$(\alpha\beta|\alpha'\beta')_{\text{Lab}} = \sum_{\alpha'',\beta''}(\alpha\beta|\alpha''\beta'')$$
$$\times \mathscr{R}^y_{\alpha''\alpha'}(-\alpha_C)\mathscr{R}^y_{\beta''\beta'}(-\alpha_D) \qquad (5.6.8)$$

where, for any angle ω,

$$\mathscr{R}^y(\omega) = \begin{pmatrix} 1 & 0 \quad 0 \quad 0 \\ 0 & \\ 0 & R[r_y(\omega)] \\ 0 & \end{pmatrix} = \begin{pmatrix} 1 & 0 & 0 & 0 \\ 0 & \cos\omega & 0 & \sin\omega \\ 0 & 0 & 1 & 0 \\ 0 & -\sin\omega & 0 & \cos\omega \end{pmatrix}. \tag{5.6.9}$$

It follows that the azimuthal ϕ-dependence of the Lab reaction parameters is also given by (5.6.5).

It should be remembered that for $NN \to NN$ one has $\alpha_C = \theta_{\mathrm{L}}$ and $\alpha_D = \theta_{\mathrm{R}}$, where θ_{L} is the Lab scattering angle and θ_{R} is the Lab recoil angle; however for a reaction like $\Lambda p \to \Lambda p$ one will have $\alpha_D = \theta_{\mathrm{R}}$ but $\alpha_C \neq \theta_{\mathrm{L}}$ (see subsection 2.2.4).

Some of the results of the most exciting experiments on spin dependence in NN scattering carried out at Argonne have been reported using a slightly different choice of Lab reference frame for each particle. The Argonne Lab frames $S_{\mathrm{L}K}^{\mathrm{ARG}}$ are

$$\begin{aligned} S_{\mathrm{L}A}^{\mathrm{ARG}} &= S_{\mathrm{L}A} \\ S_{\mathrm{L}C}^{\mathrm{ARG}} &= S_{\mathrm{L}C} \\ S_{\mathrm{L}B}^{\mathrm{ARG}} &= S_{\mathrm{L}} = r_z(-\pi) r_y(-\pi) S_{\mathrm{L}B} \\ S_{\mathrm{L}D}^{\mathrm{ARG}} &= r_z(-\pi) S_{\mathrm{L}D} \end{aligned} \tag{5.6.10}$$

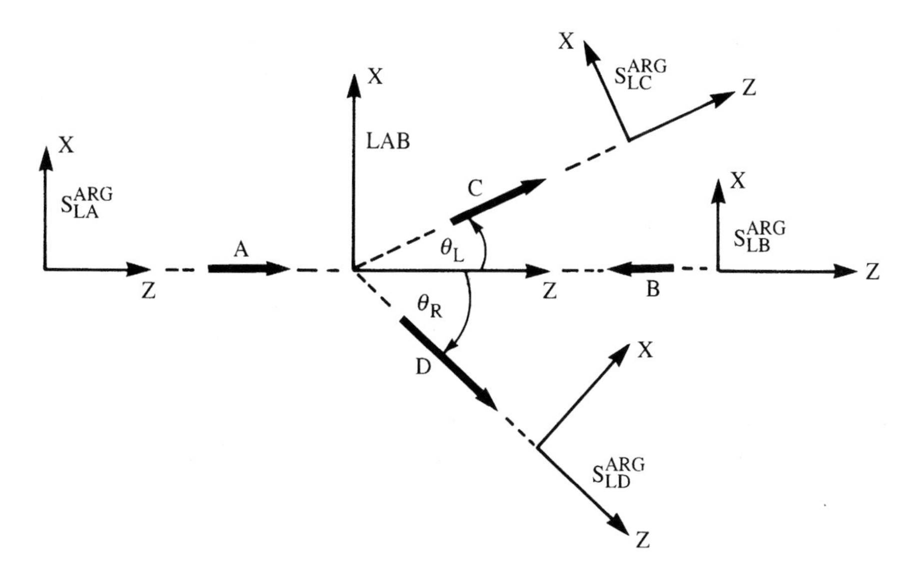

Fig. 5.2 The Argonne Lab frames for $A + B \to C + D$ as used in the reporting of several experiments.

and, as shown in Fig. 5.2, all have their Y-axes in the same direction. The Argonne frames *seem* to be a simple, sensible choice. Their drawback is that they ignore the fact that both B and D have azimuthal angle $\phi = \pi$.

By convention the directions in *each* Argonne Lab frame are not referred to as X, Y, Z but by the symbols

$$\begin{aligned} L &= \text{longitudinal} = \text{along } OZ \\ N &= \text{normal} = \text{along } OY \\ S &= \text{sideways} = \text{along } OX \end{aligned}$$

and the Argonne final state measurements are related to their initial state ones by an equation almost identical to (5.6.7) but involving the Argonne reaction parameters

$$(\alpha\beta|\alpha'\beta')^{\text{ARG}}_{\text{Lab}} = \epsilon_\beta(\alpha\beta|\alpha'\beta')_{\text{Lab}}\nu_{\beta'} \tag{5.6.11}$$

where

$$\begin{aligned} \epsilon_0 = \epsilon_X = \nu_0 = \nu_Z &= +1 \\ \epsilon_Y = \epsilon_Z = \nu_X = \nu_Y &= -1 \end{aligned}$$

The Argonne reaction parameters are connected to the CM parameters by an obvious change in (5.6.8). Clearly they have essentially the same ϕ-dependence as the Lab reaction parameters.

The detailed symmetry properties of the CM Cartesian reaction parameters and of the Argonne Lab reaction parameters are given in Appendix 8, both for $NN \to NN$ and for the more general case of reactions like $\Lambda N \to \Lambda N$.

In Appendix 9 we list the 'shorthand' notation and the nomenclature commonly used for the Argonne Lab parameters, both for $NN \to NN$ and for $\Lambda N \to \Lambda N$.

Let us look at an example of the use of (5.6.3) or (5.6.7) for a parity-conserving reaction $A + B \to A + B$ where both beam and target may be polarized and one measures the differential cross-section. Let

$$\begin{aligned} \boldsymbol{\mathcal{P}}^A &= \left(\mathscr{P}^A_x, \mathscr{P}^A_y, \mathscr{P}^A_z\right) \\ \boldsymbol{\mathcal{P}}^B &= \left(\mathscr{P}^B_x, \mathscr{P}^B_y, \mathscr{P}^B_z\right) \end{aligned}$$

be the components of the spin-polarization vectors relative to the CM or Lab frames, in which A moves along OZ.

Then, according to subsection 3.3.1 (see also eqn (5.1.7)) the spin-polarization vector that must be used for B in (5.6.3) is

$$\tilde{\boldsymbol{\mathcal{P}}}^B = \left(\mathscr{P}^B_x, -\mathscr{P}^B_y, -\mathscr{P}^B_z\right).$$

Thus, in either the CM or the Lab, using (5.6.5) in (5.6.3) or (5.6.7), one has

$$\begin{aligned}
\frac{d^2\sigma}{dtd\phi} = \frac{1}{2\pi}\frac{d\sigma}{dt}\Big\{ & 1 + A^{(A)}\left(\mathscr{P}_y^A \cos\varphi - \mathscr{P}_x^A \sin\phi\right) \\
& - A^{(B)}\left(\mathscr{P}_y^B \cos\varphi + \mathscr{P}_x^B \sin\phi\right) \\
& + A_{xx}\Big[\cos^2\varphi\, \mathscr{P}_x^A \mathscr{P}_x^B - \sin^2\phi\, \mathscr{P}_y^A \mathscr{P}_y^B \\
& \qquad + \cos\phi \sin\phi \left(-\mathscr{P}_x^A \mathscr{P}_y^B + \mathscr{P}_y^A \mathscr{P}_x^B\right)\Big] \\
& + A_{yy}\Big[\sin^2\varphi\, \mathscr{P}_x^A \mathscr{P}_x^B - \cos^2\phi\, \mathscr{P}_y^A \mathscr{P}_y^B \\
& \qquad + \cos\phi \sin\phi \left(\mathscr{P}_x^A \mathscr{P}_y^B - \mathscr{P}_y^A \mathscr{P}_x^B\right)\Big] \\
& - A_{zz}\mathscr{P}_z^A \mathscr{P}_z^B + A_{zx}\mathscr{P}_z^A\left(\cos\varphi\, \mathscr{P}_x^B - \sin\varphi\, \mathscr{P}_y^B\right) \\
& - A_{xz}\mathscr{P}_z^B\left(\cos\phi\, \mathscr{P}_x^A + \sin\phi\, \mathscr{P}_y^A\right)\Big\}
\end{aligned} \tag{5.6.12}$$

where we have used the following abbreviations for the various analysing powers:

$$\begin{aligned}
A^{(A)} &\equiv (Y0|00)_{\mathrm{CM}} = (Y0|00)_{\mathrm{Lab}} \\
A^{(B)} &\equiv (0Y|00)_{\mathrm{CM}} = (0Y|00)_{\mathrm{Lab}} \\
A_{ij} &\equiv (ij|00)_{\mathrm{CM}} = (ij|00)_{\mathrm{Lab}}.
\end{aligned} \tag{5.6.13}$$

The equality of the Lab and CM generalized analysing powers follows from (5.6.8).

Note that for identical fermions, e.g. for $pp \to pp$, $A_{zx} = -A_{xz}$ and $A^{(A)} = -A^{(B)}$. Conventionally one writes $A^{(A)} = -A^{(B)} = A_N$.

Equation (5.6.12) indicates how the analysing powers could be measured from a study of the azimuthal dependence or by comparing 'left' ($\phi = 0$) and 'right' ($\phi = \pi$) scattering, with various settings of the spin-polarization vectors.

Alternatively, if the analysing powers are known, the ϕ-dependence can be used to get some information about the spin-polarization vector of beam and target, an important issue in 'polarimetry'.

If the spin-polarization vector refers to the Argonne choice of reference frames then in the Argonne notation (5.6.12) will hold with the following substitutions:

$$\begin{aligned}
& A_y^{(A)} \to A^{(A)} \qquad A_y^{(B)} \to -A^{(B)} \\
& A_{xx} \to A_{SS} \quad A_{yy} \to -A_{NN} \qquad A_{zz} \to -A_{LL} \\
& A_{xz} \to -A_{SL} \quad A_{zx} \to A_{LS} \qquad A_{yx} \to A_{NS}
\end{aligned} \tag{5.6.14}$$

5.6.2 *The reactions* spin $0+$ spin $1/2 \rightarrow$ spin $0+$ spin $1/2$ *and* spin $1/2+$ spin $1/2 \rightarrow$ spin $0+$ spin 0

We have in mind here processes like $\pi N \rightarrow \pi N$ and $\overline{N}N \rightarrow \pi\pi$.

(*i*) $0+1/2 \rightarrow 0+1/2$. The formulae (5.6.3), (5.6.4) and (5.6.7) apply provided the following simplifications are made.

(a) Suppress completely the labels α and α'.
(b) Replace the factor $1/4$ by $1/2$ in the relation (5.6.4) for $(\beta|\beta')_\phi$ in terms of the trace over H.

(*ii*) $1/2+1/2 \rightarrow 0+0$. Again (5.6.3), (5.6.4) and (5.6.7) apply provided that one suppresses the labels α' and β' everywhere.

5.6.3 *The reactions* spin $1/2+$ spin $1/2 \rightarrow$ arbitrary-spin particles

Undoubtedly many of the most interesting experiments in the next decade will consist of the production of high-spin particles from collisions of spin-1/2 particles. We therefore recast our general results (5.3.1) and (5.5.1) into a hybrid form that takes advantage of the Cartesian formalism for the initial particles but retains the multipole description for the final particles. We get in the CM

$$t^{lL}_{mM}(C,D)\frac{d^2\sigma}{dtd\phi} = \frac{1}{2\pi}\frac{d\sigma}{dt}\sum_{\alpha,\beta}\langle\sigma_\alpha(A)\rangle\langle\sigma_\beta(B)\rangle \times(\alpha\beta|l,m;L,M)_\phi, \tag{5.6.15}$$

where the *hybrid CM reaction parameters* are given by

$$(\alpha\beta|l,m;L,M)_\phi = \frac{1}{4}\left(\frac{d\sigma}{dt}\right)^{-1} \operatorname{Tr}\left(\sigma_\alpha\sigma_\beta H^\dagger T^l_m T^L_M H\right). \tag{5.6.16}$$

For $\phi=0$ one finds that the parameters are real or imaginary according as $l+L+\delta_{\alpha 0}+\delta_{\beta 0}$ is even or odd.

The Lab version of (5.6.15), using the natural Lab analysing frames, is

$$\left[t^{lL}_{mM}(C,D)\frac{d^2\sigma}{dtd\phi}\right]_{S_{LC}S_{LD}} = \frac{1}{2\pi}\frac{d\sigma}{dt}\sum_{\alpha,\beta}\langle\sigma_\alpha(A)\rangle_{S_{LA}}\langle\sigma_\beta(B)\rangle_{S_{LB}} \times(\alpha\beta|l,m;L,M)^\phi_{\text{Lab}} \tag{5.6.17}$$

where the directions α,β for A and B refer to the frames S_{LA}, S_{LB}, with

$$(\alpha,\beta|l,m;L,M)_{\text{Lab}} = \sum_{m',M'}(\alpha\beta|l,m';L,M') \times d^l_{m'm}(-\alpha_C)d^l_{M'M}(-\alpha_D). \tag{5.6.18}$$

If the Argonne Lab frames are used, the analogue of (5.6.17) has in it

$$(\alpha\beta|l,m;L,M)_{\text{Lab}}^{\text{ARG}} = \epsilon_\beta(\alpha\beta|l,m;L,M)_{\text{Lab}}(-1)^M \tag{5.6.19}$$

where, as earlier,

$$\epsilon_0 = \epsilon_X = 1 \qquad \epsilon_Y = \epsilon_Z = -1.$$

The ϕ-dependence of the hybrid CM or Lab reaction parameters is still given by (5.6.5) and the ϕ-dependence of the Argonne parameters then follows from (5.6.19).

Consider now the most general possible experiment for a $2 \to 2$ reaction with polarized spin-1/2 beam and target. Let $\boldsymbol{\mathcal{P}}^{(A)} = \left(\mathscr{P}_x^A, \mathscr{P}_y^A, \mathscr{P}_z^A\right)$ and $\tilde{\boldsymbol{\mathcal{P}}}^{(B)} = \left(\tilde{\mathscr{P}}_x^B, \tilde{\mathscr{P}}_y^B, \tilde{\mathscr{P}}_z^B\right)$ be the components of the spin-polarization vectors of beam and target specified in the correct helicity frames for A and B (see subsection 3.3.1) using either the CM or natural Lab analysing frames. Recall (see eqn (5.1.7)) that if one specifies the components of the initial spin-polarization vector in the CM or Lab frames where A moves along OZ then, for B, $\tilde{\boldsymbol{\mathcal{P}}}^{(B)} = (\mathscr{P}_x^B, -\mathscr{P}_y^B, -\mathscr{P}_z^B)$. Then, *with f standing for the final state labels $l,m;L,M$*, one has for $AB \to CD$

$$\begin{aligned}
t_{mM}^{lL}&(C,D)\frac{d^2\sigma}{dt\,d\phi} \\
&= \frac{1}{2\pi}\left[t_{mM}^{lL}(C,D)\frac{d\sigma}{dt}\right]_{\text{unpol.}} \\
&\quad + \frac{1}{2\pi}\frac{d\sigma}{dt}\Big\{\mathscr{P}_x^A\left[\cos\phi\,(X0|f) - \sin\phi\,(Y0|f)\right] \\
&\quad + \tilde{\mathscr{P}}_x^B\left[\cos\phi\,(0X|f) - \sin\phi\,(0Y|f)\right] \\
&\quad + \mathscr{P}_y^A\left[\cos\phi\,(Y0|f) + \sin\phi\,(X0|f)\right] \\
&\quad + \tilde{\mathscr{P}}_y^B\left[\cos\phi\,(0Y|f) + \sin\phi\,(0X|f)\right] \\
&\quad + \mathscr{P}_z^A(Z0|f) + \tilde{\mathscr{P}}_z^B(0Z|f) + \mathscr{P}_z^A\tilde{\mathscr{P}}_z^B(ZZ|f) \\
&\quad + \mathscr{P}_x^A\tilde{\mathscr{P}}_x^B\Big[\cos^2\phi\,(XX|f) + \sin^2\phi\,(YY|f) \\
&\qquad\qquad - \cos\phi\sin\phi\,\langle(XY|f) + (YX|f)\rangle\Big] \\
&\quad + \mathscr{P}_y^A\tilde{\mathscr{P}}_y^B\Big[\sin^2\phi\,(XX|f) + \cos^2\phi\,(YY|f) \\
&\qquad\qquad + \cos\phi\sin\phi\,\langle(XY|f) + (YX|f)\rangle\Big] \\
&\quad + \mathscr{P}_x^A\tilde{\mathscr{P}}_y^B\Big[\cos^2\phi\,(XY|f) - \sin^2\phi\,(YX|f) \\
&\qquad\qquad + \cos\phi\sin\phi\,\langle(XX|f) - (YY|f)\rangle\Big] +
\end{aligned}$$

$$+\mathscr{P}_y^A\tilde{\mathscr{P}}_x^B\Big[\cos^2\phi\ (YX|f)-\sin^2\phi\ (XY|f)$$
$$+\cos\phi\sin\phi\ \langle(XX|f)-(YY|f)\rangle\Big]$$
$$+\mathscr{P}_x^A\tilde{\mathscr{P}}_z^B\left[\cos\phi\ (XZ|f)-\sin\phi\ (YZ|f)\right]$$
$$+\mathscr{P}_z^A\tilde{\mathscr{P}}_x^B\left[\cos\phi\ (ZX|f)-\sin\phi\ (ZY|f)\right]$$
$$+\mathscr{P}_y^A\tilde{\mathscr{P}}_z^B\left[\cos\phi\ (YZ|f)+\sin\phi\ (XZ|f)\right]$$
$$+\mathscr{P}_z^A\tilde{\mathscr{P}}_y^B\left[\cos\phi\ (ZY|f)+\sin\phi\ (ZX|f)\right]\Big\} \tag{5.6.20}$$

where the reaction parameters and the final state multipole parameters should carry an appropriate label to indicate which set of reference frames is implied.

The above is completely general in the sense that no discrete symmetries have been assumed.

We mention some of the simpler properties of the hybrid parameters.

(i) *Reality*. Using $T_m^{l\dagger}=T_{-m}^l(-1)^m$ and $\sigma_\alpha^\dagger=\sigma_\alpha$ in (5.6.16), one finds

$$(\alpha\beta|l,m;L,M)^*=(-1)^{m+M}(\alpha\beta|l,-m;L,-M). \tag{5.6.21}$$

(ii) *Parity*. For *any* one set of the above reference frames one has

$$(\alpha\beta|l,m;L,M)=\xi_\alpha^{\mathscr{P}}\xi_\beta^{\mathscr{P}}(-1)^{l+L+m+M}(\alpha\beta|l,-m;L,-M) \tag{5.6.22}$$

where $\xi_0^{\mathscr{P}}=\xi_Y^{\mathscr{P}}=+1$ and $\xi_X^{\mathscr{P}}=\xi_Z^{\mathscr{P}}=-1$. Thus

$$(\alpha\beta|l,0;L,0)=0 \quad \text{if } \xi_\alpha^{\mathscr{P}}\xi_\beta^{\mathscr{P}}(-1)^{l+L}=-1. \tag{5.6.23}$$

Combining (5.6.22) and (5.6.21) we have

$$(\alpha\beta|l,m;L,M) \text{ is } \left\{\begin{matrix}\text{real}\\ \text{imaginary}\end{matrix}\right\} \text{ as } \xi_\alpha^{\mathscr{P}}\xi_\beta^{\mathscr{P}}(-1)^{l+L}=\pm1. \tag{5.6.24}$$

(iii) *Identical particles*. If $A=B$, then for the CM reaction parameters

$$(\alpha\beta|l,m;L,M)^\theta=(-1)^{m+M}(\beta\alpha|l,m;L,M)^{\pi-\theta}. \tag{5.6.25}$$

As regards the properties of the final state multipole parameters they are of course no different from those discussed in subsections 5.4.1(ii) and 5.4.2(iv), provided that they are measured in the correct reference frames of the set being used.

Equations (5.6.20), (5.6.23), and (5.6.24) give a complete description of the states of polarization of the final particles that are possible for various choices of the initial state polarizations with and without the imposition of parity invariance.

5.6.4 Connection between photon and spin-1/2 *induced reactions*

We saw in subsection 3.1.12 that because of the absence of states $|m = 0\rangle$ for photons, the helicity density matrix for a photon is essentially a 2×2 matrix and can therefore be expressed in terms of the Pauli matrices.

Upon comparing (3.1.85) and (3.1.88) with the result (3.1.23) for spin-1/2 particles, it is clear that we can map any formulae for the observables in the reaction

$$\text{spin } 1/2 + B \to X,$$

where the spin-1/2 particles have a spin-polarization vector $\boldsymbol{\mathcal{P}}$, into the corresponding formula for the observable for the photon-induced reaction

$$\gamma + B \to X.$$

This is done by making the following replacements:

(a)
$$\mathscr{P}_x \to -\cos 2\gamma\ \mathscr{P}_{\text{lin}} \qquad \mathscr{P}_y \to -\sin 2\gamma\ \mathscr{P}_{\text{lin}}$$

corresponding to a photon linearly polarized in the XY-plane at angle γ to OX, with degree of linear polarization $\mathscr{P}_{\text{lin}}$ (see subsection 3.1.12(ii));

(b)
$$\mathscr{P}_z \to \mathscr{P}_{\text{circ}}$$

corresponding to a circularly polarized photon with circular polarization $\mathscr{P}_{\text{circ}}$ (see subsection 3.1.12(ii));

(c)
$$H_{X;\lambda\lambda_B} \to H_{X;\lambda_\gamma=2\lambda,\lambda_B}$$

for the helicity amplitudes, where λ refers to the spin-1/2 particle.

5.7 Non-linear relations amongst the observables

Consider a reaction $A + B \to C + D$ for which, after application of all the symmetries of the situation, there are found to exist n independent helicity amplitudes. Let us label these H_j with $j = 1, \ldots, n$. Since all the observables are quadratic in the H_j there will clearly exist n^2 linearly independent observables O_α of the form

$$O_\alpha = \sum_{j,k} a^\alpha_{jk} H_j H^*_k \tag{5.7.1}$$

with known coefficients a^α_{jk}. Knowing the value of the n^2 observables O_α is tantamount to knowing the value of the n^2 quantities

$$O_{jk} \equiv H_j H^*_k. \tag{5.7.2}$$

However, the number of experiments that can be carried out, i.e. the number of reaction parameters that exist, is

$$N = (2s_A + 1)(2s_B + 1)(2s_C + 1)(2s_D + 1),$$

a number that is generally much larger than n^2. For example, in elastic nucleon–nucleon scattering $n^2 = 25$ whereas $N = 256$. The symmetry properties of the reaction parameters given in subsection 5.3.1 yield linear relations amongst them just such as to reduce their number to n^2 independent parameters. The case of nucleon–nucleon scattering is displayed in some detail in Appendix 8.

To start with, though, there are only $2n$ independent real functions, the real and imaginary parts of the H_j. Moreover because the observables are quadratic functions of the H_j, one overall phase is irrelevant and can never be determined experimentally. Thus in fact all experiments must be describable in terms of $2n-1$ real functions. This implies that there must exist $n^2 - (2n-1) = (n-1)^2$ relations amongst the n^2 observables O_α. As will be seen, they are *non-linear* relations. The method for finding them is due to Klepikov, Kogan and Shamanin (1967) and Bourrely and Soffer (1975).

Consider the matrix O whose elements are the O_{jk} of (5.7.2). It is an $n \times n$ hermitian, positive matrix of rank 1. It is clear from (5.7.2) that

$$O_{jk}O_{lm} = O_{jm}O_{lk} \tag{5.7.3}$$

from which it follows that

$$O^2 = O \ \text{Tr} \ O. \tag{5.7.4}$$

Conversely one can show that if a given square matrix has elements O_{jk} such that (with no summation over repeated indices)

$$(O_{ik})^2 = O_{ii}O_{kk} \tag{5.7.5}$$

for all $i \neq k$ and

$$O_{ij}O_{jk} = O_{ik}O_{jj} \tag{5.7.6}$$

for all $i \neq k$ and any *one* value of j ($\neq i$ or k), then (5.7.3) and (5.7.4) follow. Equations (5.7.5) and (5.7.6) are the necessary and sufficient conditions for O_{jk} to be of the form (5.7.2). They constitute the desired non-linear relations amongst the observables.

The best-known example occurs in $\pi N \to \pi N$ where the reaction parameters

$$P \equiv (00|0N)^{\text{ARG}}_{\text{Lab}}$$

$$A \equiv (0L|0S)^{\text{ARG}}_{\text{Lab}}$$

$$R \equiv (0S|0S)^{\text{ARG}}_{\text{Lab}}$$

(the nomenclature P, A, R is historical) satisfy

$$P^2 + R^2 + A^2 = 1. \tag{5.7.7}$$

For nucleon–nucleon scattering the 16 non-linear relations can be found in Bourrely and Soffer (1975).

5.8 Multiparticle and inclusive reactions

We consider the simplest kind of multiparticle production process

$$A + B \to C + D_1 + D_2 + D_3 + \cdots$$

where all the variables specifying the multiparticle

$$X = D_1 + D_2 + D_3 + \cdots$$

are integrated over (except its mass M_X), i.e. we consider the *single-particle inclusive reaction*

$$A + B \to C + X$$

where A, B, C can have arbitrary spins. It does not matter, in what follows, whether X contains a fixed number of particles or whether we sum over different numbers of particles.

5.8.1 CM reaction parameters and final state density matrix

For each fixed number of particles, X can be considered as a composite 'particle', with many internal degrees of freedom and with a definite momentum $P = p_A + p_B - p_C$. It has a variable spin S_X and helicity Λ. In summing over all possible configurations of the particles that make up X we also sum over all the values of Λ incoherently.

It is then clear that, in so far as helicity dependence is concerned, the unnormalized final state density matrix for C, $\rho'(C)$, is given, in analogy with (5.2.1), by

$$\rho'_{cc'}(C; s, t, M_X^2, \phi) = \sum_{\substack{\text{internal}\\ \text{variables}}} \sum_{\Lambda} \sum_{\substack{ab\\ a'b'}} H_{c\Lambda;ab}(s, t, M_X^2, \phi) \times \rho_{\mathrm{iab;a'b'}}(A, B) H^*_{c'\Lambda;a'b'}(s, t, M_X^2, \phi) \tag{5.8.1}$$

where the H are generalized helicity amplitudes the knowledge of whose detailed properties is not necessary for our discussion.

We normalize the H in such a way that

$$\mathrm{Tr}\, \rho'(C) = 2\pi s \frac{d^3\sigma}{dtd\phi dM_X^2}. \tag{5.8.2}$$

From now on $\rho(C)$ will mean the properly normalized density matrix. It is clear that those symmetry properties of $\rho(C)$ in the $2 \to 2$ process

$A+B \to C+D$ which arise from parity conservation (for example (5.4.27)) and in which the spin s_X and the intrinsic parity η_X do not appear explicitly will continue to hold for $\rho(C)$ in $A+B \to C+X$.

In complete analogy with (5.3.1) for $2 \to 2$ scattering, we now have for the final state multipole parameter of C, in the CM,

$$\frac{d^3\sigma}{dtd\phi dM_X^2} t^{l'}_{m'}(C) = \frac{1}{2\pi}\frac{d^2\sigma}{dtdM_X^2} \sum_{\substack{lm\\LM}} (2l+1)(2L+1) t^l_m(A) t^L_M(B) \times (l,m;L,M|l',m')^{\text{inc}}_\phi \tag{5.8.3}$$

where

$$(l,m;L,M|l',m')^{\text{inc}}_\phi \equiv \sum_{\substack{\text{internal}\\ \text{variables}}} (l,m;L,M|l',m';L'=0,M'=0)^{AB\to CX}_\phi \tag{5.8.4}$$

and now depends on ϕ, t and M_X^2 as well, and where L', M' refer to 'particle' X.

Because the sum is incoherent, $(l,m;L,M|l',m')^{\text{inc}}_\phi$ has the same properties (5.3.3), (5.3.5), (5.3.6), (5.3.7), (5.3.8) and (5.3.11) as it would have had if X were a single spinless particle. It does not enjoy those properties like (5.3.16) that depend upon the intrinsic parity of 'X'.

As a consequence the properties of $t^l_m(C)$ exactly mimic those given in subsections 5.4.1(ii) and 5.4.2(iv) for $A+B \to C+D$, if s_D is put equal to zero.

In particular, for an unpolarized initial state and a parity-conserving reaction, the polarization vector of C must be perpendicular to the plane defined by $\mathbf{p}_A$ and $\mathbf{p}_C$.

If A and B have spins $1/2$ then (5.8.3) can be re-cast in an obvious way (Section 5.6) into a pure Cartesian or a hybrid Cartesian–spherical form. (See (5.6.3), (5.6.12) in which s_D would be put equal to zero.)

The CM reaction parameters can, in the present case, be related to discontinuities across the cut in M_X^2 of the forward $3 \to 3$ amplitudes for the process $A+B+\overline{C} \to A+B+\overline{C}$ (see Fig. 5.3). The original work, relating only to the unpolarized case, is due to Mueller (1970). For the generalization to spin-dependent terms see the work of Goldstein and Owens (1976). In the notation of the latter, our reaction parameter is

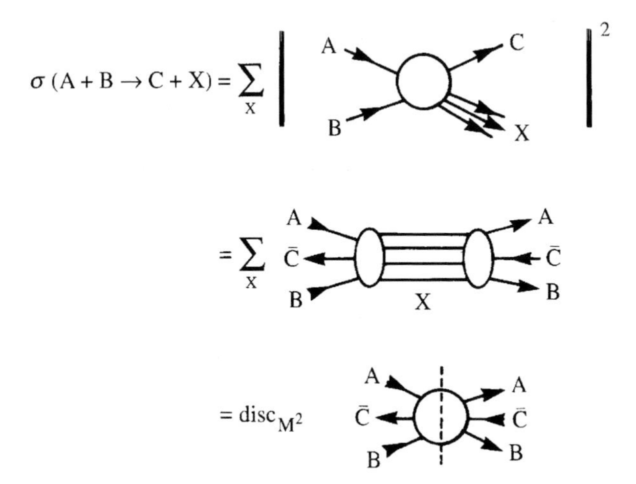

Fig. 5.3. Schematic form of the Mueller formula for inclusive cross-sections: see the text for a discussion of the quantities.

given by

$$(l,m;L,M|l',m')_{\phi}^{\text{inc}} = \sum_{\substack{ab\bar{c}\\a'b'\bar{c}'}} (T_{m'}^{l'\dagger})_{\bar{c}'\bar{c}}(T_{m}^{l})_{a'a}(T_{M}^{L})_{b'b} \times \frac{\text{disc}_{M_X^2}\, g_{a'b'\bar{c}';ab\bar{c}}}{\sum_{ab\bar{c}} \text{disc}_{M_X^2}\, g_{ab\bar{c};ab\bar{c}}} \tag{5.8.5}$$

where $g_{a'b'\bar{c}';ab\bar{c}}$ is the forward $3 \to 3$ amplitude and 'disc' refers to the discontinuities across the cut in M_X^2.

Near the forward and backward directions for C, i.e. $\theta = 0, \pi$, the $3 \to 3$ amplitude has the behaviour

$$g_{a'b'\bar{c}';ab\bar{c}} \propto (\sin\theta/2)^{\zeta_1}\,(\cos\theta/2)^{\zeta_2} \tag{5.8.6}$$

where $\zeta_1 = |a-b-\bar{c}-a'+b'+\bar{c}'|$ and $\zeta_2 = |a-b+\bar{c}-a'+b'-\bar{c}'|$, which implies for the reaction parameters the behaviour

$$(l,m;L,M|l',m')^{\text{inc}} \propto (\sin\theta/2)^{|m-M+m'|}\,(\cos\theta/2)^{|m-M-m'|}. \tag{5.8.7}$$

We end this section with a brief comment on *two-particle semi-inclusive reactions* of the type

$$A + B \to C + D + E_1 + E_2 + \cdots$$

where all variables specifying the multiparticle state $X = E_1 + E_2 + \cdots$ are integrated over, except its mass M_X.

The properties of the joint density matrix for C and D will then be analogous to those of $C + D$ in the $2 \to 3$ process

$$A + B \to C + D + X$$

where X is considered to be a particle of spin zero, but of indefinite parity.

The outcome of an experiment will be controlled by the reaction parameters

$$(l, m; L, M|l', m'; L', M')^{\text{inc}} = \sum_{\substack{\text{internal}\\ \text{variables}\\ \text{of } X}} (l, m; L, M|l', m'; L', M'; 0, 0)^{AB \to CDX}. \tag{5.8.8}$$

The most important new element is that the polarization vectors of C and D need not be perpendicular to the reaction planes ABC or ABD respectively, even for unpolarized initial beam and target.

6
The production of polarized hadrons

Crucial to all the preceding chapters is the assumption that we are able to produce beams and targets of polarized particles and that we are able to analyse the state of polarization of these particles.

In the production of targets and beams we are dealing with stable particles (or at least particles stable on the time scale involved) and the physics involved is basically a mixture of classical and quantum mechanics.

There has been extraordinary progress in the design and construction of polarized proton sources at Argonne and Brookhaven and in the development of highly polarized, radiation-resistant targets of various materials by workers at CERN, Fermilab, HERA, Basel, Virginia, SLAC and Ann Arbor.

Great advances have been made in the resolution of problems involved in the acceleration of polarized protons by groups at Bloomington and at Brookhaven. The electron beams at LEP and at HERA have been successfully polarized and a superb polarized electron source is in use at SLAC.

Also quite remarkable has been the building of secondary and tertiary beams of polarized hyperons at Fermilab. Who would have believed it possible that one can measure the magnetic moment of the Ω^-?!

Firstly we shall provide a brief discussion of the physical principles of polarized proton sources and targets and of the problems involved in accelerating beams of polarized protons without loss of polarization.

We also discuss a relatively new development, the attempt to polarize protons and antiprotons via the Stern–Gerlach effect.

Finally we consider the construction and functioning of the secondary and tertiary hyperon beams at Fermilab.

In Chapter 7 we shall study electron sources and the beautiful phenomenon whereby electron beams in a circular accelerator acquire a natural polarization (which in a perfect machine is $\approx 92\%$!) as a consequence of synchroton radiation. Polarimetry, the measurement of polarization, is taken up in Chapter 8.

The 'real-life' physics of sources, accelerators and targets is, of course, highly technical, even, to some extent, an 'art', so of necessity our treatment will emphasize only the main physical principles. For further information and access to the literature consult the series of proceedings of the International Symposium on High Energy Spin Physics (ISHESP, 1996).

6.1 Polarized proton sources

All polarized proton sources have as their aim the production of a beam of polarized ions ready for acceleration. There are many types and the produced ions may be positively or negatively charged. We shall discuss only the production of a beam of polarized protons in an atomic-beam type of source.

To understand the physics one must bear in mind the following.

(i) The S-wave ground state of the hydrogen atom is split by the hyperfine interaction into eigenstates $|s\,m\rangle$ of total spin: the spin singlet

$$|0;0\rangle = \frac{1}{\sqrt{2}}\{|\uparrow_e\downarrow_p\rangle - |\downarrow_e\uparrow_p\rangle\} \tag{6.1.1}$$

and the spin triplet

$$\begin{aligned} |1;1\rangle = |\uparrow_e\uparrow_p\rangle \qquad & |1;-1\rangle = |\downarrow_e\downarrow_p\rangle \\ |1;0\rangle = \frac{1}{\sqrt{2}}&\{|\uparrow_e\downarrow_p\rangle + |\downarrow_e\uparrow_p\rangle\} \end{aligned} \tag{6.1.2}$$

where the arrows indicate the spin projections of electron and proton.

(ii) The triplet-state energy is

$$\Delta \approx 6 \times 10^{-6}\ \text{eV} \tag{6.1.3}$$

above the singlet.

(iii) By comparison, the spin–orbit splitting is orders of magnitude greater ($\approx$ eV).

(iv) The interaction with an external magnetic field B is largely controlled by the electron, since for the magnitude of the magnetic moments

$$\mu_e \sim 660\mu_p \approx 5.8 \times 10^{-5}\ \text{eV/T}. \tag{6.1.4}$$

(v) Boltzmann's constant is

$$k \approx 0.86 \times 10^{-4}\ \text{eV/K}. \tag{6.1.5}$$

As a consequence, for a hydrogen atom in a field B there is a transition region around $B \approx 0.1$ T below which the hyperfine interaction dominates and above which the interaction between the electron and the external field takes over. The energy levels are shown in Fig. 6.1 as a function of B in tesla.

The labelling of the states $|a\rangle$, $|b\rangle$, $|c\rangle$, $|d\rangle$ has become conventional. One has

$$
\begin{aligned}
|b\rangle &= |\downarrow_e\downarrow_p\rangle \qquad |d\rangle = |\uparrow_e\uparrow_p\rangle \\
|a\rangle &= \cos\theta\;|\downarrow_e\uparrow_p\rangle - \sin\theta\;|\uparrow_e\downarrow_p\rangle \\
|c\rangle &= \sin\theta\;|\downarrow_e\uparrow_p\rangle + \cos\theta\;|\uparrow_e\downarrow_p\rangle
\end{aligned}
\tag{6.1.6}
$$

where $\tan 2\theta \approx \Delta/(2\mu_e B)$.

Note that for strong fields

$$|a\rangle \to |\downarrow_e\uparrow_p\rangle$$

and

$$|c\rangle \to |\uparrow_e\downarrow_p\rangle.$$

An example of a polarized proton source is shown in Fig. 6.2. A beam of thermal hydrogen molecules passes through an intense rf field, which dissociates the molecules into hydrogen atoms. At the temperatures involved, the hydrogen atoms populate all the hyperfine states equally, so the beam is essentially unpolarized; it is then passed through a strong

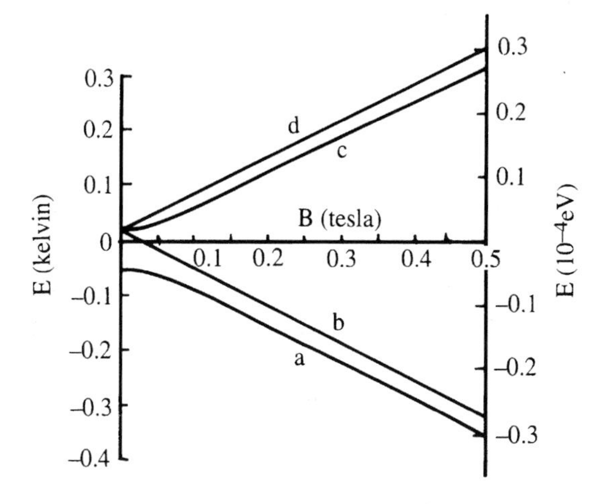

Fig. 6.1 Lowest energy levels of a hydrogen atom in a magnetic field B. $B = 0$ corresponds to the hyperfine splitting.

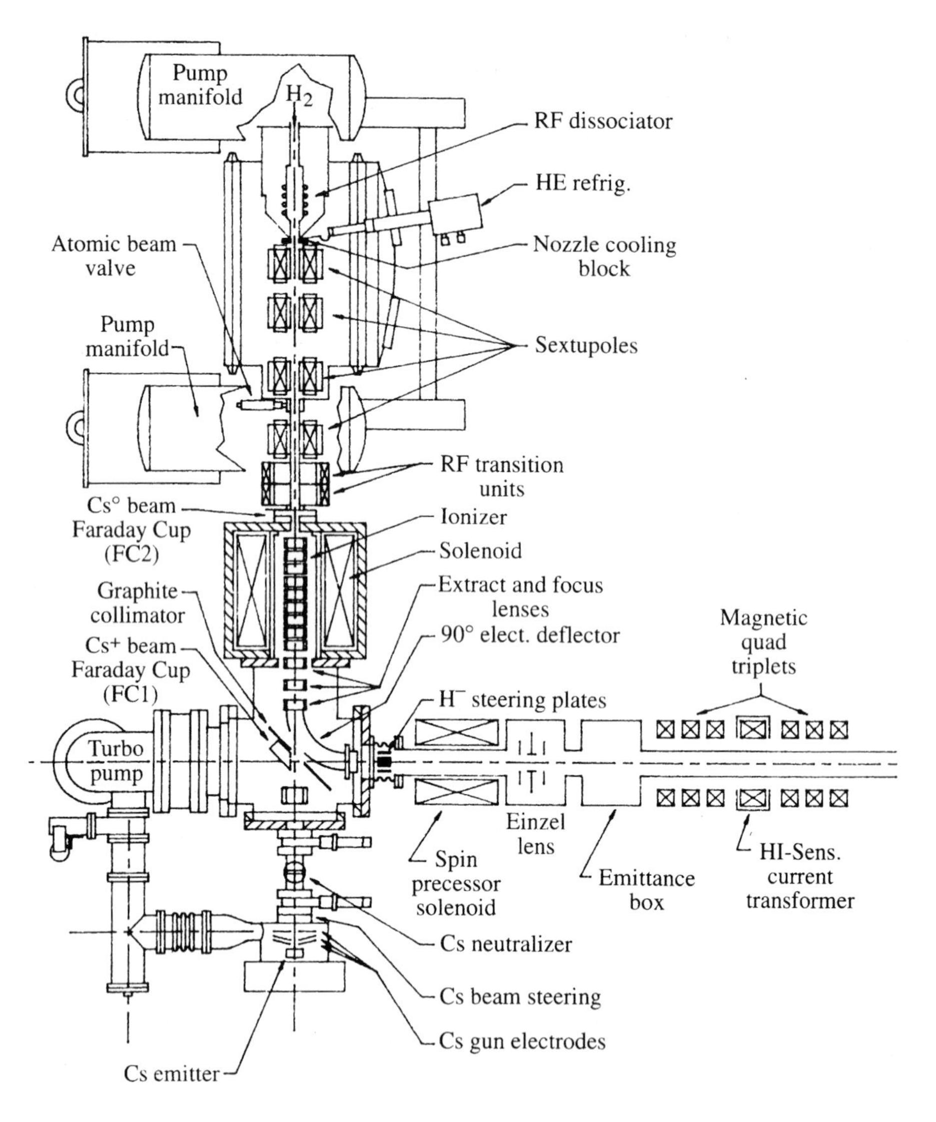

Fig. 6.2 Schematic diagram of the Brookhaven AGS polarized beam source (courtesy of Y. Makdisi).

sextupole magnet. The inhomogeneous field acts as a Stern–Gerlach apparatus separating the atoms in the states $|\uparrow_e\uparrow_p\rangle$ and $|\uparrow_e\downarrow_p\rangle$ from those in the states $|\downarrow_e\downarrow_p\rangle$ and $|\downarrow_e\uparrow_p\rangle$. The point of using a sextupole to provide the inhomogeneous field is that it *focusses* the atoms in the one pair of states while *defocussing* the others. Thus the beam that emerges contains the states $|\uparrow_e\uparrow_p\rangle$ and $|\uparrow_e\downarrow_p\rangle$ i.e. the *electron* spin, is totally polarized. The beam then passes through a uniform magnetic field and an rf field that induces transitions from $|\uparrow_e\downarrow_p\rangle$ to $|\downarrow_e\uparrow_p\rangle$. The electrons are thereby depolarized but the protons are now completely polarized in the state $\uparrow_p$. The atoms pass through an ionizing field that strips off the electrons and one is left with a beam of polarized protons.

The angle of the mean spin relative to the direction of motion can be altered by using the fact that, at *non-relativistic* energies, when a proton is deflected by an electric field the mean spin essentially does not change direction. This is most easily seen from the discussion of the dynamics of the relativistic mean spin vector given in subsection 6.3.1 below, in particular from eqn (6.3.22). In this way one finally obtains a source of polarized protons whose mean spin vector is perpendicular to the plane of the accelerator into which the beam is fed.

The most ambitious series of spin measurements ever undertaken has just begun at the RHIC accelerator at Brookhaven, where proton–proton collisions up to a CM energy of 500 GeV will be possible, with both beams polarized and with a high luminosity $\mathscr{L} = 2 \times 10^{32}$ cm^{-2} s^{-1}. To achieve this, use is made of a new source which yields 500 μA of current.

6.2 Polarized proton targets

Two developments seem to offer the most promise for high energy physics: (1) the construction of targets using frozen ammonia, and (2) the use of gas jets or cells of polarized protons. We shall briefly discuss their essential features.

6.2.1 Frozen targets

In many polarization experiments one requires a high-intensity beam. This causes problems in the target for two reasons. It creates a heat load that requires a high-powered cryogenic system and it causes radiation damage to the material, resulting in a fall-off in polarization. Ammonia is found to be more resistant to radiation damage than other target materials and its ability to become polarized can be recovered by a process of annealing. In addition the fraction of the target material that becomes polarized is higher ($\approx 18\%$) than in other materials.

Consider a diamagnetic solid into which are embedded some paramagnetic impurities, which, for our purposes, can be considered as localized electron spins. The embedding can be done by chemical doping or by irradiation in an electron beam.

For temperatures of the order of 0.5 K and for magnetic fields of the order of 2.5 T the electron spins will be almost completely aligned whereas the protons spins will be unpolarized. The polarization of the electrons is transferred to the protons via a process known as 'dynamical nuclear polarization'.

Let us suppose that all electron spins are in the thermal equilibrium state $\uparrow_e$ and that there is a dipole–dipole coupling between the electron spins and the neighbouring proton spins. Under a microwave field of the correct frequency, transitions $|\uparrow_e\downarrow_p\rangle \to |\downarrow_e\uparrow_p\rangle$ can be induced while $|\uparrow_e\uparrow_p\rangle \not\to |\downarrow_e\downarrow_p\rangle$. Of course the reverse transition $|\downarrow_e\uparrow_p\rangle \to |\uparrow_e\downarrow_p\rangle$ also takes place, so that there would be no change in the proton polarization if the only mechanism were the electron spin resonance effect. But the electron spins, via coupling to the lattice, have a relaxation time that is orders of magnitude shorter than that of the proton spins, so that the flipped electron spin $\downarrow_e$ rapidly returns to its thermal equilibrium state $\uparrow_e$, from which it can once again induce a proton spin-flip $\downarrow_p\to\uparrow_p$. In this fashion the electron spin polarization is transferred to the proton spins.

With this method, polarizations of about 70% could be reached in two hours and the target could handle beams of about 10^{10} protons per second. A major improvement was achieved at Ann Arbor when Crabb *et al.* (1990), working at 1 K and using a 5 T field, succeeded in obtaining a polarization of 96% in about 25 minutes! To what extent such polarizations can be achieved in a beam depends critically on the power of the cooling system.

The above frozen ammonia target has been further developed by a Virginia, Basel, SLAC group (Crabb and Day, 1995) and used very successfully in the polarized electron beam at SLAC in the E143 experiment. Using a ^{4}He evaporation refrigerator with a large pumping system, it was operated at temperatures ≤ 1 K in a beam of 5×10^{11} electrons per second while retaining substantial polarization.

Experiments were done using normal ammonia, $^{14}NH_3$, and deuterated ammonia, $^{14}ND_3$, as well as with $^{15}NH_3$ and $^{15}ND_3$, since in the isotope ^{15}N the unpaired polarizable neutron in ^{14}N is paired with a second neutron. This leads to smaller corrections when trying to interpret a measured asymmetry as a proton asymmetry. Proton polarizations of about 80% and deuteron polarizations of about 40% were achieved under actual experimental running conditions.

Figures 6.3 and 6.4 give some idea of the complexity of an ethylene glycol target and a frozen ammonia target. For a review of solid polarized targets, see Crabb and Meyer (1997).

One of the most influential experiments of the 1980s was the European Muon Collaboration (EMC) measurement of deep inelastic lepton–proton scattering. This used longitudinally polarized muons incident upon a

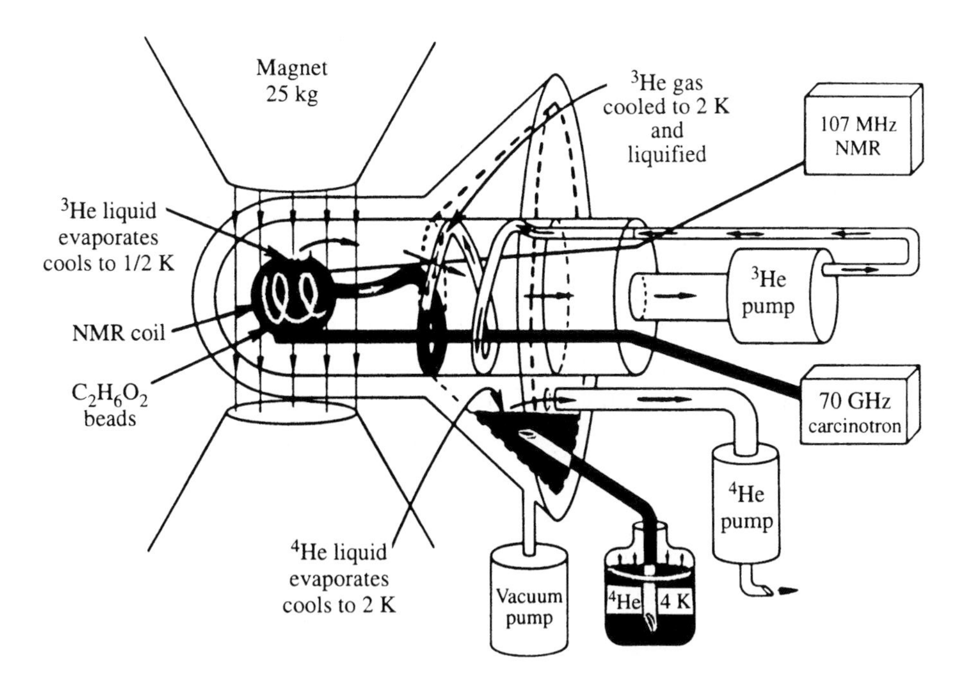

Fig. 6.3 Ethylene glycol target using ^{3}He evaporation cryostat (courtesy of A.D. Krisch).

longitudinally polarized proton target (Ashman *et al.*, 1988). The startling results of this collaboration, which are discussed in Section 11.5, stimulated enormous theoretical and experimental interest and their work has been continued by the Spin Muon Collaboration (SMC) at CERN. Building upon the experience of the EMC, the apparatus and especially the targets used by the SMC reached a remarkable level of refinement. For proton data a frozen ammonia target was used and for deuteron data the target material was deuterated butanol. The target material was contained in two identical thin cylindrical cells, one behind the other, with the beam traversing them longitudinally, as shown in Fig. 6.5. The material in the upstream and downstream cells had opposite polarizations – of great help in eliminating systematic errors.

A typical polarization build-up induced by the microwave field takes several hours, but to obtain maximum polarization takes several days. When the microwave power is switched off the target is operated in the 'frozen-spin' mode at extremely low temperatures, of 30–50 mK. This is only possible because of the relatively low intensity of the muon beam, about 2×10^7 per second.

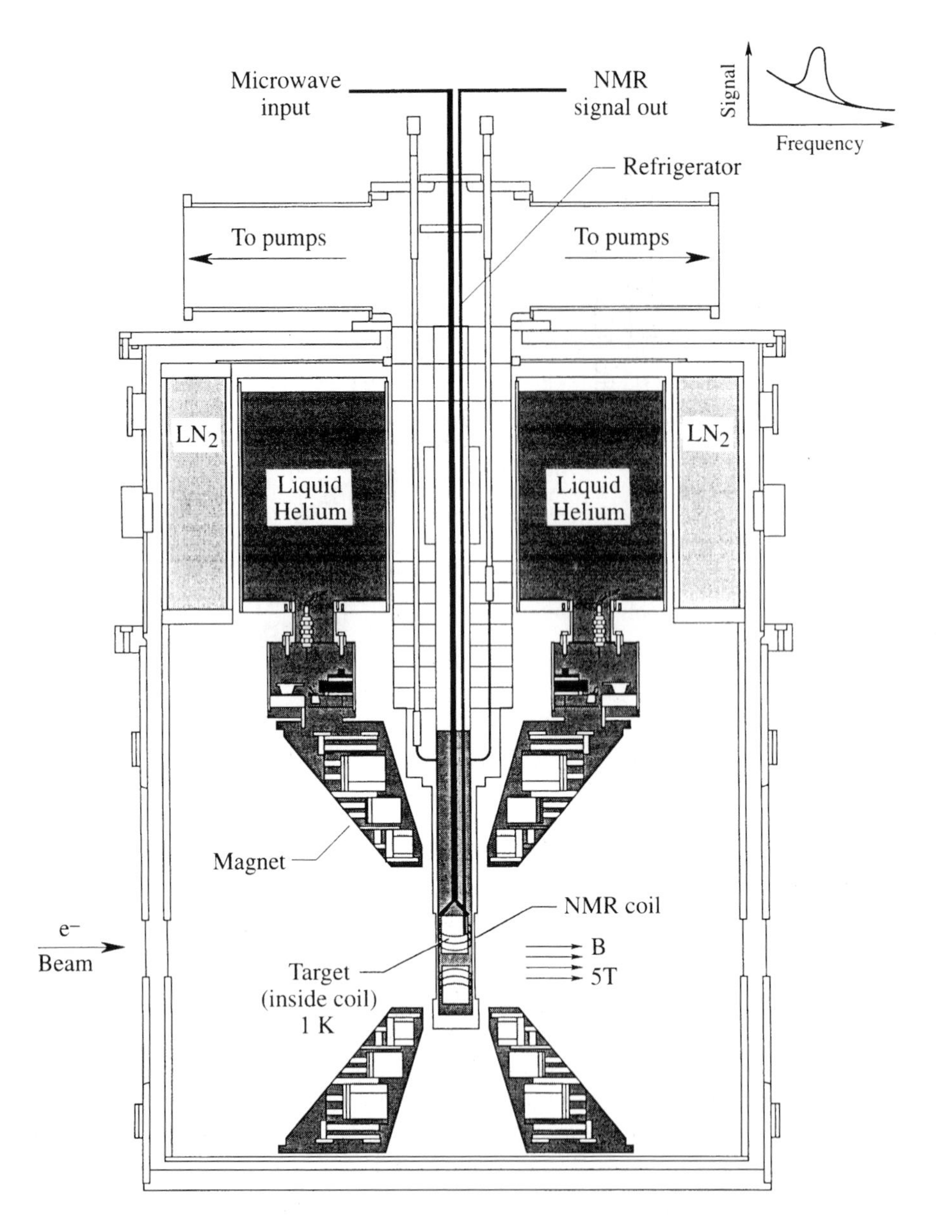

Fig. 6.4 Virginia, Basel, SLAC frozen ammonia target using ^{4}He refrigerator (courtesy of D.G. Crabb).

Once in a frozen-spin mode the direction of the longitudinal polarization can be reversed relatively quickly by varying the fields in the various magnets that surround the target. Typically the polarization was reversed five times per day. It was also possible to rotate the direction of polarization into a direction transverse to the beam.

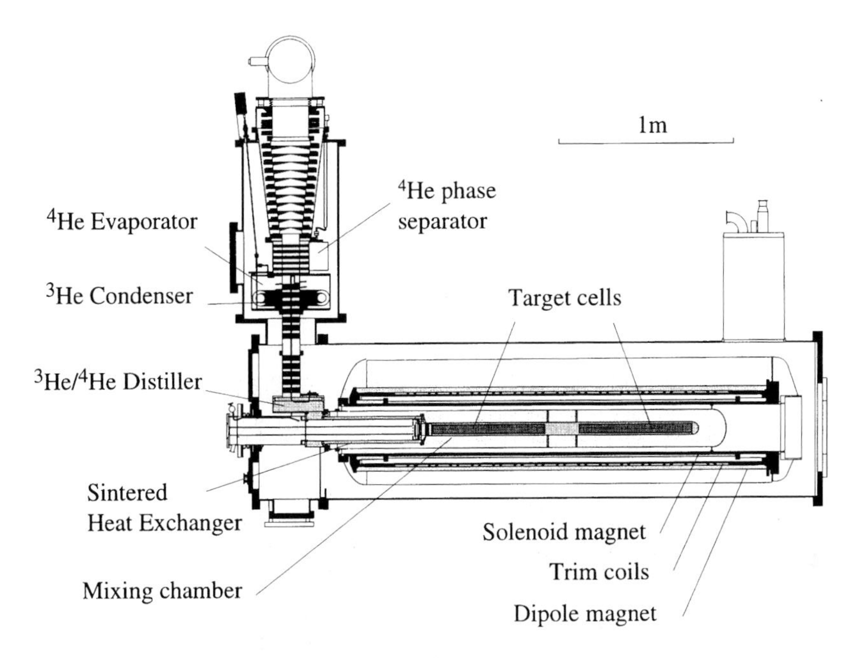

Fig. 6.5 Polarized target used by the Spin Muon Collaboration at CERN. On the left, the dilution refrigerator; on the right the target cells within a system of superconducting magnets. (Courtesy of G. Mallot.)

The average polarizations achieved and the accuracy to which they were known are impressive:

$$\langle \mathscr{P}_p \rangle = 0.86 \pm 0.02 \qquad \langle \mathscr{P}_d \rangle = 0.50 \pm 0.02.$$

6.2.2 Gas-jet targets

Unpolarized gas jets have been used for some time in hadronic physics. The jet is fired across the circulating beam in the accelerator and the low density of the jet compared with a solid target is compensated by the large number of times the beam bunch traverses the jet. The construction of a *polarized* jet has been discussed for a long time (Dick *et al.*, 1980). A great advantage over the solid polarized targets discussed in subsection 6.2.1 is that in an atomic hydrogen jet all the material of the jet is polarized, not just a small fraction of it.

We shall discuss three types of polarized-gas-jet target:

(i) the Mark-II ultra-cold polarized-hydrogen-jet target being developed at Ann Arbor for use as an internal target to study proton–proton collisions at the 400 GeV UNK proton accelerator under construction at Protvino;

(ii) the HERMES polarized-proton gas cell already in use at HERA in measurements of deep inelastic electron–proton scattering; and

(iii) the SLAC high-density gaseous polarized ^{3}He target in use at SLC for electron–^{3}He scattering, which yields very direct information on electron–neutron scattering.

(i) The Mark II ultra-cold polarized-hydrogen-jet target

The basic idea, which was suggested many years ago (Niinikoski, 1980; Kleppner and Greytak, 1982) seems at last to be on the point of becoming a practical tool; see Fig. 6.6(a). The aim is to have a low-velocity jet of intensity of about 4×10^{17} atoms per second, giving an equivalent target thickness of 10^{13} polarized protons per cm^2, a figure more than an order of magnitude better than achieved with conventional atomic beam sources (for a status report see Luppov *et al.*, 1996).

One of the challenging problems is to avoid spoiling the very high vacuum (10^{-9} torr) required in the regime of the jet. Thus the vertical jet must be captured very efficiently after the jet passes across the accelerator beam. For this purpose a so-called 'cryocondensation catcher' pump has been developed with an extremely high pumping speed, for atomic hydrogen, of about 1.2×10^7 litres per second!

The basic idea is to accumulate electron-polarized atomic hydrogen in a magnetic storage bottle at densities greater than 10^{18} atoms/cm^3 and then produce the beam by electron spin resonance microwave pumping. The crucial point is that unpolarized hydrogen atoms at the above density will rapidly combine to form molecular hydrogen, whereas atoms whose electron spins are parallel experience a weak repulsion.

Hydrogen atoms are produced in an rf dissociator and then rapidly cooled to ≈ 0.3 K. They flow into a cell lying inside a 12 T superconducting solenoid. The nature of the cell walls is critical since atomic hydrogen is absorbed strongly on most surfaces. To avoid this, the surfaces are coated with liquid ^{4}He. For such high fields the spin states given in eqn (6.1.6) become

$$\begin{aligned} |b\rangle &= |\downarrow_e \downarrow_p\rangle \qquad & |a\rangle &= |\downarrow_e \uparrow_p\rangle \\ |d\rangle &= |\uparrow_e \uparrow_p\rangle \qquad & |c\rangle &= |\uparrow_e \downarrow_p\rangle \end{aligned} \tag{6.2.1}$$

and atoms in states $|a\rangle$ and $|b\rangle$ are accelerated by the field gradient towards the high-field region and escape from the cell. The atoms in states $|c\rangle$ and $|d\rangle$ are repelled towards the low-field region and effuse from the exit aperture with the electron spin polarized. (Other forces are, of course, totally dominated by the magnetic moment of the electrons.)

The beam then passes through an rf transition unit (see Fig. 6.6(a)), where microwaves are injected at a suitable frequency to induce, say, the transition $|c\rangle \to |a\rangle$. The resulting beam with all protons polarized in the state $\uparrow_p$ passes through a superconducting sextupole magnet that focusses the atoms in state $|a\rangle$ into the interaction region while defocussing the atoms in state $|d\rangle$; the latter are cryopumped away.

Recently it has been discovered that a quasi-parabolic copper mirror, coated with superfluid ^{4}He, gives enhanced focussing of the emerging beam and might lead to an order-of-magnitude increase in the beam intensity (Fig. 6.6(b)).

This whole development is extremely attractive and could lead to jets with essentially 100% polarization and with sufficient density to be able to study large-momentum-transfer reactions, where severe tests of QCD may be possible (see Chapters 12 and 13).

(ii) The HERMES polarized-gas cell

This is not strictly a gas-jet target, in the sense that instead of simply crossing the circulating electron beam the gas flows through a T-shaped open ended pipe or storage cell, as shown in Fig. 6.7, so that the gas forms a longitudinal tube along the direction of the electron beam. The incoming gas jet is produced in conventional fashion from an atomic-beam source, as described in Section 6.1, and without the longitudinal cell would produce an effective target thickness of a few times 10^{11} protons per cm^2. With the cell, effective thicknesses of about 10^{14} protons per cm^2 have been achieved. On the negative side, the cell walls present a source of background scattering (which in a proton beam might present a severe problem) and it is necessary to collimate the electron beam so as to prevent the tail of the beam from scattering off the cell walls.

The HERMES source was first tested in 1994 using unpolarized gas. During 1995 it ran with polarized ^{3}He with stable polarizations of about 50% and in 1996 functioned successfully with hydrogen and deuterium, achieving proton polarizations of 80%–90%. The quality and stability of the ^{3}He polarization is vividly illustrated in Fig. 6.8.

The principle used to produce polarized ^{3}He in the HERMES experiment will be explained briefly. (A somewhat more detailed explanation of the SLAC ^{3}He target is given in the next subsection.) The method was first exploited by Colegrove, Schearer and Walters (1963) and the more modern variant now utilized is described in Lee *et al.* (1993). Helium-3, with a nuclear spin of 1/2, possesses a 1^1S_0 ground state and a 2^3S_1 metastable excited state, which is easily excited by a weak electrical discharge. Placed in a weak magnetic field these levels undergo a hyperfine splitting. If, using a tunable laser, the gas is now optically pumped with, say, left-circularly-polarized light propagating in the direction of the magnetic field, transitions can be excited out of the $J_z = -3/2$ and $-1/2$

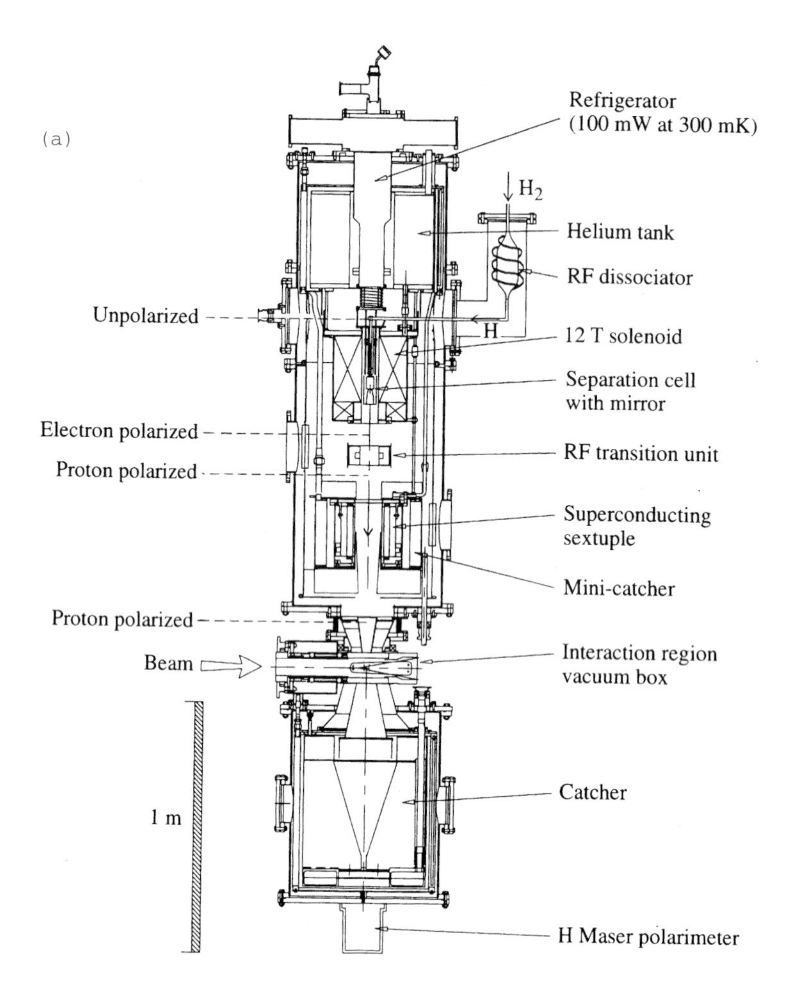

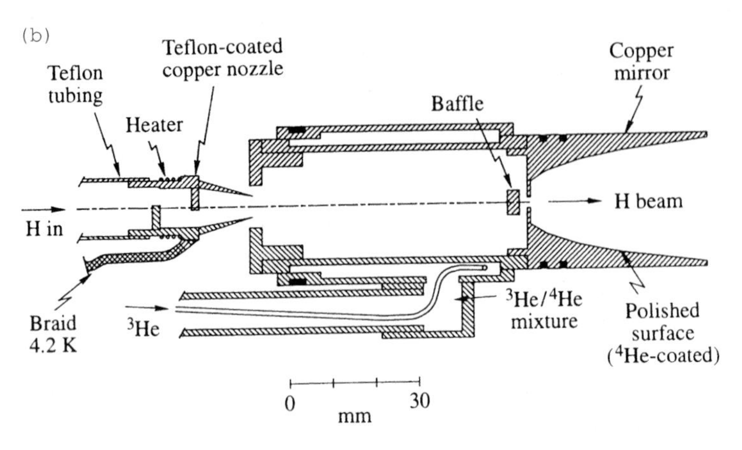

Fig. 6.6 (a) Mark-II ultra-cold polarized-hydrogen-jet target (courtesy of A. D. Krisch); (b) mechanism for improved focussing.

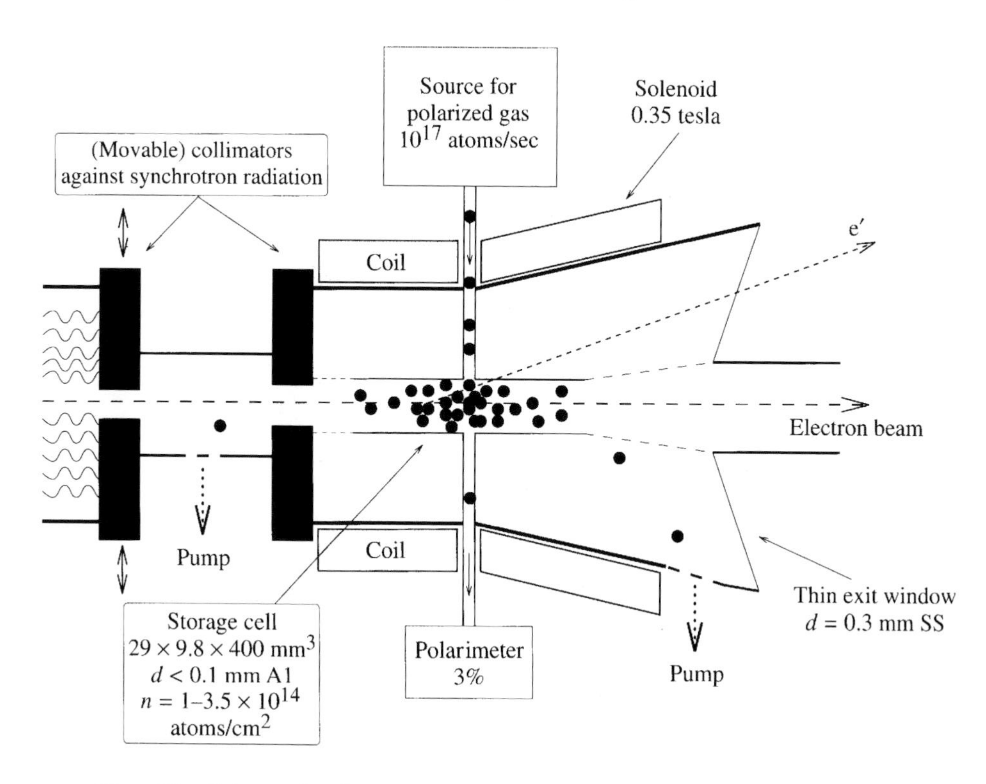

Fig. 6.7 Schematic view of the HERMES polarized gas cell (courtesy of W. Lorenzon).

hyperfine sub-levels of the $^3S_1$ state to certain of the $^3P_0$ sub-levels. Collision mixing then populates all the $^3P_0$ sub-levels and these decay back to the $^3S_1$ sub-levels with equal probability. There is thus a net depletion of the metastable $J_z = -3/2$ and $-1/2$ sub-levels. In collisions between the polarized metastable atoms and atoms in either sub-level of the ground state there is some probability of a transfer of angular momentum, leading to the polarization of the ground state atoms, i.e. to the polarization of the ^{3}He nuclei, since the electrons in the ground state are in a $^1S_0$ state.

(iii) The SLAC high-density gaseous polarized ^{3}He *target*

Polarized ^{3}He is an ideal target for studying electron–neutron interactions because, it is believed, the nucleus is essentially in a spatially symmetric S-state where the spins of the two protons must be in opposite directions. Thus the spin of the nucleus is almost entirely provided by the neutron and only very small corrections are required to extract the 'true' polarized neutron results from the ^{3}He results.

The idea of a ^{3}He target is not new (Bouchiat *et al.*, 1960), but a high-density target capable of operating effectively in the presence of an

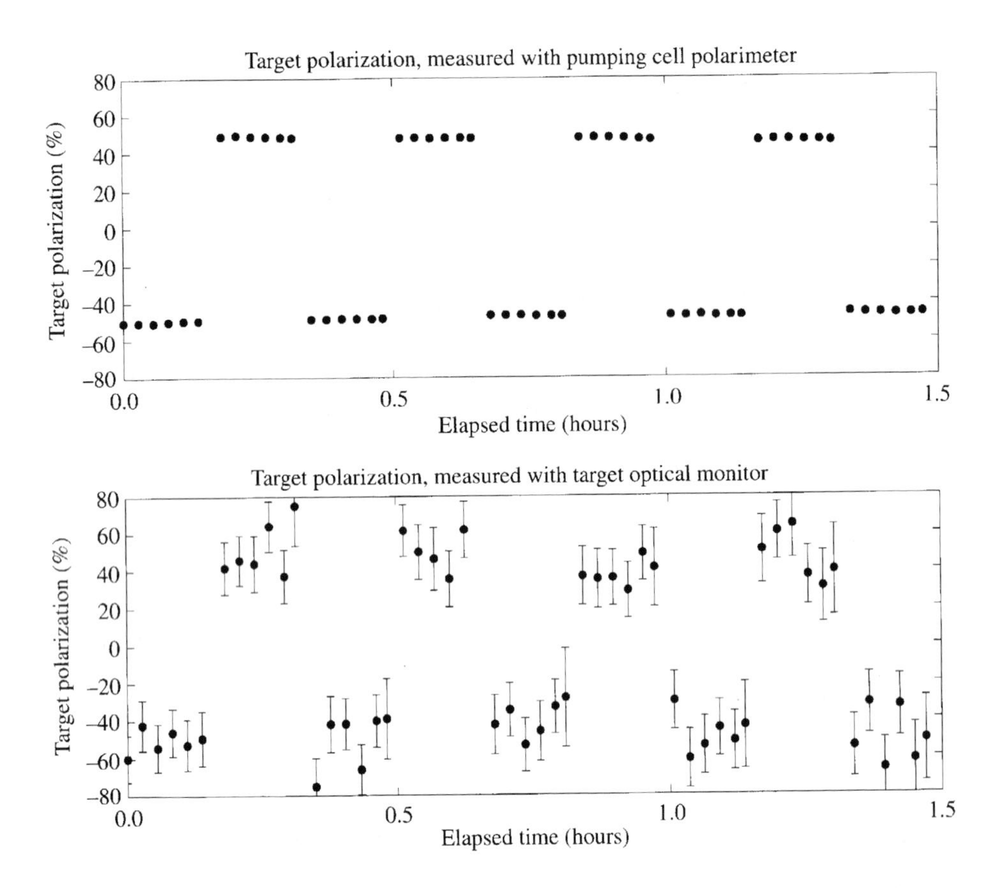

Fig. 6.8 Polarization of ^{3}He in the HERMES gas cell as a function of time, measured by two methods (courtesy of W. Lorenzon).

intense electron beam is a non-trivial matter and a prototype of the SLAC target was first developed in 1992 by a Harvard–MIT team (Chupp *et al.*, 1992).

The essential physical mechanism involved in the polarizing of the ^{3}He is reasonably straightforward and is based on spin transfer in collisions between ^{3}He and highly polarized Rb. Rubidium ($Z = 37$), an alkali metal, has one valence electron outside a closed shell, so that in the ground state it is in a $5S_{1/2}$ state. The Rb vapour lies in a uniform static magnetic field $\mathbf{B}_0$, which defines the Z-direction, and is optically pumped using circularly polarized (σ^+) light of wavelength 794.7 nm from a Ti: sapphire laser, with the beam directed along $\mathbf{B}_0$. In this process the electron, in the ground state with $J_z = -1/2$ is excited to the $^5P_{1/2}$ state, with $J_z = 1/2$. The low-density Rb vapour, with about 10^{14} atoms/cm^3, is immersed in ^{3}He gas of a much higher density ($> 10^{20}$ atoms/cm^3) so that as a result of the high collision rate and the very small energy differences between the various P states, the Rb P-states become equally populated and the radiative decay back to the ground state feeds the $J_z = \pm 1/2$ states with equal probability. There is thus a net depletion of the ground state with $J_z = -1/2$ until eventually the Rb vapour becomes electron-spin-polarized along $\mathbf{B}_0$, in the ground state $J_z = 1/2$, at which point it is transparent to the σ^+ light.

In the collisions between the Rb and the ^{3}He, a very tiny part of the interaction is due to a hyperfine interaction between the Rb valence electron and the ^{3}He nucleus and thus leads to an extremely small, but non-zero, probability for spin exchange. Provided that mechanisms for depolarization are weak enough, there will then be a slow transfer of the valence-electron polarization to the ^{3}He nucleus.

Herein lies the technical challenge. The polarization build-up time is very long, about 10 hours! So one has to construct a target in which the many potential sources of depolarization (or spin relaxation) are minimized to a fantastic degree. Foremost amongst these dangers are:

(a) in an intense electron beam the rubidium will be depolarized through ionization;
(b) radiation damage will darken the glass cell walls and prevent optical repumping of the cell.

The two-cell structure shown in Fig. 6.9 was a brilliant solution to these problems (Chupp *et al.*, 1992; Johnson *et al.*, 1995). The upper cell, lying in an oven, contains the Rb vapour and the ^{3}He and is connected via a transfer tube to the sealed target cell, which lies in the electron beam. The ^{3}He is polarized in the upper cell and diffuses through to the target cell with a time constant of about 10 minutes (much smaller than the characteristic times for spin relaxation). The target cell remains at a much lower temperature than the upper cell so that the Rb vapour density therein is negligible. The actual physical construction of such a two-chamber cell is incredibly difficult and it is a triumph that it has been possible to produce targets with effective densities of about 7×10^{21} ^{3}He nuclei per cm^2 with a stable polarization of 40% over periods of several weeks!

6.3 The acceleration of polarized particles

We discuss here the problems that arise when one tries to accelerate polarized particles, namely, the mechanisms that tend to depolarize the particles and how these can be overcome. The motion of a particle in an accelerator is largely a question of classical dynamics and the behaviour of the spin is best understood in terms of the covariant mean spin vector

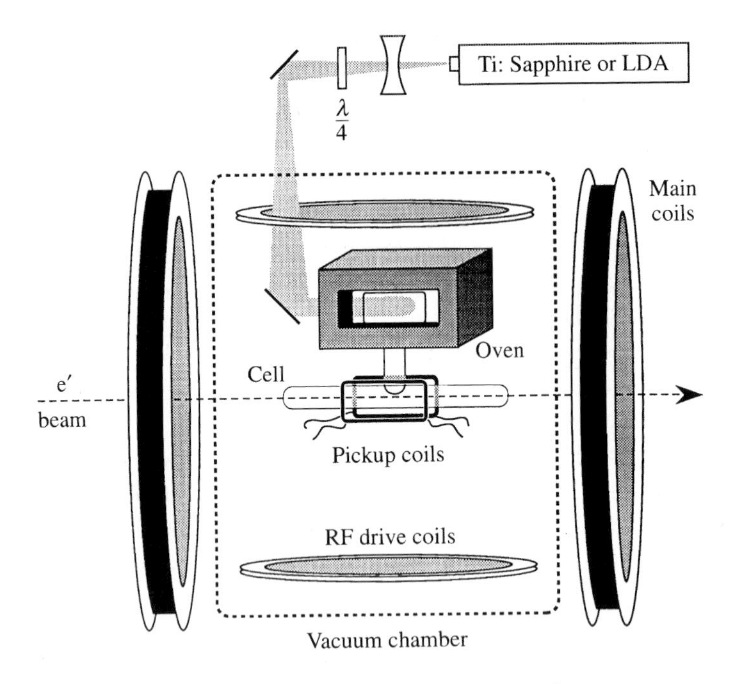

Fig. 6.9 The SLAC high-density gaseous polarized ^{3}He target (courtesy of J.R. Johnson).

$\mathscr{S}^\mu$, which was introduced in Section 3.4. Thus we begin with a study of the behaviour of $\mathscr{S}^\mu$ in the presence of macroscopic electric and magnetic fields.

6.3.1 *Dynamics of the relativistic mean spin vector*

Ehrenfest's theorem assures us that the motion of $\mathscr{S}^\mu$ is controlled in a classical fashion by the mean value of the interaction energy, which will involve the electric and magnetic fields.

Being macroscopic these fields do not vary on the scale of the particle's wave packet and so may be taken out of the mean value expressions. Thus the mean interaction energy may be constructed from the *given* classical field interacting with the *mean* spin vector $\mathscr{S}^\mu$.

In this section we shall deal exclusively with the *canonical* mean spin vector

$$\overset{\circ}{\mathscr{S}}{}^{\mu} = \frac{1}{s}(0, m\mathbf{s}) \tag{6.3.1}$$

where $\mathbf{s}$ is the mean spin vector in the canonical (comoving) rest frame reached from the frame where the particle of spin s has velocity $\mathbf{v}$ by the pure boost $l(\mathbf{v})$ (see Section 1.2). Then

$$\mathscr{S}^{\mu} = \Lambda^{\mu}{}_{\nu}[l(\mathbf{v})]\overset{\circ}{\mathscr{S}}{}^{\nu} = (\mathscr{S}^0, \boldsymbol{\mathcal{S}}) \tag{6.3.2}$$

so that $\mathscr{S}$ has the form

$$\begin{aligned} \boldsymbol{\mathcal{S}} &= \frac{1}{s}\left[\mathbf{s} + \frac{\gamma^2}{\gamma+1}(\boldsymbol{\beta}\cdot\mathbf{s})\,\boldsymbol{\beta}\right] \\ \mathscr{S}^0 &= \boldsymbol{\beta}\cdot\boldsymbol{\mathcal{S}} = \frac{\gamma}{s}(\boldsymbol{\beta}\cdot\mathbf{s}), \end{aligned} \tag{6.3.3}$$

the latter following from

$$p\cdot\mathscr{S} = 0. \tag{6.3.4}$$

We now seek a classical covariant equation for the rate of change of $\mathscr{S}^{\mu}$, with respect to the proper time τ, when it is acted upon by an arbitrary electromagnetic field specified by $F^{\mu\nu}$ in which the **E** and **B** fields are given by

$$\begin{aligned} E_j &= -F^{0j} \\ \epsilon^{ijk}B_k &= -F^{ij}. \end{aligned} \tag{6.3.5}$$

(In this subsection we use Gaussian units.)

Given that $\mathscr{S}^{\mu}$ is an axial vector, and assuming that $d\mathscr{S}^{\mu}/d\tau$ is linear in the fields and that it does not depend upon kinematic variables other than the 4-velocity

$$U^{\mu} = (\gamma c, \gamma\mathbf{v}) \tag{6.3.6}$$

and $\dot{U}^{\mu} \equiv dU^{\mu}/d\tau$, the most general form possible is

$$\frac{d\mathscr{S}^{\mu}}{d\tau} = aF^{\mu\nu}\mathscr{S}_{\nu} + b(\mathscr{S}_{\alpha}F^{\alpha\beta}U_{\beta})U^{\mu} + d(\mathscr{S}_{\alpha}\dot{U}_{\alpha})U^{\mu} \tag{6.3.7}$$

where a, b, d are constants.

The condition that $\mathscr{S}\cdot p = \mathscr{S}\cdot U = 0$ for all τ requires that

$$a = bc^2 \qquad d = -1/c^2 \tag{6.3.8}$$

if it is assumed that (6.3.7) holds no matter what sort of force causes the acceleration $\dot{U}^{\mu}$.

Finally we assume that for a particle at rest in a uniform magnetic field $\overset{\circ}{\mathbf{B}}$

$$\frac{d\mathbf{s}}{d\tau} = \text{torque} = \boldsymbol{\mu}\times\overset{\circ}{\mathbf{B}}. \tag{6.3.9}$$

We write, for the magnetic moment,

$$\boldsymbol{\mu} = \frac{g\mu_0}{\hbar s}\mathbf{s} \tag{6.3.10}$$

where for a particle of charge q

$$\mu_0 = \frac{q\hbar s}{2mc} \tag{6.3.11}$$

and where g is the gyromagnetic ratio.

Then

$$\frac{d\mathbf{s}}{d\tau} = \frac{g\mu_0}{\hbar s}\mathbf{s} \times \overset{\circ}{\mathbf{B}}. \tag{6.3.12}$$

Upon writing eqn (6.3.7) in the canonical rest frame, we recover (6.3.12) provided

$$a = g\mu_0. \tag{6.3.13}$$

Thus

$$\frac{d\mathscr{S}^\mu}{d\tau} = g\mu_0 \left[F^{\mu\nu}\mathscr{S}_\nu + \frac{1}{c^2}(\mathscr{S}_\alpha F^{\alpha\beta}U_\beta)U^\mu\right] - \frac{1}{c^2}(\mathscr{S}_\alpha \dot{\mathbf{U}}_\alpha)U^\mu. \tag{6.3.14}$$

If the particle is charged and if, in addition, its motion is largely controlled by the Lorentz-force interaction with the electromagnetic fields then

$$\frac{d\mathbf{p}}{dt} = q[\mathbf{E} + (\boldsymbol{\beta} \times \mathbf{B})] \tag{6.3.15}$$

and

$$\frac{dp_0}{dt} = q\boldsymbol{\beta} \cdot \mathbf{E}, \tag{6.3.16}$$

which, since $p^\mu = mU^\mu$ (p^μ is the kinetic, not the canonical, 4-momentum), can be written as

$$\dot{U}^\mu \equiv \frac{dU^\mu}{d\tau} = \frac{q}{mc}F^{\mu\nu}U_\nu. \tag{6.3.17}$$

In this case (6.3.14) becomes the celebrated Thomas–BMT equation:

$$\frac{d\mathscr{S}^\mu}{d\tau} = \frac{q}{mc}\left[\frac{g}{2}F^{\mu\nu}\mathscr{S}_\nu + \frac{1}{c^2}\left(\frac{g}{2} - 1\right)(\mathscr{S}_\alpha F^{\alpha\beta}U_\beta)U^\mu\right] \tag{6.3.18}$$

(Thomas, 1927; Bargmann, Michel and Telegdi, 1959). In (6.3.18) it is clear that $g = 2$ is a very special value.

Before turning to the question of the spin motion in an accelerator it is instructive to use (6.3.14) to rederive the Thomas precession dealt with in subsection 2.2.8.

We are interested in the rate of change of the canonical rest system $\mathbf{s}$ with laboratory time t. Substitution of (6.3.3) into (6.3.14) followed by a daunting piece of algebra leads to

$$\frac{d\mathbf{s}}{dt} = \mathbf{s} \times \left\{ \frac{g\mu_0}{\hbar s} \left[\mathbf{B} - \frac{\gamma}{\gamma+1}(\boldsymbol{\beta} \cdot \mathbf{B})\boldsymbol{\beta} - \boldsymbol{\beta} \times \mathbf{E} \right] + \frac{\gamma^2}{\gamma+1} \boldsymbol{\beta} \times \frac{d\boldsymbol{\beta}}{dt} \right\}. \tag{6.3.19}$$

For the case of a pure electric field (the Coulomb field of the nucleus) and with $\mu_0 = -e\hbar s/(2mc)$ for an electron, (6.3.19) becomes

$$\frac{d\mathbf{s}}{dt} = \frac{ge}{2mc^2} \mathbf{s} \times (\mathbf{v} \times \mathbf{E}) + \boldsymbol{\omega}_{\mathrm{T}} \times \mathbf{s} \tag{6.3.20}$$

where, in terms of the acceleration $\mathbf{a}$,

$$\boldsymbol{\omega}_{\mathrm{T}} = \frac{1}{c^2} \left(\frac{\gamma^2}{\gamma+1} \right) (\mathbf{a} \times \mathbf{v})$$

in agreement with eqns (2.2.34), (2.2.35).

When the motion of a charged particle is controlled by the Lorentz force, (6.3.15) and (6.3.16) yield for the acceleration

$$\frac{d\boldsymbol{\beta}}{dt} = \frac{q}{\gamma mc} [\mathbf{E} + \boldsymbol{\beta} \times \mathbf{B} - (\boldsymbol{\beta} \cdot \mathbf{E})\boldsymbol{\beta}], \tag{6.3.21}$$

which, substituted into (6.3.19), gives

$$\frac{d\mathbf{s}}{dt} = \frac{q}{mc} \mathbf{s} \times \left[\left(\frac{g}{2} - 1 + \frac{1}{\gamma} \right) \mathbf{B} - \left(\frac{g}{2} - 1 \right) \frac{\gamma}{\gamma+1} (\boldsymbol{\beta} \cdot \mathbf{B})\boldsymbol{\beta} - \left(\frac{g}{2} - 1 + \frac{1}{\gamma+1} \right) \boldsymbol{\beta} \times \mathbf{E} \right]. \tag{6.3.22}$$

6.3.2 *Difficulties in the acceleration of polarized particles*

We shall utilize eqn (6.3.22) to give a brief explanation of the problems that occur when one attempts to accelerate polarized particles. We consider protons being accelerated in a planar circular accelerator whose guide field $\mathbf{B}_0$ is in the OZ direction, so that the equilibrium orbit lies in the XY-plane. In this case, in (6.3.22) $\mathbf{E} = 0$ and $\boldsymbol{\beta} \perp \mathbf{B}_0$, so that

$$\frac{d\mathbf{s}}{dt} = -\frac{q}{mc} \left(\frac{g}{2} - 1 + \frac{1}{\gamma} \right) \mathbf{B}_0 \times \mathbf{s} \tag{6.3.23}$$

and (6.3.21) becomes

$$\frac{d\boldsymbol{\beta}}{dt} = -\frac{q}{\gamma mc} \mathbf{B}_0 \times \boldsymbol{\beta}. \tag{6.3.24}$$

Since $g/2 > 1$ both **s** and $\boldsymbol{\beta}$ rotate about $\mathbf{B}_0$ in the same sense, with angular frequences as follows.

$$\begin{aligned} &\text{For } \boldsymbol{\beta}: \quad \Omega_c = \text{relativistic cyclotron frequency} = -\frac{q}{\gamma mc} B_0. \\ &\text{For } \mathbf{s}: \quad \Omega_s = -\frac{q}{mc}\left(G + \frac{1}{\gamma}\right) B_0 = (\gamma G + 1)\Omega_c, \end{aligned} \tag{6.3.25}$$

where

$$G = g/2 - 1 \tag{6.3.26}$$

is called the *gyromagnetic anomaly.*

The difference in angular frequencies is then

$$\Omega \equiv \Omega_s - \Omega_c = -\frac{qG}{mc} B_0 = \Omega_c G\gamma. \tag{6.3.27}$$

Thus the quantity $G\gamma$ measures how much bigger Ω is than Ω_c in a frame that rotates with the velocity vector. It thus measures the number of complete spin precessions in this frame per revolution of the particle in its orbit. It is known as the *spin tune* and is often written v_s.

Note that for protons $G = 1.79$, so that in a high energy accelerator $G\gamma$ is a large number. For electrons G is exceedingly small, $G \approx 1.16 \times 10^{-3}$, but at LEP or HERA γ is very large, so that again $G\gamma$ is large (≈ 103 at the Z^0 mass at LEP, ≈ 63 at HERA). This means, via (6.3.27), that if the beam direction is altered by a small angle as a consequence of passing through some field **B**, then the component of the mean spin vector perpendicular to **B** will rotate through a much larger angle, namely $1 + G\gamma$ times larger.

In a perfect machine, with all particles moving along the equilibrium orbit of radius R and with uniform $\mathbf{B} = B_0\mathbf{e}_z$ the mean spin **s** would simply precess about the Z-axis.

It should be remembered that the Ωs are not constant but are functions of time. For example, the equilibrium orbit is given by

$$r = R = \frac{v\gamma mc}{qB_0} \tag{6.3.28}$$

and is constant in a synchrotron, so that

$$|\Omega_c| = v/R \tag{6.3.29}$$

increases as the particle accelerates.

There are three main effects that disturb the ideal situation in a synchrotron:

(i) the quadrupole fields that focus the beam have a component of **B** in the horizontal plane and the component B_z varies in the radial direction;
(ii) there are imperfections in the fields due to misaligment of magnets, field errors etc.;
(iii) spin-flip occurs as a result of synchrotron radiation.

The latter is very important for electron machines and will be dealt with in Section 7.1. We shall ignore it in discussing proton accelerators.

(i) Consider protons injected into an accelerator with all spins parallel to OZ, so that initially the beam is 100% polarized. A particle that is not moving along the equilibrium orbit $r = R$, $z = 0$ experiences vertical and horizontal focussing forces that cause it to oscillate about the equilibrium orbit. These vertical and horizontal *betatron oscillations* have frequencies that depend upon the *field index* n, which, to first order, describes how $B_z(r)$ varies near $r = R$. That is, n is defined by writing

$$B_z(R + \delta r) = B_0 \left(1 - \frac{n\delta r}{R}\right) \tag{6.3.30}$$

where $B_0 \equiv B_z(R)$.

Because of these perturbations, after each revolution **s** will differ slightly from the value given by eqn (6.3.23). Since **s** is initially along OZ it is not affected by the fact that B_z varies with r. However, it will precess around the horizontal components of **B** during its vertical betatron oscillations.

In a system of perfect magnets these precessions would average to zero. But in reality there are always stray horizontal fields at the end of any magnet or group of magnets and the strength of these fields will vary with z. Thus when a particle traverses the gap between such magnets while undergoing vertical betatron oscillations, **s** will pick up a net non-zero precession about the horizontal axes along and perpendicular to the beam.

Because of the weakness of the horizontal fields, the effect, per revolution, is very small. But large resonant results can build up if the small perturbations are in phase.

Let Q_z be the *vertical tune*, i.e. the number of vertical betatron oscillations per revolution, and let the accelerator have a K-fold symmetry. Then, using eqn (6.3.22) with the horizontal fields included, putting $\mathbf{s} = \mathbf{s}_0 + \Delta\mathbf{s}$, where $\mathbf{s}_0(t)$ is the mean spin vector in the presence of the uniform field $\mathbf{B}_0$, and keeping only terms linear in $\Delta\mathbf{s}$ in the resulting equation one finds (Froissart and Stora, 1959) that the condition for resonance is

$$\nu_s \equiv \gamma G = mK \pm Q_z \qquad m = \text{ integer.} \tag{6.3.31}$$

Equation (6.3.31) can be understood intuitively as follows. The horizontal components of **s** involve the oscillatory functions $\cos\Omega t$, $\sin\Omega t$. These are subject to a perturbation that depends upon the vertical betatron oscillations and thus involves $\cos(Q_z\Omega_c t)$, $\sin(Q_z\Omega_c t)$. The angular frequencies involved in the product of such terms are then $\Omega \pm Q_z\Omega_c$, and the times at which the perturbation acts are $t_N = N\tau/K$, where τ, the period of revolution, is given by $2\pi/\Omega_c$. For a resonant build-up we require that the phases of the resultant angular functions change by $2m\pi$ when $N \to N+1$. Thus we require

$$(\Omega \pm Q_z\Omega_c)\frac{2\pi}{K\Omega_c} = 2m\pi,$$

which yields (6.3.31) immediately.

The above resonances are known as *intrinsic resonances.*

(ii) Because of imperfections in the magnetic field even the equilibrium orbit will not be the idealized circle we have been assuming. However, since the real orbit is closed, the path must retrace itself each revolution. The imperfection causes a small change in the expected **s** and once again this will build up resonantly if the spin precession is in phase with the time of encountering the imperfection. This will occur whenever $G\gamma$ is an integer. These resonances are the *imperfection resonances.* They occur at energies given by

$$E = m\gamma c^2 = \left(\frac{mc^2}{G}\right) G\gamma = \left(\frac{mc^2}{G}\right) \times \text{ integer} \qquad (6.3.32)$$

so that the spacing is

$$\Delta E = \left(\frac{mc^2}{G}\right), \qquad (6.3.33)$$

which for protons ($G = 1.79$) is ≈ 523 MeV.

In both the above mechanisms, the resonance conditions depend upon the energy of the particle. Of course the actual consequence of a given resonance will depend upon the details of the particular accelerator. But what appears inevitable in the above is that in the process of accelerating to higher and higher energies the particle will encounter an ever growing number of depolarizing resonances of varying strength. The challenge is to find a way to 'jump' or 'cross' these resonances with as little loss of polarization as possible.

A traditional method, pioneered at the Argonne ZGS, is to use pulsed quadrupole magnets to induce a rapid change in Q_z while the beam energy is in the vicinity of a resonance value. The choice of pulse duration, timing and strength seems to be an art. Figure 6.10 shows the degree of polarization as the beam momentum increases in the ZGS with and without the pulsed quadrupoles. The results are dramatic!

However, each resonance has to be studied separately so that, as the accelerator output energy goes up and the number of resonances encountered increases more or less proportionately to the energy, this becomes an extremely difficult task, with thousands of resonances to be manipulated.

A wonderful solution to the above was suggested many years ago (in 1976!) but only now appears to have become a practical tool: the Siberian snake, which we now discuss.

6.3.3 The Siberian snake

A radical solution to the above problem is provided by the Siberian snake (Derbenev and Kondratenko, 1976), in which **s** is rotated through 180° around a horizontal axis each revolution, with a consequent cancelling-out of the depolarizing effects.

Moreover the arrangement can be made independent of energy so that the spin tune ν_s is a fixed number, a half-integer, and all integer spin resonances have disappeared!

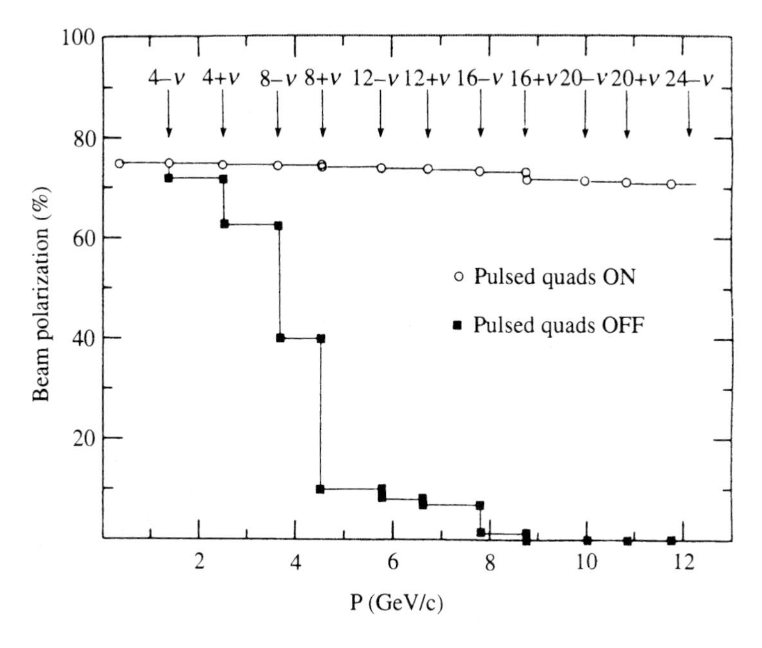

Fig. 6.10 Beam polarization as a function of momentum, as various resonances are crossed at the ZGS (from Fernow and Krisch, 1981).

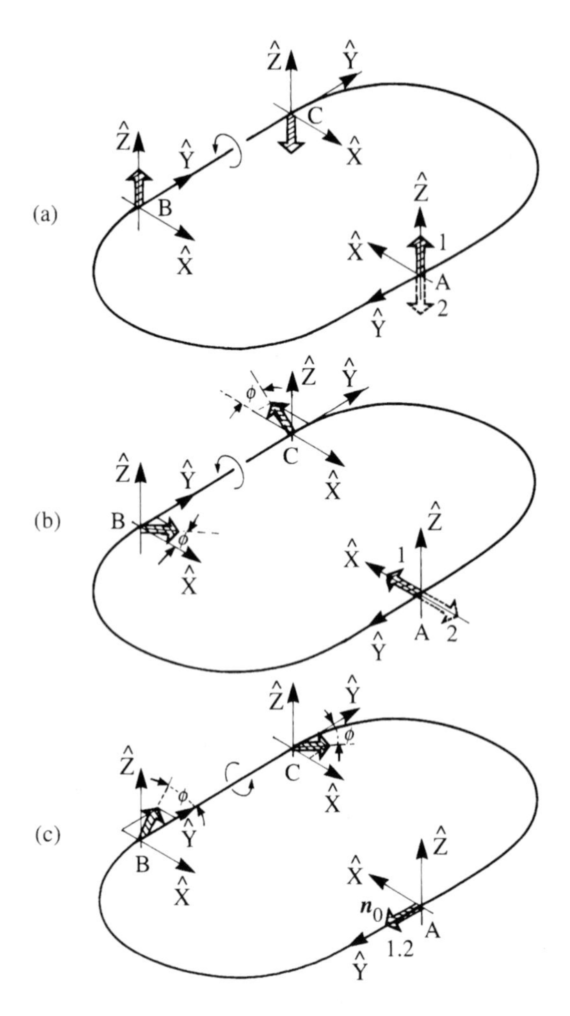

Fig. 6.11 Precession of various spin vectors through a Siberian snake on a particle's path $A \to B \to C \to A$: (a) $\mathbf{s}_A^z$; (b) $\mathbf{s}_A^x$; (c) $\mathbf{s}_A^y$.

The functioning of a Siberian snake can be understood pictorially in Fig. 6.11, which has been adapted from the review article of Montague (1984). We refer everything to the frame $\hat{X}\hat{Y}\hat{Z}$ attached to the particle, with $O\hat{Y}$ along the velocity and $O\hat{Z}$ perpendicular to the plane of the accelerator, and we consider what happens to the mean spin vector $\mathbf{s}_A$ at A as the particle makes one revolution on its orbit. The Siberian snake 'mechanism' is installed in the section BC and, in this version, consists of a longitudinal magnetic field, i.e. a field along the particle's motion. We assume a vertical guide field $B_0\mathbf{e}_z$ along the semicircular sections of the orbit.

At A we resolve $\mathbf{s}_A$ into orthogonal vectors $\mathbf{s}_A^j$, lying along the axes at A, and follow the precession of each component vector (shown hatched) separately. These precessions along the path $A \to B \to C \to A$ are shown in Fig. 6.11(a), (b), (c) respectively. To understand what happens between B and C let us return to eqn (6.3.22) for the case where $\mathbf{B} = \mathbf{b}_\parallel$ is along the motion. Then

$$\frac{d\mathbf{s}}{dt} = \left(\frac{q}{mc\gamma}\right)\frac{g}{2}\mathbf{s} \times \mathbf{b}_\parallel \tag{6.3.34}$$

and $\mathbf{s}$ precesses about its velocity vector $\boldsymbol{\beta}$ with angular frequency

$$\Omega_\parallel = -\frac{qgb_\parallel}{2mc\gamma}. \tag{6.3.35}$$

We suppose that $b_\parallel$ is adjusted so that the rotation angle about $\boldsymbol{\beta}$ between B and C is 180°.

Consider, in Fig. 6.11(a), the precession of $\mathbf{s}_A^z$. Along $A \to B$ it is parallel to the guide field, so reaches B unaltered. Along $B \to C$ it rotates about $O\hat{Y}$ by 180° then returns to A antiparallel to $O\hat{Z}$. In (b) we follow $\mathbf{s}_A^x$. Along $A \to B$ it rotates about $O\hat{Z}$ at angular frequency $\Omega = \Omega_c G\gamma$; see (6.3.27). The $O\hat{X}$ axis has turned through angle $\Omega_c t = \pi$, so the precession angle is $G\gamma\pi$ which we write as $2N\pi - \phi$, so that at B $\mathbf{s}_A^x$ makes an angle ϕ with $O\hat{X}$ as shown. On $B \to C$ it rotates about $\boldsymbol{\beta}$ by 180°, ending up at C with the orientation shown. From $C \to A$ it precesses through $2N\pi - \phi$ degrees so that at A it is antiparallel to $O\hat{X}$. Finally in (c) we follow $\mathbf{s}_A^y$ and find that it ends up parallel to $O\hat{Y}$ at A. Thus $\mathbf{s}_A$ rotates about its velocity by 180° per revolution.

Although suggested in 1976, the practical construction of a Siberian snake is a non-trivial matter and the first ever was tested at Indiana only a few years ago (Krisch *et al.*, 1989). The results shown in Fig. 6.12 plot the beam degree of polarization against a quantity that controls the strength of an imperfection resonance. Without the snake the degree of polarization drops rapidly as a function of the strength of the imperfection. With the snake on, the polarization remains essentially unchanged. The above experiment was carried out at fixed beam momentum.

It seems that to overcome imperfection resonances a 'partial' snake, which does not rotate the spin through a full 180°, is adequate, but the full snake is required for intrinsic resonances. Experiments at Indiana have continued with great success, including the overcoming of the depolarization from overlapping depolarizing resonances. A Siberian snake has also been built at the Brookhaven AGS with very encouraging results (Huang and Roser, 1994) and is being incorporated in the chain that produces polarized protons in RHIC. A partial snake will overcome all

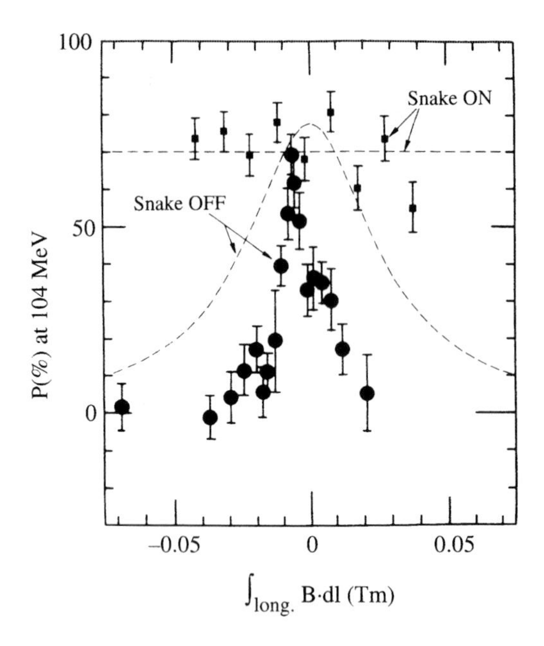

Fig. 6.12 First test of the Siberian snake. See text; the curve gives the theoretical resonance shape (from Krisch *et al.*, 1989).

imperfection resonances during acceleration up to 24 GeV in the AGS, while the six main intrinsic resonances encountered will be corrected by the pulsed quadrupole method. After transfer to the RHIC accelerator the beam will pass through four split Siberian snakes, as shown in Fig. 6.13. This configuration should preserve the polarization up to an energy per beam of 250 GeV. The expected beam polarization is 70%.

6.3.4 Stern–Gerlach polarization of protons and antiprotons

Finally we give a brief outline of a totally new approach to the production of polarized protons and antiprotons, which, however, has not yet been shown to be a practical tool but which seems promising.

The idea is to avoid all the problems encountered when accelerating polarized particles in a circular accelerator, by first accelerating unpolarized particles and then polarizing them while they circulate at fixed energy in a storage ring. It would permit the fantastic possibility of a high energy polarized antiproton beam.

The basic idea is the following (Niinikoski and Rossmanith, 1985). The beam in a storage ring is focussed by alternate quadrupoles of opposite polarity. Normally one considers only the effect on the motion due to the charge, but each quadrupole will, in addition, give rise to a tiny Stern–

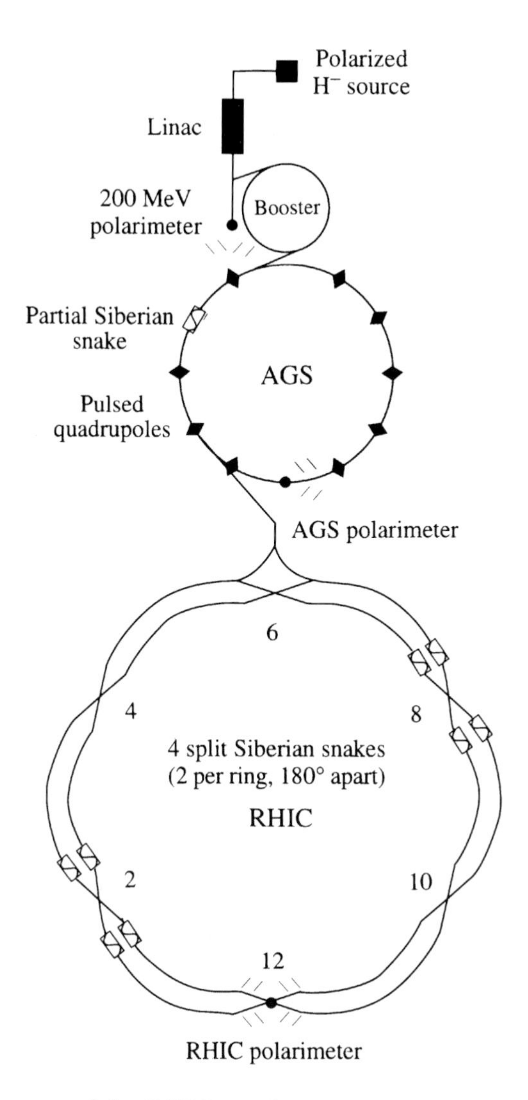

Fig. 6.13 Lay-out of the RHIC accelerator showing the Siberian snakes (from Bunce *et al.*, 1992).

Gerlach splitting of the beam. (For this reason the phenomenon has been christened the *spin-splitter* effect.)

With the system of axes indicated, the field of the quadrupole shown in Fig. 6.14 near its central axis is given approximately by

$$B_x = by \qquad B_y = bx \tag{6.3.36}$$

where b is called the radial field gradient. The beam is travelling in the OZ direction and the force on an arbitrary magnetic moment $\boldsymbol{\mu}$ has

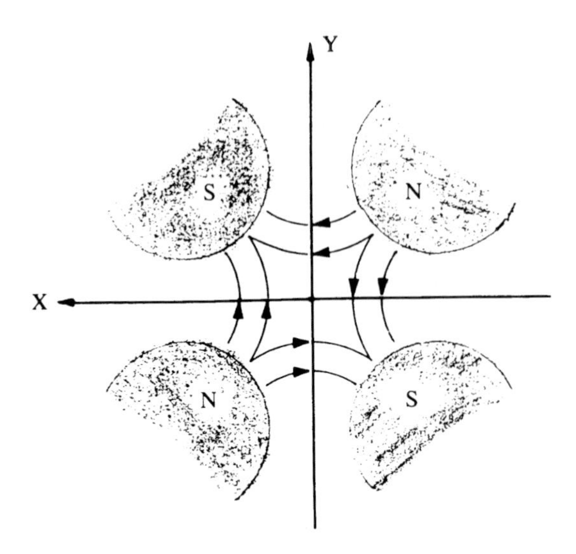

Fig. 6.14. Schematic picture of the field of a magnetic quadrupole.

components

$$F_x = b\mu_y \qquad F_y = b\mu_x. \tag{6.3.37}$$

For alternate quadrupoles b changes sign, so that the direction of the force on $\boldsymbol{\mu}$ alternates from one quadrupole to the next.

Each quadrupole of length L will generate an additional angular deflection δ of order

$$\delta \approx \frac{b\mu L}{E}. \tag{6.3.38}$$

For typical gradients ≈ 10 T/m and $L \approx 1$ m one has

$$\delta \approx \frac{10^{-13}}{E(\text{GeV}) \text{ radians}}. \tag{6.3.39}$$

This seems a hopelessly small angle. However, if the sequence of deflections can be phased to add up coherently then it has been suggested that it might be possible to build up a spatial separation in the beam, between particles of opposite spin direction, of the order of a few millimetres in a period of about 1 hour.

If the spin did not precess between the quadrupoles of opposite polarity then there would be an effective cancelling-out of deflections. Thus we require that $\boldsymbol{\mu}$ precesses by a rotation angle that is an odd number of π radians during the passage from one quadrupole to the next. This could be the normal precession around the guide field between quadrupoles, which would give rise to a reversal of μ_x and hence, by (6.3.37), to a coherent build-up of a vertical separation between particles of opposite μ_x, i.e. of opposite transverse, *horizontal*, components of spin.

We are trying to build up a displacement vertically by means of periodic impulse from the quadrupoles. At the same time the particle is undergoing vertical betatron oscillations. Clearly to ensure a resonant build-up, revolution by revolution, we require the spin precession to be in phase with the vertical betatron oscillations. This is just the condition (6.3.31), which, ironically, as discussed in subsection 6.3.2 is the condition for an intrinsic resonance causing depolarization of the *vertical* component of the spin.

A slightly modified scheme to polarize antiprotons at the low energy CERN LEAR collider was proposed by Onel, Penzo and Rossmanith (1986) in which the spin precession between quadrupoles was to be caused by a Siberian snake in which $\boldsymbol{\mu}$ rotates about a longitudinal **B**-field. For a momentum of 200 MeV/c and $b = 20$ T/m it was estimated that a separation of 2.5 mm per hour could be built up. Unfortunately this scheme was never brought to fruition. It seems unrealistic in practice to try to have complete separation of the beams, especially given that the effect decreases with energy; see (6.3.39). In the time involved (hours) several other effects could destroy the build-up of polarization, e.g. depolarization due to imperfections, intrabeam scattering, fluctuations in power supplies etc. Thus the original idea does not seem practicable.

However, there are two new developments that suggest that a practicable scheme may be possible. The first idea does not directly resolve the difficulties but it does allow for a more flexible approach to the type of spin-based separation created.

Conte, Penzo and Pusterla (Conte *et al.*, 1995) argued that the use of longitudinal magnetic fields, with a field gradient along the particle's trajectory, induces tiny longitudinal forces that will result in minute changes ΔE of the kinetic energy of the particle. These changes will be negative or positive, depending on the sign of the longitudinal component of $\boldsymbol{\mu}$, and could be used to create a longitudinal separation in the beam that is correlated to the spin direction. In this case the effect does not decrease with energy, but, like the vertical or horizontal separation discussed earlier, it may be impractical to generate a utilizable separation in a reasonable time span.

The second idea is due to Derbenev (1990) and relies on a concept of resonant enhancement of the small splitting effects. For the vertical or horizontal splitting the particle is undergoing betatron oscillations whose phase is related to the spin direction. For the case of longitudinal separation the magnetic gradient could be provided by passing the beam through a transverse electric (TE) cavity, thus generating an oscillatory energy change ΔE whose phase is again linked to the spin direction. This would give rise to small induced coherent synchrotron oscillations in addition to the usual incoherent ones. If now a detector is constructed to measure either kind of oscillation it can be used to drive a feedback system that will apply a suitable rf field, leading to a resonant enhancement of the effect. Estimates suggest that significant polarization could be achieved in a few minutes. A practical scheme is discussed in Akchurin *et al.* (1996).

6.4 Polarized secondary and tertiary beams

It is a mysterious empirical fact that hyperons produced with medium momentum transfer in the collision of an unpolarized proton beam with an unpolarized hydrogen or nuclear target emerge with a significant degree of polarization. Moreover, the polarization is largely independent of the

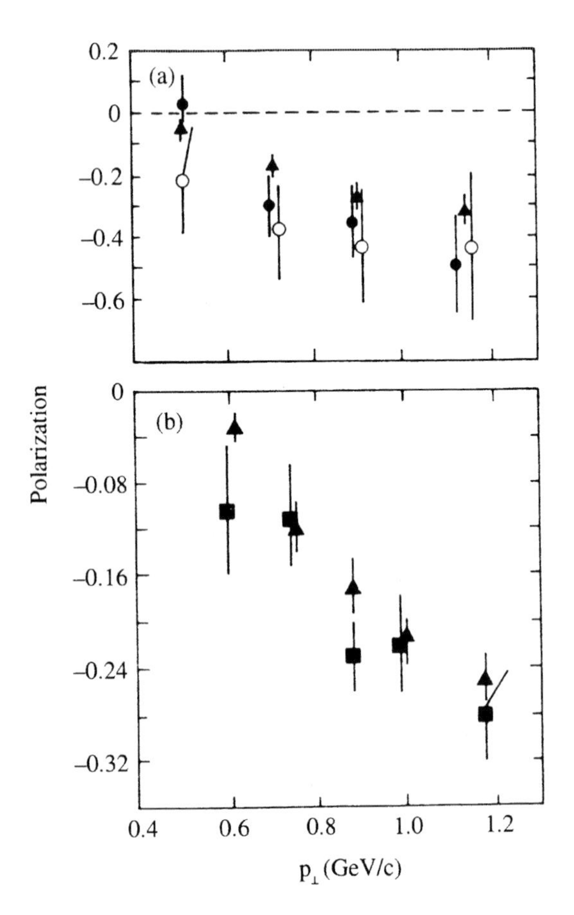

Fig. 6.15. Polarization of inclusive Λ production at several energies vs. $p_{\perp}$. (a) $p+N \rightarrow \Lambda^0+X$: ▲, 400 GeV; •, 1500 GeV; ∘, 2000 GeV. (b) ▲, 400 GeV H_2; ■, 28 GeV H_2 or D_2. The five sets of points in (b) correspond to x_F-values, from left to right, of 0.36, 0.45, 0.52, 0.59 and 0.69.

collision energy over a wide range, as shown for Λ particles in Fig. 6.15. The dependence of the polarization upon the type of hyperon produced, for the $p - Be$ reaction at $P_{\text{Lab}} = 400$ GeV/c and at a fixed Lab angle of 5 mrad, is shown in Fig. 6.16 plotted against the Lab momentum of the produced hyperon. It is seen that there is a significant polarization for Λ, Σ^+, Ξ^-. There is no convincing theoretical explanation for this polarization, as will be discussed in Chapter 13.

It has proved possible at Fermilab to utilize these secondary polarized hyperons in further interactions in which tertiary hyperons are produced. By analysing the decay distribution of the latter, one can measure the depolarization or polarization transfer parameters for reaction of the type

$$\text{hyperon} + \text{nucleus} \rightarrow \text{hyperon} + X.$$

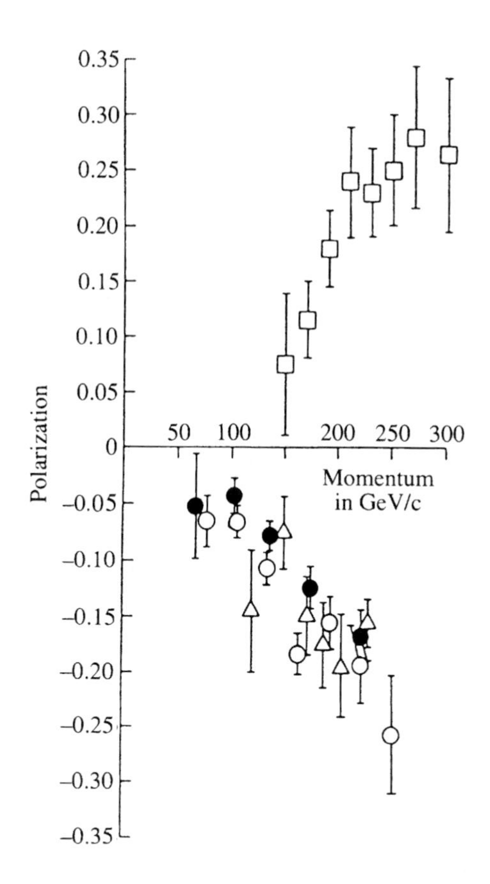

Fig. 6.16 Polarization in inclusive hyperon production at 400 GeV/c at a fixed Lab angle of 5 mrad: $\circ$, $p + \text{Be} \rightarrow \Lambda^0 + X$; $\bullet$, $p + p \rightarrow \Lambda^0 + X$; $\triangle$, $p + \text{Be} \rightarrow \Xi^- + X$; $\square$, $p + \text{Be} \rightarrow \Sigma^+ + X$.

Almost at the boundary of science fiction are experiments at Fermilab on the production of Ω^- (Longo *et al.*, 1989). Using 800 GeV/c unpolarized protons on a beryllium target, a sample of about 100 000 Ωs was obtained and was found to have essentially zero polarization; see Fig. 6.17.

When instead a secondary beam consisting of a mixture of polarized Λs and Ξ^0s strikes a copper target the produced Ωs are found to be significantly polarized! (See Fig. 6.18.) At the time of the Minneapolis Conference (1988) some 20 000 polarized Ωs had been produced in this way and had been allowed to precess in a magnetic field so as to allow the first ever measurement of their magnetic moment. By now (the year 2001) the magnetic moment is known with some precision,

$$\mu(\Omega) = (-2.02 \pm 0.05)\mu_{\rm N},$$

in nuclear magnetons; this is a result of importance in testing the constituent quark model of the hadrons.

One of the most beautiful developments in recent years has been the construction of very energetic, highly polarized, *tertiary* proton and an-

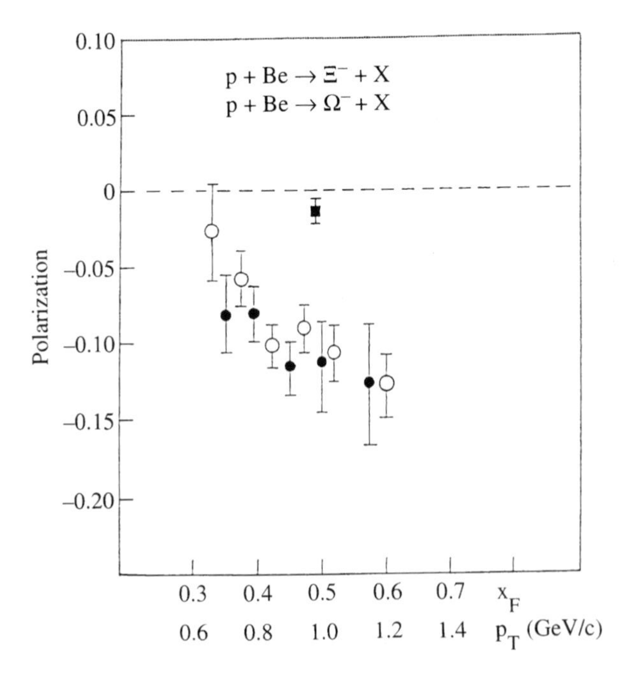

Fig. 6.17 Comparison of Ξ and Ω polarization in inclusive hyperon production on beryllium (from Longo *et al.*, 1989): ∘, Ξ_- at 400 GeV, 5 mrad; •, Ξ^- at 800 GeV, 2.5 mrad; , Ω^- at 800 GeV, 2.5 mrad.

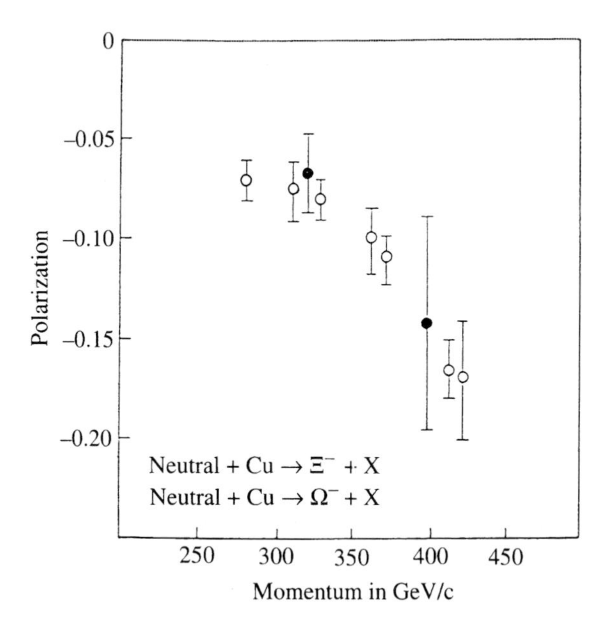

Fig. 6.18 Polarization of Ωs produced when a polarized hyperon beam strikes a copper target (from Longo *et al.*, 1989): ∘, Ξ^-; •, Ω^-.

tiproton beams at Fermilab (Grosnick *et al.*, 1990). To be specific, we shall discuss the proton beam; the antiproton case is analogous.

Protons of momentum 800 GeV/c from the Fermilab Tevatron strike a beryllium target and produce copious numbers of approximately forward-going Λs. The forward-going Λs must be unpolarized on account of eqns (3.1.35), (5.4.2) and (5.3.18b). However, their decay in flight into $p\pi^-$ is parity violating and the protons are produced with a longitudinal polarization $\mathscr{P} \approx 64\%$ (see subsection 8.2.1).

The spin-polarization vector $\boldsymbol{\mathcal{P}}$ is shown in Fig. 6.19 in the helicity rest frame S_p of the proton as reached from the helicity rest frame S_Λ of the Λ particle. The proton is produced at an angle θ in S_Λ, as shown, and with energy

$$E = \frac{m_\Lambda^2 + m^2 - \mu^2}{2m_\Lambda} \approx 943 \text{ MeV}$$

and magnitude of momentum

$$p = \frac{\left\{\left[(m_\Lambda + m)^2 - \mu^2\right]\left[(m_\Lambda - m)^2 - \mu^2\right]\right\}^{1/2}}{2m_\Lambda} \approx 101 \text{ MeV}/c.$$

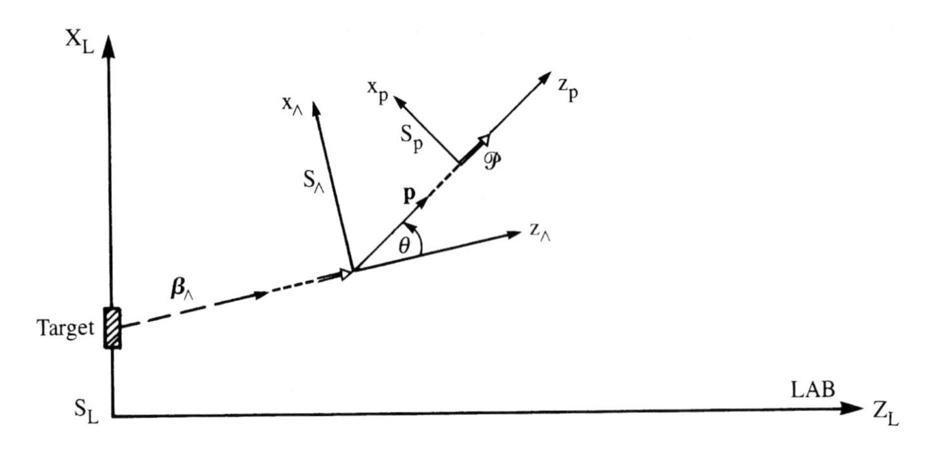

Fig. 6.19 The decay $\Lambda \to p\pi^-$ showing the Λ helicity rest frame S_Λ and the proton helicity rest frame S_p as reached from S_Λ. The proton has momentum $\mathbf{p}$ in S_Λ. $\mathscr{P}$ is the spin-polarization vector.

The spin-polarization vector $\mathcal{P}$ lies along OZ_p in the proton's helicity rest frame. Note that the proton is almost completely non-relativistic in the rest frame S_Λ.

Viewed in the Lab we have then the picture shown in Fig. 6.20, in which S_p' is the proton's helicity rest frame as reached from the laboratory frame. One has

$$p_L \sin\theta' = p\sin\theta \tag{6.4.1}$$

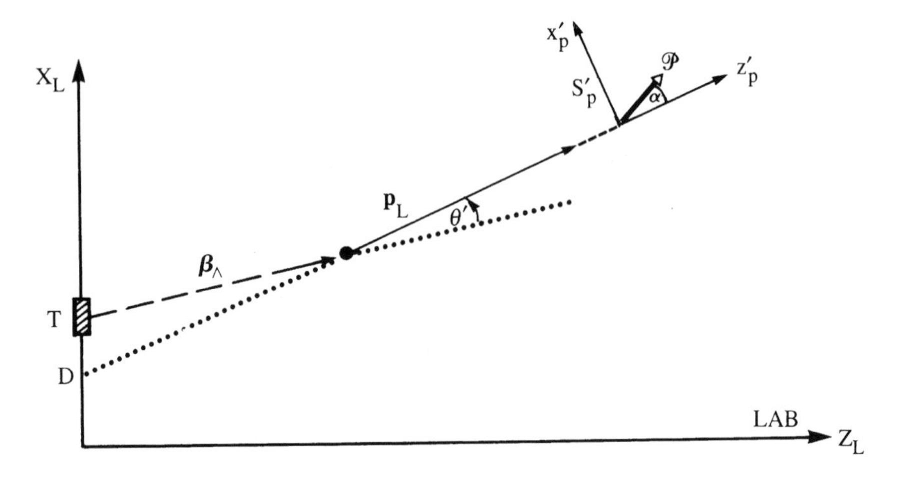

Fig. 6.20 The decay $\Lambda \to p\pi^-$ in the Lab. S_p' is the proton helicity rest frame reached from the Lab, where the proton has momentum $\mathbf{p}_L$. $\mathscr{P}$ makes an angle α with OZ_p'.

where p_{L} is the magnitude of the proton's Lab momentum and is ≈ 185 GeV/c.

The boost of axes $l(\boldsymbol{\beta})$ with $\boldsymbol{\beta} = -\boldsymbol{\beta}_\Lambda$ that takes us to the Lab induces a Wick helicity rotation on the spin-polarization vector so that $\boldsymbol{\mathcal{P}}$, which was along OZ_p in S_p is now at an angle α to OZ'_p of S'_p; from (2.2.9), (2.2.6) and (3.2.10) $\alpha = -\theta_{\mathrm{Wick}}$ is given by

$$\begin{aligned}\sin\alpha &= \frac{m}{p_{\mathrm{L}}}\gamma_\Lambda\beta_\Lambda \sin\theta \\ &= \frac{m}{m_\Lambda}\frac{p_\Lambda}{p_{\mathrm{L}}}\sin\theta\end{aligned}$$

and using (6.4.1)

$$\sin\alpha = \frac{m}{m_\Lambda}\frac{p_\Lambda}{p}\sin\theta'. \qquad (6.4.2)$$

The component of the spin-polarization vector *transverse* to the proton's motion is

$$\begin{aligned}\mathscr{P}_T &= \mathscr{P}\sin\alpha \\ &= \mathscr{P}\frac{m}{m_\Lambda}\frac{p_\Lambda}{p}\sin\theta'. \end{aligned} \qquad (6.4.3)$$

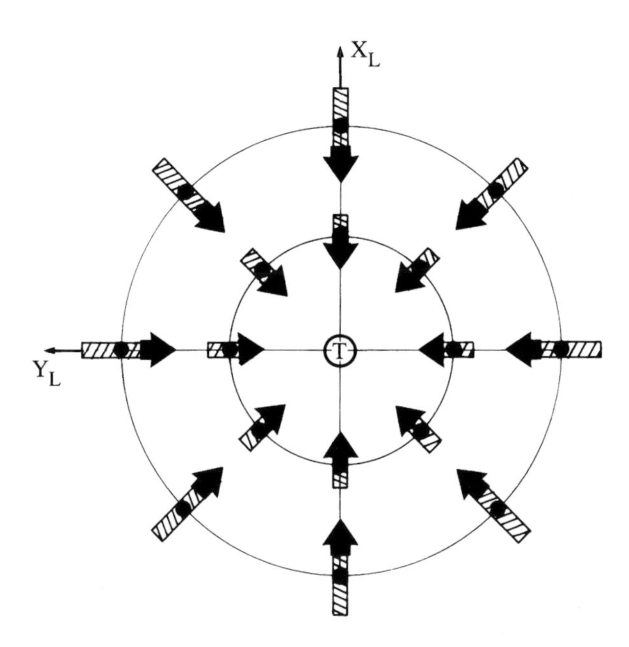

Fig. 6.21 Correlation between the magnitude and direction of the transverse component of the proton's spin-polarization vector and the position of the virtual source in the $X_{\mathrm{L}}Y_{\mathrm{L}}$-plane. The solid circles give the positions to which the spin-polarization vectors refer.

Moreover the geometry is such that θ' is an extremely small angle, so that if the line of motion of the proton is projected backwards until it reaches the plane of the target at D, as shown in Fig. 6.20, then

$$DT = L\theta' \tag{6.4.4}$$

where L is the distance covered by the Λs before decaying.

Thus we see that $\mathcal{P}_T$ is proportional to the distance from the target to the 'virtual source' D of the protons in the Lab. Moreover, if $\mathbf{p}$ lies below OZ_Λ, i.e. $\mathbf{p} = (p, \theta, \pi)$, then $\alpha = +\theta_{\text{Wick}}$ and $\mathcal{P}_T$ points in the opposite direction to the case $\mathbf{p} = (p, \theta, 0)$. Since, in addition, there is cylindrical symmetry about OZ_L we end up with a correlation between the position of the virtual source in the $X_\text{L}Y_\text{L}$-plane and the magnitude and direction of $\mathcal{P}_T$. This is shown qualitatively in Fig. 6.21.

The construction of these tertiary proton and antiproton beams was completed in 1989 and a major experimental programme was begun. Many of the asymmetries, e.g. in inclusive π^0 production, originally discovered at much lower energies, persist at higher energies, and a rich harvest of results has emerged. These are discussed in Chapter 13.

7
The production of polarized $e^{\pm}$

Quite dramatic progress has been made in the production and utilization of polarized $e^{\pm}$ beams at CERN's LEP, at HERA at DESY and at the Stanford linear collider SLC. The motivation for trying to overcome the tremendous technical problems involved derives from two sources:

(i) the realization that longitudinally polarized electrons permit extremely accurate measurement of the fundamental parameters of the Standard Model of electroweak interactions;
(ii) the discovery in 1987, by the European Muon Collaboration (Ashman *et al.*, 1988), that only a very small fraction of the proton's spin appeared to be carried by its quarks, leading to what was characterized as a 'crisis in the parton model' (Leader and Anselmino, 1988). This made it important to carry out further studies of deep inelastic lepton–hadron scattering using longitudinally polarized leptons colliding with a longitudinally polarized proton target.

Though not a primary impetus, it turns out also that polarized $e^{\pm}$ permit an exceedingly accurate calibration of the beam energy at LEP and HERA.

The problems involved in having stable polarized beams are quite different in circular storage rings and in linear accelerators. Hence we shall discuss the two cases separately.

7.1 The natural polarization of electrons circulating in a perfect storage ring

As mentioned in the introduction to Chapter 6, in principle a circulating electron beam gradually acquires a natural polarization in which its magnetic moment $\boldsymbol{\mu}_e$ becomes aligned parallel to the guide field **B**. Ultimately a maximum degree of polarization

$$\mathscr{P}_0 = \frac{8}{5\sqrt{3}} \approx 92\% \tag{7.1.1}$$

is attained (Sokolov and Ternov, 1963). This is known as the Sokolov–Ternov effect. For electrons, with $\boldsymbol{\mu}_e$ opposite in direction to **s**, the mean spin vector will be polarized antiparallel to **B**.

We consider an idealized perfectly circular ring of radius R with a uniform guide field $B_0\mathbf{e}_z$ and with all particles having a fixed energy E. Our discussion relies heavily on a very illuminating treatment by Jackson (1976).

At first sight the closeness of $\mathscr{P}_0$ to 100% together with the fact that $\boldsymbol{\mu}_e$ lines up along **B** suggests that the phenomenon is trivial, namely that radiative transitions cause the system to populate the state of lowest energy in the hamiltonian $H = -\boldsymbol{\mu} \cdot \mathbf{B}$. This would be true for an isolated spin system, but to regard an electron in a storage ring as an isolated spin system is self-contradictory. This can be seen as follows.

In this picture, in its canonical rest frame the electron would see a magnetic field $\gamma B_0\mathbf{e}_z$, so that the energy levels, in this frame, would have separation

$$\Delta\overset{\circ}{E}_{\text{spin}} = \frac{g}{2}\left(\frac{e\hbar\gamma B_0}{mc}\right). \tag{7.1.2}$$

This separation, in the Lab, is

$$\Delta E_{\text{spin}} = \gamma\Delta\overset{\circ}{E}_{\text{spin}} = \left(\frac{g\gamma^3}{2}\right)\hbar\Omega_{\text{c}}, \tag{7.1.3}$$

where, for an electron,

$$\Omega_{\text{c}} = \frac{eB_0}{\gamma mc}. \tag{7.1.4}$$

Now consider the orbital angular momentum (kinetic, not canonical) of the electron: it is

$$l\hbar \approx Rp = mR\gamma v$$

so that

$$l = \frac{cmR\gamma\beta}{\hbar} = \left(\frac{R}{\lambda\!\!\!^{-}_e}\right)\gamma\beta \tag{7.1.5}$$

where $\lambda\!\!\!^{-}_e$, the Compton wavelength of the electron, $\approx 4 \times 10^{-13}$ m. For an ultra-relativistic electron with, say, $R \approx 1000$ m, $\gamma \approx 10^5$ and $\beta \approx 1$ we have

$$l \approx 4 \times 10^{21}. \tag{7.1.6}$$

The orbital levels with different values of l_z will be separated by

$$\Delta E_{\text{orbital}} = \hbar\Omega_{\text{c}}. \tag{7.1.7}$$

Thus

$$\frac{\Delta E_{\text{orbital}}}{\Delta E_{\text{spin}}} = \frac{2}{g\gamma^3} \ll 1. \tag{7.1.8}$$

Hence any radiative transition involving energies of order ΔE_{spin} will involve huge changes in l_z and there will exist a strong coupling between the spin and orbital degrees of freedom. In other words the spin system is not at all isolated!

Incidentally, this does not imply that the motion is non-classical, because for an emitted photon whose energy is given by (7.1.3) the electron recoil will imply a change in momentum, using (6.3.29),

$$\Delta p \simeq \frac{g}{2}\left(\frac{\gamma^3\hbar\Omega_{\text{c}}}{c}\right) \simeq \frac{g\gamma^3}{2R}.$$

This yields a large change in l of order

$$\delta l \approx \frac{g}{2}\gamma^3 \approx \frac{g}{2} \times 10^{15}. \tag{7.1.9}$$

Nonetheless,

$$\frac{\delta l}{l} \simeq \frac{g}{2}\left(\frac{\lambda\!\!\!^{-}_e}{R}\right)\gamma^2 \approx 10^{-6} \tag{7.1.10}$$

and so remains very small for our example.

We shall now outline how the effect can be understood in the framework of quasi-classical radiation theory, on the basis of the relativistic motion of the spin vector.

Recall that, in the usual quasi-classical radiation theory, the spontaneous emission of a photon with momentum $\mathbf{k}$ and polarization vector $\boldsymbol{\varepsilon}$ arises from a time-dependent perturbation engendered by the coupling of the charge of the particle to a classical electromagnetic vector potential $\mathbf{A}'(\mathbf{r}, t)$. This is chosen to correspond, in intensity, to having one photon present. Thus one takes

$$H'_{\text{charge}} = -\frac{q}{mc}\mathbf{p}\cdot\mathbf{A}' \tag{7.1.11}$$

with

$$\mathbf{A}' = c\sqrt{\frac{2\pi\hbar}{\omega}}\boldsymbol{\varepsilon}^* e^{i(\omega t - \mathbf{k}\cdot\mathbf{r})}. \tag{7.1.12}$$

To obtain the radiative transitions due to the spin we note that the effective hamiltonian giving rise to the spin motion (6.3.22) is

$$H_{\text{spin}} = -\frac{q}{mc}\mathbf{s}\cdot\left[\left(\frac{g}{2}-1+\frac{1}{\gamma}\right)\mathbf{B}-\left(\frac{g}{2}-1\right)\frac{\gamma}{\gamma+1}(\boldsymbol{\beta}\cdot\mathbf{B})\boldsymbol{\beta}\right.$$
$$\left.-\left(\frac{g}{2}-1+\frac{1}{\gamma+1}\right)\boldsymbol{\beta}\times\mathbf{E}\right] \tag{7.1.13}$$

and we produce a spontaneous transition by taking the electric and magnetic fields to correspond to $\mathbf{A}'$ in (7.1.12), i.e.

$$\begin{aligned}\mathbf{E}' &= -i\sqrt{2\pi\hbar\omega}\boldsymbol{\varepsilon}^* e^{i(\omega t-\mathbf{k}\cdot\mathbf{r})} \equiv \mathscr{E}\boldsymbol{\varepsilon}^*\\ \mathbf{B}' &= -i\sqrt{2\pi\hbar\omega}(\hat{\mathbf{k}}\times\boldsymbol{\varepsilon}^*)e^{i(\omega t-\mathbf{k}\cdot\mathbf{r})} = \mathscr{E}(\hat{\mathbf{k}}\times\boldsymbol{\varepsilon}^*).\end{aligned} \tag{7.1.14}$$

We shall simplify life by dealing only with electrons from now on, and so take $g/2-1=0$. Thus the perturbing hamiltonian becomes

$$H'_{\text{spin}} = \frac{e}{\gamma mc}\mathbf{s}\cdot\left(\mathbf{B}'-\frac{\gamma}{\gamma+1}\boldsymbol{\beta}\times\mathbf{E}'\right). \tag{7.1.15}$$

The total hamiltonian is then

$$H = H^0 + H'_{\text{charge}} + H'_{\text{spin}} \equiv H^0 + H' \tag{7.1.16}$$

where H^0 includes the interaction with the guide field $\mathbf{B}_0 = B_0\mathbf{e}_z$, which can be taken to come from a vector potential

$$\mathbf{A}_0 = -\tfrac{1}{2}\mathbf{r}\times\mathbf{B}_0. \tag{7.1.17}$$

Thus, using (7.1.13),

$$\begin{aligned}H^0 &= \sqrt{(c\mathbf{P}^2+e\mathbf{A}_0)^2+m^2c^2}+\frac{1}{\gamma}\frac{e}{mc}\mathbf{s}\cdot\mathbf{B}_0\\ &\equiv H^0_{\text{charge}}+H^0_{\text{spin}}\end{aligned} \tag{7.1.18}$$

where $\mathbf{P}$ is the *canonical* momentum.

The problem is solved hierarchically as follows.

(1) The usual classical motion is obtained from H^0_{charge}, ignoring the influence of the spin upon the orbit (s is explicitly of order $\hbar$).
(2) The unperturbed motion of the spin is then controlled by H^0_{spin}. The influence of the orbital motion has been taken into account in going from (7.1.13) to (7.1.18) (the appearance of $g/2-1+1/\gamma=1/\gamma$ instead of $g/2$). Because $g=2$ the mean spin vector $\mathbf{s}(t)$ rotates about $\mathbf{B}_0$ at the same angular frequency as $\boldsymbol{\beta}(t)$ (see eqn (6.3.24)), i.e. with the

relativistic cyclotron frequency

$$\omega_c \equiv \Omega_c^{\text{electron}} = \frac{eB_0}{\gamma mc}. \tag{7.1.19}$$

(3) Time-dependent perturbation theory tells us that the probability for the spontaneous emission of a photon with $\mathbf{k}$ in $d^3\mathbf{k}$ $(= \omega^2 d\omega d\Omega/c^3)$ during the time interval $t_1 \to t_2$ is

$$dp = \left| \frac{1}{i\hbar} \int_{t_1}^{t_2} \langle f|H'(t)|i\rangle dt \right|^2 \frac{\omega^2 d\omega d\Omega}{(2\pi c)^3}. \tag{7.1.20}$$

The term H'_{charge} in (7.1.11) gives rise to the usual synchrotron radiation and is of no interest to us here. We thus utilize (7.1.15) for H', in which $\mathbf{s}$ is now the unperturbed spin operator.

It is simplest to work in the Heisenberg picture because there the time-dependent operators obey equations of motion that are formally the same as those governing the motion of the mean values of the operators. Thus, in (7.1.15)

$$\mathbf{s} \to \hat{\mathbf{s}}(t) \equiv (\hbar/2)\, \boldsymbol{\sigma}(t) \tag{7.1.21}$$

where $\boldsymbol{\sigma}(t)$ rotates about $\mathbf{B}_0$ at angular freqency ω_c. Thus, we can take

$$\begin{aligned} \sigma_x(t) &= \sigma_x \cos\omega_c t - \sigma_y \sin\omega_c t \\ \sigma_y(t) &= \sigma_x \sin\omega_c t + \sigma_y \cos\omega_c t \\ \sigma_z(t) &= \sigma_z. \end{aligned} \tag{7.1.22}$$

The perturbation H'_{spin} will give rise to both non-flip and spin-flip emission. Here we are only interested in the latter: if we quantize our states along OZ then spin-flip can only arise from the matrices $\sigma_x(t)$, $\sigma_y(t)$ or, more precisely, from

$$\sigma_\pm(t) = \tfrac{1}{2}\left[\sigma_x(t) \pm i\sigma_y(t)\right] = \sigma_\pm e^{\pm i\omega_c t} \tag{7.1.23}$$

where $\sigma_\pm = (\sigma_x \pm \sigma_y)/2$ are the usual spin-raising and spin-lowering matrices.

Using the fact that for any two vectors $\mathbf{C}$, $\mathbf{D}$,

$$\mathbf{C} \cdot \mathbf{D} = 2(C_+ D_- + C_- D_+) + C_z D_z$$

we see that the relevant, spin-flip, part of (7.1.15), is

$$\frac{e\hbar}{\gamma mc}\left[\sigma_+(t)\left(\mathbf{B}' - \frac{\gamma}{\gamma+1}\boldsymbol{\beta}\times\mathbf{E}\right)_- + \sigma_-(t)\left(\mathbf{B}' - \frac{\gamma}{\gamma+1}\boldsymbol{\beta}\times\mathbf{E}\right)_+\right]. \tag{7.1.24}$$

Now, it is known that the radiation from a relativistic particle whose acceleration is perpendicular to its velocity is confined to a narrow cone about $\mathbf{v}$ of opening angle $\theta \approx 1/\gamma$. So we may, for simplicity, consider a linearly polarized photon with $\mathbf{k}$ in the XY-plane, say along OY. Then we can put

$$\boldsymbol{\varepsilon}^* = (\sin\alpha, 0, \cos\alpha) = (\epsilon_x, 0, \epsilon_z) \tag{7.1.25}$$

and take

$$\boldsymbol{\beta}(t) = \beta(-\sin\omega_c t, \cos\omega_c t, 0). \tag{7.1.26}$$

Then (7.1.24) involves, from (7.1.14),

$$\begin{aligned}\left(\mathbf{B}' - \frac{\gamma}{\gamma+1}\boldsymbol{\beta}\times\mathbf{E}'\right) &= \mathscr{E}\left[(\hat{\mathbf{k}}\times\boldsymbol{\varepsilon}^*) - \frac{\gamma}{\gamma+1}\boldsymbol{\beta}\times\boldsymbol{\varepsilon}^*\right] \\ &= \mathscr{E}\left[(\epsilon_z, 0, -\epsilon_x)\right. \\ &\quad \left. - \left(\frac{\beta\gamma}{\gamma+1}\right)(\epsilon_z\cos\omega_c t, \epsilon_z\sin\omega_c t, -\epsilon_x\cos\omega_c t)\right]\end{aligned}$$

so that

$$\left(\mathbf{B}' - \frac{\gamma}{\gamma+1}\boldsymbol{\beta}\times\mathbf{E}'\right)_{\pm} = \tfrac{1}{2}\mathscr{E}\cos\alpha\left(1 - \frac{\beta\gamma}{\gamma+1}e^{\pm i\omega_c t}\right).$$

Finally, then, (7.1.24) becomes

$$\frac{e\hbar\mathscr{E}\cos\alpha}{2\gamma mc}\left[\left(e^{i\omega_c t} - \frac{\beta\gamma}{\gamma+1}\right)\sigma_+ + \left(e^{-i\omega_c t} - \frac{\beta\gamma}{\gamma+1}\right)\sigma_-\right]. \tag{7.1.27}$$

This is the key result. It shows that the spin-raising and spin-lowering parts of the perturbing hamiltonian are different.

Now because, as mentioned, the radiation cone has opening angle $\approx 1/\gamma$, for our choice of $\mathbf{k}$ along OY the relevant times will be those for which $\boldsymbol{\beta}$ lies in such a cone, i.e. $|\omega_c t| \lesssim 1/\gamma$. Thus we can expand the exponentials in (7.1.27) and use the fact that $1-\beta$ is of order $1/\gamma^2$ for $\gamma \gg 1$, to obtain

$$\begin{aligned}H'_{\text{spin-flip}} &= \frac{e\hbar\mathscr{E}\cos\alpha}{2\gamma mc}\left\{\left(1 - i\omega_c t - \beta + \frac{\beta}{\gamma+1}\right)\sigma_+\right. \\ &\quad \left. + \left(1 + i\omega_c t - \beta + \frac{\beta}{\gamma+1}\right)\sigma_-\right\} \\ &\simeq \frac{e\hbar\mathscr{E}\cos\alpha}{2\gamma^2 mc}\{(1-iu)\sigma_+ + (1+iu)\sigma_-\}\end{aligned} \tag{7.1.28}$$

where

$$u \equiv \gamma\omega_c t. \tag{7.1.29}$$

Substituting for $\mathscr{E}$ from (7.1.14), the time integral in (7.1.20) is thus of the form

$$\frac{1}{i\hbar}\int \langle f|H'_{\text{spin-flip}}(t)|i\rangle dt = -\sqrt{\frac{2\pi\omega}{\hbar}}\frac{e\hbar\cos\alpha}{2\gamma^2 mc}\langle f|\sigma_\pm|i\rangle \times \int_{-1/\gamma_c}^{1/\gamma_c}(1\mp iu)e^{i[\omega t-\mathbf{k}\cdot\mathbf{R}(t)]}dt \qquad (7.1.30)$$

where $\mathbf{R}(t)$ is the position vector along the trajectory.

The integrals are similar to those that occur in ordinary synchrotron radiation. They are approximated in a standard fashion, which we will not reproduce here. However, we shall at least show how the various powers of γ enter the transition rate. The exponent in (7.1.30), for our case of $\mathbf{k}$ along OY, and bearing in mind $|t|\lesssim(\gamma\omega_c)^{-1}$, is

$$\omega t - kR\sin\omega_c t = \omega\left(t - \frac{\beta}{\omega_c}\sin\omega_c t\right) = \omega\left(t - \beta t + \frac{\beta\omega_c^2 t^3}{6} + \cdots\right)$$
$$\cong \omega\left(\frac{t}{2\gamma^2} + \frac{\beta\omega_c^2 t^3}{6} + \cdots\right) \quad \text{for } \beta \simeq 1.$$

Both terms are of order $\omega_c^{-1}\gamma^{-3}$ whereas the terms left out are of order $\omega_c^{-1}\gamma^{-5}$.

The wave factor in (7.1.30) then becomes

$$\exp\left[i\frac{3}{2}\left(\frac{\omega}{\omega_{\text{cr}}}\right)u\left(1+\frac{u^2}{3}\right)\right] \qquad (7.1.31)$$

where, conventionally,

$$\omega_{\text{cr}} \equiv 3\gamma^3\omega_c \qquad (7.1.32)$$

is the characteristic frequency of synchrotron radiation. The integrals in (7.1.30) yield a result of the form

$$\frac{1}{\gamma\omega_c}\left[f_1\left(\frac{\omega}{\omega_{\text{cr}}}\right) \pm f_2\left(\frac{\omega}{\omega_{\text{cr}}}\right)\right]$$

where $f_{1,2}$ are in fact Bessel-type functions.

Gathering all factors from (7.1.14), (7.1.15) and (7.1.18) into (7.1.20), we have for the two spin-flip probabilities, per revolution,

$$\frac{dp^{\downarrow\uparrow/\uparrow\downarrow}}{d\Omega d\omega} = \frac{e^2\hbar\omega^3}{32\pi^2 m^2 c^5\gamma^6\omega_c^2}\left[f_1\left(\frac{\omega}{\omega_{\text{cr}}}\right) \pm f_2\left(\frac{\omega}{\omega_{\text{cr}}}\right)\right]^2. \qquad (7.1.33)$$

We have cheated in (7.1.33) in not taking into account any angular dependence when $\mathbf{k}$ points outside the XY-plane. We account for this dependence roughly by taking $d\Omega \sim 2\pi/\gamma$. Then we divide by the period of revolution $2\pi/\omega_c$ to get a transition rate and integrate over ω, changing to the variable $\xi = \omega/\omega_{cr}$. The result is of the form

$$w^{\downarrow\uparrow/\uparrow\downarrow} \simeq \frac{e^2\hbar\omega_{cr}^4}{32\pi^2 m^2 c^5 \gamma^7 \omega_c}(a \pm b)$$

where $w^{\downarrow\uparrow}$ is the rate for the transition $\downarrow$ to $\uparrow$, $w^{\uparrow\downarrow}$ is the rate for the transition $\downarrow$ to $\uparrow$ and a, b are numbers of order unity. Substituting for ω_{cr} gives

$$w^{\downarrow\uparrow/\uparrow\downarrow} \approx \left(\frac{81}{32\pi^2}\right)\frac{e^2\hbar\gamma^5}{m^2c^2R^3}(a \pm b). \tag{7.1.34}$$

The precise numerical values of a and b depend upon a careful integration over angles and ω, but the essential kinematic dependence is correctly given by (7.1.34). An accurate treatment yields

$$w^{\uparrow\downarrow/\downarrow\uparrow} = \left(\frac{5\sqrt{3}}{16}\right)\frac{e^2\hbar\gamma^5}{m^2c^2R^3}\left(1 \pm \frac{8}{5\sqrt{3}}\right) \tag{7.1.35}$$

$$\equiv \frac{1}{2\tau_{ST}}\left(1 \pm \frac{8}{5\sqrt{3}}\right) \tag{7.1.36}$$

where τ_{ST}, as will be seen, is the characteristic rise time for the Sokolov–Ternov polarization to build up from an unpolarized state: one has

$$\tau_{ST} = \frac{8}{5\sqrt{3}}\frac{m^2c^2R^3}{e^2\hbar\gamma^5}. \tag{7.1.37}$$

For $R \sim 1000$ m, $\gamma \sim 10^5$, $\tau_{ST} \sim 5$ minutes. For LEP, running near the Z^0 mass, $\tau_{ST} \approx 310$ minutes.

Consider now the numbers of particles with spin up, $n_\uparrow(t)$, or down, $n_\downarrow(t)$, assuming that at time $t = 0$ $n_\uparrow = n_\downarrow$, i.e. the system is unpolarized. We have

$$\begin{aligned} \frac{dn_\uparrow}{dt} &= w^{\uparrow\downarrow}n_\downarrow - w^{\downarrow\uparrow}n_\uparrow = nw^{\uparrow\downarrow} - n_\uparrow w \\ \frac{dn_\downarrow}{dt} &= w^{\downarrow\uparrow}n_\uparrow - w^{\uparrow\downarrow}n_\downarrow = nw^{\downarrow\uparrow} - n_\downarrow w \end{aligned} \tag{7.1.38}$$

where

$$w \equiv w^{\uparrow\downarrow} + w^{\downarrow\uparrow} = 1/\tau_{ST} \tag{7.1.39}$$

and

$$n = n_\uparrow + n_\downarrow = \text{ constant.}$$

The degree of polarization along OZ is

$$\mathscr{P}(t) = \frac{n_\uparrow(t) - n_\downarrow(t)}{n}. \tag{7.1.40}$$

So

$$\frac{d\mathscr{P}}{dt} = \frac{1}{n}\left(\frac{dn_\uparrow}{dt} - \frac{dn_\downarrow}{dt}\right) = \left(w^{\uparrow\downarrow} - w^{\downarrow\uparrow}\right) - w\mathscr{P}.$$

Thus

$$\mathscr{P}(t) = \frac{w^{\uparrow\downarrow} - w^{\downarrow\uparrow}}{w}\left(1 - e^{-t/\tau_{\mathrm{ST}}}\right) \tag{7.1.41}$$

The ultimate polarization, in a perfect machine, due to the Sokolov–Ternov mechanism is thus

$$\mathscr{P}_{\mathrm{ST}} = \frac{8}{5\sqrt{3}}. \tag{7.1.42}$$

It seems clear that the precise value of $\mathscr{P}_{\mathrm{ST}}$ is *not* related to some simple physical fact. It emerges from integrals over Bessel functions. Moreover, Jackson (1976) studied the situation for arbitrary values of g, and for a certain range of *positive* g-values, $0 < g < 1.2$, finds that $\mathscr{P}_{\mathrm{ST}}$ even has the opposite sign to (7.1.42)!

7.1.1 Imperfect storage rings

In the previous section we dealt with a perfect storage ring, i.e. one that is absolutely planar with its guide field perpendicular to the orbit plane and with no magnetic imperfections.

In that case there is a unit vector $\mathbf{n}$, along or opposite to the guide field, such that a mean spin vector initially along $\mathbf{n}$ will remain so, independent of the azimuthal angle θ that specifies the position of the particle on its orbit. A general spin vector, not along $\mathbf{n}$, will precess around $\mathbf{n}$ as the particle moves in its orbit.

In the case of an imperfect machine there is no such fixed direction, but for a particle on a closed orbit there does exist a direction $\mathbf{n}(\theta)$, varying with the particle position, which is periodic, i.e.

$$\mathbf{n}(\theta + 2\pi) = \mathbf{n}(\theta) \tag{7.1.43}$$

and such that if the mean spin vector $\mathbf{s}(\theta)$ is initially along $\mathbf{n}(\theta)$ at some angle θ it will continue to point along $\mathbf{n}(\theta)$ as θ changes. Thus $\mathbf{n}(\theta)$ represents a periodic solution to the equations of spin motion (6.3.23).

For a closed orbit the magnetic fields experienced by the particle are periodic i.e. $\mathbf{B}(\theta + 2\pi) = \mathbf{B}(\theta)$, from which it is easy to show that any solution to the equation of spin motion must satisfy

$$\mathbf{s}(\theta + 2\pi) = R_\theta \mathbf{s}(\theta) \tag{7.1.44}$$

where R_θ is some rotation, which depends on θ and is itself periodic, $R_{\theta+2\pi} = R_\theta$.

If, now, we resolve an arbitrary $\mathbf{s}(\theta)$ into components along $\mathbf{n}(\theta)$ and orthogonal to it, then (7.1.43) and (7.1.44) clearly imply that the components of $\mathbf{s}(\theta)$ orthogonal to $\mathbf{n}(\theta)$ rotate around $\mathbf{n}(\theta)$ by a fixed number of radians per revolution. Moreover, since eqns (6.3.23)–(6.3.27) hold for an arbitrary field $\mathbf{B}$, the angle involved is simply $2\pi\nu_s$ where ν_s is the spin tune; $\nu_s = G\gamma$, introduced in (6.3.27). In short, in an imperfect machine the mean spin vector precesses about a periodic solution $\mathbf{n}(\theta)$ instead of about the unique guide field in a perfect machine. To the extent that one is only dealing with very small imperfections and there are no special spin rotator magnets in the ring, $\mathbf{n}(\theta)$ should deviate only slightly from the guide field direction except in the vicinity of the spin resonances discussed in subsection 6.3.2.

The mechanism of the natural Sokolov–Ternov polarization discussed in Section 7.1 continues to operate in the case of non-uniform fields but the direction of the equilibrium polarization is along $\mathbf{n}(\theta)$ rather than the guide field. The problem of imperfection and intrinsic resonances, which bedevils the acceleration of polarized protons (where the growing energy implies a growing spin tune which thus continually intercepts resonance values) ought not, ideally, to affect a storage ring, where the particles are circulating at a fixed energy chosen so that ν_s is well clear of a resonance value. In reality however, there may be a significant spread of energies so that electrons far from the central value may hit a depolarizing resonance.

The main mechanism for the spread in energies is discrete photon emission in addition to the usual classical synchrotron radiation. It is important for electrons, but totally negligible for protons. And, as we now explain, it gives rise to an important depolarizing effect. The probabilities for spin-flip in the emission of the photon ($w^{\uparrow\downarrow}$ and $w^{\downarrow\uparrow}$ given in (7.1.34)) are orders of magnitude smaller than the non-flip probabilities, so the emission may be considered to take place without spin-flip.

Photon emission is a random process, the time scale for which is minute in comparison with changes in orbit position or direction of the mean spin vector. The only significant effect on an electron following a closed orbit is thus its energy loss, so that it finds itself with too little energy to remain on its original orbit. It thus begins to execute horizontal betatron oscillations, and these lead to vertical oscillations as well.

Some indication of the mechanisms at work can be elicited by supposing that the electron was originally on the central closed orbit at an energy well clear of depolarizing resonances, with its mean spin vector along the associated periodic solution $\mathbf{n}_0(\theta)$. After emission it is on an orbit for which $\mathbf{n}(\theta) \neq \mathbf{n}_0(\theta)$, so its spin vector begins to precess about $\mathbf{n}(\theta)$. As the electron gradually picks up RF energy its orbit oscillations are damped out, its orbit approaches the central orbit and $\mathbf{n}(\theta) \to \mathbf{n}_0(\theta)$. In this relatively slow process the mean spin vector continues to precess about $\mathbf{n}(\theta)$, adiabatically following its change until it ends up precessing about $\mathbf{n}_0(\theta)$. Since it was originally *along* $\mathbf{n}_0(\theta)$ its component along $\mathbf{n}_0(\theta)$ has decreased.

In fact each electron emits many photons and, since the emissions are uncorrelated both in time and energy, the perturbations give rise to a random walk of the mean spin vector superposed on any coherent precession motion. The stochastic nature of the photon emission results in a diffusion of the spin vectors and hence to a depolarization of the beam.

Bearing in mind that only a tiny fraction of emissions involve spin-flip and thus contribute to the Sokolov–Ternov mechanism, it is clear that the spin diffusion is potentially a very strong effect and the achievable polarization $\mathscr{P}_{\max}$ may be much less than $\mathscr{P}_{\mathrm{ST}}$.

The strength of the depolarizing process can be characterized by a diffusion time τ_{D}. The competition between the Sokolov–Ternov and diffusion mechanisms then results in an asymptotic maximal polarization

$$\mathscr{P}_{\max} = \left(\frac{\tau_{\mathrm{D}}}{\tau_{\mathrm{ST}} + \tau_{\mathrm{D}}}\right) \mathscr{P}_{\mathrm{ST}}. \tag{7.1.45}$$

The polarization build-up time is reduced to

$$\tau = \left(\frac{\mathscr{P}_{\max}}{\mathscr{P}_{\mathrm{ST}}}\right) \tau_{\mathrm{ST}}. \tag{7.1.46}$$

Without special precautions, τ_{D} can be quite small compared with τ_{ST}, leading to a serious loss of beam polarization. It is thus essential to take steps to counteract the depolarizing mechanism.

The horizontal and vertical orbit oscillations are not purely simple harmonic. However, they may be expanded in a Fourier series with frequencies per revolution specified by integers k. The actual associated integer depolarization resonances then occur at $\nu_{\mathrm{s}} = k$, but even for $\nu_{\mathrm{s}} \neq k$ they have an influence that depends upon their strength and their proximity to ν_{s}.

The vertical orbit oscillations are the most damaging for the polarization since the result is that the spin-vector is rotated away from the essentially vertical direction $\mathbf{n}_0(\theta)$. It is possible to compensate for these by the method of *harmonic spin matching* (Rossmanith and Schmidt, 1985; Barber *et al.*, 1994). Additional vertical distortions ('bumps') are introduced at strategic positions along the orbit and tuned to correspond to those harmonic components of the vertical oscillations closest to the spin tune.

Initially such corrections were implemented empirically by varying the bump amplitudes and monitoring the resultant polarization. With improved accuracy in monitoring the beam position and a suitable feedback system the optimal corrections can be applied automatically.

These techniques have been used with great success at LEP at CERN and at PETRA and HERA in Hamburg. A detailed discussion of the approach used at HERA can be found in Barber (1994). A general summary of the state of the art in this field is given in Barber (1996).

7.2 Polarization at LEP and HERA

Ever since the mid-1980s there have been studies of the possibility of having polarized e^+e^- beams in LEP, this *transverse* polarization arising from the Sokolov–Ternov effect discussed in Section 7.1. These studies were catalysed by the realization that experiments with *longitudinally* polarized leptons would allow very accurate measurements of the parameters of the Standard Model of electroweak interactions, as will be discussed in Chapter 9. It was envisaged that the transverse polarization would be rotated into the longitudinal direction by special magnets without difficulty.

But there were major problems and doubts. The near-miraculous natural polarization $\mathscr{P}_{ST} \approx 92\%$ derived in Section 7.1 assumes a perfect machine. All the difficulties that beset the acceleration of polarized protons (subsection 6.3.2) in an imperfect machine apply to electrons as well; in addition, the greater synchrotron radiation leads to bigger problems with synchrotron oscillations.

Nonetheless an extraordinary collaboration of accelerator and particle physicists at CERN and HERA has succeeded in mastering many of the difficulties.

7.2.1 Polarization at LEP

An early attempt to calculate theoretically the expected behaviour of the polarization as a function of beam energy is shown in Fig. 7.1 (Koutchouk and Limberg, 1988). The first observation of a stable transverse polarization of $(9.1 \pm 0.3 \pm 1.8)\%$ at LEP was reported in 1991 (Knudsen *et al.*, 1991) and despite its smallness soon led to an improvement in our knowledge of the parameters of the Standard Model, albeit in an indirect way – by an improved calibration of the beam energy in LEP and thereby of the mass and width of the Z^0.

The idea is to use *resonant depolarization.* A frequency-controlled radial rf magnetic field causes the spin vector of the particle to precess away from its transverse (vertical) direction. An artificial depolarizing resonance occurs when the frequency ν_{dep} of the oscillatory magnetic field equals the spin precession frequency, i.e. when (see eqn (6.3.27))

$$\nu_{dep} = \nu_{dep}^{res} \equiv G\gamma f_{rev} = \frac{GE}{m_e c^2} f_{rev} \tag{7.2.1}$$

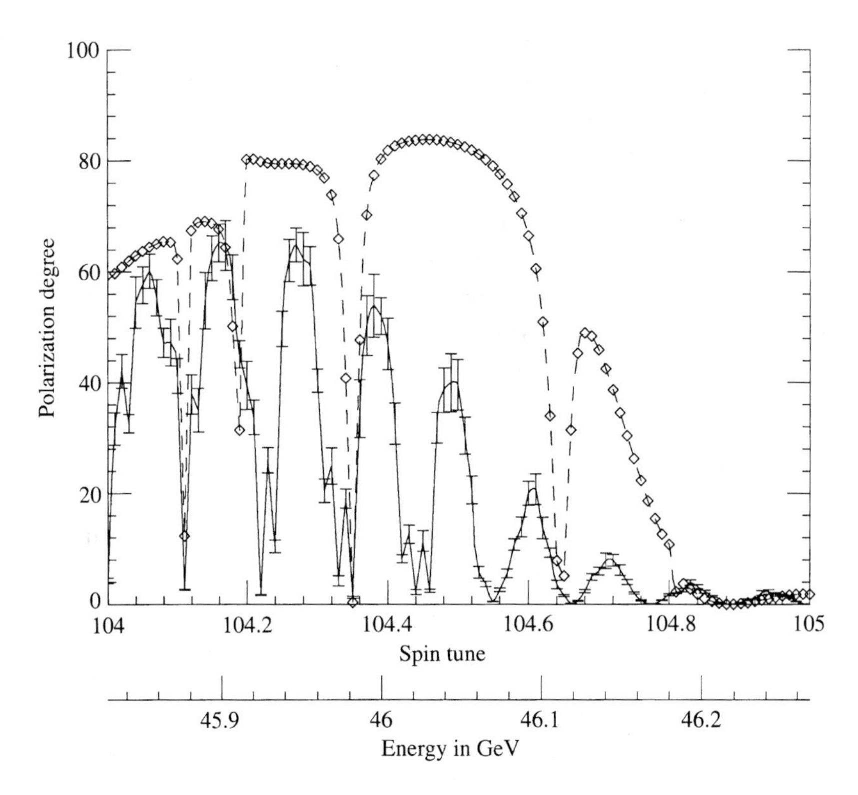

Fig. 7.1 Early theoretical estimate, in two models, of the beam polarization as a function of beam energy at LEP (courtesy of J.P. Koutchouk): diamonds on broken line, linear calculation; solid line with error bars, spin tracking.

where f_{rev} is the particle's frequency of revolution at energy E and, for electrons, G is known to fantastic accuracy:

$$G = \frac{g}{2} - 1 = 1.159652188 \times 10^{-3}. \tag{7.2.2}$$

The spin tune $\nu_s \equiv G\gamma$ can be written as

$$\nu_s = N_s + \delta\nu_s, \tag{7.2.3}$$

where N_s is an integer ($N_s = 103$ at the Z^0 mass), and it is then sufficient to measure $\delta\nu_{\text{dep}}^{\text{res}} \equiv \delta\nu_s f_{\text{rev}}$ at resonance. One ends up with a formula for the beam energy,

$$E_{\text{beam}} = 0.4406486 \left(N_s + \frac{\delta\nu_{\text{dep}}^{\text{res}}}{f_{\text{res}}} \right) \text{GeV}, \tag{7.2.4}$$

that led, in 1991, to a LEP beam energy calibration to an accuracy of $\lesssim 1$ MeV i.e. about one part in 10^5!

In fact this method had already been used with great success in lower-energy electron machines, VEPP2 and VEPP4 at Novosibirsk, DORIS in Hamburg and CESR at Cornell, leading to greatly improved precision in the measurements of the masses of the vector mesons $\omega, \phi, J/\psi, \psi'$, and the upsilon family Υ, Υ' and Υ'.

In a perfect LEP machine the rise time τ_{ST} to reach the polarization $\mathscr{P}_{\mathrm{ST}} = 92\%$ is found from (7.1.37) to be about 5 hours at the Z^0 mass, an enormously long time during which, in an imperfect machine, all kinds of depolarizing effects will operate. The situation, as explained earlier, can be summarized by introducing a characteristic depolarization time τ_{D}, in which case the asymptotic polarization is reduced to

$$\mathscr{P}_{\infty} = \frac{1}{1 + \tau_{\mathrm{ST}}/\tau_{\mathrm{D}}} \mathscr{P}_{\mathrm{ST}} \tag{7.2.5}$$

and, in parallel, the rise time is reduced to

$$\tau = \frac{1}{1 + \tau_{\mathrm{ST}}/\tau_{\mathrm{D}}} \tau_{\mathrm{ST}}. \tag{7.2.6}$$

The original polarization of about 9% at LEP has been steadily improved upon, using the method of harmonic spin matching. In this way polarizations of about 60% have been achieved for non-interacting beams.

Another problem stems from the solenoids used by the experimental groups at LEP, which have strong longitudinal fields that cause the mean spin vector to rotate about a longitudinal axis. This has been solved by introducing additional bumps before and after each solenoid to compensate for the longitudinal rotation.

More recently studies have begun of the effect of interactions on the polarization. It has been possible to attain a stable transverse polarization of about 40% with one interaction region and with a high luminosity of about 1.5×10^{30} cm^{-2} s^{-1} (Assmann *et al.*, 1995), but no comprehensive spin physics programme was undertaken.

Given that extremely precise measurements of the electroweak parameters are envisaged, it is important to try to eliminate sources of systematic error, principally in the measurements of the polarization of the beams and in the normalization of data samples taken with different settings of the e^+e^- helicities. A very clever trick (Blondel, 1998; Placidi and Rossmanith, 1985) permits the elimination of both these errors. The transverse polarization of the e^+ and e^- will be in opposite directions and after rotation to the longitudinal direction this will still be true. Thus the (longitudinal) spins of e^+ and e^- will be opposite, so that the *helicities* of the e^+ and e^- will be the same. It is relatively easy to depolarize a beam.

Moreover, this can be done to the individual bunches in the beam, so that one can have a pattern of bunch–bunch collisions with various settings of the spins, as shown in Fig. 7.2. For the four types of collision indicated, the total cross-sections depend upon the degree of longitudinal polarization $\mathscr{P}_e$, $\mathscr{P}_{\bar{e}}$ of the electron and positron beams and upon an *asymmetry parameter* $A_{\rm LR}$. As explained in Chapter 9, an accurate measurement of $A_{\rm LR}$ sheds valuable light on the electroweak parameters. One has

$$\begin{aligned}
\sigma_1 &= \sigma(1 + \mathscr{P}_{\bar{e}} A_{\rm LR}) \\
\sigma_2 &= \sigma(1 - \mathscr{P}_e A_{\rm LR}) \\
\sigma_3 &= \sigma \\
\sigma_4 &= \sigma\,[1 - \mathscr{P}_e \mathscr{P}_{\bar{e}} + (\mathscr{P}_{\bar{e}} - \mathscr{P}_e) A_{\rm LR}]
\end{aligned} \tag{7.2.7}$$

where σ is the unpolarized cross-section.

Remarkably, these four measurements permit us to deduce the values of $\mathscr{P}_e$, $\mathscr{P}_{\bar{e}}$ and $A_{\rm LR}$! It should be noted that this is a fairly miraculous situation. It happens only because in the Standard Model we are able to show that the coefficient of $\mathscr{P}_{\bar{e}}\mathscr{P}_e$ is -1.

7.2.2 *Polarization at HERA*

Ever since its conception there have been plans to polarize the leptons in the $e^{\pm}$–proton collider HERA, one objective being the study of polarized deep inelastic lepton–proton scattering by the HERMES collaboration utilizing a polarized-proton gas cell (see subsection 6.2.2) as target. The project gained much impetus from the startling results of the 1988 European Muon Collaboration experiment involving longitudinally polarized muons colliding with a polarized proton target (this is discussed in Chapter 11).

The HERMES collaboration began its first data-taking in 1995. Consideration is now being given to the possibility of polarizing the 820

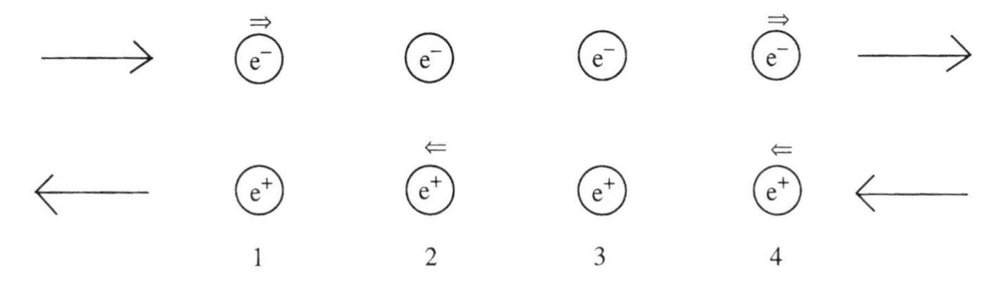

Fig. 7.2 A possible setting of the spins in successive e^+ and e^- bunches at LEP: the number 1–4 show four kinds of collision. Absence of an arrow $\Rightarrow$ indicates an unpolarized bunch.

GeV proton beam as well. Such a facility would allow very interesting experiments on the polarized structure function $g_1(x, Q^2)$ at very small x and large Q^2 and also could provide much needed information about the polarization of gluons in polarized protons (see Section 11.6).

Unlike LEP the natural rise time for the transverse Sokolov–Ternov polarization, assuming a perfect machine, is quite short: for $e^{\pm}$ 40 minutes at 27 GeV and 11 minutes at 35 GeV. When the machine was optimized for polarization using empirical harmonic orbit corrections, as discussed in subsection 7.1.1, the depolarization time τ_D could be made as long as 2 hours. Consequently, stable transverse polarizations of electrons or positrons of 60–70% were achieved routinely during 1995, well above the 50% design goal of the HERMES experiment.

In May 1994 the spin rotators were brought into operation and HERA became the first high energy electron machine to achieve *longitudinal* polarization. The so called 'Mini-rotators' (Buon and Steffen, 1986) consist of a sequence of dipole magnets designed to deflect the beam sequentially in the vertical and horizontal directions, as shown in Fig. 7.3. At each small angular deflection of the beam the component of the mean spin vector perpendicular to the field of the bending magnet precesses through an angle which is $1 + G\gamma = 63.5$ times bigger than the deflection angle; see (6.3.27).

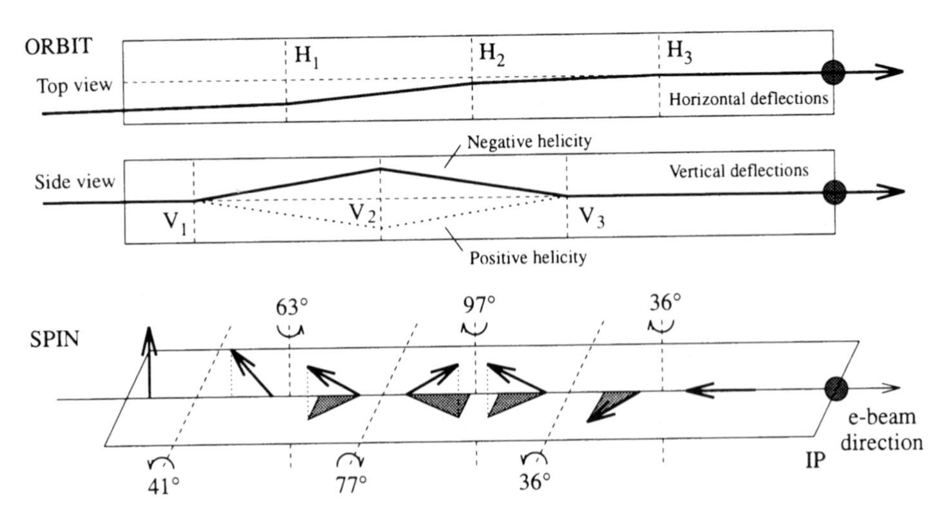

Fig. 7.3 Schematic diagram of a 'mini-rotator', showing horizontal and vertical beam deflections at the points $H_{1,2,3}$ and $V_{1,2,3}$ and the corresponding precession of the spin vector (courtesy of M. Düren).

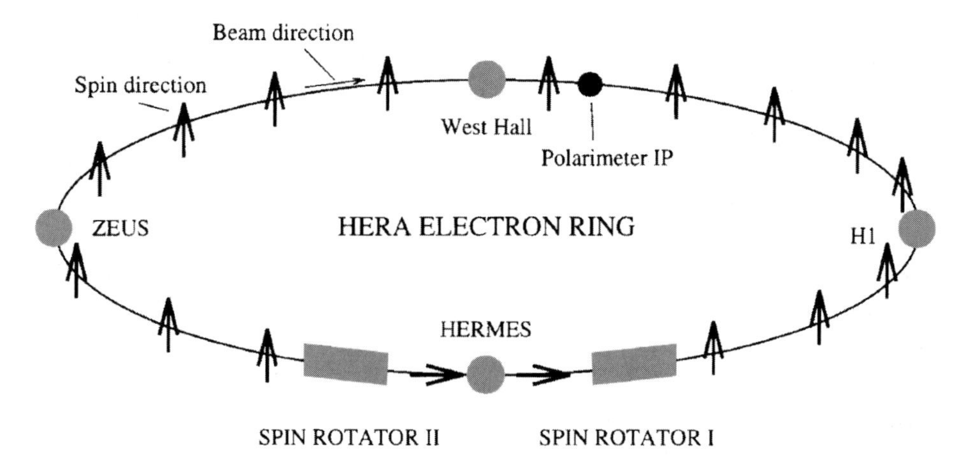

Fig. 7.4 Layout at the HERA ring showing the spin rotators and an idealized picture of the mean spin directions (courtesy of M. Düren).

Of course, once past the HERMES experimental region the longitudinal polarization must be rotated back to the transverse direction. The layout of rotators, spin directions etc. around the HERA ring is shown in Fig. 7.4.

The system has worked outstandingly well and stable longitudinal polarizations of about 70% are routinely achieved. An exciting and challenging investigation of the spin structure of the nucleon is in full swing and many interesting results have already emerged.

7.3 Polarization at SLC

In the Stanford linear collider the acceleration of the $e^{\pm}$ beams takes place along straight sections of the accelerator, so there is no Sokolov–Ternov effect in operation and the polarization must be produced at the electron source. The $e^{\pm}$ beams are brought together for collision along circular arms of the accelerator, but, since they only transverse these arcs once, there is no danger of resonant build-up of the depolarization effects that plague circular accelerators.

The principal challenge, then, is to produce a source of polarized electrons with a stable high degree of polarization and with a high output intensity. There is a long history of attempts to construct these sources. In more recent times, the desire to study the spin structure of the proton at SLAC in the 1970s led to the development of a photoionization source that played an essential rôle in the first experiments on the seminal process of polarized deep inelastic scattering (Chapter 11). However, the need for the much higher current required for the SLC led to the development of a new polarized source based upon photoemission from gallium arsenide (GaAs). The use of molecular beam epitaxy to grow thin layers of strained GaAs on wafers of bulk GaAs led ultimately to the achievement of polarizations above 80% with short bunch currents of a few amperes. The electrons are photoexcited by a pulsed, tunable laser. A comprehensive description of the SLAC polarized electron source, shown in Fig. 7.5, can be found in the review article by Alley *et al.* (1995).

The physical mechanism responsible for the polarization of the photoelectrons can be understood from the energy level diagrams (Fig. 7.6) for ordinary GaAs (top figure) and strained GaAs (lower figure). The solid and broken lines correspond respectively to transitions induced by right circularly polarized light (σ^+), and left circularly polarized light (σ^-). E_g is the band gap. The numbers in circles are the relative transition probabilities for the transitions.

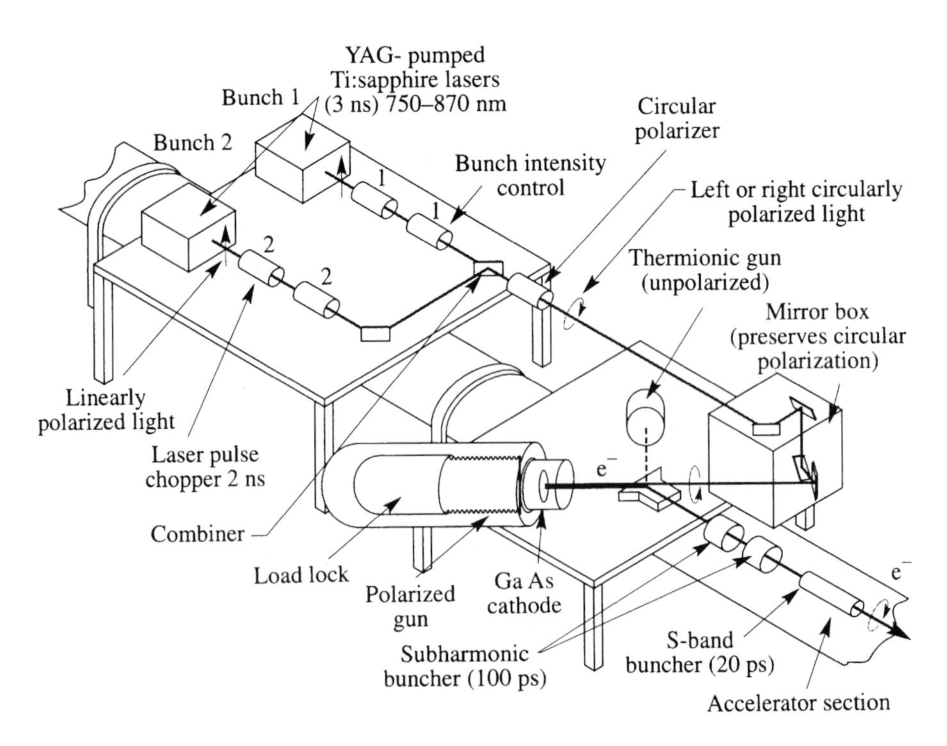

Fig. 7.5 The Stanford linear accelerator polarized electron source (courtesy of J. Clendenin and L. Piemontese).

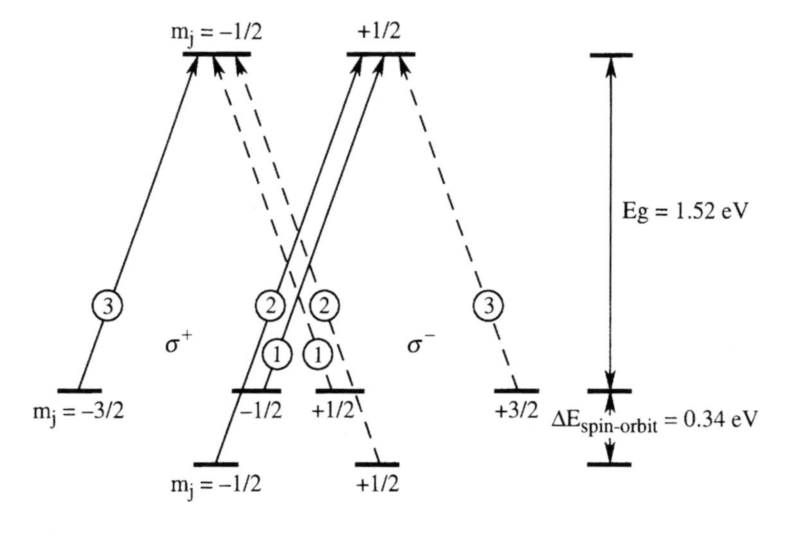

Strain-induced splitting ⇓ Strain axis parallel to incident photon axis

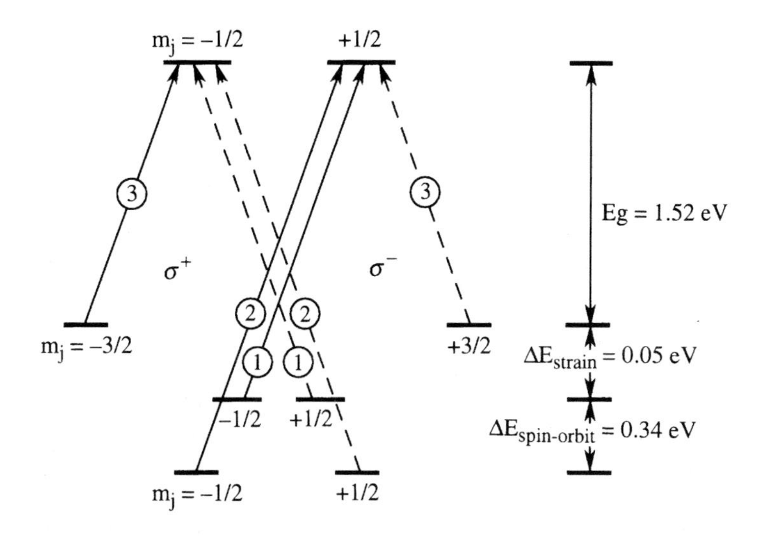

Fig. 7.6 Energy level diagram for ordinary GaAs (upper figure) and strained GaAs (lower figure) (courtesy of J. Clendenin and L. Piemontese).

Let us concentrate on the case of left circularly polarized light. In the unstrained case, if the light frequency is adjusted so that

$$E_{\mathrm{g}} < h\nu < E_{\mathrm{g}} + \Delta E_{\text{spin-orbit}}$$

then the only transitions into the conduction band are $m_j = 1/2 \to m_j = -1/2$, with relative transition rate 1, and $m_j = 3/2 \to m_j = 1/2$, with relative transition rate 3. The polarization is then

$$\mathscr{P}_{\sigma^-} = \frac{N(m_j = 1/2) - N(m_j = -1/2)}{N(m_j = 1/2) + N(m_j = -1/2)} = \frac{1}{2} \tag{7.3.1}$$

with a similar argument giving $\mathscr{P}_{\sigma^+} = -1/2$.

In the strained case the degenerate valence band levels are split. Thus by choosing the light frequency such that

$$E_{\mathrm{g}} < h\nu < E_{\mathrm{g}} + \Delta E_{\mathrm{strain}}$$

one can eliminate the transition $m_j = 1/2 \rightarrow m_j = -1/2$, leaving only the transition to the $m_j = 1/2$ state. This yields, in principle, 100% positive polarization, $\mathscr{P}_{\sigma^-} = 1$, and similarly for right circularly polarized light, $\mathscr{P}_{\sigma^+} = -1$.

The SLAC polarized electron source has functioned extremely efficiently. It has played an important rôle in testing the Standard Model of electroweak interaction via e^+e^- collisions in the SLC (Chapter 9) and is, at present, providing data of extraordinary accuracy in polarized deep inelastic scattering, where the polarized electron beam collides with polarized fixed targets of hydrogen, deuterium and helium-3 (Chapter 11).

8
Analysis of polarized states: polarimetry

In the previous chapters we have dealt with the production of the polarized states that serve as initial states in reactions. Here we turn to the measurement of the state of polarization of an ensemble of particles, i.e. to polarimetry.

In the analysis of the state of polarization we may be dealing with stable or unstable particles. If the particles are stable it may be possible to rely on well-understood reactions, such as those of QED, to achieve the polarization analysis, via, e.g. Coulomb interference or scattering off a laser beam. Or, if this is impracticable, it is sometimes possible to use a double-scattering technique even if the reaction mechanism is unknown. The only assumption needed for this is time-reversal invariance. If the particles are unstable their decay angular distribution gives information on their state of polarization prior to decay. This is not surprising if the decay is electromagnetic, so that the decay amplitudes are precisely known. What is remarkable, however, is that even when the decay mechanism is *not known* certain decays are 'magic' and still provide information on the polarization state of the decaying particle. Examples are $\rho \to \pi\pi$, $\omega \to \gamma\pi$, $D^* \to \gamma D$, $\psi \to \rho\pi$, $a_2 \to \rho\pi$ etc.

For electron beams, where we can rely on QED, it has been possible to construct very accurate and rapidly acting polarimeters.

One of the most interesting challenges at the moment is to construct efficient high energy proton polarimeters for use at RHIC, UNK and possibly at Fermilab. We shall discuss some of the current ideas in this field.

We shall also give a general treatment of the measurement of the density matrix from sequential scattering and resonance decays. The approach is remarkably simple and powerful and applies to the decay of a resonance of arbitrary spin.

8.1 Stable particles

Here we are primarily concerned with spin-1/2 particles, electrons, protons and neutrons. We consider separately the following cases.

(a) The reaction mechanism is known or essentially understood; this inevitably means electromagnetic or electroweak interactions.
(b) The reaction amplitudes cannot be calculated from first principles, as is the case for strong interactions.

The fundamental ingredient is the fact that differential cross-sections display azimuthal asymmetries if the initial state is polarized. We consider a fixed axis system with a polarized beam A moving along OZ and unpolarized target B. Then for a $2 \to 2$ reaction $AB \to CD$ one has from subsection 5.4.2

$$\frac{d^2\sigma}{dtd\phi} = \frac{1}{2\pi}\frac{d\sigma}{dt}\sum_{l,m}(2l+1)t^l_m(A)(lm;0,0|0,0;0,0)e^{-im\phi} \tag{8.1.1}$$

where C has polar angles θ, ϕ.

For an inclusive reaction $AB \to CX$, from Section 5.8 one has

$$\frac{d^3\sigma}{dtd\phi dM_X^2} = \frac{1}{2\pi}\frac{d^2\sigma}{dtdM_X^2}\sum_{l,m}(2l+1)t^l_m(A) \times (lm;0,0|0,0;0,0)^{\text{inc}}e^{-im\phi} \tag{8.1.2}$$

Thus if the 'lm analysing powers' $(l,m;0,0|0,0;0,0)$ for the reaction are known or can be calculated one can learn about the polarization state of the beam from the ϕ-dependence of the differential cross-section.

For spin-1/2 *beam* particles, with spin-polarization vector $\boldsymbol{\mathcal{P}}$ and with parity-conserving reactions, (8.1.1) and (8.1.2) simplify to (using (5.6.5) and (3.1.35))

$$\frac{d^2\sigma}{dtd\phi} = \frac{1}{2\pi}\frac{d\sigma}{dt}\left[1 + A(t)(\mathscr{P}_y\cos\phi - \mathscr{P}_x\sin\phi)\right] \tag{8.1.3}$$

$$\frac{d^3\sigma}{dtd\phi dM_X^2} = \frac{1}{2\pi}\frac{d^2\sigma}{dtdM_X^2}\left[1 + A_{\text{inc}}(t)(\mathscr{P}_y\cos\phi - \mathscr{P}_x\sin\phi)\right] \tag{8.1.4}$$

where the As are the analysing powers of the reactions for particle A.

In the above we have chosen an arbitrary reference frame with the beam A arriving along OZ and B either at rest or moving along the negative Z-axis. The combination $\mathscr{P}_y\cos\phi - \mathscr{P}_x\sin\phi$ is just $\boldsymbol{\mathcal{P}}\cdot\hat{\mathbf{n}}$ where $\hat{\mathbf{n}}$ is a unit normal to the scattering plane, i.e. $\hat{\mathbf{n}}$ is along $\mathbf{p}_A \times \mathbf{p}_C$. Thus a measurement of the ϕ-dependence gives us the components of $\boldsymbol{\mathcal{P}}$ perpendicular to the collision plane.

For the case of the collision of two spin-1/2 particles whose spin-polarization vectors are unknown, the more general results (5.6.12) or (5.6.20) can be used to measure their polarizations, providing, of course, that we know the values of the various generalized analysing powers.

8.1.1 Reaction mechanism understood

When two hadrons interact, their interaction is controlled by a mixture of strong (nuclear) and electromagnetic forces, and in an exact treatment one would add together the nuclear and electromagnetic hamiltonians. Generally the nuclear forces totally dominate, but there are certain kinematical regions where the long range of the electromagnetic forces leads to transition amplitudes that grow rapidly and eventually exceed the nuclear amplitudes. Of particular interest is the region of small momentum transfer, where, for example, the one-photon exchange amplitudes diverge as $t \to 0$.

There is thus a region of very small t (typically $\approx 10^{-3}$ $(\mathrm{GeV}/c)^2$ at high energies) where the known electromagnetic and the nuclear amplitudes are comparable. Although the nuclear amplitudes cannot be calculated from first principles they are expected to have smooth finite limits as $t \to 0$. Moreover any significant variation with t is expected to occur only for scales of order of a typical hadron mass squared, so that, to a first approximation, we can use just their values in the forward direction $t = 0$.

In summary a knowledge of the electromagnetic amplitudes together with some limited information on the forward nuclear amplitudes may yield enough information to estimate the analysing power of the reaction, at least for very small t. However, it will be seen that in situations involving hadrons it is perhaps an overstatement to claim that the reactions are truly understood.

(i) Electromagnetic–hadronic interference in proton–proton scattering

Interference at very small angles between the electromagnetic (EM) and hadronic contributions to the scattering amplitudes has long been used as a tool in the study of the phase of the hadronic amplitude. This only utilizes the interference between the hadronic forces and the longest-range part of the EM interaction, namely the Coulomb force. But at high energies magnetic effects become important and we expect to find that EM contributions to helicity-flip amplitudes gives rise to spin-dependent interference phenomena.

Here we shall focus only on the most dominant effects, and we shall approximate the amplitudes as a sum of the one-photon exchange and nuclear amplitudes, as shown in Fig. 8.1. For a detailed treatment and a

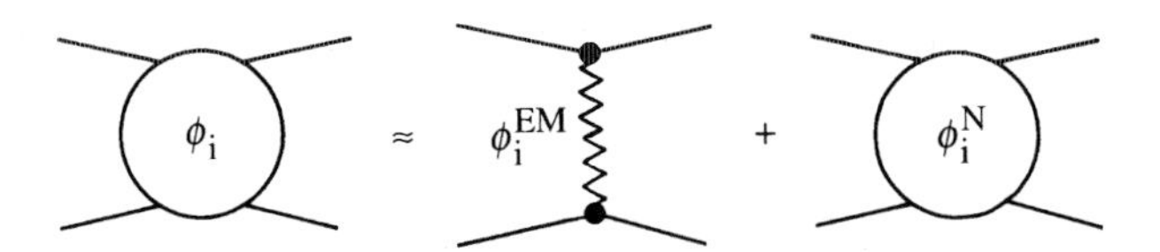

Fig. 8.1 Approximate form of the proton–proton amplitude as a sum of one-photon-exchange and nuclear amplitudes.

discussion of possible inaccuracies in this simple approach, the reader is referred to Buttimore, Gotsman and Leader (1978) and to Leader (1997).

With our normalization the most singular EM contributions to the $pp \to pp$ amplitudes are, for $s \gg m^2$, $t \to 0$,

$$\phi_1^{EM} \equiv H_{++;++}^{EM} \approx \frac{\sqrt{4\pi\alpha}}{t} \qquad \phi_3^{EM} \equiv H_{+-;+-}^{EM} \approx \frac{\sqrt{4\pi\alpha}}{t}$$
$$\phi_5^{EM} \equiv H_{++;+-}^{EM} \approx -\frac{\sqrt{4\pi\alpha}}{\sqrt{-t}}\frac{\alpha\kappa}{2m} \tag{8.1.5}$$

where κ is the *anomalous* magnetic moment of the proton in units of the proton magneton. The EM contributions to

$$\phi_2 \equiv H_{++;--} \qquad \text{and} \qquad \phi_4 \equiv H_{+-;-+} \tag{8.1.6}$$

are non-singular as $t \to 0$.

The nuclear (N) contributions to all ϕ_i are non-singular as $t \to 0$. Indeed from (4.3.1) we expect

$$\phi_{1,2,3}^N \approx \text{constant} \qquad \phi_5^N \approx \sqrt{-t} \qquad \phi_4^N \approx t. \tag{8.1.7}$$

It is generally supposed, upon the basis of models and some rather sparse low energy data, that the double-flip amplitude ϕ_2^N is negligible at high energies. Moreover, to a good approximation the non-flip amplitudes are imaginary, so can be estimated for *very* small t via the optical theorem (see eqn (5.1.4)). Thus, for very small t,

$$\begin{aligned}\phi_1(t) + \phi_3(t) &\approx \phi_1(0) + \phi_3(0) \\ &\approx i\,\mathrm{Im}\,[\phi_1(0) + \phi_3(0)] = \frac{i}{\sqrt{4\pi}}\sigma_{\text{tot}}.\end{aligned} \tag{8.1.8}$$

With the above approximations the differential cross-section is given by (see eqn (4.1.4))

$$\frac{d\sigma}{dt} \approx \frac{1}{2}\left(|\phi_1|^2 + |\phi_3|^2\right) \approx 4\pi\left[\frac{\alpha^2}{t^2} + \left(\frac{\sigma_{\text{tot}}}{8\pi}\right)^2\right]. \tag{8.1.9}$$

We see that the nuclear and electromagnetic contributions are comparable for

$$|t| \approx |t_C| \equiv \frac{8\pi\alpha}{\sigma_{tot}}. \tag{8.1.10}$$

For $\sigma_{tot} \geq 40$ mb we get electromagnetic dominance for

$$|t| < 2 \times 10^{-3}\ (\text{GeV}/c)^2. \tag{8.1.11}$$

The analysing power to be used in (8.1.3) is given by (see Table A10.5)

$$A \approx -\frac{2\ \text{Im}\left[\phi_5^*(\phi_1 + \phi_3)\right]}{|\phi_1|^2 + |\phi_3|^2} \tag{8.1.12}$$

Using (8.1.5), (8.1.7) and (8.1.9) we can eventually write, for very small t,

$$A(t) \approx A_{max}\left(\frac{4\left(t/t_{max}\right)^{3/2}}{3\left(t/t_{max}\right)^2 + 1}\right) \tag{8.1.13}$$

which has a maximum value $A = A_{max}$ at

$$t = t_{max} = -\frac{8\sqrt{3}\pi\alpha}{\sigma_{tot}} = -\sqrt{3}|t_C|. \tag{8.1.14}$$

The maximum value is

$$A_{max} = \frac{\kappa\sqrt{-3t_{max}}}{4m}. \tag{8.1.15}$$

For a proton beam with Lab momentum $p_L \approx 200$ GeV/c and taking $\sigma_{tot} \sim 100($ GeV$/c)^2$ one has $t_{max} \approx -3\times10^{-3}($ GeV$/c)^2$ and $A_{max} \approx 4.6\%$. $A(t)$ and $d\sigma/d\Omega$ are shown in Fig. 8.2.

We see that $A(t)$ is generally small and decreases rapidly with t. Outside the interference region it might well grow owing to purely hadronic effects, but of course we cannot calculate it. Indeed it is somewhat miraculous that we can estimate A for small t by lumping all our ignorance of the strong interactions into a few qualitative features plus the value of σ_{tot}.

At high energies the range of t where $A(t)$ is a few per-cent corresponds to extremely small laboratory scattering angles, so that it is immensely difficult to carry out the asymmetry measurement. Nonetheless work has progressed at Brookhaven on a CNI (Coulomb nuclear interference) polarimeter for use with RHIC and the method was tested at Fermilab (Grosnick *et al.*, 1990).

It should be noted that though the analysing power is small, it would be totally negligible if the proton had no anomalous magnetic moment. For then the helicity-flip amplitude ϕ_5^{EM} in (8.1.5) would have arisen from γ^μ

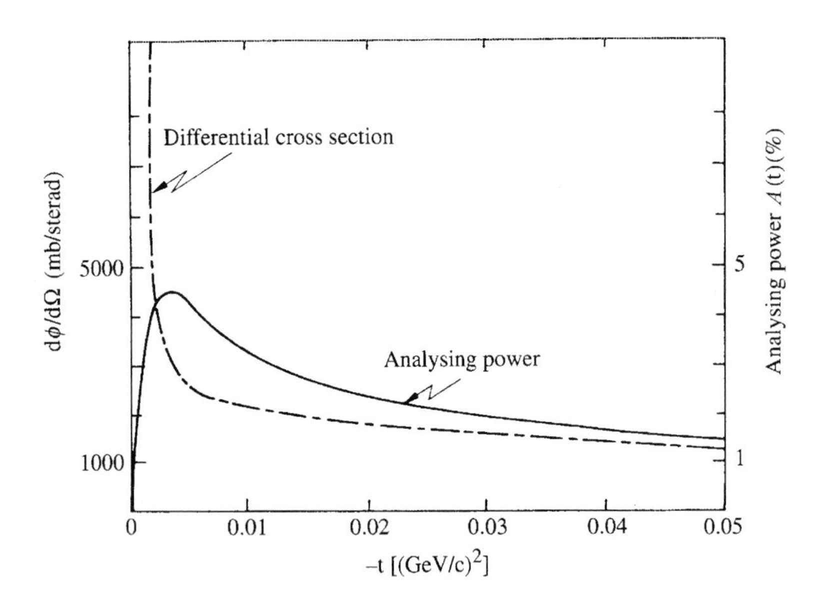

Fig. 8.2 Differential cross-section and analysing power $A(t)$ for $pp \rightarrow pp$ at $p_{\mathrm{L}} = 200$ GeV$/c$.

coupling, which, at high energies, conserves helicity (see subsection 4.6.2), so that we would have found an extra factor of $m/\sqrt{s}$ in ϕ_5^{EM}.

At the time of writing there is great interest in being able to measure beam polarizations at high energies to an accuracy of about 5%. On the theoretical side, attempts have been made to test or improve the accuracy of (8.1.13) by inclusion of hadronic helicity-flip amplitudes (Jakob and Kroll, 1992; Trueman, 1996). On the experimental side attempts have been made to measure A in the Coulomb interference region in pp elastic scattering at 200 GeV$/c$ (Akchurin *et al.*, 1993).

The experiment is exceedingly difficult and the data points, with large errors, are compatible with the result (8.1.13) but do not really test it to any significant degree of accuracy.

It is not clear at present whether one will be able to calculate A to an accuracy of 5%, though a somewhat optimistic conclusion was reached at the RHIC–Brookhaven workshop on CNI polarimetry (Leader, 1997; Leader and Trueman, 1997). This was based upon a new analysis of the magnitude and phase expected for the part of the hadronic-flip amplitudes that might survive at asymptotically high energies. It was suggested that a more accurate expression than (8.1.13), valid for $|t| \lesssim 0.01 (\mathrm{GeV}/c)^2$, is

$$A\frac{d\sigma}{dt} = \frac{\alpha\sigma_{\mathrm{tot}}}{m\sqrt{-t}}\left(\frac{\kappa}{2} - \mathrm{Im}\, r_5\right) \tag{8.1.16}$$

where r_5 is defined by

$$\phi_5 = r_5 \left(\frac{\sqrt{-t}}{m} \right) \text{ Im } \left(\frac{\phi_1 + \phi_3}{2} \right).$$

The unknown parameter r_5 is in principle a function of both energy and t, but it is argued that Im r_5 in (8.1.16) can be taken as a constant in the RHIC energy range and for $|t|$ as specified above.

Very recently it has been discovered that pp elastic scattering in the CNI region is self-calibrating, in the sense that if enough spin-dependent observables are measured then one can determine not only the values of the various helicity amplitudes but also, most surprisingly, the value of the polarizations of the initial protons (Buttimore *et al.*, 1999). Thus in effect one has an absolute polarimeter for which the theoretical error in the expression for the analysing power is of the order of the fine structure constant α. This will be discussed in Chapter 14.

(ii) Primakoff-type reactions

In this variant of the original Primakoff effect a π^0 is diffractively produced in the interaction of a proton with the Coulomb field of a heavy nucleus Z:

$$p + Z \to p + \pi^0 + Z.$$

When the final state $p\pi^0$ is moving almost forwards, i.e. at very small momentum transfer to the nucleus, the reaction is dominated by one-photon exchange, as shown in Fig. 8.3.

The Feynman diagram involves the amplitude for the 'reaction'

$$\text{virtual photon } + p \to \pi + p$$

and for very small momentum transfers in the Primakoff process, say $|k^2| \approx 10^{-3}(\text{ GeV}/c)^2$, the virtual photon is almost on mass shell. Thus to a very good approximation we should be able to consider the amplitude involved as the physical amplitude for genuine photoproduction $\gamma + p \to \pi + p$, a reaction which has been well studied at low and medium energies.

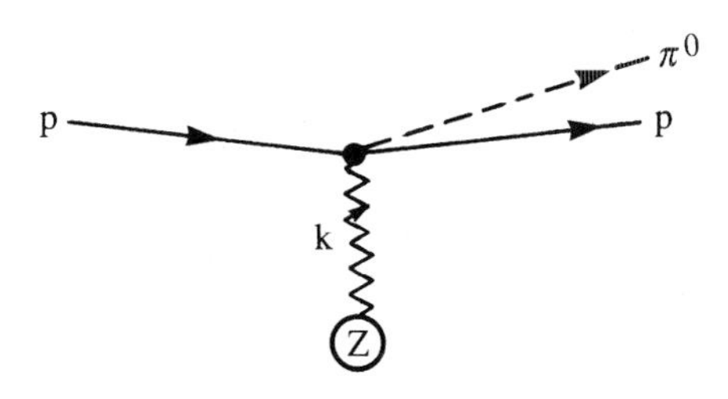

Fig. 8.3. Feynman diagram for Primakoff effect.

The beautiful and subtle point is that, even for a high energy initial proton, the CM energy of the photoproduction reaction (let us call it $M_{\pi p}$) is small, as we shall show, and at low energies it is known empirically that the photoproduction analysing power is large.

Consider a high energy proton, mass m and with magnitude of momentum p_{L} incident along OZ upon a fixed target in the Lab made up of heavy nuclei of mass $M \gg m$. If we focus only on reactions in which $t \equiv k^2$ is very small in modulus, $|t| \lesssim 10^{-3} (\,\mathrm{GeV}/c)^2$, then one can show that the maximum value of $M_{\pi p}$ is given by

$$\left(M_{\pi p}^{\max}\right)^2 = m^2 + 2p_{\mathrm{L}}\sqrt{-t}. \tag{8.1.17}$$

Thus even for $p_{\mathrm{L}} = 300\ \mathrm{GeV}/c$, $M_{\pi p} \lesssim 4.5\ \mathrm{GeV}/c^2$ and we are dealing with a relatively low energy reaction, which has been well studied experimentally.

The relevant analysing power, usually denoted $T(\theta)$, of the $\gamma p \to \pi^0 p$ reaction varies both with energy and CM scattering angle θ. It is large in the region $1.36 \leq M_{\pi p} \leq 1.52\ \mathrm{GeV}/c^2$ and has a maximum magnitude of about 90%.

The realization that a measurement of the proton polarization at high energies can be linked to low energy photoproduction is due to Underwood (1979). The basic theory was developed by Margolis and Thomas (1978) and a practical feasibility analysis was presented by Kuroda (1982). The experimental possibilities of the approach were finally demonstrated at Fermilab in 1989 by *measuring* the analysing power of the Primakoff reaction using a 185 GeV/c proton beam of *known* polarization and demonstrating that it is in accord with the theoretical expectation. (Carey *et al.*, 1990).

Referring to Fig. 8.4, let p^μ be the 4-momentum of the initial proton and let

$$P^\mu = q^\mu + p'^\mu \qquad (P^2 = M_{\pi p}^2) \tag{8.1.18}$$

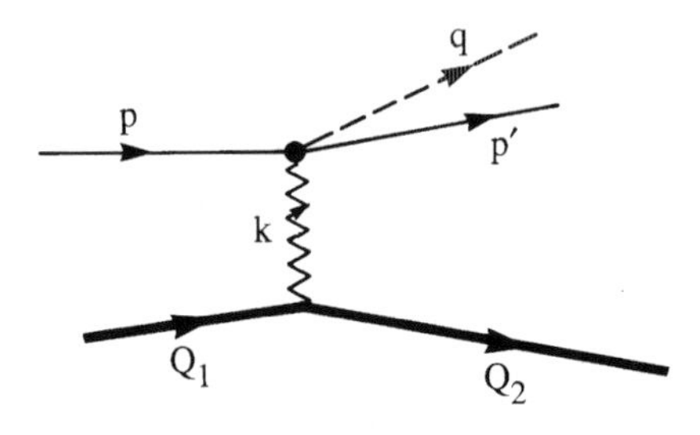

Fig. 8.4. Kinematics for Primakoff effect. Q_1, Q_2 are the initial and final momenta of the heavy nucleus Z.

be the total 4-momentum of the final πp pair. Define the invariant momentum transfer t for the following reactions:

$$p + Z \rightarrow (\pi p) + Z: \qquad t = (P - p)^2 = k^2 \tag{8.1.19}$$

$$\gamma + p \rightarrow \pi + p: \qquad t_0 = (p' - p)^2, \tag{8.1.20}$$

Then, assuming that only the diagram in Fig. 8.3 contributes, with a spinless target nucleus of charge Ze, one can show (Margolis and Thomas, 1978) that

$$\frac{d\sigma}{dM_{\pi p}^2 dt dt_0 d\phi_\pi} = \frac{\alpha Z^2}{\pi} \frac{|F(t)|^2}{M_{\pi p}^2 - m^2} \left(\frac{P_\perp^2}{t^2}\right) \frac{d\sigma}{dt_0 d\phi_\pi}(\vec{\gamma}\vec{p} \rightarrow \pi^0 p) \tag{8.1.21}$$

where the arrow overbars indicate a polarized particle, $F(t)$ is an unknown nuclear electromagnetic form factor and $\mathbf{P}_\perp$ is the transverse momentum vector of the πp system in the Lab:

$$P_\perp^2 = -t - \frac{\left(M_{\pi p}^2 - m^2\right)^2}{4p_{\mathrm{L}}^2}. \tag{8.1.22}$$

The cross-section $d\sigma/dt_0 d\phi_\pi$ is the differential cross-section for $\gamma p \rightarrow \pi p$ with polarized photon and polarized initial proton. The angle ϕ_π is, strictly speaking, the azimuthal angle of the π in the πp rest system, i.e. in the $\gamma p \rightarrow \pi p$ CM, with Z-axis along $\mathbf{P}$ in the Lab and some fixed Y-axis. Because the direction of $\mathbf{P}$ differs only infinitesimally from the direction of $\mathbf{p}_{\mathrm{L}}$, for the kinematic region under study, ϕ_π is then also simply the azimuthal angle of the produced pion in the Lab reference frame.

Margolis and Thomas (1978) showed that the almost-real photon is linearly polarized. Then if ϕ_γ is the angle between the polarization vector $\boldsymbol{\varepsilon}$ and the reaction plane, and if $\mathcal{P}$ is the spin-polarization vector for the initial proton in the CM, one has (see Storrow, 1978)

$$\frac{d\sigma}{dt_0 d\phi_\pi} = \frac{1}{2\pi} \left.\frac{d\sigma}{dt_0}\right|_{\text{unpol.}} \left\{1 - \mathscr{P}_{\text{lin}} \left[\Sigma(\theta) \cos 2\phi_\gamma + \mathscr{P}_x H(\theta) \sin 2\phi_\gamma \right.\right.$$
$$\left.\left. + \mathscr{P}_y P(\theta) \cos 2\phi_\gamma - \mathscr{P}_z G(\theta) \sin 2\phi_\gamma\right] + \mathscr{P}_y T(\theta)\right\} \tag{8.1.23}$$

where θ is the CM scattering angle, $\mathscr{P}_{\text{lin}}$ is the linear polarization and here the direction OY is along the normal to the reaction plane, i.e. along $\mathbf{k} \times \mathbf{q}$ in the CM. The various functions $\Sigma(\theta)$, $H(\theta)$, $P(\theta)$, $G(\theta)$ and $T(\theta)$ are dynamics-dependent reaction parameters that also depend upon the energy of the $\gamma p \rightarrow \pi p$ reaction.

Margolis and Thomas (1978) also showed that $\boldsymbol{\varepsilon}$ lies along the direction of the vector $\mathbf{P}_\perp$ as seen in the $\gamma p \rightarrow \pi p$ CM. Moreover the cross-section (8.1.21) is independent of the azimuthal angle Φ of $\mathbf{P}$ in the Lab. Hence if for fixed ϕ_π we average over the direction of $\mathbf{P}_\perp$ we are in effect averaging over ϕ_γ. In this case almost all terms in (8.1.23) average to zero and we are left with

$$\left\langle \frac{d\sigma}{dM_{\pi p}^2 dt dt_0 d\phi_\pi} \right\rangle = \frac{\alpha Z^2}{\pi} \frac{|F(t)|^2}{M_{\pi p}^2 - m^2} \left(\frac{P_\perp^2}{t^2} \right) \times \frac{1}{2\pi} \left. \frac{d\sigma}{dt_0} \right|_{\text{unpol.}} [1 + \boldsymbol{\mathcal{P}} \cdot \mathbf{n} T(\theta)] \tag{8.1.24}$$

where $\mathbf{n}$ is the unit normal to the reaction plane and the angle brackets imply an average over Φ.

Since we are using the reaction as an analyser and we do not know the direction of $\boldsymbol{\mathcal{P}}$ it is perhaps simplest to discuss (8.1.24) in the CM reference frame with fixed X- and Y-axes such that the pion has azimuthal angle ϕ_π. Then the polarization-dependent term in (8.1.24) is just

$$1 + (\mathscr{P}_y \cos\phi_\pi - \mathscr{P}_x \sin\phi_\pi)\, T(\theta)$$

where $T(\theta)$ is supposed known.

A study of the ϕ_π-dependence of the cross-section thus gives information on $\mathscr{P}_x$ and $\mathscr{P}_y$. Ideally this could be done for values of θ where $T(\theta)$ is large, but it may be necessary in practice to integrate over θ to increase the statistics.

Unfortunately the very beautiful result (8.1.24) cannot be used directly for polarimetry, because we have ignored, in the above, all contributions arising from the purely hadronic diffraction production of the πp system. It is usually assumed that the hadronic amplitude is due to Pomeron exchange, does not depend on helicity and is essentially imaginary. Typically it is taken to be of the form $iC\exp(-bP_\perp^2)$ for very small $P_\perp^2$, with C real. The slope b should reflect the 'size' R of the nucleus $Z: b \propto 1/R^2$. For Pb one estimates $b \sim 250\ (\text{GeV}/c)^2$, so that the hadronic differential cross-section has a slope of about 500 $(\text{GeV}/c)^2$. However, the cross-section in (8.1.24) has a sharp peak at

$$P_\perp^2 = \frac{\left(M_{\pi p}^2 - m^2\right)^2}{2p_\text{L}^2},$$

which, for $M_{\pi p} \simeq 1.23\ \text{GeV}/c^2$ and $p_\text{L} = 200\ \text{GeV}/c$, corresponds to the tiny value $P_\perp^2 = 1.5 \times 10^{-5}\ (\text{GeV}/c)^2$. Thus a fit to the $P_\perp^2$ distribution can help to estimate the hadronic part of the cross-section.

For the region of such small values of $P_\perp^2$ and $|t|$, the form factor $F(t)$ can safely taken to be $F(0) = 1$. The observed $P_\perp^2$ distribution can then be fitted by

$$\left.\frac{d\sigma}{dP_\perp^2}\right|_{\text{Expt}} = \left.\frac{d\sigma}{dP_\perp^2}\right|_{\text{Primakoff}} + C^2 e^{-bP_\perp^2} \tag{8.1.25}$$

from which C and b could be determined, in principle.

The net effect would then be that the ϕ_π-dependent part of (8.1.24) becomes

$$1 + \boldsymbol{\mathcal{P}} \cdot \mathbf{n} T(\theta) f(P_\perp^2) \tag{8.1.26}$$

where

$$f(P_\perp^2) = \left.\frac{d\sigma}{dP_\perp^2}\right|_{\text{Primakoff}} \Bigg/ \left.\frac{d\sigma}{dP_\perp^2}\right|_{\text{expt}} \tag{8.1.27}$$

is a dilution factor such that the effective analysing power is $T(\theta)f(P_\perp^2)$.

In the experiment of Carey *et al.* (1990) mentioned earlier, where a proton beam of *known* polarization was used, a somewhat similar approach was taken to the analysis, and f, averaged over the $P_\perp^2$ of the experiment, was measured to be 0.55 with an error of about ± 0.18.

Clearly there is a significant influence from the hadronic amplitude and (8.1.24) cannot be used as an absolute polarimeter as it stands. But the dilution factor is not catastrophic and (8.1.26) seems to offer a practicable approach to high energy polarimetry provided f can be measured accurately.

The argument that led to (8.1.26) is actually flawed. The photoproduction analysing power $T(\theta)$ would be zero if the photoproduction amplitudes were all real! (see Appendix section A10.4).

Thus there could be important interference effects between the Primakoff and hadronic amplitudes. However, as we shall now explain, this cannot change the *form* of (8.1.26) – only the physical interpretation of $f(P_\perp^2)$ changes.

Using methods based on (5.4.31) and on the analysis of resonance decay to be dealt with in Section 8.2, in which parity conservation is assumed, it is possible to show that the cross-section for

$$p + Z \to \pi + p + Z,$$

averaged over $\mathbf{P}_\perp$, depends on the initial proton polarization only via a factor

$$1 + A(p_{\text{L}}, P_\perp^2, M_{\pi p}^2, \theta) \boldsymbol{\mathcal{P}} \cdot \mathbf{n} \tag{8.1.28}$$

where $A(p_{\text{L}}, P_\perp^2, M_{\pi p}^2, \theta)$ is the proton analysing power of the reaction

$$p + Z \to \pi + p + Z,$$

irrespective of the dynamical mechanism. It has to be determined experimentally.

This can be seen in a simpler way. The cross-section must be invariant under space inversion. Thus the pseudovector $\boldsymbol{\mathcal{P}}$ must occur in a scalar product with some other pseudovectors. *A priori* the latter could be $\mathbf{k} \times \mathbf{q}$ and $\mathbf{k} \times \mathbf{p}'$. However, at fixed $\mathbf{q}$, it follows that $\mathbf{p}' = \mathbf{P} - \mathbf{q}$ and so averaging over $\mathbf{P}_\perp$ causes $\boldsymbol{\mathcal{P}} \cdot (\mathbf{k} \times \mathbf{p}')$ to reduce to $-\boldsymbol{\mathcal{P}} \cdot (\mathbf{k} \times \mathbf{q})$. Thus the only possibility is $\boldsymbol{\mathcal{P}} \cdot \hat{\mathbf{n}}$ and (8.1.28) is the most general form possible. From this more general point of view the Primakoff analysis simply tells us in what kinematic regime we can expect to find a significant analysing power for the reaction

$$p + Z \to \pi + p + Z.$$

Despite these rather disappointing conclusions, it could still be possible to have an absolute Primakoff polarimeter if one had enough events to be able to restrict oneself to really very small $P_\perp^2$ i.e. of order 10^{-5} $(\text{GeV}/c)^2$. For, from the study of Carey *et al.* (1990), one can deduce that in this kinematic region $d\sigma|_{\text{Primakoff}} \gtrsim 100 d\sigma|_{\text{Hadronic}}$.

(iii) *Møller scattering*

The reaction

$$e + e \to e + e$$

is, from a spin point of view, formally identical to elastic proton–proton scattering, so that all the formula relating CM reaction parameters to helicity amplitudes may be taken over from Table A10.3.

Being an electromagnetic interaction we treat it in the Born, i.e. the one-photon-exchange, approximation, in which case all the helicity amplitudes are real. From the first two entries of Table A10.5 we see that the standard analysing power is then zero. However, the initial state spin correlation parameters $A_{\alpha\beta}$ will be non-zero and the reaction can be used to measure the polarization of the beam provided that we use a polarized target with known polarization. This can be achieved by using very thin magnetized ferromagnetic foils, in which degrees of polarization of about 8% are attained. The direction of the spin-polarization is easily reversed.

To begin with we work in the CM of the reaction. We choose our Y-axis such that the known target spin-polarization vector lies in the YZ-plane. Let the beam and target spin-polarization vectors be specified in the CM frame by

$$\boldsymbol{\mathcal{P}} = (\mathscr{P}_x, \mathscr{P}_y, \mathscr{P}_z) \qquad \text{and} \qquad \boldsymbol{\mathcal{P}}^{\mathrm{T}} = (0, \mathscr{P}_y^{\mathrm{T}}, \mathscr{P}_z^{\mathrm{T}}). \tag{8.1.29}$$

The general form of the differential cross-section $d^2\sigma/dtd\phi$ is given in (5.6.12), in which particle B is the target. We must now calculate the reaction parameters A_{ij} for our process.

We are particularly interested in high energy electrons, so we may greatly simplify the calculation by going to the CM, where all the electrons are highly relativistic, and making use of the result (see subsection 4.6.2) that helicity is conserved for vector coupling in this situation. Thus neglecting terms of order m^2/s, where m is the electron mass, in our normalization, the only non-negligible, correctly symmetrized (see (4.2.16)) helicity amplitudes are (Buttimore, Gotsman and Leader, 1978, Appendix A):

$$\begin{aligned}\phi_1 &= \sqrt{4\pi\alpha}\left(\frac{1}{t}+\frac{1}{u}\right)\\ \phi_3 &= \sqrt{4\pi\alpha}\frac{1}{t}\left(1+\frac{t}{s}\right)e^{i\phi}\\ \phi_4 &= -\sqrt{4\pi\alpha}\frac{1}{u}\left(1+\frac{u}{s}\right)e^{i\phi}\end{aligned} \tag{8.1.30}$$

where we have used (4.1.7) to generalize the Buttimore *et al.* results to $\phi \neq 0$. Using these in the expressions given in Table A10.5 we find

$$\begin{aligned}A^{(A)} = A^{(B)} &= A_{XZ} = A_{ZX} = 0\\ A_{XX} = A_{YY} &= -\frac{\sin^4\theta}{(4-\sin^2\theta)^2}\\ A_{ZZ} &= \frac{\sin^2\theta(8-\sin^2\theta)}{(4-\sin^2\theta)^2}.\end{aligned} \tag{8.1.31}$$

Then (5.6.12) becomes

$$\frac{d^2\sigma}{dtd\phi} = \frac{1}{2\pi}\frac{d\sigma}{dt}\left[1+\frac{\sin^4\theta}{(4-\sin^2\theta)^2}\mathscr{P}_y\mathscr{P}_y^{\mathrm{T}} - \frac{\sin^2\theta(8-\sin^2\theta)}{(4-\sin^2\theta)^2}\mathscr{P}_z\mathscr{P}_z^{\mathrm{T}}\right] \tag{8.1.32}$$

Note that there is no azimuthal dependence. Under reversal of $\boldsymbol{\mathcal{P}}^{\mathrm{T}}$ we then have the asymmetry:

$$\frac{\frac{d\sigma}{dt}(\boldsymbol{\mathcal{P}},\boldsymbol{\mathcal{P}}^{\mathrm{T}}) - \frac{d\sigma}{dt}(\boldsymbol{\mathcal{P}},-\boldsymbol{\mathcal{P}}^{\mathrm{T}})}{\frac{d\sigma}{dt}(\boldsymbol{\mathcal{P}},\boldsymbol{\mathcal{P}}^{\mathrm{T}}) + \frac{d\sigma}{dt}(\boldsymbol{\mathcal{P}},-\boldsymbol{\mathcal{P}}^{\mathrm{T}})} = \frac{\sin^2\theta}{(4-\sin^2\theta)^2} \times\left[\sin^2\theta\,\mathscr{P}_y\mathscr{P}_y^{\mathrm{T}} - (8-\sin^2\theta)\mathscr{P}_z\mathscr{P}_z^{\mathrm{T}}\right] \tag{8.1.33}$$

so that $\mathscr{P}_y$ and $\mathscr{P}_z$ can be determined by suitably varying $\mathscr{P}_y^{\mathrm{T}}$ and $\mathscr{P}_z^{\mathrm{T}}$. Note that the 'analysing powers' vary strongly with CM angle θ and are large near $\theta = \pi/2$. Also the cross-section in this region is relatively large so that an efficient polarimeter is feasible.

The technique of Møller polarimetry has been used successfully in a range of experiments, most recently in the SLAC E142, E143, E154 and E155 experiments (Feltham and Steiner, 1997; Band, 1997) on polarized deep inelastic scattering. A statistical precision of 1%–2% is achieved in typically 15 minutes!

For access to the experimental aspects of Møller polarimetry the reader is referred to the *Proceedings of the 12th International Symposium on High-Energy Spin Physics, Amsterdam, 1996* (World Scientific, Singapore, 1997).

(iv) Compton scattering

The reaction

$$\gamma + e \to \gamma + e$$

was studied in great detail in lowest-order QED by Lipps and Tolhoek (1954). For an unpolarized initial photon the cross-section is independent of the polarization of the electron. However, it does depend upon any linear polarization of the photon and for a circularly polarized photon depends also upon the polarization of the electron. Hence by scattering a laser beam of *known* circular polarization off the electrons and measuring the differential cross-section, one can learn about the spin-polarization vector of the electrons.

This technique has been used with great success at several e^+e^- storage rings, most recently at SLAC and HERA.

At the SLC/SLD at SLAC (see Fig. 8.5), a 532 nm YAG laser beam, corresponding to circularly polarized photons of energy 2.33 eV in the Lab, collides almost head-on with a high energy (45.6 GeV) polarized electron beam. The electrons are scattered into a narrow forward cone and are detected in a Čerenkov detector. The photons are backscattered but are not detected. The polarization measurement involves a comparison of the detection rates when the circular polarization of the laser beam is reversed. The Compton polarimeter is capable of achieving a statistical accuracy of 1% in 3 minutes and polarizations are now quoted with a total accuracy of $\pm 0.67\%$!

At HERA there are two Compton polarimeters in operation, one using a pulsed Nd:YAG laser, the other using a continuous argon-ion laser. In both it is the backscattered photons that are detected. Statistical errors of 0.4% are achieved in 10 minutes and an overall accuracy of about 3% is expected.

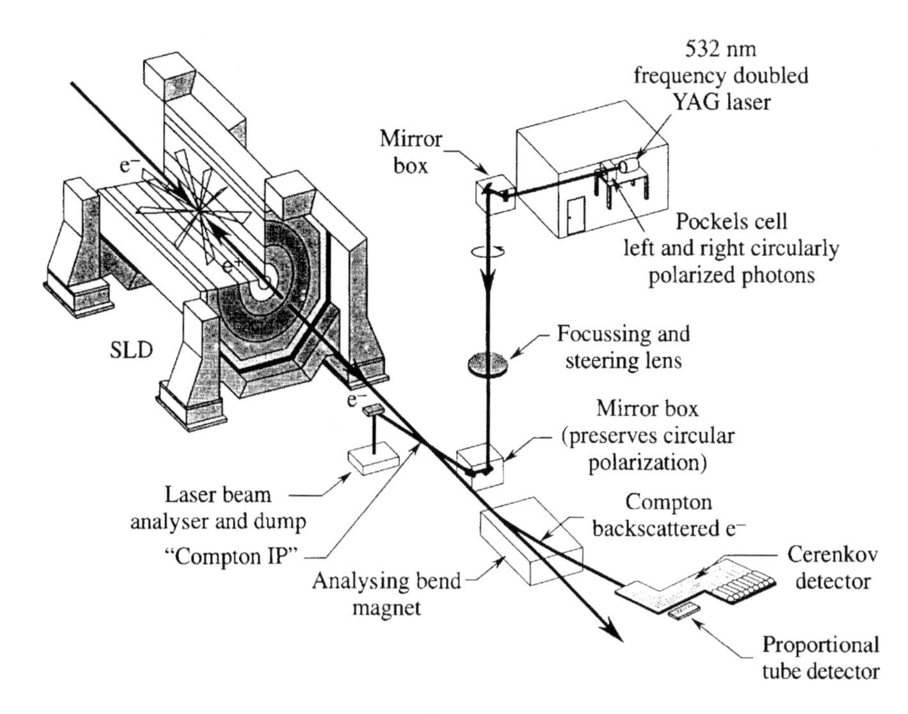

Fig. 8.5 Compton polarimeter at SLC/SLD (courtesy of J. Clendenin, L. Piemontese and M. Swartz).

Consider now the kinematics of Compton scattering. Usually, in the literature, the reaction is discussed in what is called the laboratory frame, meaning the frame where the target electron is at rest. Here, the collision in the laboratory frame is between the laser beam and the fast-moving electron beam. So we shall, for clarity, refer to these frames as the *electron rest frame* and the *Lab collision frame*. The kinematics in these frames is shown in Fig. 8.6. The angle between the incoming laser beam and the electron beam is so small that it can be ignored, so that we have, in effect, a head-on collision and the two frames are related by a simple boost along the Z-axis. We assume some fixed axis system with the electron beam incoming along OZ. The photon is scattered into polar angles $\theta = \pi - \theta_\gamma$, $\phi = \phi_\gamma$ in the electron rest frame, as shown.

Note that in the Lab collision frame a very high energy electron collides with a very low energy photon, so that the final state particles are largely swept forward along OZ. On the contrary, in the electron rest frame a very high energy photon collides with the stationary electron and the final state particles are largely swept along the negative OZ direction.

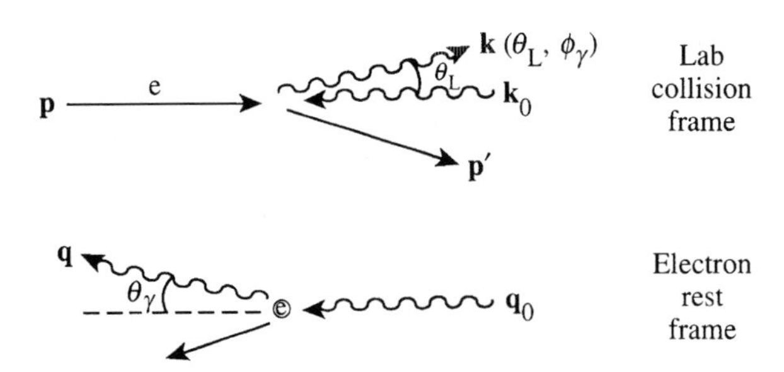

Fig. 8.6 Kinematics of Compton scattering. Initially the electron moves along OZ in the Lab collision frame; in both parts of the figure the Z-axis is to the right.

In the electron rest frame we have the famous Compton relation

$$\frac{m_e}{q} - \frac{m_e}{q_0} = 1 - \cos\theta_\gamma. \tag{8.1.34}$$

The connection between the electron rest frame variables and the given Lab collision frame variables, the photon energy k_0, the electron energy E_e and the angle θ_L as shown, are (neglecting m_e in comparison with E_e),

$$q_0 = \frac{2E_e k_0}{m_e} \qquad \tan\theta_L = \left(\frac{m_e}{E_e}\right)\frac{\sin\theta_\gamma}{1-\cos\theta_\gamma}. \tag{8.1.35}$$

Some care must be exercised in taking over the result of the Lipps–Tolhoek papers. Firstly, their results are presented in an electron rest frame with the Y-axis perpendicular to the reaction plane whereas we wish to analyse the electron's spin-polarization vector with respect to some fixed-axis system. Secondly, their paper was written prior to the invention of helicity states, so their photon density matrix is given in a basis that utilizes the states of linear polarization $|\mathbf{e}_{(x)}\rangle$ and $|\mathbf{e}_{(y)}\rangle$, given in (3.1.75), rather than in the helicity basis. The necessary alterations to the Lipps–Tolhoek results can easily be made by use of the results in subsection 3.1.12 and Section 5.4.

We suppose that a laser beam moving along the negative OZ direction contains a fraction $1-f$ of its photons with linear polarization $\mathscr{P}_{\text{lin}}$ along an axis at angle γ to OX and a fraction f of its photons with circular polarization $\mathscr{P}_{\text{circ}}$, where $\mathscr{P}_{\text{circ}} > 0$ corresponds to positive helicity and to left-circular polarization in classical optics. If the beam collides with an electron whose spin-polarization vector is $\boldsymbol{\mathcal{P}} = (\mathscr{P}_x, \mathscr{P}_y, \mathscr{P}_z)$ in the fixed reference frame then, the invariant differential cross-section is given, in the electron rest frame, by

$$\frac{d\sigma}{dt d\phi} = \frac{m_e^2 r_0^2}{s^2}\left(\frac{q}{q_0}\right)\left\{1+\cos^2\theta_\gamma + \frac{1}{m_e}(q_0-q)(1-\cos\theta_\gamma)\right.$$
$$+(1-f)\mathcal{P}_{\text{lin}}\cos(2\gamma-2\phi_\gamma)\sin^2\theta_\gamma + f\frac{\mathcal{P}_{\text{circ}}}{m_e}\left[\mathcal{P}_z(q_0+q)\cos\theta_\gamma\right.$$
$$\left.\left. -\ (\mathcal{P}_x\cos\phi_\gamma+\mathcal{P}_y\sin\phi_\gamma)q\sin\theta_\gamma\right](1-\cos\theta_\gamma)\right\} \tag{8.1.36}$$

where r_0 is the 'classical electron radius' $e^2/(m_e c^2)$.

If we consider a purely circularly polarized laser beam ($f = 1$) and we analyse the data in the electron rest frame then a study of the ϕ_γ-dependence will, in principle, yield all the components of the electron's spin-polarization vector $\boldsymbol{\mathcal{P}}$.

In practice the polarization of the laser beam is known with great precision, whereas the cross-section measurement suffers from significant normalization errors. It is therefore better to study the asymmetry under reversal of the circularity of the photon polarization, i.e. to study

$$\frac{d\sigma(\mathcal{P}_{\text{circ}})-d\sigma(-\mathcal{P}_{\text{circ}})}{d\sigma(\mathcal{P}_{\text{circ}})+d\sigma(-\mathcal{P}_{\text{circ}})}$$
$$=\mathcal{P}_{\text{circ}}\frac{\left[\mathcal{P}_z(q_0+q)\cos\theta_\gamma-(\mathcal{P}_x\cos\phi_\gamma+\mathcal{P}_y\sin\phi_\gamma)q\sin\theta_\gamma\right](1-\cos\theta_\gamma)}{m_e(1+\cos^2\theta_\gamma)+(q_0-q)(1-\cos\theta_\gamma)}. \tag{8.1.37}$$

For longitudinally polarized electrons, where we attempt to measure $\mathcal{P}_z$, one uses cross-sections integrated over the azimuthal angle ϕ_γ, in which case the measured asymmetry becomes

$$\frac{\langle d\sigma(\mathcal{P}_{\text{circ}})-d\sigma(-\mathcal{P}_{\text{circ}})\rangle}{\langle d\sigma(\mathcal{P}_{\text{circ}})+d\sigma(-\mathcal{P}_{\text{circ}})\rangle} = -\mathcal{P}_{\text{circ}}\mathcal{P}_z A_{\text{C}}(\theta_\gamma) \tag{8.1.38}$$

where the underlying Compton asymmetry is

$$A_{\text{C}}(\theta_\gamma) = \frac{-(q_0+q)\cos\theta_\gamma(1-\cos\theta_\gamma)}{m_e(1+\cos^2\theta_\gamma)+(q_0-q)(1-\cos\theta_\gamma)}. \tag{8.1.39}$$

For 100% right (R) or left (L) circularly polarized light the measured asymmetry is thus

$$\frac{\langle d\sigma(\text{R})-d\sigma(\text{L})\rangle}{\langle d\sigma(\text{R})+d\sigma(\text{L})\rangle} = \mathcal{P}_z A_{\text{C}}(\theta_\gamma). \tag{8.1.40}$$

The basic asymmetry $A_{\text{C}}(\theta_\gamma)$ is shown as a function of $\cos\theta_\gamma$ in Fig. 8.7 for $E_e = 47$ GeV.

Note that (8.1.39) can be written as a function of E'_e/E_e, where E'_e is the electron recoil energy in the Lab collision frame. Figure 8.8 shows the

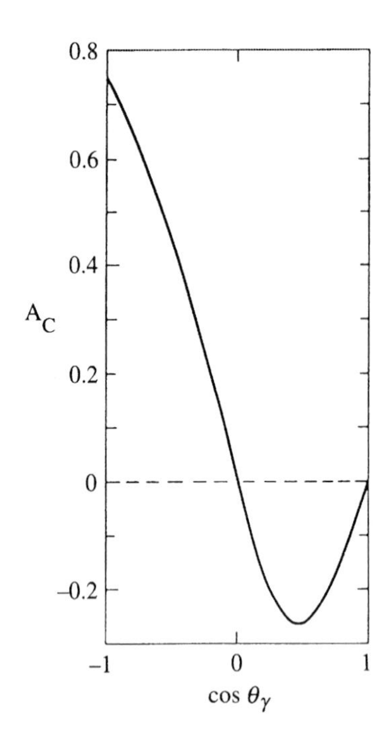

Fig. 8.7 The underlying Compton asymmetry $A_C(\theta_\gamma)$ as a function of $\cos\theta_\gamma$, for $E_e = 47$ GeV (from Alexander *et al.*, 1996).

measured and theoretical asymmetry (8.1.40) at SLAC, for $\mathscr{P}_z = 77.5\%$, as a function of E'_e.

It can be seen from these figures that the asymmetries are very large for θ_γ near π, which corresponds to electrons with smaller recoil energy E'_e.

(v) Polarimetry via decay distributions

A seminal series of experiments on polarized deep inelastic lepton hadron scattering has been in progress at CERN for more than a decade, utilizing a high energy polarized muon beam (the European Muon Collaboration, Ashman *et al.* 1989; the Spin Muon Collaboration, Adams *et al.*, 1997).

The muons, say μ^+, are produced by the decay in flight of high energy pions, $\pi^+ \to \mu^+ + \nu_\mu$. In the Standard Model the neutrinos are entirely left-handed i.e. $\lambda_\nu = -1/2$. Consequentially, in the rest frame of the pion, where the muon and neutrino are produced back to back, the μ^+ can only be produced with $\lambda_\mu = -1/2$, because the π is spinless. Thus in the pion rest system (more correctly in the helicity rest frame of the muon reached from the pion rest frame) the muons are -100% longitudinally polarized.

In the Lab frame (more correctly in the muon helicity rest frame reached from the Lab frame) the muon spin-polarization vector will be different on account of the Wick helicity rotation discussed in subsection 2.2.2. The

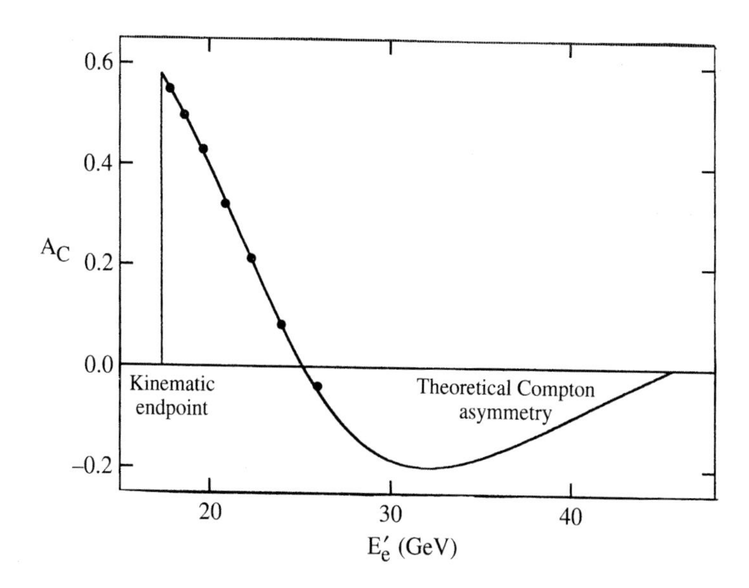

Fig. 8.8 The measured and theoretical Compton asymmetry $\mathscr{P}_z A_C(E'_e)$ at SLAC for $E_e = 45.6$ GeV and for $\mathscr{P}_z = 77.5\%$, as a function of electron recoil energy E'_e. The solid line gives the theory; the points are the measurements, made using a Čerenkov detector.

Wick angle θ_{Wick} will be given by (2.2.6), in which $\delta = \theta \equiv \theta^*$ is the angle between the muon and pion directions of flight, in the pion rest system and $\beta = -\beta_\pi \approx -1$. The rotation is about an axis perpendicular to the plane containing the muon and pion momenta, the latter taken to be along OZ in the Lab.

As a consequence the purely longitudinal mean spin vector of a given muon in the pion rest frame will appear in the Lab frame to have z-component

$$\mathscr{P}_z^{(\mu)}(\theta^*) = -\frac{E^* \cos\theta^* + p^*}{E^* + p^* \cos\theta^*}, \tag{8.1.41}$$

where E^* and p^* are variables in the pion rest system, and will also have a component of $\boldsymbol{\mathcal{P}}$ perpendicular to OZ. Averaging over the azimuthal angles of all muons with a given value of θ^* yields $\mathscr{P}_\perp^{(\mu)}(\theta^*) = 0$ for the ensemble.

Only for strictly forward-going muons $\theta^* = 0$, i.e. the most energetic ones in the Lab, will $\mathscr{P}_z = -1$. (Indeed for $\theta^* = \pi$, $\mathscr{P}_z = +1$!) But the muon beam, of necessity, contains a cone of particles with a range of momenta. It will thus be an ensemble that is longitudinally polarized, with $\boldsymbol{\mathcal{P}}^{(\mu)} = \mathscr{P}_z^{(\mu)} \mathbf{e}_{(z)}$, where $|\mathscr{P}_z^{(\mu)}| < 1$. For accurate work it is therefore necessary to measure $\mathscr{P}_z^{(\mu)}$ for the beam.

Consider the concrete case of positively charged muons. The μ^+ eventually undergo β-decay in flight, $\mu^+ \to e^+\nu_e\bar{\nu}_\mu$, and the shape of the positron energy spectrum is sensitive to the polarization of the muon and is controlled by the so-called Michel parameter (see Commins and Bucksbaum, 1983, Chapter 3.2; for the transformation to the Lab see Combley and Picasso, 1974). One has

$$\frac{dN}{dy} = N\left[\frac{5}{3} - 3y^2 + \frac{4y^3}{3} - \mathscr{P}_z^{(\mu)}\left(\frac{1}{3} - 3y^2 + \frac{8y^3}{3}\right)\right] \tag{8.1.42}$$

where $y = E_e/E_\mu$ is the ratio of the Lab energies of the e^+ and μ^+ and N is the total number of muon decays.

The spectrum dN/dy is shown as a function of y in Fig. 8.9. In practice QED corrections, which have a small but non-negligible effect on the spectrum, are also taken into account (Adeva *et al.*, 1994).

Using this approach the SMC were able to determine that $\mathscr{P}_z = -0.82$ with an error of $\pm 3\%$ for their 100–200 GeV muon beam.

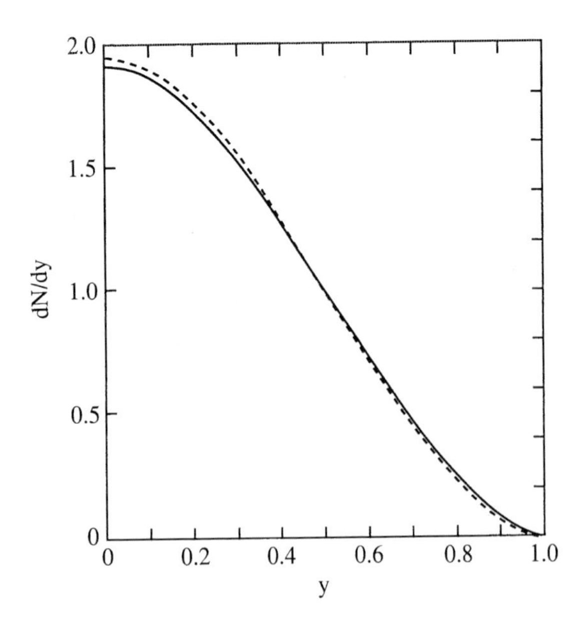

Fig. 8.9 Positron energy spectrum in $\mu^+ \to e^+\nu_e\bar{\nu}_\mu$ as a function of $y = E_e/E_\mu$ with (- - -) and without (—) QED corrections (from Adeva *et al.*, 1994.)

8.1.2 Reaction mechanism not known

It is sometimes possible to use a reaction as a polarization analyser even though its analysing powers are not calculable from first principles. Often, some ingenuity is required in order to do so.

(i) Beam and/or target polarization adjustable

If the target and/or the beam can be put into a state of known polarization, either by electromagnetic means or by utilizing a parity-violating decay to produce the beam (see Section 6.4), so that the initial t^l_m or t^L_M are known, then by measuring the ϕ-dependence of $d^2\sigma/dtd\phi$ one can measure the generalized analysing powers $(l, m; 0, 0|0, 0; 0, 0)$ and/or $(0, 0; L, M|0, 0; 0, 0)$. This is discussed at length in Section 5.4.

(ii) Double-scattering experiments

Another approach is to rely on time-reversal invariance and to do a double-scattering experiment $A + B \to C + D$ followed by $C + D \to A + B$. The initial beam is scattered through an angle θ off one target and the scattered beam is re-scattered off an identical target through the same angle θ. Suppose that both targets are unpolarized and that the initial beam is also unpolarized. The scattered beam C will be polarized, in general, so an analysis of the ϕ-dependent cross-section asymmetries for the *second reaction*, in the natural Lab analysing frame S_{LC}, will yield the combination, see (5.4.14),

$$C_m(\theta) \equiv \sum_{l \geq m} (2l + 1) \left[t^l_m(C; \theta) \right]_{S_{LC}} \times (l, m; 0, 0|0, 0; 0, 0)^{CD \to AB}_{\theta; \phi=0}. \tag{8.1.43}$$

In deriving (8.1.43) we have used a form of (5.4.11) in which C is now the beam particle and we have stressed the θ-dependence. C is produced with multipole parameters $t^l_m(C; \theta)$ in the CM of the first reaction, but, as discussed in subsection 3.3.2(ii), it is the multipole parameters in the appropriate frame S_{LC}, where C is the incoming beam particle, that must be used in (8.1.43). From eqn (3.3.14),

$$\left[t^l_m(C; \theta) \right]_{S_{LC}} = \sum_{m'} d^l_{mm'}(\alpha_C) t^l_{m'}(C; \theta). \tag{8.1.44}$$

Now from (5.4.2), for an unpolarized initial state,

$$t^l_m(C; \theta) = (0, 0; 0, 0|l, m; 0, 0)^{AB \to CD}_{\theta; \phi=0} \tag{8.1.45}$$

and assuming time-reversal invariance we have, from (5.3.9) and (5.3.3),

$$\begin{aligned} (0, 0; 0, 0|l, m; 0, 0)^{AB \to CD}_{\theta; \phi=0} &= (l, m; 0, 0|0, 0; 0, 0)^{CD \to AB}_{\theta; \phi=\pi} \\ &= e^{-im\pi} (l, m; 0, 0|0, 0; 0, 0)^{CD \to AB}_{\theta; \phi=0} \end{aligned} \tag{8.1.46}$$

Thus

$$\left[t^l_m(C;\theta)\right]_{S_{LC}} = \sum_{m'} d^l_{mm'}(\alpha_C)e^{-im'\pi} \times (l,m';0,0|0,0;0,0)^{CD\to AB}. \tag{8.1.47}$$

We see from (8.1.43) and (8.1.47) that what is measured is a bilinear combination of the analysing powers of the reaction $CD \to AB$. If the beam can be passed through a magnetic field between the two scatterings (see subsection 5.4.2(ii)) then for the same angle θ several combinations $C_m(\theta, \delta_i)$ can be measured, involving different bilinear combinations, with known coefficients, of the analysing powers. A sufficient number of measurements with various values of the magnetic field may allow one to solve (8.1.42) and (8.1.47) for the $(l,m;0,0|0,0;0,0)$, though sign ambiguities may remain that must be settled by other means. A beautiful example of this technique can be found in Button and Mermod's experiment on non-relativistic deuteron–deuteron scattering (Button and Mermod, 1960).

A word of caution is necessary, however. What is hidden in the usual discussion (and in the above) is the fact that the reaction parameters one is trying to measure also depend, in principle, on the CM energy of each collision. Since the scattered particle from the first reaction loses energy in the Lab frame, especially if it scatters through a large angle, the CM energy of the second reaction will be less than for the first. Care should be taken to assess in any given situation whether this is a relevant consideration.

We consider some interesting practical examples.

(a) *$p + p \to p + p$ followed by $p + p \to p + p$*

After the first reaction, taken to lie in the XZ-plane, proton C emerges with spin-polarization vector $\boldsymbol{\mathcal{P}} = (0, A(\theta), 0)$ where $A(\theta)$ is the analysing power. There is no Wick helicity rotation in this case (as can be seen using (8.1.44) or from the fact that the rotation is about OY), so that the same vector $\boldsymbol{\mathcal{P}}$ specifies the spin polarization for the incident nucleon in the second reaction. For the second reaction the Y-axis of the S_{LC} is in the same direction as the Y-axis of the Lab frame so we may use (5.6.12) to get

$$\begin{aligned}\frac{d^2\sigma}{dtd\phi} &= \frac{1}{2\pi}\frac{d\sigma}{dt}\left[1 + \cos\phi\ A(\theta)\mathscr{P}_y\right] \\ &= \frac{1}{2\pi}\frac{d\sigma}{dt}\left[1 + \cos\phi\ A^2(\theta)\right]\end{aligned} \tag{8.1.48}$$

so that the ϕ-dependence yields $A(\theta)$ up to a sign. The method is relatively simple for spin-1/2 particles. The practical problem is that $A(\theta)$ gets very small at high energies, at least for moderate t-values.

(b) $A+B \to \gamma + D$ *followed by* $\gamma + D \to A + B$
Consider the state of polarization of the γ produced in the first reaction from an unpolarized initial state. From (5.4.2)

$$t^l_m(\gamma,\theta) = (0,0;0,0|l,m;0,0)^{A+B\to\gamma+D}. \tag{8.1.49}$$

The absence of helicity zero for photons implies, from the properties of the matrices $T^l_m(s=1)$, see (3.1.26), that the only non-zero reaction parameters, see (5.3.2), have $m=0$ or $m=2$. Also, from (5.3.7), $(0,0;0,0|1,0;0,0)=0$. Thus (3.1.84) and (3.1.87) imply that the photon can only be linearly polarized. Moreover, from (5.3.8), for a parity-conserving reaction $(0,0;0,0|2,2;0,0)$ is real. Thus in (3.1.87) $\gamma=0$ and from (3.1.86) the photon is linearly polarized along OX in its standard helicity frame.

Because we are dealing with a photon there is no Wick helicity rotation and the $t^l_m(\gamma;\theta)$ given by (8.1.49) are the correct multipole parameters for the photon as the beam particle in the second reaction. (The second reaction is described in a frame whose Y-axis is in the same direction as the Lab Y-axis of the first reaction.)

For the time-reversed reaction $\gamma + D \to A + B$, with linearly polarized γ, we have from (5.4.11)

$$\frac{d^2\sigma}{dtd\phi} = \left(\frac{2}{3}\right)\frac{1}{2\pi}\frac{d\sigma}{dt}\sum_{l,m}(2l+1)t^l_m(\gamma;\theta) \times (l,m;0,0|0,0;0,0)^{\gamma D\to AB}_{\theta,\phi} \tag{8.1.50}$$

If conventionally we define the photon tensor analysing power of the above reaction $\gamma D \to AB$ by

$$\Sigma(\theta) \equiv \frac{2\sqrt{5}}{\sqrt{3}}(2,2;0,0|0,0;0,0)^{\gamma D\to AB}_{\theta,\phi=0} \tag{8.1.51}$$

and use the fact that for photons (as can be deduced from (5.3.2) and (3.1.26) or from (5.4.2) and (3.1.87))

$$(2,0;0,0|0,0;0,0) = \frac{1}{\sqrt{10}} \tag{8.1.52}$$

and that

$$t^2_2(\gamma;\theta) = (0,0;0,0|2,2;0,0)^{AB\to\gamma D} = (2,2;0,0|0,0;0,0)^{\gamma D\to AB}_{\phi=\pi}$$

by time reversal, and so by (5.3.3)

$$t^2_2(\gamma;\theta) = (2,2;0,0|0,0;0,0)^{\gamma D\to AB}_{\phi=0} = \frac{\sqrt{3}}{2\sqrt{5}}\Sigma(\theta),$$

then we eventually end up, for the second reaction, with

$$\frac{d^2\sigma}{dtd\phi} = \frac{1}{2\pi}\frac{d\sigma}{dt}\left\{1 + [\Sigma(\theta)]^2 \cos 2\phi\right\} \tag{8.1.53}$$

so that $\Sigma(\theta)$ can be found, up to a sign, from the azimuthal dependence.

Of course if one has a photon beam of *known* linear polarization, $\Sigma(\theta)$ can be measured in a much simpler fashion: see Appendix 10.4.

(iii) *Asymmetries in inclusive reactions*

It is an empirical fact that many high energy inclusive hadronic reactions have a significant polarizing power (e.g. $pp \to \Lambda X$) and/or a large analysing power (e.g. $\vec{p}p \to \pi^{\pm}X$). These polarizing and analysing powers appear to remain big even for large-momentum-transfer reactions, in contradiction to naive perturbative QCD expectations. Some possible mechanisms for these are discussed in Chapter 13.

There is, to date, no generally agreed dynamical explanation for the observed behaviour. However, the event rates are high and provided the analysing power can be *measured empirically* to sufficient accuracy they can be utilized as polarimeters.

Perhaps the most promising reaction for polarization analysis of proton beams is

$$\vec{p} + p \to \pi^{+,-,0} + X. \tag{8.1.54}$$

For an unpolarized target, let the unknown spin-polarization vector for the incoming proton beam be $\boldsymbol{\mathcal{P}} = (\mathscr{P}_x, \mathscr{P}_y, \mathscr{P}_z)$ with respect to some fixed Lab or CM reference frame. Then, according to the arguments given in Section 5.8, the differential cross-section will be given by eqn (5.6.12) with the target spin-polarization vector $\boldsymbol{\mathcal{P}}^B$ put to zero and the various analysing powers now referring to the reaction (8.1.54). One has then

$$\frac{d^2\sigma}{dtd\phi} = \frac{1}{2\pi}\frac{d\sigma}{dt}\left[1 + A_N\left(\mathscr{P}_y \cos\phi - \mathscr{P}_x \sin\phi\right)\right] \tag{8.1.55}$$

where we have followed convention and utilized (see (5.6.13) and discussion thereafter)

$$A_N \equiv A^{(A)} \tag{8.1.56}$$

for the proton analysing power of the reaction.

For an unpolarized beam and polarized target exactly the same formula holds with $A_N \to -A_N$ (since $A^{(B)} = -A^{(A)}$ in this case) provided the spin-polarization vector of the target is specified in its natural helicity rest frame reached from the CM, as explained in subsection 3.3.1.

The azimuthal dependence in (8.1.55) then allows us to determine the component of $\boldsymbol{\mathcal{P}}$ perpendicular to the collision axis, provided $A_N(\theta)$ is known.

Measurements of $A_N(\theta)$ were originally carried out at Argonne, Brookhaven and CERN, but the experiments of most relevance to high energy polarimetry have been those carried out at Fermilab using the 200 GeV/c tertiary polarized proton beam described in Section 6.4.

8.2 Unstable particles

There is a vast literature concerning the derivation of the properties, especially the spin and parity, of a resonance from an analysis of its decay, stemming from the heyday of the 1960s when vast numbers of new particles were discovered (Jackson, 1965). Recently this kind of analysis has again come into vogue with the study of the charm and bottom families and the search for glueballs. Our concern is with cases where the resonance or unstable particle is well known and we wish to use its decay purely to learn about its density matrix as it emerges from the production reaction.

The resonance typically decays into two or three particles and their angular distribution is presented in their CM, i.e. in a rest frame of the decaying particle. As mentioned in subsection 3.3.2 several choices of frame are popular – helicity, Gottfried–Jackson, Adair, transversity. The density matrix in one of these frames then differs from the density matrix in the CM of the production reaction by at most a rotation.

We shall discuss decay distributions almost entirely in the helicity rest frame of the decaying particle. Details about other frames can be found in the review article by Bourrely, Leader and Soffer (1980). The treatment we shall give is both general and straightforward, and the use of multipole parameters is far simpler and clearer than density matrix elements.

8.2.1 Two-particle decay of spin-J resonance

We consider the decay of particle C of arbitrary spin J, $C \to E+F$, where E, F are also of arbitrary spin. We consider this decay in the *helicity rest frame* S_C of C, where E emerges with momentum $\mathbf{p}_E = (p_E, \theta_E, \phi_E)$. The initial state of C is then described by the CM helicity density matrix $\rho(C)$ or the CM multipole parameters $t^l_m(C)$ of the *production reaction.*

The decay amplitude is a special case of (4.1.8) in which the initial state has a unique value of J:

$$H_{ef;c}(\theta_E, \phi_E) \propto \langle ef|T^J|c\rangle e^{ic\phi_E} d^J_{c\mu}(\theta_E)$$

where e, f, c refer to the helicities of E, F and C and $\mu = e - f$. However, for a single particle C we must have $c \equiv \lambda_C = J_z$. Then by rotational invariance the matrix element cannot depend on c.

We thus write

$$H_{ef;c}(\theta_E, \phi_E) \propto M_C(e, f) e^{ic\phi_E} d^J_{c\mu}(\theta_E) \qquad (8.2.1)$$

where the $M_C(e, f)$, the *reduced helicity amplitudes*, are dynamics dependent parameters describing the decay.

We shall normalize the $M_C(e, f)$ so that

$$\sum_{e,f} |M_C(e, f)|^2 = 1. \tag{8.2.2}$$

(i) The decay angular distribution

From (5.2.1) and (5.2.3) the *normalized* angular distribution of E is given by

$$W(\theta_E, \phi_E) = \frac{2J+1}{4\pi} \sum_{\substack{e,f \\ c,c'}} |M_C(e, f)|^2 \times e^{i\phi_E(c-c')} d^J_{c\mu}(\theta_E) \rho_{cc'}(C) d^J_{c'\mu}(\theta_E). \tag{8.2.3}$$

Since ρ is hermitian (8.2.3) can be rewritten as

$$W(\theta_E, \phi_E) = \frac{2J+1}{4\pi} \sum_{\substack{e,f \\ c,c'}} |M_C(e, f)|^2 \left\{ \cos\left[\phi_E(c-c')\right] \mathrm{Re}\ \rho_{cc'} - \sin\left[\phi_E(c-c')\right] \mathrm{Im}\ \rho_{cc'} \right\} d^J_{c\mu}(\theta_E) d^J_{c'\mu}(\theta_E). \tag{8.2.4}$$

If parity is conserved in the *decay* then from (4.2.3)

$$M_C(e, f) = \frac{\eta_E \eta_F}{\eta_C} (-1)^{J-s_E-s_F} M_C(-e, -f) \tag{8.2.5}$$

and one obtains, for any production process,

$$W(\theta_E, \phi_E) = W(\pi - \theta_E, \pi + \phi_E), \tag{8.2.6}$$

for a parity-conserving decay, W is symmetric under reflection through the origin of S_C.

If parity is conserved in the *production* reaction and if either

(1) the initial state in that reaction is unpolarized, or
(2) it is polarized, the state of polarization satisfies the experimental conditions (s_1)–(s_3) of subsection 5.4.2(iv) or (s_4) of subsection 5.4.4 and C emerges in the beam-containing plane perpendicular to the quantization plane

then $\rho(C)$ will satisfy (5.4.8) and for *any* decay mechanism, one has

$$W(\theta_E, \phi_E) = W(\pi - \theta_E, \pi - \phi_E), \tag{8.2.7}$$

i.e. W is symmetric under reflection through the Y-axis of S_C.

If the above holds for the production reaction *and* the decay conserves parity then (8.2.6) and (8.2.7) give

$$W(\theta_E, \phi_E) = W(\theta_E, -\phi_E), \tag{8.2.8}$$

which implies that the part of (8.2.4) that depends on Im $\rho_{cc'}$ must vanish in this case, and one has

$$W(\theta_E, \phi_E) = \frac{2J+1}{4\pi} \sum_{\substack{e,f \\ c,c'}} |M_C(e,f)|^2 \cos\left[\phi_E(c-c')\right] \times \mathrm{Re}\, \rho_{cc'}\, d^J_{c\mu}(\theta_E) d^J_{c'\mu}(\theta_E). \tag{8.2.9}$$

Hence the decay distribution in this case yields information only on Re $\rho_{cc'}$. Since for this case (5.4.8) holds, this is equivalent to obtaining information only on the *even-polarization* part of ρ. In fact, this is a special case of a completely general result (see the next section) that in any parity-conserving decay, no matter what the production reaction is, $W(\theta_E, \phi_E)$ depends only on the *even* multipole parameters $t^l_m(C)$ of C.

(ii) Distribution of the final state multipole parameters

The reaction formalism developed in Section 5.3 simplifies enormously when the initial state consists of a single particle. With the multipole parameters all specified in the helicity rest frame of C, the final state multipole parameters are controlled by reaction parameters $(l,m|l',m';L',M')_{\phi_E}$ in which the dependence on θ_E is now explicit:

$$(l,m|l',m';L',M')_{\phi_E} = \frac{1}{4\pi} \frac{1}{\sqrt{2l+1}} C_l(l',m';L',M') \times \mathscr{D}^{(l)}_{m,m'-M'}(\phi_E, \theta_E, 0) \tag{8.2.10}$$

where the C_l, the *decay parameters* dependent on the decay amplitude, are constants and the angular functions are the well-known representation functions of the rotation group (Rose, 1957). In fact

$$\mathscr{D}^{(l)}_{m,m'-M'}(\phi, \theta, 0) = e^{-im\phi} d^l_{m,m'-M'}(\theta). \tag{8.2.11}$$

Explicitly, with our normalization convention (8.2.2) and when the spin of C is J,

$$\begin{aligned} C_l(l',m';L',M') = {} & \left[\frac{(2J+1)(2s_E+1)(2s_F+1)}{(2l'+1)(2L'+1)}\right]^{1/2} (-1)^{J+s_F-s_E} \\ & \times \sum_{e,f} M^*_C(e,f) M_C(e-m', f-M') \langle l',m'|s_E,e;s_E,m'-e\rangle \\ & \times \langle L',M'|s_F,f;s_F,M'-f\rangle \\ & \times \langle l,m'-M'|J,e-f;J,m'-M'-e+f\rangle . \end{aligned} \tag{8.2.12}$$

The relation between the initial and final state multipole parameters is then

$$W(\theta_E,\phi_E)t^{l'L'}_{m'M'}(E,F) = \frac{1}{4\pi}\sum_{l\geq|m'-M'|}^{2J}\sqrt{2l+1}\;C_l(l',m';L',M')$$
$$\times\sum_m t^l_m(C)\mathscr{D}^{(l)}_{m,m'-M'}(\phi_E,\theta_E,0). \qquad (8.2.13)$$

The decay parameters enjoy the following properties:

(α) Normalization:

$$C_0(0,0;0,0) = \sum_{e,f}|M_C(e,f)|^2 = 1. \qquad (8.2.14)$$

(β) l-value constraint:

$$C_l(l',m';L',M') = 0 \qquad \text{if } |m'-M'| > l. \qquad (8.2.15)$$

(γ) Reality:

$$C^*_l(l',m';L',M') = C_l(l',-m';L',-M'). \qquad (8.2.16)$$

(δ) Parity: if parity is conserved in the decay then

$$C_l(l',-m';L',-M') = (-1)^{l+l'+L'}C_l(l',m';L',M'). \qquad (8.2.17)$$

Thus

$$C_l(l',0;L',0) = 0 \quad \text{if } l+l'+L' \text{ is odd.} \qquad (8.2.18)$$

When (8.2.18) is combined with (8.2.16) we have, for a parity conserving decay

$$C_l(l',m';L',M') \text{ is } \left\{\begin{matrix}\text{real}\\ \text{imaginary}\end{matrix}\right\} \text{ if } l+l'+L' \text{ is } \left\{\begin{matrix}\text{even}\\ \text{odd}\end{matrix}\right\}. \qquad (8.2.19)$$

It is clear from (8.2.12) that there will be linear relationships amongst the different $C_l(l',m';L',M')$ for a *fixed* set of values of m' and M', since they are all expressed in terms of the same product $M^*_C(e,f)M_C(e-m',f-M')$ of matrix elements. An example will be given in subsection 8.2.1(ix).

As a special case of (8.2.13) the decay distribution is

$$W(\theta_E,\phi_E) = \frac{1}{4\pi}\sum_{l\geq 0}^{2J}C_l(0,0;0,0)\sum_m t^{l*}_m(C)Y_{lm}(\theta_E,\phi_E) \qquad (8.2.20)$$

where we have used the relation for the spherical harmonics

$$\mathscr{D}^{(l)}_{m,0}(\phi,\theta,0) = \sqrt{\frac{4\pi}{2l+1}}\,Y^*_{lm}(\theta,\phi) \qquad (8.2.21)$$

as well as

$$t^{l*}_m = (-1)^m t^l_{-m} \qquad \text{and} \qquad Y^*_{lm} = (-1)^m Y_{l-m}.$$

A crucial feature is that for a parity-conserving decay only *even* values of l appear on the right-hand side of (8.2.20).

It is important to note that the $t^{l'L'}_{m'M'}(E,F)$ are the multipole parameters in the helicity rest frame S_C of C. They are also the correct multipole parameters in the helicity rest frames S_E, S_F of E and F reached from S_C.

The generalization of (8.2.20) to the two-body decays of unstable particles C_1, C_2 of spin J_1 and J_2 created in some reaction, i.e.

$$A + B \to C_1 + C_2$$
$$C_1 \to E_1 + F_1$$
$$C_2 \to E_2 + F_2,$$

where the decays are *independent* of each other, is straightforward:

$$W(\theta_1,\phi_1;\theta_2,\phi_2) = \frac{1}{4\pi}\sum_{l_1\geq 0}^{2J_1}\sum_{l_2\geq 0}^{2J_2} C_{l_1}(0,0;0,0)C_{l_2}(0,0;0,0) \times \sum_{m_1,m_2} t^{l_1l_2}_{m_1m_2}(C_1,C_2)Y_{l_1m_1}(\theta_1,\phi_1)Y_{l_2m_2}(\theta_2,\phi_2) \tag{8.2.22}$$

where the $t^{l_1l_2}_{m_1m_2}(C_1,C_2)$ are the joint multipole parameters for the produced C_1 and C_2 particles. Of course all angles in (8.2.22), i.e. θ_1,ϕ_1 for E_1 and θ_2,ϕ_2 for E_2, refer to the helicity rest frames S_{C_1} and S_{C_2} of C_1 and C_2. (There is also an obvious generalization of (8.2.13).) A nice application of this formalism to $e^-e^+ \to \tau^-\tau^+$ followed by $\tau^- \to \pi^-\nu_\tau$ and $\tau^+ \to \pi^+\bar{\nu}_\tau$ will be given in subsection 9.2.1(iii).

In the following we shall deal just with the simpler case (8.2.20).

We note the following general properties that follow from (8.2.20) or (8.2.13).

(α) If parity is conserved in the decay then (8.2.18) implies that $W(\theta_E,\phi_E)$ depends only on those $t^l_m(C)$ with l even.

(β) For an unpolarized initial state, i.e. $t^l_m(C) = 0$ for $l \geq 1$,

$$t^{l'L'}_{m'M'}(E,F) = 0 \quad \text{if } m' \neq M'. \tag{8.2.23}$$

In particular, for the effective multipole parameters of E (or F)

$$t^{l'}_{m'}(E) = 0 \quad \text{if } m' \neq 0. \tag{8.2.24}$$

Note that (8.2.23) and (8.2.24) are much stronger results than in the $2 \to 2$ reaction case.

(γ) If parity is conserved in the decay then

$$t^{l'L'}_{m'M'}(E,F;\theta_E,\phi_E) = (-1)^{l'+L'} t^{l'L'}_{-m'-M'}(E,F;\pi-\theta_E,\pi+\phi_E). \tag{8.2.25}$$

For an unpolarized initial state the only non-zero effective multipole parameters for E (or F) are the $t^{l'}_0$ with l' even.

(δ) If the production reaction gives a $\rho(C)$ that satisfies (5.4.8) (see the discussion after eqn (8.2.6)) and if the decay conserves parity then

$$t^{l'L'}_{m'M'}(\theta_E,\phi_E) = (-1)^{l'+L'+m'+M'} t^{l'L'}_{-m'-M'}(\theta_E,-\phi_E). \tag{8.2.26}$$

Equation (8.2.8) for $W(\theta_E,\phi_E)$ is just a special case of this general relation.

(iii) *Moments of the experimental distributions*

Let $G(\theta_E,\phi_E)$ be any function of the polar angles of E in S_C. We denote by $\langle G\rangle$ the average, over all angles, of G weighted by the normalized decay distribution $W(\theta_E,\phi_E)$, i.e.

$$\langle G\rangle \equiv \int_0^{2\pi} d\phi_E \int_0^{\pi} \sin\theta_E \; d\theta_E W(\theta_E,\phi_E)G(\theta_E,\phi_E). \tag{8.2.27}$$

By taking moments of suitable multipole parameters of the decay products one can isolate individual multipoles $t^l_m(C)$ of the initial state. In complete generality one has, from (8.2.13),

$$\left\langle t^{l'L'}_{m'M'} \mathscr{D}^{(l)^*}_{m,m'-M'} \right\rangle = \frac{1}{\sqrt{2l+1}} C_l(l',m';L',M') t^l_m(C) \qquad (l \le 2J). \tag{8.2.28}$$

The simplest example, where no spin properties of E or F are measured, is (see (8.2.13) and (8.2.21))

$$\langle Y_{lm}\rangle = \frac{1}{\sqrt{4\pi}} C_l(0,0;0,0) t^l_m(C) \qquad (l \le 2J). \tag{8.2.29}$$

To use (8.2.28) in practice one requires a table of $\mathscr{D}$-functions. These can be obtained from (8.2.11) and the explicit table of $d^l_{\lambda\mu}$ in Appendix 1. For $l \ge 3$ one must resort to recursion relations. A new and simplified form of these is given in Appendix 1, where also the detailed symmetry properties of the $d^l_{\lambda\mu}$ are stated.

Note that:

(α) In principle, a particular $t^l_m(C)$ can be found from many different moments, i.e. from all those with $|m'-M'| \le l$. In practice, however, one will not know the $C_l(l',m';L',M')$ needed on the right-hand side of (8.2.28) for arbitrary values of its arguments and one will not, in any case, be able to measure joint multipole parameters very easily. The simplest way to obtain the $t^l_m(C)$ is discussed below.

(β) For a parity-conserving decay the right-hand side of (8.2.29) is zero for l odd. The equation is nevertheless useful, and the left-hand side should be calculated from the data also for odd l as a check on experimental biases.

(γ) Moments with $l > 2J$ must give zero. (This is a general result on the maximal angular complexity in the decay of a spin-J particle.) If they do not there are either experimental biases or C is not what it was thought to be!

We see in general that the moments of the angular distribution of the particles, or of their multipole parameters, give us information only about the *product* of $t^l_m(C)$ and certain decay parameters whose value depends upon the decay mechanism.

For a *parity-violating* decay, the decay distribution functions as a complete analyser of the polarization state of C, so all we require to know are the decay parameters $C_l(0,0;0,0)$. Of these,

$$N \equiv \min\{(2s_E + 1)(2s_F + 1) - 1; 2J\}$$

are independent.

For a *parity-conserving* decay we need the decay distribution and the distribution of any one of the odd multipole parameters of either E or F in order to get a complete analysis of the state of polarization of C. We thus require to know the $C_l(0,0;0,0)$ and, say, the $C_l(1,0;0,0)$. In total these depend upon

$$N_{\mathscr{P}} \equiv \begin{cases} \min\left\{\dfrac{(2s_E + 1)(2s_F + 1)}{2} - 1;\ 2J\right\} \\ \text{or} \\ \min\left\{\dfrac{(2s_E + 1)(2s_F + 1)}{2} - \dfrac{1}{2};\ 2J\right\} \end{cases}$$

real parameters, for $(2s_E + 1)(2s_F + 1)$ even or odd, respectively.

If we cannot calculate the C_l from first principles, then in order to use a decay as an analyser we must first carry out N or $N_{\mathscr{P}}$ measurements on the decay products in such a way that the polarization state of C is irrelevant, i.e. by measuring moments with $l = m = 0$ (see eqn (8.2.28)). This gives us the minimum required number of C_l and thereafter the decay can function as a complete analyser for arbitrary states of polarization of C.

Fortunately many decays in Nature have $N_{\mathscr{P}} = 0$ or $N = 1$ so the whole problem of finding the C_l disappears or at least becomes relatively simple. We shall refer to decays with $N_{\mathscr{P}} = 0$ and which thus automatically function as analysers as *magic* decays.

Moreover, many high-spin resonances undergo a 'decay chain', i.e. a sequence of two-body decays of which the last, at least, has $N_{\mathscr{P}} = 0$ or $N = 1$, and we shall show later how this fact can be used to bypass the problem of finding the C_l for the intermediate links of the chain.

(iv) Magic decays

We now list the magic decays. The decay parameters are given in terms of vector addition coefficients, which are fully tabulated in Appel (1968) or can be extracted from the table of Clebsch–Gordan coefficients in any biennial 'Review of particle properties' in *Phys Rev D*, for example Barnett *et al.* (1996). The following results are for the case where the particle with non-zero spin has decay angles θ_E, ϕ_E.

(a) $J \to 0+0$ (e.g. $\rho \to \pi\pi$, $f \to \pi\pi$, $D \to K\pi$):

$$C_l(0,0;0,0) = (-1)^J\sqrt{2J+1}\,\langle l,0|J,0;J,0\rangle \qquad l \text{ even.} \tag{8.2.30}$$

(b) $J \to 1/2+0$ with parity conserved and $J \to \nu+0$ with maximal parity violation as in the Standard Model (e.g. $\Delta \to N\pi$, $\tau^- \to \nu\pi$):

$$\begin{aligned} &C_l(0,0;0,0) = (-1)^{J-1/2}\sqrt{2J+1}\,\langle l,0|J,1/2;J,-1/2\rangle \qquad l \text{ even.}\\ &C_l(1,0;0,0) = \tfrac{1}{\sqrt{3}}(-1)^{J-1/2}\sqrt{2J+1}\,\langle l,0|J,1/2;J,-1/2\rangle \qquad l \text{ odd.}\\ &C_l(1,1;0,0) = 0. \end{aligned} \tag{8.2.31}$$

(c) $J \to$ photon $+0$, parity conserved (e.g. $\omega \to \gamma\pi$, $D^{*+} \to \gamma D$):

$$\begin{aligned} &C_l(0,0;0,0) = (-1)^{J-1}\sqrt{2J+1}\,\langle l,0|J,1;J,-1\rangle \qquad l \text{ even}\\ &C_l(1,0;0,0) = (-1)^{J}\sqrt{2J+1}\,\langle l,0|J,1;J,-1\rangle \qquad l \text{ odd}\\ &C_l(2,0;0,0) = (-1)^{J-1}\sqrt{\tfrac{1}{2}}\sqrt{2J+1}\,\langle l,0|J,1;J,-1\rangle \qquad l \text{ even}\\ &C_l(2,2;0,0) = (-1)^{J-1}\sqrt{\tfrac{3}{5}}\sqrt{2J+1}\,\langle l,2|J,1;J,1\rangle \qquad l \text{ even}\\ &C_l(1,1;0,0) = C_l(2,1;0,0) = 0. \end{aligned} \tag{8.2.32}$$

(d) $J \to 1+0$, parity conserved *and* intrinsic parities satisfy $\eta_0\eta_1 = (-1)^J\eta_J$ (e.g. $\psi \to \rho\pi$, $a_2 \to \rho\pi$, $\omega(1670) \to \rho\pi$): same as (c).

(v) The decay $C(\text{spin } J) \to E(\text{spin } 1/2) + F(\text{spin } 0)$

The decay $J \to 1/2+0$ with or without parity conservation is of particular interest, so we study it in some detail. We split each decay amplitude into a parity-conserving (PC) and a parity-violating (PV) piece,

$$M(\lambda) = M_{\text{PC}}(\lambda) + M_{\text{PV}}(\lambda) \qquad \lambda = \pm 1/2, \tag{8.2.33}$$

such that

$$M_{\mathrm{PC/PV}}(-\lambda) = \pm\eta M_{\mathrm{PC/PV}}(\lambda)$$

where the plus and minus apply to M_{PC} and M_{PV} respectively and where η is the phase factor in eqn (8.2.5).

There are three real linearly independent parameters controlling the decay,

$$\alpha = \frac{2\,\mathrm{Re}\,(M_{\mathrm{PC}}M^*_{\mathrm{PV}})}{|M_{\mathrm{PC}}|^2 + |M_{\mathrm{PV}}|^2}, \qquad \beta = \frac{2\,\mathrm{Im}\,(M_{\mathrm{PC}}M^*_{\mathrm{PV}})}{|M_{\mathrm{PC}}|^2 + |M_{\mathrm{PV}}|^2},$$
$$\gamma = -\frac{|M_{\mathrm{PC}}|^2 - |M_{\mathrm{PV}}|^2}{|M_{\mathrm{PC}}|^2 + |M_{\mathrm{PV}}|^2} \tag{8.2.34}$$

which satisfy $\alpha^2+\beta^2+\gamma^2 = 1$. Our definitions of α, β, γ have been chosen so as to agree with those used in the *Review of Particle Properties* (Particle Data Group, 1978) for the case $\Lambda \to N\pi$. As a consequence our γ is opposite in sign to the γ defined in Jackson (1965).

If parity is conserved one has $\alpha = \beta = 0$, $\gamma = -1$.

The decay parameters are given in terms of these by

$$C_l(0,0;0,0) = (-1)^{J-1/2}\sqrt{2J+1}\,\langle l,0|J,1/2;J,-1/2\rangle \times \begin{Bmatrix} 1 \\ \alpha \end{Bmatrix} \quad \text{for } l \begin{Bmatrix} \text{even} \\ \text{odd} \end{Bmatrix}$$

$$C_l(1,0;0,0) = (-1)^{J-1/2}\sqrt{\tfrac{1}{3}}\sqrt{2J+1}\,\langle l,0|J,1/2;J,-1/2\rangle \times \begin{Bmatrix} \alpha \\ 1 \end{Bmatrix} \quad \text{for } l \begin{Bmatrix} \text{even} \\ \text{odd} \end{Bmatrix}$$

$$C_l(1,1;0,0) = \bar{\eta}\sqrt{\tfrac{1}{6}}\sqrt{2J+1}\,\langle l,1|J,1/2;J,1/2\rangle\,(-\gamma + i\beta) \qquad \text{for } l \text{ odd}$$
$$= 0 \qquad \text{for } l \text{ even} \tag{8.2.35}$$

with $\bar{\eta} = \eta_C\eta_E\eta_F$ in terms of intrinsic parities.

From these and eqns (8.2.13), (8.2.20) and (8.2.28) we learn the following.

(a) If parity is conserved we have a magic decay; thus a measurement of $W(\theta_E, \phi_E)$ and the distribution of any one component of the polarization of E will yield a complete analysis of the $t^l_m(C)$. (Recall that $\mathscr{P}_z = \sqrt{3}t^1_0(E)$, $\mathscr{P}_x = \sqrt{3/2}(t^1_{-1} - t^1_1)$ and $\mathscr{P}_y = i\sqrt{3/2}(t^1_{-1} + t^1_1)$, the components referring to the axes in the helicity rest frame of E.)

(b) If parity is not conserved a measurement of the decay distribution $W(\theta_E, \phi_E)$ yields the $t^l_m(C)$ for l even and $\alpha t^l_m(C)$ for l odd. The asymmetry parameter α can be found most simply from the average longitudinal polarization of E:

$$\langle \mathscr{P}_z(E) \rangle = \sqrt{3} C_0(1,0;0,0) = \alpha. \tag{8.2.36}$$

The most common decays of this type are e.g. $\Lambda \to N\pi$, $\Sigma \to N\pi$.

Let us assume C is produced in a parity-conserving reaction in a situation such that eqn (5.4.8) is satisfied. Then $t^1_{-1}(C) = t^1_1(C)$, $t^1_0(C) = 0$ and the polarization vector of C is normal to the production plane, i.e. along the Y-axis in the helicity rest frame of C. Explicitly then

$$W(\theta_E, \phi_E) = \frac{1}{4\pi} \left(1 + \alpha \mathscr{P}^C_y \sin\theta_E \sin\phi_E\right) \tag{8.2.37}$$

giving rise to the well-known up–down symmetry with respect to the production plane.

If $(\mathscr{P}^E_{x'}, \mathscr{P}^E_{y'}, \mathscr{P}^E_{z'})$ are the components of the polarization vector for E in its helicity rest frame S_E, one has

$$\begin{aligned}
W(\theta_E, \phi_E)\mathscr{P}^E_{z'} &= \frac{1}{4\pi}\left(\alpha + \mathscr{P}^C_y \sin\theta_E \sin\phi_E\right) \\
W(\theta_E, \phi_E)\mathscr{P}^E_{x'} &= -\frac{\eta \mathscr{P}^C_y}{4\pi}(\gamma \cos\theta_E \sin\phi_E + \beta\cos\phi_E) \\
W(\theta_E, \phi_E)\mathscr{P}^E_{y'} &= \frac{\eta \mathscr{P}^C_y}{4\pi}(\beta \cos\theta_E \sin\phi_E - \gamma\cos\phi_E)
\end{aligned} \tag{8.2.38}$$

where η is the phase factor in eqn (8.2.5).

If we define a 3-vector $\bar{\mathscr{P}}^E$ in the rest frame S_C of C, such that its components $(\bar{\mathscr{P}}^E_x, \bar{\mathscr{P}}^E_y, \bar{\mathscr{P}}^E_z)$ are numerically equal to the components of $\mathscr{P}^E$ along the axes of the Adair rest frame of E (the Adair rest frame is reached from S_C by a *pure* Lorentz transformation along the direction of motion of E), then (8.2.38) can be written as a 3-vector equation in S_C:

$$\begin{aligned}
4\pi W(\theta_E, \phi_E)\bar{\mathscr{P}}^E &= (\alpha + \mathbf{e}\cdot\boldsymbol{\mathcal{P}}^C)\mathbf{e} + \eta\gamma\left[\mathbf{e}\times(\mathbf{e}\times\boldsymbol{\mathcal{P}}^C)\right] \\
&\quad + \eta\beta(\mathbf{e}\times\boldsymbol{\mathcal{P}}^C)
\end{aligned} \tag{8.2.39}$$

where $\mathbf{e}$ is a unit vector along the momentum of E. Note that our formula has explicit reference to the relative parities of the particles in it (through η) and also to the spin J of C. It will thus only agree with the formula given by the Particle Data Group in reactions like $\Lambda \to N\pi$ where $\eta = -1$.

(vi) Fermionic decay $C(J) \to E(J_1) + 0$, $J \geq 3/2$

It is now necessary to measure the average values for each of the $t^{l'}_0(E)$, which, from (8.2.28) yield

$$\left\langle t^{l'}_0(E) \right\rangle = C_0(l', 0; 0, 0). \tag{8.2.40}$$

(If parity is conserved in the decay, the right-hand side $= 0$ for l' odd.)

Then from (8.2.12) we obtain

$$C_l(0,0;0,0) = \sqrt{2l+1}C_0(l,0;0,0), \tag{8.2.41}$$

which is another example of a reaction's power to analyse being related to its power to polarize.

There are then two possibilities.

(a) If parity is not conserved, all the $C_l(0,0;0,0)$ being now known the distribution of E functions as a complete analyser for the state of polarization of C.

(b) If parity is conserved the $C_l(0,0;0,0)$ with odd l are zero. The decay distribution yields the $t^l_m(C)$ only for l even.

To find the $t^l_m(C)$ with odd l, one studies the distribution of $t^{l'}_0(E)$ for any odd value of l'. One thus requires the decay parameters $C_l(l',0;0,0)$, which are calculated as follows.

One solves (8.2.12) for the moduli squared of the decay matrix elements

$$|M_C(\lambda_1)|^2 = \frac{1}{2J_1+1}\sum_l(2l+1)\,\langle J_1,\lambda_1|J_1,\lambda_1;l,0\rangle \\ \times C_0(l,0;0,0) \tag{8.2.42}$$

(note that the left-hand side is just $\rho_{\lambda_1\lambda_1}(E_1)$ if C is *unpolarized*) and then substitutes in

$$C_l(l',0;0,0) = \sqrt{2l+1}\sum_{\lambda_1}\langle J_1,\lambda_1|J_1,\lambda_1;l',0\rangle \\ \times \langle J,\lambda_1|J,\lambda_1;l,0\rangle\,|M_C(\lambda_1)|^2. \tag{8.2.43}$$

In electroweak interactions, which violate parity, it often happens that the decaying particle C is produced with longitudinal polarization $\mathscr{P}^C_z$. In this case the angular distribution of E is given by

$$W(\theta_E,\phi_E) = \frac{1}{4\pi}\left(1+\alpha\mathscr{P}^C_z\cos\theta_E\right). \tag{8.2.44}$$

For many resonance decays the values of the decay parameter α are quite well known, as shown in Table 8.1. Equation (8.2.44) will be of use in discussing reactions like $e^+e^- \to \Lambda + X$ where, in the Standard Model, one expects the Λs to be highly polarized longitudinally (see subsection 9.2.2(ii)).

Table 8.1. Values of the decay parameter α for various hyperon decays

Decay	Value of α
$\Lambda \to p\pi^-$	0.642 ± 0.013
$\Sigma^+ \to p\pi^0$	-0.980 ± 0.016
$\Sigma^+ \to n\pi^+$	0.068 ± 0.013
$\Sigma^- \to n\pi^-$	-0.068 ± 0.008
$\Xi^0 \to \Lambda\pi^0$	-0.411 ± 0.022
$\Xi^- \to \Lambda\pi^-$	-0.456 ± 0.014
$\Lambda_c^+ \to \Lambda\pi^+$	-0.98 ± 0.19
$\Lambda_c^+ \to \Sigma^+\pi^0$	-0.45 ± 0.32

(vii) Fermionic decay chains

A common situation is that the produced fermion C decays sequentially, emitting a spin-0 meson at each step. We thus consider

$$\begin{array}{l} C(J) \to E_1(J_1) + 0 \\ \qquad\quad \searrow E_2(J_2) + 0 \\ \qquad\qquad\quad \searrow \\ \qquad\qquad\qquad \cdots \\ \qquad\qquad\qquad\quad \searrow E_n(J_n) + 0 \\ \qquad\qquad\qquad\qquad\quad \searrow E(1/2) + 0. \end{array} \tag{8.2.45}$$

We denote by $t^l_m(E_k)$ the multipole parameters of E_k in its helicity rest system as reached from the helicity rest frame of its parent E_{k-1}.

From the discussion in (iv) above, a study of the distribution of E in the frame S_{E_n}, and if necessary of the distribution of one of its polarized components, will always yield all the $t^l_m(E_n)$ whether or not parity is conserved. From (iv) or (v) the distribution of E_n and one of its odd multipoles $t^{l_n}_0(E_n)$ is enough to give all the $t^l_m(E_{n-1})$ etc.

In principle, therefore, one can work one's way back up the chain until one obtains all the $t^l_m(C)$.

(viii) Bosonic decay chains

We consider the decay of a heavy boson and suppose that it is dominated by a sequence of two-body decay as in (8.2.45), but with various possibilities for the final decay. We assume that parity is conserved in each decay.

There are three cases of interest, depending on the form of the last decay in the chain.

(a) If the last decay is of the form $E_n(J_n) \to E(0)+0$ then the distribution of E in S_{E_n} gives the *even-l* $t^l_m(E_n)$, but there is no way to find the $t^l_m(E_n)$ with odd l. Proceeding up the chain one ends up with only the even-multipole parameters of the original resonance C.

(b) If the last reaction is of the form $E_n(J_n) \to$ photon $+\ 0$ and if the state of polarization of the photon can be measured (see subsection 3.1.12) then since the decay is magic all the required decay parameters are known (see eqn (8.2.32)) and one can obtain all the $t^l_m(E_n)$. One can then work back up the chain to obtain all $t^l_m(C)$ of the original resonance.

(c) If it is possible to detect decay into a stable fermion–antifermion pair, e.g. $\rho \to \mu^-\mu^+$, and if the longitudinal polarization of the fermions can be measured then all $t^l_m(E_n)$ can be found and, proceeding up the chain, eventually all $t^l_m(C)$.

Suppose the final decay is $E_n(J_n) \to E(1/2)+\bar{E}(1/2)$. The decay parameters that appear in the angular distribution of E and of its longitudinal polarization are, from (8.2.12),

$$\begin{aligned} C_l(0,0;0,0) &= (-1)^{J_n}\sqrt{2J_n+1}\,\big[(1-\epsilon)\,\langle l,0|J_n,0;J_n,0\rangle \\ &\qquad -\epsilon\,\langle l,0|J_n,1;J_n,-1\rangle\big] \qquad \text{for } l \text{ even} \\ C_l(1,0;0,0) &= \epsilon(-1)^{J_n+1}\sqrt{\frac{2J_n+1}{3}} \\ &\qquad \times \langle l,0|J_n,1;J_n,-1\rangle \qquad \text{for } l \text{ odd,} \end{aligned} \tag{8.2.46}$$

where ϵ is a measure of the relative decay probabilities into helicities $++$ or $+-$. Specifically,

$$|M_{E_n}(+-)|^2/|M_{E_n}(++)|^2 = \epsilon/(1-\epsilon). \tag{8.2.47}$$

Thus a measurement of the moments of the distribution of E and of its longitudinal polarization will give all the $t^l_m(E_n)$ as functions of one parameter ϵ. To find ϵ requires the measurement of a correlation between the spins of E and $\bar{E}$. If the spin projections for E and $\bar{E}$ are referred to the axes of their respective helicity rest frames, then

$$\int_0^{2\pi} d\phi_E \int_0^{\pi} \sin\theta_E\, d\theta_E\, \langle \sigma_z(E)\sigma_z(\bar{E})\rangle_{\theta_E,\phi_E}\, W(\theta_E,\phi_E) = 1-2\epsilon. \tag{8.2.48}$$

For $J_n = 1$, and where the decay is into a lepton–anti-lepton pair coupled purely through a minimal electromagnetic-type γ^μ coupling (e.g. $\rho \to \mu^-\mu^+$) one finds that

$$\epsilon = \left(1+2m_l^2/M_{E_n}^2\right)^{-1}$$

where m_l and M_{E_n} are the lepton and resonance masses. Thus $\epsilon \sim 1$ and the decay functions as a very efficient analyser. In this case, clearly one need not measure ϵ.

(ix) Decay of W-boson

As discussed in Chapter 9 one of the most remarkable developments in particle physics has been the prediction and discovery of the massive vector bosons $W^{\pm}$, Z^0 associated with the unification of the weak and electromagnetic forces. Polarization phenomena in the decay of the Z^0 are treated in detail in subsection 9.2.1. Here we describe a fundamental test to show that the W is really a spin-1 boson.

Suppose that the W had spin J, and consider its decay

$$W \to E + F$$

where E, F have spins S_E, S_F respectively.

Using the properties of the vector addition coefficients (Rose, 1957), one can show that

$$\sqrt{\frac{J(J+1)}{3}}C_1(0,0;0,0) = \sqrt{S_E(S_E+1)}C_0(1,0;0,0) - \sqrt{S_F(S_F+1)}C_0(0,0;1,0). \quad (8.2.49)$$

To turn this into a relation amongst measured quantities we note that from (8.2.29)

$$\begin{aligned}\langle \cos\theta_E\rangle &\equiv \langle P_1(\theta_E)\rangle = \sqrt{\frac{4\pi}{3}}\,\langle Y_{10}\rangle \\ &= \sqrt{\frac{1}{3}}\sqrt{\frac{J}{(J+1)}}C_1(0,0;0,0)\mathscr{P}_z(W) \\ &= \sqrt{\frac{1}{3}}\frac{1}{\sqrt{J(J+1)}}C_1(0,0;0,0)\,\langle \hat{s}_z\rangle_W\,. \end{aligned} \quad (8.2.50)$$

Also, from (8.2.28),

$$\left\langle t^1_0(E)\right\rangle = C_0(1,0;0,0) \qquad \left\langle t^1_0(F)\right\rangle = C_0(0,0;1,0). \quad (8.2.51)$$

Thus, substituting in (8.2.49) and using (3.1.35) and (1.1.27)

$$\langle \cos\theta_E\rangle = \frac{1}{J(J+1)}\left[\langle\lambda_E\rangle - \langle\lambda_F\rangle\right]\langle \hat{s}_z\rangle_W \quad (8.2.52)$$

where $\langle\lambda_i\rangle$ is the mean helicity of particle i. This result, clearly, is quite general and does not depend specifically on the initial particle's being a W. Indeed (8.2.52) was first derived by Jacob (1958) in the context of strange particle decay, but its most dramatic success was in connection with the original discovery of $W^+ \to e^+\nu_e$, where the value $\langle\cos\theta_{e^+}\rangle = 0.5$ was found. This 'killed three birds with one stone'! It required $J_W = 1$, $\langle\lambda_{e^+}\rangle = -\langle\lambda_{\nu_e}\rangle = 1/2$ and $\langle\hat{s}_z\rangle_W = 1$, all in perfect agreement with the predictions of the standard model of electroweak interactions.

8.2.2 *Three-particle decay of a spin-J resonance*

We consider the decay of a particle C of arbitrary spin J and mass m_C,

$$C \to E_1 + E_2 + E_3,$$

where the particles E_i have arbitrary spin. The most common decays, in practice, are those in which all E_i have spin zero, e.g. 3π, or in which one particle, say E_1, has spin 1/2 and the others are pions. We shall thus not discuss the most general case but limit ourselves to the situation where the polarization, or the multipole parameters of, at most *one* of the decay products is observed. We shall always refer to this particle as E_1.

(i) *Decay amplitudes*

There are many ways to specify the configuration of a three-particle state. Werle (1963) used the polar angles θ_1, ϕ_1 of E_1 and one further angle to specify the orientation of the *decay triangle*, i.e. the triangle formed by the momenta $\mathbf{p}_i$ of the final particles in the rest system of C. Berman and Jacob (1965) characterized the state by the polar angles of the *normal* to the decay triangle and an angle specifying the orientation of the triangle once the normal is fixed. We will utilize the latter only, since we have found that it leads to simpler results.

Let S_C with axes X, Y, Z be the helicity rest frame of C as reached from the CM of the production reaction.[1] Let ω_i, $\mathbf{p}_i$ be the energies and momenta of the particles E_i, with

$$\omega_1 + \omega_2 + \omega_3 = m_C. \tag{8.2.53}$$

An arbitrary state in which the E_i have helicities λ_i is written as

$$|\omega_1\lambda_1, \omega_2\lambda_2, \omega_3\lambda_3; \phi_n, \theta_n, \gamma\rangle,$$

where $\theta = \theta_n$, $\phi = \phi_n$ are the polar angles of the normal $\mathbf{n}$ to the decay triangle, the direction of $\mathbf{n}$ being along $\mathbf{p}_1 \times \mathbf{p}_2$.

The significance of γ is best seen by noting that the above state can be obtained from a 'standard' one by a rotation through Euler angles ϕ_n, θ_n, γ:

$$|\omega_i\lambda_i; \phi_n, \theta_n, \gamma\rangle = U[r(\phi_n, \theta_n, \gamma)]|\omega_i\lambda_i; 0, 0, 0\rangle \tag{8.2.54}$$

[1] Any other rest frame of C is equally good provided the correct density matrix for C is used in the formulae that follow.

where, in $|\omega_i\lambda_i; 0, 0, 0\rangle$, $\mathbf{p}_1$ lies along OX and $\mathbf{p}_2$ lies in the XY-plane with $p_2 > 0$, so that $\mathbf{n}$ lies along OZ.

For a given event, once the polar angles θ_n, ϕ_n of the normal are determined, the angle γ can be found from the polar angles θ_1, ϕ_1 of E_1 as follows:

$$\begin{aligned}\cos\gamma &= \cos\theta_1/\sin\theta_n \\ \sin\gamma &= \sin\theta_1(\sin\phi_1\cos\phi_n - \cos\phi_1\sin\phi_n).\end{aligned} \tag{8.2.55}$$

The decay amplitude is of the form

$$\begin{aligned}&H_{\lambda_1\lambda_2\lambda_3;\lambda_C}(\omega_1, \omega_2, \omega_3; \phi_n, \theta_n, \gamma) \\ &\qquad \propto \sum_{\mathcal{M}=-J}^{J} F_{\mathcal{M}}(\omega_1\lambda_1, \omega_2\lambda_2, \omega_3\lambda_3)\mathcal{D}^{(J)^*}_{\lambda_C\mathcal{M}}(\phi_n, \theta_n, \gamma)\end{aligned}$$

where, because $\lambda_C = J_z$, $F_{\mathcal{M}}$ is independent of λ_C by rotational invariance. The physical significance of $\mathcal{M}$ is that it is the projection of $\mathbf{J}$ along the normal to the decay plane; the dependence on $\mathcal{M}$ is a new feature compared with the two-body decay situation.

We normalize $F_{\mathcal{M}}$ in such a way that

$$\iint d\omega_1 d\omega_2 \sum_{\mathcal{M}}\sum_{\lambda_i} |F_{\mathcal{M}}(\omega_i\lambda_i)|^2 = 1 \tag{8.2.56}$$

where the integration over ω_1, ω_2 corresponds to summing over the Dalitz plot.

The most detailed distribution we consider involves measurement of the helicity multipole moments of particle E_1 and of their dependence on ϕ_n, θ_n, γ. (These are the multipole moments in the helicity rest frame S_{E_1} of E_1 reached from S_C. S_{E_1} has its Z-axis along $\mathbf{p}_1$ and OX opposite to the normal to the decay plane.) It is assumed that a summation over the Dalitz plot is performed.[1] Then

$$W(\phi_n, \theta_n, \gamma)t^l_m(E_1) = \frac{1}{8\pi^2}\sum_{L,M}(2L+1)t^L_M(C)[LM|lm]_{\phi_n\theta_n\gamma} \tag{8.2.57}$$

where $t^L_M(C)$ are the helicity multipole parameters of C, $W(\phi_n, \theta_n, \gamma)$ is the normalized decay distribution and

$$\begin{aligned}[LM|lm]_{\phi_n\theta_n\gamma} \equiv &\sum_{\mathcal{M}\mathcal{M}'} R_{\mathcal{M}\mathcal{M}'}(l, m)\langle J, \mathcal{M}'; L, \mathcal{M} - \mathcal{M}'|J, \mathcal{M}\rangle \\ &\times \mathcal{D}^{(L)}_{M,\mathcal{M}-\mathcal{M}'}(\phi_n, \theta_n, \gamma)\end{aligned} \tag{8.2.58}$$

[1] When particles E_2 and E_3 are identical we relax this assumption, as discussed later.

in which the *decay parameters* are

$$R_{\mathscr{M}\mathscr{M}'}(l,m) = \int\int d\omega_2 d\omega_3 \sum_{\lambda_i} \langle s_1,\lambda_1|s_1,\lambda_1-m;l,m\rangle \times F^*_{\mathscr{M}}(\omega_2,\omega_3;\lambda_1,\lambda_2,\lambda_3)F_{\mathscr{M}'}(\omega_2,\omega_3;\lambda_1-m,\lambda_2,\lambda_3). \tag{8.2.59}$$

Note that

$$R^*_{\mathscr{M}\mathscr{M}'}(l,m) = (-1)^m R_{\mathscr{M}'\mathscr{M}}(l,-m). \tag{8.2.60}$$

For simplicity we shall write $R_{\mathscr{M}'\mathscr{M}}$ for $R_{\mathscr{M}'\mathscr{M}}(0,0)$.

(ii) The angular distribution of the normal to the decay plane
If we take $l=m=0$ in (8.2.57) and integrate over γ we are left with the normalized angular distribution for the normal to the decay triangle:

$$\begin{aligned}\bar{W}(\phi_n,\theta_n) &\equiv \int_0^{2\pi} d\gamma\, W(\phi_n,\theta_n,\gamma)\\ &= \frac{1}{\sqrt{4\pi}}\sum_{L,M}\sqrt{2L+1}\, t^{L\,*}_M(C) R_L Y_{LM}(\theta_n,\phi_n)\end{aligned} \tag{8.2.61}$$

where the real constants R_L are given by

$$R_L = \sum_{\mathscr{M}} R_{\mathscr{M}\mathscr{M}}\langle J,\mathscr{M};L,0|J,\mathscr{M}\rangle . \tag{8.2.62}$$

The normalization condition (8.2.56) implies that

$$R_0 = \sum_{\mathscr{M}} R_{\mathscr{M}\mathscr{M}} = 1.$$

Note that (8.2.61) is exactly analogous to the two-particle decay distribution (8.2.20), with the R_L, which depend on the dynamics of the process, playing the rôle of the decay parameters $C_L(0,0;0,0)$.

(a) *Consequences of parity conservation*
If parity is conserved in the decay process, one finds that

$$F_{\mathscr{M}}(\omega_2,\omega_3;\lambda_1,\lambda_2,\lambda_3) = e^{i\pi\mathscr{M}}\eta_C\eta_1\eta_2\eta_3(-1)^{s_1-\lambda_1+s_2-\lambda_2+s_3-\lambda_3} \times F_{\mathscr{M}}(\omega_2,\omega_3;-\lambda_1,-\lambda_2,-\lambda_3) \tag{8.2.63}$$

where the η are intrinsic parities. If the E_i are spinless particles then (8.2.63) will cause the vanishing of either the even-$\mathscr{M}$ or the odd-$\mathscr{M}$ amplitudes, depending on the intrinsic parities. In the general spin case one has, from (8.2.59),

$$R_{\mathscr{M}\mathscr{M}'}(l,-m) = (-1)^{l+\mathscr{M}-\mathscr{M}'} R_{\mathscr{M}\mathscr{M}'}(l,m). \tag{8.2.64}$$

Thus

$$R_{\mathscr{M}\mathscr{M}'}(l,0)=0 \qquad \text{if } l+\mathscr{M}-\mathscr{M}' \text{ is odd.} \tag{8.2.65}$$

In contrast to the two-body case, (8.2.18), parity invariance here does not make R_L vanish for odd L in general, and $\bar{W}(\phi_n,\theta_n)$ serves as an analyser of both the even and odd parts of the polarization of C.[1]

We see from (8.2.61) that the ratios $t^L_M(C)/t^L_{M'}(C)$ for any L and arbitrary M, M' can be obtained directly from $\bar{W}(\phi_n,\theta_n)$ without requiring any knowledge of the R_L. To obtain the dependence on L does require a knowledge of the R_L.

(b) *Identical particles*

If particles E_2 and E_3 are identical, or if they are in an eigenstate of isotopic spin, the decay amplitude must be made either symmetric or antisymmetric under the exchange of the space and spin labels of E_2 and E_3. The correctly symmetrized amplitude $F^{\mathscr{S}}$ is then found to satisfy

$$F^{\mathscr{S}}_{\mathscr{M}}(\omega_1\lambda_1,\omega_2\lambda_2,\omega_3\lambda_3)=\pm(-1)^{J+\lambda_1+\lambda_2+\lambda_3} \times F^{\mathscr{S}}_{-\mathscr{M}}(\omega_1\lambda_1,\omega_3\lambda_3,\omega_2\lambda_2). \tag{8.2.66}$$

Introducing the quantity $R_{\mathscr{M}\mathscr{M}'}(l,m;\omega_2,\omega_3)$ as the unintegrated analogue of $R_{\mathscr{M}\mathscr{M}'}(l,m)$ (see eqn (8.2.59)), one obtains, for the symmetrized form,

$$R^{\mathscr{S}}_{\mathscr{M}\mathscr{M}'}(l,m;\omega_2,\omega_3)=e^{-i\pi m}R^{\mathscr{S}}_{-\mathscr{M}-\mathscr{M}'}(l,m;\omega_3,\omega_2). \tag{8.2.67}$$

If we also introduce the unintegrated version of R_L, i.e.

$$R_L(\omega_2,\omega_3)=\sum_{\mathscr{M}}R_{\mathscr{M}\mathscr{M}}(0,0;\omega_2,\omega_3)\,\langle J,\mathscr{M};L,0|J,\mathscr{M}\rangle \tag{8.2.68}$$

then we obtain

$$R^{\mathscr{S}}_L(\omega_2,\omega_3)=\tfrac{1}{2}\sum_{\mathscr{M}}\langle J,\mathscr{M};L,0|J,\mathscr{M}\rangle \times\left[R^{\mathscr{S}}_{\mathscr{M}\mathscr{M}}(0,0;\omega_2,\omega_3)+(-1)^L R^{\mathscr{S}}_{\mathscr{M}\mathscr{M}}(0,0;\omega_3,\omega_2)\right\} \tag{8.2.69}$$

and we see that after integration over the whole Dalitz plot we will find

$$R^{\mathscr{S}}_L=0 \qquad \text{for } L \text{ odd.}$$

To avoid this loss of analysing efficiency when $E_2=E_3$ or when E_2 and E_3 are in eigenstates of isotopic spin, one should restrict the Dalitz-plot integration region to, say, $\omega_2>\omega_3$.

---

[1] The one exception to the above occurs when C has spin 1, all the decay particles have spin zero and $\eta_C=\eta_1\eta_2\eta_3$. In this case $\mathscr{M}=\pm1,0$ only and, by (8.2.63), $F_{\pm1}=0$. Then $R_L=0$ for odd L by (8.2.62) and only the even polarization of C is analysed.

Several examples for $J \leq 2$ are worked out in detail in Berman and Jacob (1965) and in the following section.

(c) *Moments of the experimental distributions*
Let $G(\phi_n, \theta_n, \gamma)$ be any function of the angles ϕ_n, θ_n, γ that specify the configuration of the three-particle final state in the rest frame S_C. We denote by $\langle G \rangle$ the angle average $(\phi_n, \theta_n, \gamma)$ of G weighted by the normalized decay distribution:

$$\langle G \rangle \equiv \int_0^{2\pi} d\phi_n \int_0^{\pi} \sin\theta_n \, d\theta_n \int_0^{2\pi} d\gamma \, W(\phi_n, \theta_n, \gamma) G(\phi_n, \theta_n, \gamma). \tag{8.2.70}$$

The most general case we consider involves the taking of moments of the angular distribution, or of the distribution of helicity multipole parameters of E_1. From (8.2.57) and (8.2.58) one has immediately:

$$\left\langle t^l_m(E_1) \mathscr{D}^{(L)^*}_{M\mu} \right\rangle = t^L_M(C) \sum_{\mathscr{M}} \langle J, \mathscr{M} - \mu; L, \mu | J, \mathscr{M} \rangle \times R_{\mathscr{M}, \mathscr{M}-\mu}(l, m). \tag{8.2.71}$$

Note that moments with $\mu = 0$ just correspond to integrating over the angle γ.

We see that information about a specific $t^L_M(C)$ can be obtained from many different moments, by choosing different values of μ, or from studying the distribution of the different $t^l_m(E_1)$.

In general one may need to know some of the decay parameters in order to extract the t^L_M. Just as in the two-body decay case, some reactions are *magic* and yield a complete analysis of the $t^L_M(C)$. Other reactions may yield only ratios of t^L_M at fixed M as L varies, and these reactions require the knowledge of certain of the dynamic-dependent decay parameters in order to get actual values of the $t^L_M(C)$. These parameters can be obtained in a model-independent fashion only if the resonance C can be prepared in a definite state of polarization, and that seems to be impossible if $J \geq 1$.

We shall look at several cases of interest.

(c.1) *Decay into spinless particles.* We are here limited to moments of the angular distribution only, i.e. to $l = m = 0$ in (8.2.71). From (8.2.63), if parity is conserved in the decay then $F_{\mathscr{M}} = 0$ for $\mathscr{M}$ odd or even according to whether $\eta_C \eta_1 \eta_2 \eta_3 = \pm 1$. In both cases $R_{\mathscr{M}, \mathscr{M}-\mu}$ will vanish for μ odd. Thus only moments with μ even are non-zero. There are then several possibilities.

- $J = 1$ and $\eta_C \eta_1 \eta_2 \eta_3 = +1$. In this case only $F_0 \neq 0$ and we are left with just R_{00}, which is equal to unity by the normalization condition (8.2.56). Then from (8.2.71)

$$\left\langle \mathscr{D}^{(1)^*}_{M0} \right\rangle = 0 \qquad \left\langle \mathscr{D}^{(2)^*}_{M0} \right\rangle = \langle 1, 0; 2, 0 | 1, 0 \rangle \, t^2_M(C) = -\sqrt{\tfrac{2}{5}} t^2_M(C) \tag{8.2.72}$$

so no information is obtained about the $t^1_M(C)$ but the $t^2_M(C)$ are fully determined.

• $J=2$ and $\eta_C\eta_1\eta_2\eta_3=+1$. Now we have R_{22}, R_{00}, $R_{-2-2}=1-R_{22}-R_{00}$, $R_{20}=R^*_{02}$, $R_{-20}=R^*_{0-2}$, R_{2-2} and from (8.2.71)

$$\left\langle \mathscr{D}^{(L)^*}_{M0}\right\rangle = t^L_M(C)\{\langle 2,2;L,0|2,2\rangle + R_{00}\left[\langle 2,0;L,0|2,0\rangle - \langle 2,2;L,0|2,2\rangle\right]\} \quad \text{for } L=0,2,4 \quad (8.2.73)$$

$$\left\langle \mathscr{D}^{(L)^*}_{M0}\right\rangle = t^L_M(C)\,\langle 2,2;L,0|2,2\rangle \times (R_{00}+2R_{22}-1) \quad \text{for } L=1,3 \quad (8.2.74)$$

$$\left\langle \mathscr{D}^{(L)^*}_{M2}\right\rangle = t^L_M(C)\,\langle 2,0;L,2|2,2\rangle \times \left[R_{20}+(-1)^L R_{0-2}\right] \quad \text{for } L=2,3,4 \quad (8.2.75)$$

$$\left\langle \mathscr{D}^{(L)^*}_{M4}\right\rangle = t^L_M(C)\,\langle 2,-2;L,4|2,2\rangle\, R_{2-2} \quad \text{for } L=4. \quad (8.2.76)$$

From (8.2.75) we can obtain the ratio $t^2_M(C)/t^4_M(C)$. The same ratio is also obtained from (8.2.73), as a function of R_{00}, which can thus be determined. Hence we obtain explicitly the even-rank multipoles. From (8.2.74) we can obtain the ratio $t^1_M(C)/t^3_M(C)$ but not the explicit values of the odd multipoles.

• Arbitrary integer J and $\eta_C\eta_1\eta_2\eta_3=+1$. The previous method can be generalized, and one obtains the even multipoles explicitly and the ratios only of the odd multipoles. Briefly, choosing $\mu=2J-2$ yields $t^{2J-2}_M(C)/t^{2J}_M(C)$. Using this in the moment with $\mu=2J-4$ allows the elimination of $R_{J,-J+4}+R_{J-4,-J}$, after which $t^{2J-4}_M(C)/t^{2J-2}_M(C)$ can be evaluated. Proceeding thus, one ends up with explicit values for all the even multipoles. For the odd multipoles, $\mu=2J-4$ yields $t^{2J-3}_M(C)/t^{2J-1}_M(C)$, which, used in the moment with $\mu=2J-6$, yields $t^{2J-5}_M(C)/t^{2J-3}_M(C)$. Proceeding thus, one ends up with the ratios of all the odd multipoles but not their explicit value.

• $J=1$ and $\eta_C\eta_1\eta_2\eta_3=-1$. The non-zero parameters are R_{11} (real); $R_{-1-1}=1-R_{11}$ and $R_{1-1}=R^*_{-11}$. Then from (8.2.71)

$$\left\langle \mathscr{D}^{(1)^*}_{M0}\right\rangle = \frac{1}{\sqrt{2}}\left[2R_{11}-1\right] t^1_M(C)$$
$$\left\langle \mathscr{D}^{(2)^*}_{M0}\right\rangle = \frac{1}{\sqrt{10}} t^2_M(C) \qquad \left\langle \mathscr{D}^{(2)^*}_{M2}\right\rangle = \sqrt{\tfrac{3}{5}}R_{1-1}t^2_M(C). \quad (8.2.77)$$

Again the $t^2_M(C)$ are fully determined, and now the $t^1_M(C)$ are also determined, up to one overall factor.

• $J = 2$ and $\eta_C\eta_1\eta_2\eta_3 = -1$. We have R_{11}, $R_{-1-1} = 1-R_{11}$, $R_{1-1} = R^*_{-11}$ and from (8.2.71)

$$\begin{aligned}
\left\langle \mathscr{D}^{(L)^*}_{M0} \right\rangle &= t^L_M(C)\,\langle 2,1;L,0|2,1\rangle \qquad \text{for } L = 0,2,4 \\
\left\langle \mathscr{D}^{(L)^*}_{M0} \right\rangle &= t^L_M(C)\,\langle 2,1;L,0|2,1\rangle\,(2R_{11}-1) \qquad \text{for } L = 1,3 \\
\left\langle \mathscr{D}^{(L)^*}_{M2} \right\rangle &= t^L_M(C)\,\langle 2,-1;L,2|2,1\rangle\, R_{1-1} \\
\left\langle \mathscr{D}^{(L)^*}_{M3} \right\rangle &= \left\langle \mathscr{D}^{(L)^*}_{M4} \right\rangle = 0.
\end{aligned} \tag{8.2.78}$$

Thus the even-L multipoles are fully determined but for the odd-L ones only their ratio is determined.

• Arbitrary integer J and $\eta_C\eta_1\eta_2\eta_3 = -1$. The method described above for the case arbitrary J and $\eta_C\eta_1\eta_2\eta_3 = +1$ is applicable and yields the explicit value of the even multipoles and the ratios of the odd multipoles.

In summary, for arbitrary J and $\eta_C\eta_1\eta_2\eta_3 = \pm 1$ all even multipoles are determined but only the ratios of the odd multipoles. It should be noted that moments of the γ distribution are essential for this to be possible, once $J \geq 2$.

(c.2) *Decay into one spin-1/2 and two spinless particles.* Here, in principle, we can utilize moments of the angular distribution and of the distribution of the spin components of E_1. The decay then functions as a *magic* analyser and the *complete* density matrix of the decaying particle can be determined.

We consider the decay of a spin-3/2 resonance in detail and outline the approach for the case of arbitrary half-integer spin.

• $C(3/2) \to E_1(1/2) + E_2(0) + E_3(0)$. Because of parity invariance and because E_1 can have only two values for its helicity, it turns out that all $R_{\mathscr{M}\mathscr{M}'}(l,m)$ can be written in terms of one combination of the decay amplitudes, namely

$$Q_{\mathscr{M}\mathscr{M}'} \equiv \int d\omega_2 d\omega_3 F^*_{\mathscr{M}}(\omega_2,\omega_3;1/2)F_{\mathscr{M}'}(\omega_2,\omega_3;1/2). \tag{8.2.79}$$

One has then from (8.2.59)

$$\begin{aligned}
R_{\mathscr{M}\mathscr{M}'} \equiv R_{\mathscr{M}\mathscr{M}'}(0,0) &= \left[1+(-1)^{\mathscr{M}-\mathscr{M}'}\right] Q_{\mathscr{M}\mathscr{M}'} \\
R_{\mathscr{M}\mathscr{M}'}(1,0) &= \tfrac{1}{\sqrt{3}}\left[1-(-1)^{\mathscr{M}-\mathscr{M}'}\right] Q_{\mathscr{M}\mathscr{M}'} \\
R_{\mathscr{M}\mathscr{M}'}(1,1) &= -(-1)^{\mathscr{M}-\mathscr{M}'} R_{\mathscr{M}\mathscr{M}'}(1,-1) \\
&= \eta_C\eta_1\eta_2\eta_3 e^{i\pi\mathscr{M}'}\sqrt{\tfrac{2}{3}}Q_{\mathscr{M}\mathscr{M}'}.
\end{aligned} \tag{8.2.80}$$

Finally, it will turn out convenient to define the combinations

$$Q^{\pm}_{\mathcal{M}\mathcal{M}'} = Q_{\mathcal{M}\mathcal{M}'} \pm Q_{-\mathcal{M}'-\mathcal{M}}. \tag{8.2.81}$$

For the $\mu = 0$ moments of the angular distribution one then finds

$$\begin{aligned}\left\langle \mathscr{D}^{(L)^*}_{M0} \right\rangle &= t^L_M(C)\Big\{\langle 3/2, 3/2; L, 0|3/2, 3/2\rangle \\ &\qquad + 2Q^{+}_{1/2\ 1/2}\left[\langle 3/2, 1/2; L, 0|3/2, 1/2\rangle\right. \\ &\qquad \left.- \langle 3/2, 3/2; L, 0|3/2, 3/2\rangle\right]\Big\} \qquad \text{for } L = 0, 2 \\ \left\langle \mathscr{D}^{(L)^*}_{M0} \right\rangle &= 2t^L_M(C)\Big\{\langle 3/2, 3/2; L, 0|3/2, 3/2\rangle\, Q^{-}_{3/2\ 3/2} \\ &\qquad + \langle 3/2, 1/2; L, 0|3/2, 1/2\rangle\, Q^{-}_{1/2\ 1/2}\Big\} \\ &\qquad\qquad \text{for } L = 1, 3.\end{aligned} \tag{8.2.82}$$

For the $\mu = 0$ moments of the distribution of the longitudinal component of the spin of E_1 one finds

$$\left\langle \mathscr{P}_z(E_1)\mathscr{D}^{(L)^*}_{M0} \right\rangle = \sqrt{3}\left\langle t^1_0(E_1)\mathscr{D}^{(L)^*}_{M0} \right\rangle = 0 \tag{8.2.83}$$

where the subscript refers to the axis OZ of the helicity rest frame S_{E_1}.

In particular the mean longitudinal polarization is zero after integration over γ.

For the $\mu = 0$ moments of the transverse polarization, along OY of S_{E_1}, one finds similarly

$$\left\langle \mathscr{P}_y(E_1)\mathscr{D}^{(L)^*}_{M0} \right\rangle = i\sqrt{\tfrac{3}{2}}\left\langle \left[t^1_1(E_1) + t^1_{-1}(E_1)\right] \mathscr{D}^{(L)^*}_{M0} \right\rangle = 0. \tag{8.2.84}$$

Again the mean transverse polarization, in the decay plane, is zero.

For the component of spin $\mathscr{P}_n$ along the normal to the decay plane, since OX in the helicity rest frame S_{E_1} is opposite to $\mathbf{n}$ we have (with $\eta = \eta_C\eta_1\eta_2\eta_3$)

$$\begin{aligned}\left\langle \mathscr{P}_n(E_1)\mathscr{D}^{(L)^*}_{M0} \right\rangle &= \sqrt{\tfrac{3}{2}}\left\langle \left[t^1_1(E_1) - t^1_{-1}(E_1)\right] \mathscr{D}^{(L)^*}_{M0} \right\rangle \\ &= -2i\eta t^L_M(C)\Big\{\langle 3/2, 3/2; L, 0|3/2, 3/2\rangle\, Q^{\mp}_{3/2, 3/2} \\ &\qquad - \langle 3/2, 1/2; L, 0|3/2, 1/2\rangle\, Q^{\mp}_{1/2\ 1/2}\Big\} \\ &\qquad\qquad \text{for } L \text{ even/odd}\end{aligned} \tag{8.2.85}$$

We note in particular that from the moment with $L = M = 0$ we get

$$\langle \mathscr{P}_n(E_1)\rangle = -2i\eta\left(Q^{-}_{3/2\ 3/2} - Q^{-}_{1/2\ 1/2}\right). \tag{8.2.86}$$

We now show how all the multipole moments $t^L_M(C)$ can be obtained explicitly from just the $\mu = 0$ moments of the angular distribution and those of the polarization along the normal to the decay plane.

Use of (8.2.86) allows us to eliminate $Q^-_{3/2\ 3/2}$ in eqns (8.2.85) and (8.2.82). Then for $L = 1$ and 2 in (8.2.82) and (8.2.85) we have, schematically,

$$\begin{aligned}
\left\langle \mathscr{D}^{(2)^*}_{M0} \right\rangle &= t^2_M(C)\left(a + bQ^+_{1/2\ 1/2}\right) \\
\left\langle \mathscr{D}^{(1)^*}_{M0} \right\rangle &= t^1_M(C)\left(c + dQ^-_{1/2\ 1/2}\right) \\
\left\langle \mathscr{P}_n(E_1)\mathscr{D}^{(2)^*}_{M0} \right\rangle &= t^2_M(C)\left(e + fQ^-_{1/2\ 1/2}\right) \\
\left\langle \mathscr{P}_n(E_1)\mathscr{D}^{(1)^*}_{M0} \right\rangle &= t^1_M(C)\left(g + hQ^+_{1/2\ 1/2}\right)
\end{aligned} \qquad (8.2.87)$$

where a, b, c, d, e, f, g, h are *known* constants, combination of vector addition coefficients and the measured value of $\langle \mathscr{P}_n(E_1)\rangle$.

If the four quantities (for fixed M) on the left-hand side of (8.2.87) are measured then we have four equations for the four unknowns, t^1_M, t^2_M, $Q^+_{1/2\ 1/2}$ and $Q^-_{1/2\ 1/2}$. Thus we can get the explicit values of $t^1_M(C)$ and $t^2_M(C)$. With $Q^-_{1/2\ 1/2}$ now known, a measurement of $\left\langle \mathscr{D}^{(3)^*}_{M0} \right\rangle$ will yield the value of $t^3_M(C)$ via (8.2.82).

If particles E_2 and E_3 are identical, care should be taken to integrate over the region $\omega_2 > \omega_3$ only in the Dalitz plot, as discussed in subsection (b) above.

The moments described above have $\mu = 0$, so that γ is simply integrated over, and they suffice to determine all the $t^L_M(C)$. Moments with $\mu \neq 0$ are also interesting if one wishes to study the dynamics of the decay mechanisms. The following general rule holds for moments with arbitrary μ:

$$\left\langle \left[t^l_m(E_1) - (-1)^{l+\mu} t^l_{-m}(E_1) \right] \mathscr{D}^{(L)^*}_{M\mu} \right\rangle = 0. \qquad (8.2.88)$$

It follows that $\left\langle t^1_0(E_1)\mathscr{D}^{(L)^*}_{M\mu} \right\rangle = 0$ if $l + \mu$ is odd. The result (8.2.88) can be used as a check on experimental biases. The vanishing of the $\mu = 0$ moments of $\mathscr{P}_z$ and $\mathscr{P}_y$ mentioned above is a special case of (8.2.88).

• C (arbitrary half-integer J) $\to E_1(1/2) + E_2(0) + E_3(0)$. We outline briefly how the $t^L_M(C)$ can be obtained from the moments of the various distributions. For $J \geq 5/2$ it is necessary to utilize moments with $\mu > 0$ as well.

For fixed L, the dependence of the $t^L_M(C)$ upon M is trivially obtained from the ratio of moments with the same L and various M. As can be seen from (8.2.71) all the decay-dependent parameters will cancel out.

We thus concentrate on moments with one value of M, namely $M = 0$.

For simplicity let us denote by M^L_μ a moment of the type $\left\langle \mathscr{D}^{(L)^*}_{0\mu} \right\rangle$ and by N^L_μ the type $\left\langle \mathscr{P}_n(E_1)\mathscr{D}^{(L)^*}_{0\mu} \right\rangle$, and let us put t^L for $t^L_0(C)$.

For $\mu = 2J - 1$, M^{2J}_{2J-1} and N^{2J-1}_{2J-1} will be proportional to $t^{2J} Q^{-}_{J,-J+1}$ and $t^{2J-1} Q^{-}_{J,-J+1}$ respectively. Their ratio gives t^{2J-1}/t^{2J}.

For $\mu = 2J - 3$ there are four unknown decay constants, $Q^{\pm}_{J,-J+3}$ and $Q^{\pm}_{J-1,-J+2}$. By taking ratios of M^{2J}_{2J-3}, M^{2J-1}_{2J-3}, N^{2J}_{2J-3} and N^{2J-1}_{2J-3} and using the known value of t^{2J-1}/t^{2J} one is able to obtain, say, $Q^{\pm}_{J-1,-J+2}$ and $Q^{-}_{J,-J+3}$ in the form (known constant) $\times\, Q^{+}_{J,-J+3}$. Making these substitutions everywhere, $Q^{+}_{J,-J+3}$ becomes a common factor in the equations relating moments to t^L. Thus we can obtain t^{2J-2}/t^{2J-1} and t^{2J-3}/t^{2J-2}.

Proceeding in this way one finds for each choice of μ information from the measured moments and from the previously determined multipole ratios that is more than enough, to express all $Q^{\pm}_{\mathcal{M},\mathcal{M}-\mu}$ in the form (known constant) $\times Q^{+}_{J,J-\mu}$. Substitution of this into the equations yields the new ratios $t^{\mu}/t^{\mu+1}$, $t^{\mu+1}/t^{\mu+2}$.

At the last stage, $\mu = 0$, we know the ratios $t^2/t^3, t^3/t^4, \ldots, t^{2J-1}/t^{2J}$ and we can obtain all $Q^{\pm}_{\mathcal{M},\mathcal{M}}$ in the form (known constant) $\times\, Q^{+}_{J,J}$, whereafter the ratio t^1/t^2 can be found. Now, however, we have also the normalization condition $\sum_{\mathcal{M}>0} Q^{+}_{\mathcal{M},\mathcal{M}} = 1/2$, which follows from (8.2.56), (8.2.79) and (8.2.81), and this determines the actual value of $Q^{+}_{J,J}$. Then the value of t^1, say, can be found explicitly, from which follow the values of all the other t^L.

Note that we end up with not just the desired t^L_M but also the whole set of decay parameters $Q^{\pm}_{\mathcal{M},\mathcal{M}-\mu}$ (μ even).

(c.3) *Two-body resonance domination of three-body state.* If the three-body final state is dominated by resonance formation between two of the particles then we regard the decay as a two-step process

$$
\begin{array}{l}
C \to C' + E_3 \\
\quad\quad\;\; \searrow \\
\quad\quad\quad\; E_1 + E_2
\end{array}
$$

and this is discussed fully in subsections 8.2.1(vii), (viii) above.

(c.4) *Decay into photon and two spinless particles,* C(J integer) $\to$ Photon$+$ $E_2(0) + E_3(0)$. Because $\lambda = 0$ is forbidden to a photon, this case is very similar to that where E_1 has spin 1/2. Parity conservation allows all decay parameters to be related to the combination

$$Q_{\mathcal{M}\mathcal{M}'} \equiv \int d\omega_2 d\omega_3 F^*_{\mathcal{M}}(\omega_2, \omega_3; 1) F_{\mathcal{M}'}(\omega_2, \omega_3; 1). \tag{8.2.89}$$

We have

$$
\begin{aligned}
R_{\mathcal{M}\mathcal{M}'}(0,0) &= \left[1 + (-1)^{\mathcal{M}-\mathcal{M}'}\right] Q_{\mathcal{M}\mathcal{M}'} \\
R_{\mathcal{M}\mathcal{M}'}(2,0) &= \tfrac{1}{\sqrt{10}} R_{\mathcal{M}\mathcal{M}'}(00) \\
R_{\mathcal{M}\mathcal{M}'}(2,2) &= \eta_C \eta_2 \eta_3 (-1)^{\mathcal{M}'+1} \sqrt{\tfrac{3}{5}} Q_{\mathcal{M}\mathcal{M}'} \\
R_{\mathcal{M}\mathcal{M}'}(1,0) &= \tfrac{1}{\sqrt{2}} \left[1 - (-1)^{\mathcal{M}-\mathcal{M}'}\right] Q_{\mathcal{M}\mathcal{M}'}.
\end{aligned}
\tag{8.2.90}
$$

If the state of polarization of the photon can be determined then the measured moments will yield the $t^L_M(C)$ via (8.2.71) and (8.2.90) in similar manner to case (c.2) above.

9
Electroweak interactions

One of the most dramatic events in the history of elementary particle physics was the unification of the electromagnetic and the weak interactions into a single, beautiful gauge theory, which was created by Weinberg, Salam and Glashow and which is nowadays referred to as the 'Standard Model' (SM). For a detailed pedagogical account of the need for and development of such a theory, the reader is referred to Leader and Predazzi (1996). We simply recall that this tightly knit theory contains the astounding and incredible prediction of the existence of a set of three vector bosons, $W^{\pm}, Z^0$, with huge masses, $m_W \approx 80$ GeV/c^2, $m_Z \approx 90$ GeV/c^2, and that these unlikely objects were eventually discovered. (The W was identified at CERN in January 1983 and the Z^0, also at CERN, a few months later.) A test for the spin of the W is described in subsection 8.2.1(ix).

In the Standard Model the electroweak interactions are mediated by the exchange of photons, Zs and Ws, whose coupling to the basic fermions (leptons and quarks) is a mixture of vector and axial-vector. To begin with all particles are massless, and their masses are generated by spontaneous symmetry breaking. The usual mechanism of symmetry breaking requires a neutral scalar particle, the Higgs meson H, whose mass is not determined by the theory. H has not yet been detected experimentally and is the most serious missing link in the theory. But in every other respect the theory has been remarkably successful. All the first-generation experimental tests have been passed with flying colours and a new generation of more refined and demanding tests has been carried out at the two highest energy e^+e^- colliding beam machines, LEP at CERN and SLC at Stanford. It has been realized that some of the cleanest tests involve spin-dependent measurements and SLC has made excellent use of such ideas. Some work on polarized $e^{\pm}$ beams has been done at LEP, but the push for higher energies and the use of LEP in the construction of the new large hadron collider (LHC) means that a detailed spin programme was never carried out.

We shall recall the essential elements of the SM and then concentrate on the spin-dependent possibilities.

9.1 Summary of the Standard Model

There are three generations of leptons, (e^-, ν_e), (μ^-, ν_μ), and (τ^-, ν_τ) and the neutrinos are treated as massless. The charged lepton fields, which we designate by the symbol for the particle, are split into left- and right-handed parts, see eqns (4.6.53), (4.6.54), and the neutrinos are, by definition, left-handed.

The left-handed parts are grouped into *weak isospin* doublets

$$\begin{pmatrix} \nu_e \\ e \end{pmatrix} \quad \begin{pmatrix} \nu_\mu \\ \mu^- \end{pmatrix} \quad \begin{pmatrix} \nu_\tau \\ \tau^- \end{pmatrix}$$

and the charged bosons $W^\pm$ interact universally with these. Interactions involving $W^\pm$ are called *charged current* interactions and these only involve left-handed leptons. The relevant part of the interaction Lagrangian density has the following form:

$$\mathscr{L}_{W-\text{lept}} = \frac{e}{2\sqrt{2}\sin\theta_W} \left\{ \left[\bar{\nu}_e \gamma^\mu (1-\gamma_5) e W_\mu + \bar{e}\gamma^\mu (1-\gamma_5)\nu_e W_\mu^\dagger \right] + \mu^- \text{ terms} + \tau^- \text{ terms} \right\}. \tag{9.1.1}$$

Here e is the magnitude of the electron charge and θ_W, the Weinberg angle, is a crucial parameter in the unifying of the weak and electromagnetic interactions.

The above interaction gives rise to the following Feynman diagram vertices:

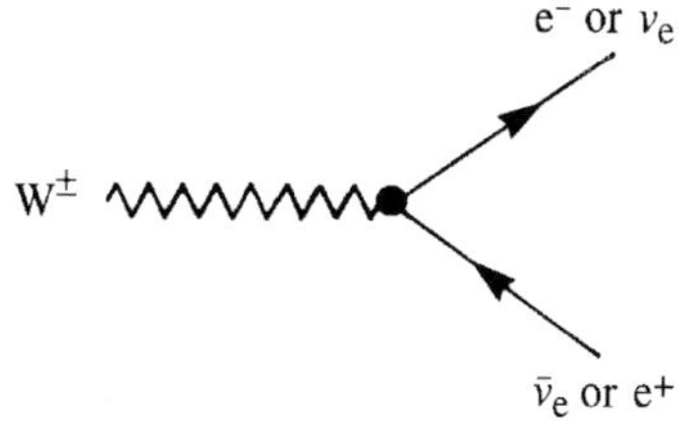

$W^\pm$ — vertex with e^- or ν_e (outgoing) and $\bar{\nu}_e$ or e^+ (incoming)

$$= \frac{ie}{2\sqrt{2}\sin\theta_W}\gamma^\mu(1-\gamma_5) \tag{9.1.2}$$

where the arrow shows the flow of fermion number or, equivalently, lepton number. Identical vertices occur for $\mu^\pm$ and $\tau^\pm$.

The form of the W-propagator depends upon the gauge choice. It is simplest in what is called the unitary gauge:

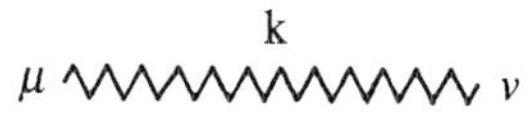

μ ~~~ k ~~~ ν

$$= \frac{i\left(-g_{\mu\nu} + k_\mu k_\nu / m_W^2\right)}{k^2 - m_W^2 + i\epsilon}. \tag{9.1.3}$$

Note, in (9.1.1), that only left-handed leptons are annihilated but also, because $\gamma^\mu(1-\gamma_5) = (1+\gamma_5)\gamma^\mu$, that only left-handed leptons are created. This follows since, from (4.6.53),

$$\bar{u}_L = u_L^\dagger \gamma^0 = \left[\tfrac{1}{2}(1-\gamma_5)u\right]^\dagger \gamma^0 = \bar{u}\tfrac{1}{2}(1+\gamma_5).$$

The Z^0, which gives rise to *neutral current* weak interactions, interacts with a superposition of left- and right-handed charged leptons. The relevant part of the interaction Lagrangian density is

$$\mathscr{L}_{Z\text{-lept}} = e\left\{\left[\bar{e}\gamma^\mu(v_e - a_e\gamma_5)eZ_\mu + \bar{v}_e\gamma^\mu(v_v - a_v\gamma_5)v_eZ_\mu\right] + \mu^- \text{ terms} + \tau^- \text{ terms}\right\}. \tag{9.1.4}$$

where, for a fermion f,

$$\begin{aligned} v_f &= \frac{I_3^f - 2Q_f\sin^2\theta_W}{2\sin\theta_W\cos\theta_W} \\ a_f &= \frac{I_3^f}{2\sin\theta_W\cos\theta_W}. \end{aligned} \tag{9.1.5}$$

Here I_3^f and Q_f are the third component of weak isospin and the charge (in units of e) of the fermion. Thus

$$v_e = \frac{-1+4\sin^2\theta_W}{4\sin\theta_W\cos\theta_W} \qquad a_e = \frac{1}{4\sin\theta_W\cos\theta_W} \tag{9.1.6}$$

and

$$v_{v_e} = \frac{1}{4\sin\theta_W\cos\theta_W} = a_{v_e}. \tag{9.1.7}$$

Note that because $\sin^2\theta_W \approx 0.23$ one finds that $v_e \ll a_e$, so that the coupling to the charged leptons is almost purely axial-vector.

The Feynman diagram vertices are

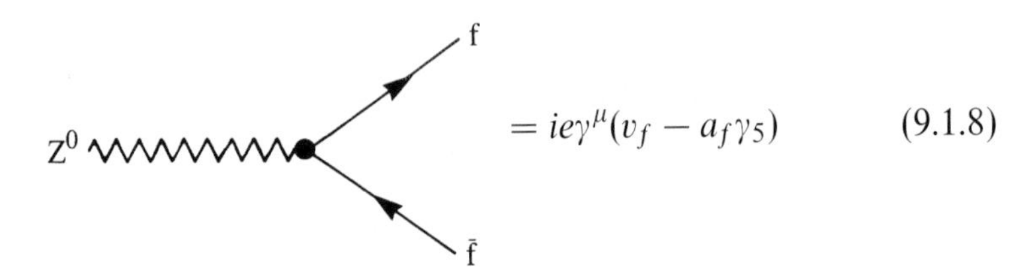

with identical vertices for the generations e, μ, τ.

A fascinating feature of the theory is the interference between Z^0 and photon exchange, so we recall that the standard QED vertex is

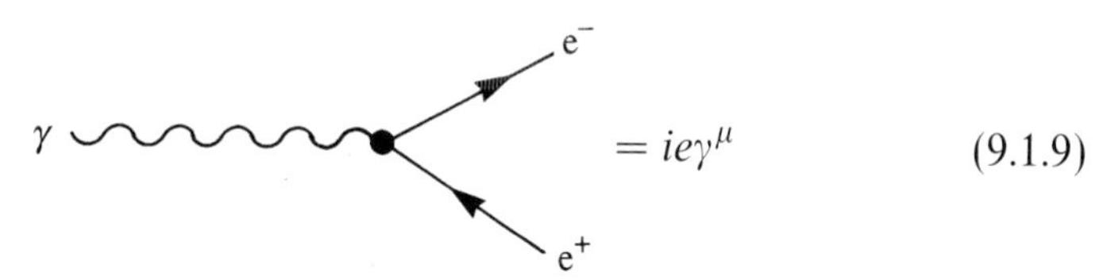

$$= ie\gamma^\mu \tag{9.1.9}$$

with identical vertices for $\mu^\pm$, $\tau^\pm$.

The Z^0 and photon propagators are

$$\mu \overset{k}{\underset{Z^0}{\text{~~~~}}} \nu = \frac{i\left(-g_{\mu\nu} + k_\mu k_\nu / m_Z^2\right)}{k^2 - m_Z^2 + i\epsilon}. \tag{9.1.10}$$

$$\mu \overset{k}{\underset{\gamma}{\text{~~~~}}} \nu = \frac{-ig_{\mu\nu}}{k^2 + i\epsilon}. \tag{9.1.11}$$

For all fermions, the propagators are

$$\underset{p}{\longrightarrow} = \frac{i}{\not{p} - m + ie} \tag{9.1.12}$$

where $\not{p}$ is the momentum flow in the direction of the fermion number-flow arrow.

Because interference effects between different diagrams are so interesting, care must be taken to allow for possible relative signs between diagrams; these signs arise from the sequential order of the fermionic operators that occur in the products of operators responsible for the diagrams.

Some of the most beautiful effects arise because (9.1.1) and (9.1.4) contain a mixture of vector and axial-vector coupling and thus do not conserve parity.

The parameter θ_W fixes the relative couplings of γ, Z, W to charged leptons, but even to the lowest order in perturbation theory, it plays several other rôles as well (see Chapter 4 of Leader and Predazzi (1996)).

The Higgs mechanism, which gives mass to $W^\pm$ and Z, results in the relation

$$m_W = m_Z \cos\theta_W \tag{9.1.13}$$

and a computation of the muon lifetime in $\mu^- \to e^- + \bar{\nu}_e + \nu_\mu$ relates θ_W and m_W to the Fermi coupling constant $G \equiv G_F \equiv G_\mu$:

$$m_W = \left(\frac{\pi\alpha}{\sqrt{2}G}\right)^{1/2} \frac{1}{\sin\theta_W} \tag{9.1.14}$$

where α is the fine structure constant.

The coupling of the vector bosons to quarks is analogous to the leptons except that Cabibbo–Kobayashi–Maskawa generation mixing takes place and there is universal coupling to the three left-handed doublets

$$\begin{pmatrix} u_1 \\ d'_1 \end{pmatrix}_L \equiv \begin{pmatrix} u \\ d' \end{pmatrix}_L \qquad \begin{pmatrix} u_2 \\ d'_2 \end{pmatrix}_L \equiv \begin{pmatrix} c \\ s' \end{pmatrix}_L \qquad \begin{pmatrix} u_3 \\ d'_3 \end{pmatrix}_L \equiv \begin{pmatrix} t \\ b' \end{pmatrix}_L$$

where

$$\begin{pmatrix} d' \\ s' \\ b' \end{pmatrix}_L = V \begin{pmatrix} d \\ s \\ b \end{pmatrix}_L \tag{9.1.15}$$

or

$$d'_i = V_{ij} d_j \tag{9.1.16}$$

and V is the 3×3 unitary Kobayashi–Maskawa matrix.

Although its existence had more or less been taken for granted, on account of its rôle in calculations that agreed with a host of data, the top quark t was discovered only in 1994, at Fermilab. Its mass has turned out to be somewhat larger than originally expected: $m_t \approx 175\ \mathrm{GeV}/c^2$.

The relevant parts of the Lagrangian density for the charged current interactions are:

$$\begin{aligned} \mathscr{L}_{W-\text{quark}} = \frac{e}{2\sqrt{2}\sin\theta_W} \Big\{ & \Big[\bar{u}\gamma^\mu(1-\gamma_5)d' W_\mu + \bar{d}'\gamma^\mu(1-\gamma_5)u W_\mu^\dagger \Big] \\ & + c, s' \text{ term} + t, b' \text{ term} \Big\}. \end{aligned} \tag{9.1.17}$$

This gives rise to the following Feynman diagram vertices

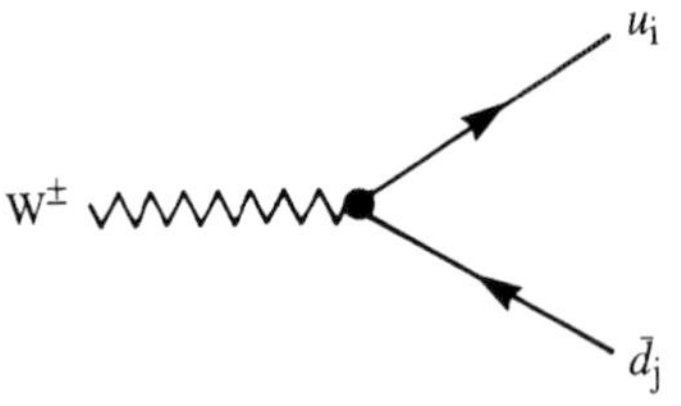

u_i

W$^\pm$ $\quad = \dfrac{ieV_{ij}}{2\sqrt{2}\sin\theta_W}\gamma^\mu(1-\gamma_5)$ (9.1.18)

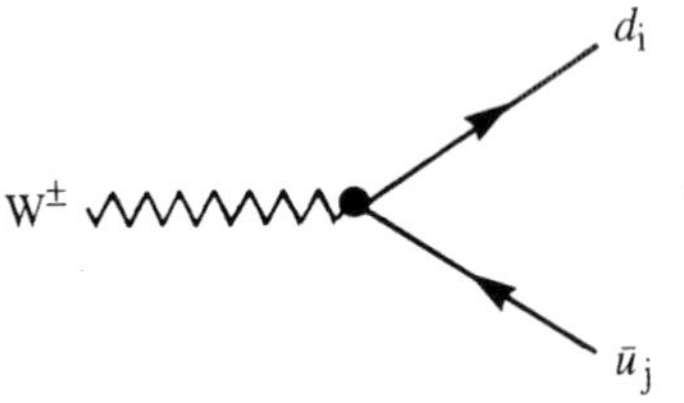

$\bar{d}_j$

d_i

W$^\pm$ $\quad = \dfrac{ieV_{ij}^\dagger}{2\sqrt{2}\sin\theta_W}\gamma^\mu(1-\gamma_5).$ (9.1.19)

$\bar{u}_j$

For the neutral current interactions there is no generation mixing and the coupling to $q\bar{q}$ has exactly the same structure as for the lepton–antilepton pairs, as given in (9.1.4) and (9.1.5), where now I_3^f and Q_f refer to the quark weak isospin and charge. The vertices are thus shown in (9.1.8).

Of course when dealing with quarks one must remember that all the above applies equally to each quark colour.

In addition to the above there are interaction terms involving the Higgs meson coupling to fermions and to the vector mesons and the self-coupling of the vector mesons. None of these is directly relevant to our study, which will deal mainly with fermionic reactions, but of course they will play a rôle in higher-order perturbative corrections. The detailed Feynman rules can be found in Appendix 2 of Leader and Predazzi (1996).

The Higgs meson does contribute to the reactions we shall consider, but its effect, in lowest order, is negligible because of the weakness of the coupling to fermions:

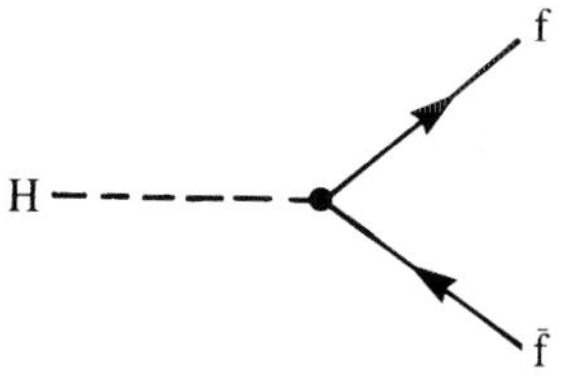

$$= -i\left(\sqrt{2}G\right)^{1/2} m_f I. \qquad (9.1.20)$$

This is especially small for reactions at LEP and SLC, where f in the initial state is always an electron. Note, from (9.1.14), that

$$\left(\sqrt{2}G\right)^{1/2} m_f = \left[\frac{(\pi\alpha)^{1/2}}{\sin\theta_W}\right]\left(\frac{m_f}{m_W}\right).$$

Higgs exchange will be ignored in the following.

9.2 Precision tests of the Standard Model

The properties of many experimental reactions have been calculated in lowest-order perturbation theory (the Born or tree approximation); they are all consistent with the results of the first generation of experiments carried out in the past few years. In particular the parameter θ_W occurs in many different situations and its various determinations are all mutually consistent.

There is now great interest in testing the deeper quantum-field-theoretic aspects of the theory by comparing precision experiments with calculations done to higher order. (Recall the seminal rôle of the Lamb shift and $g-2$ for QED!) But the procedure is not quite straightforward, given the non-discovery thus far of the Higgs, because while at the Born level we can simply deal with reactions that do not involve it, in higher orders it is unavoidable. For example, although the Higgs couples very weakly to light fermions, its coupling to the vector bosons (which is dimensionless) is effectively large, $2(\sqrt{2}G)^{1/2}m_W^2$, so that its contribution to the W

propagator,

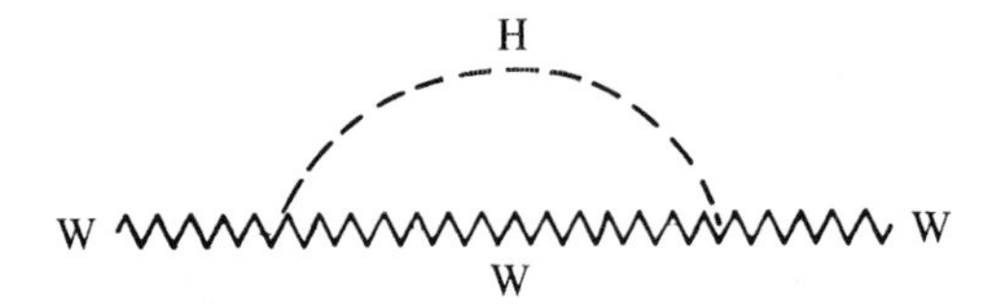

is important.

Thus the detailed higher-order corrections will depend upon the unknown parameter m_H and it becomes very interesting to look for observables that are particularly sensitive to this parameter.

Finally, in going to higher orders, because of infinite renormalization effects one has to decide more carefully what exactly the parameters are that go into the perturbative calculations.

A natural set to use would be α, m_W, m_Z and m_t, m_H, but m_W is less accurately known on account of the neutrino involved in its decay. To obviate this problem one can compute the rate for $\mu^- \to e^- + \bar{\nu}_e + \nu_\mu$ to order α^2, in which case (9.1.14) is altered to

$$m_W^2 = \left(\frac{\pi\alpha}{\sqrt{2}G}\right)\left(\frac{1}{\sin^2\theta_W}\right)\frac{1}{1-\Delta r} \tag{9.2.1}$$

where Δr is a calculated correction of order α, whose precise value depends upon the renormalization scheme used.

If one chooses a scheme where (9.1.13) holds exactly (the so-called 'on-shell' scheme), i.e. θ_W is *defined* by

$$\cos\theta_W \equiv \frac{m_W}{m_Z}, \tag{9.2.2}$$

then one can use the fact that G is known to great accuracy,

$$G = 1.66389(22) \times 10^{-5}\left(\text{GeV}/c^2\right)^2, \tag{9.2.3}$$

and take α, G and m_Z as basic parameters. Now m_W and θ_W are *calculated* from (9.2.1) and (9.2.2), i.e.

$$\cos^2\theta_W \sin^2\theta_W = \left(\frac{\pi\alpha}{\sqrt{2}Gm_Z^2}\right)\frac{1}{1-\Delta r} \tag{9.2.4}$$

and

$$m_W^2 = m_Z^2\cos^2\theta_W. \tag{9.2.5}$$

The dominant contribution to Δr is

$$\Delta r \approx \Delta\alpha - \frac{3}{8\sqrt{2}\pi^2}\frac{\cos^2\theta_W}{\sin^2\theta_W}Gm_t^2$$

where $\Delta\alpha$ is a QED correction, $\Delta\alpha \simeq 0.064$.

It is estimated that the error in the calculation of Δr arising from imperfectly controlled hadronic physics implies an uncertainty in $\sin^2\theta_W$ as calculated from (9.2.4) of

$$\delta(\sin^2\theta_W) = \pm 0.0004. \tag{9.2.6}$$

This then sets a fantastic goal for the accuracy in the new generation of measurements; thus one should look for other reactions in which $\sin^2\theta_W$ plays such a sensitive rôle that it can be measured to the accuracy (9.2.6).

The most promising approach seems to be via the measurement of the vector part of the Z coupling to fermion–antifermion pairs, i.e. of v_f, defined in (9.1.5). But because v_f is so small it is essential to look for parity-violating effects, where interference between vector and axial-vector couplings will give rise to observables proportional to $v_f a_f$ rather than to $v_f^2 + a_f^2$ as in parity-conserving quantities.

Several reactions seem possible, but by the far the most sensitive to $\sin^2\theta_W$ are those involving forward–backward asymmetries using longitudinally polarized $e^\pm$ beams, and those involving measurement of the polarization of the final state fermion. In order to optimize the event rate the e^+e^- energy should be close to the Z^0 peak.

To evaluate the dominant dependence on $\sin^2\theta_W$ it will be sufficient to discuss the reaction in the Born approximation but, clearly, in the eventual comparison between theory and experiment the theoretical predictions must include higher-order effects. (These are described in detail in Consoli and Hollik (1989).)

9.2.1 The reaction $e^-e^+ \to$ fermion–antifermion pair

Consider the process

$$e^-e^+ \to f\bar{f}$$

in the region of the Z peak,[1] so that we can ignore photon exchange, for simplicity, with longitudinally polarized electrons and positrons; here f is any *lepton*. At the huge energies involved $m/E \ll 1$ for all the fermions involved, so that, as discussed in subsection 4.6.3, helicity and chirality are indistinguishable. The lowest-order diagrams are:

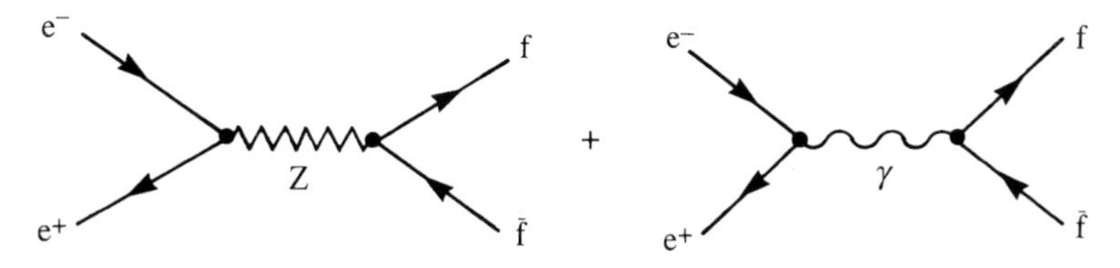

[1] Paramenters evaluated at the Z energy are sometimes called the 'pole paramenters' and are given a superscript 0

Because the energy is close to the Z^0 mass it is not adequate to use the propagator given in (9.1.10) and a more realistic version that takes into account the finite width Γ_Z of the Z must be used:

$$\mu \overset{k}{\underset{Z}{\sim\sim\sim\sim}} \nu = \frac{i\left(-g_{\mu\nu} + k_\mu k_\nu/m_Z^2\right)}{k^2 - m_Z^2 + ik^2\Gamma_Z/m_Z} \qquad (k^2 \approx m_Z^2). \tag{9.2.7}$$

The contribution from γ exchange is of order Γ_Z/m_Z compared with Z exchange, which will be neglected in our qualitative discussion. It could be included easily. In fact we will not actually evaluate the Feynman diagrams but derive the results in a fashion that highlights the physical ingredients. Thus we shall view the process as a physical process of resonance formation and decay,

$$e^- e^+ \to Z \to f\bar{f}.$$

From subsection 8.2.1, the amplitude is of the form

$$\begin{aligned} H_{f\bar{f};e\bar{e}}(\theta) &= M(f,\bar{f})M(e,\bar{e})d^1_{\lambda\mu}(\theta) \\ \lambda = e - \bar{e} &\qquad \mu = f - \bar{f}, \end{aligned} \tag{9.2.8}$$

where, because we shall be interested only in ratios of cross-sections at the same energy, we have left out a function of energy related to the behaviour of the Z propagator. The labels in (9.2.8) refer to the helicities of the relevant particles and the Ms measure the amplitudes for $e^-e^+ \to Z$ and $f\bar{f} \to Z$ respectively.

We know from subsection 4.6.2 that the fermions and antifermions must have opposite helicity, so only two decay amplitudes occur, $M(+,-)$ and $M(-,+)$. Moreover if in (9.1.8) we write

$$v - a\gamma_5 = \tfrac{1}{2}(v+a)(1-\gamma_5) + \tfrac{1}{2}(v-a)(1+\gamma_5) \tag{9.2.9}$$

we see that apart from irrelevant normalization we can take

$$M(+,-) = v - a \qquad M(-,+) = v + a. \tag{9.2.10}$$

The only helicity amplitudes are then

$$\begin{aligned} H_{+-;+-} &= (v_f - a_f)(v_e - a_e)d^1_{11}(\theta) \\ H_{-+;-+} &= (v_f + a_f)(v_e + a_e)d^1_{-1-1}(\theta) \\ H_{+-;-+} &= (v_f - a_f)(v_e + a_e)d^1_{-11}(\theta) \\ H_{-+;+-} &= (v_f + a_f)(v_e - a_e)d^1_{1-1}(\theta) \end{aligned} \tag{9.2.11}$$

where

$$d^1_{11}(\theta) = d^1_{-1-1}(\theta) = \tfrac{1}{2}(1+\cos\theta)$$

and

$$d^1_{1-1}(\theta) = d^1_{-11}(\theta) = \tfrac{1}{2}(1 - \cos\theta).$$

The unpolarized differential cross-section has the simple form, in the CM frame,

$$\frac{d\sigma}{d\Omega} = \frac{3}{16\pi}\sigma\left[1 + \cos^2\theta + 2\mathscr{A}_f\mathscr{A}_e\cos\theta\right] \tag{9.2.12}$$

where

$$\mathscr{A}_i \equiv \frac{2v_i a_i}{v_i^2 + a_i^2} \tag{9.2.13}$$

is a direct measure of the vector coupling v_i and σ is the total integrated cross-section.

The most general experiment possible in $e^-e^+ \to f\bar{f}$ is described by eqn (5.6.3). Since we are looking for parity-violating effects we consider longitudinally polarized $e^\mp$. Let $\mathscr{P}_e$, $\mathscr{P}_{\bar{e}}$ be the degree of longitudinal polarization of the e^-, e^+ beams respectively in their helicity rest frames, as depicted in Fig. 3.1 for particles A and B, so that, for both $e^\mp$,

$$\mathscr{P} = \frac{n_\mathrm{R} - n_\mathrm{L}}{n_\mathrm{R} + n_\mathrm{L}}$$

where $n_{\mathrm{R,L}}$ are the relative numbers of right- and left-handed particles. Then the initial state density matrix is determined by

$$\boldsymbol{\mathcal{P}}_e = (0, 0, \mathscr{P}_e) \qquad \text{and} \qquad \boldsymbol{\mathcal{P}}_{\bar{e}} = (0, 0, \mathscr{P}_{\bar{e}}). \tag{9.2.14}$$

Because of the simple structure of the helicity amplitudes (9.2.11), the CM reaction parameters defined in (5.6.4) are easily evaluated. Those that interest us are independent of ϕ, as follows from (5.6.2) and (5.3.3). One has:

- *the electron longitudinal analysing power,*

$$\begin{aligned} A_Z(\theta) \equiv A_Z(e^-) &= (Z0|00) \\ &= -\frac{\mathscr{A}_e(1 + \cos^2\theta) + \mathscr{A}_f(2\cos\theta)}{1 + \cos^2\theta + \mathscr{A}_e\mathscr{A}_f(2\cos\theta)} && (9.2.15) \\ &= -A_Z(e^+); && (9.2.16) \end{aligned}$$

- *the initial state correlation parameter,*

$$A_{ZZ} \equiv (ZZ|00) = -1. \tag{9.2.17}$$

Equation (9.2.17) is a direct signature of the fact that only electrons and positrons of opposite helicity can interact with each other.

Using these in (5.6.3) yields

$$\begin{aligned}\frac{d\sigma}{d\Omega}(\mathscr{P}_e,\mathscr{P}_{\bar{e}}) &= \frac{d\sigma}{d\Omega}\left[(1-\mathscr{P}_e\mathscr{P}_{\bar{e}})+(\mathscr{P}_e-\mathscr{P}_{\bar{e}})A_Z(\theta)\right] \\ &= \frac{3\sigma}{16\pi}\left\{[1-\mathscr{P}_e\mathscr{P}_{\bar{e}}+\mathscr{A}_e(\mathscr{P}_e-\mathscr{P}_{\bar{e}})]\,(1+\cos^2\theta)\right. \\ &\qquad \left.+\mathscr{A}_f\,[\mathscr{A}_e(1-\mathscr{P}_e\mathscr{P}_{\bar{e}})+\mathscr{P}_{\bar{e}}-\mathscr{P}_e]\,2\cos\theta\right\}\end{aligned} \tag{9.2.18}$$

Thus

$$\frac{d\sigma}{d\Omega}(\mathscr{P}_e,\mathscr{P}_{\bar{e}}) = \frac{3\sigma}{16\pi}\left[(1+\gamma_1)(1+\cos^2\theta)+\mathscr{A}_f(\mathscr{A}_e+\gamma_2)2\cos\theta\right] \tag{9.2.19}$$

where

$$\gamma_1 \equiv -\mathscr{P}_e\mathscr{P}_{\bar{e}}+\mathscr{A}_e(\mathscr{P}_e-\mathscr{P}_{\bar{e}}) \qquad \gamma_2 \equiv \mathscr{P}_{\bar{e}}-\mathscr{P}_e-\mathscr{P}_e\mathscr{P}_{\bar{e}}\mathscr{A}_e. \tag{9.2.20}$$

For the integrated cross-section we have

$$\sigma(\mathscr{P}_e,\mathscr{P}_{\bar{e}}) = \sigma\left\{1-\mathscr{P}_e\mathscr{P}_{\bar{e}}+\mathscr{A}_e(\mathscr{P}_{\bar{e}}-\mathscr{P}_e)\right\}. \tag{9.2.21}$$

(i) *The left–right asymmetry* A_{LR}

Let σ_L and σ_R be the integrated cross-section for the interaction of left and right-handed electrons respectively with unpolarized positrons. Then from (9.2.21) we have

$$A_{LR} \equiv \frac{\sigma_L-\sigma_R}{\sigma_L+\sigma_R} \tag{9.2.22}$$

$$= \mathscr{A}_e \qquad \text{in the Born approximation.} \tag{9.2.23}$$

Taking $\sin^2\theta_W = 0.23$ yields $\mathscr{A}_e = 0.16$.

When higher-order corrections are taken into account the relationship (9.2.23) will change only slightly, because the radiative corrections largely cancel in the asymmetry. The structure of (9.2.21) remains the same but with $\mathscr{A}_e$ replaced by A_{LR}, i.e.

$$\sigma(\mathscr{P}_e,\mathscr{P}_{\bar{e}}) = \sigma\left[1-\mathscr{P}_e\mathscr{P}_{\bar{e}}+A_{LR}(\mathscr{P}_{\bar{e}}-\mathscr{P}_e)\right]. \tag{9.2.24}$$

As discussed in subsection 7.2.1 this could have been used to *measure* $\mathscr{P}_e$, $\mathscr{P}_{\bar{e}}$ and A_{LR} by running LEP in the four polarization settings:

$$(\mathscr{P}_e,\mathscr{P}_{\bar{e}}) = (0,0),\ (\mathscr{P}_e,0),\ (0,\mathscr{P}_e),\ (\mathscr{P}_e,\mathscr{P}_{\bar{e}}).$$

If it were possible to have 50% polarization and 10^6 events the statistical precision on A_{LR} would be $\delta A_{LR} = 0.002$, leading to $\delta(\sin^2\theta_W) \approx 0.0004$.

Now that m_t is reasonably well determined, a measurement of A_{LR} to the above accuracy will quite strongly constrain the possible values of m_H.

The most advanced studies thus far have been carried out by the SLD collaboration at the SLC at Stanford (see Prepost, 1996). Although using only 93 000 events at the Z^0 mass, the beam polarization $\mathscr{P}_e$ is known with great accuracy to be $(77.23 \pm 0.52)\%$ and A_{LR} is measured with amazing precision:

$$A_{\mathrm{LR}} = 0.1543 \pm 0.0039. \tag{9.2.25}$$

Allowing for the higher-order radiative corrections, this result is expressed as a value for $\sin^2 \theta_{\mathrm{W}}^{\mathrm{eff}}$, which differs from $\sin^2 \theta_{\mathrm{W}}$ defined in (9.2.2) by small radiative corrections (see Hollik, 1990). The result is

$$\sin^2 \theta_{\mathrm{W}}^{\mathrm{eff}} = 0.23060 \pm 0.00050, \tag{9.2.26}$$

making it the world's most precise determination of $\theta_{\mathrm{W}}^{\mathrm{eff}}$ from a single experiment.

(ii) The forward–backward asymmetry A_{FB}

It is clear from (9.2.19) that a forward–backward asymmetry exists (i.e. under $\theta \to \pi - \theta$) because of the term linear in $\cos\theta$ and that this is non-zero because of interference between vector and axial-vector terms. The forward–backward asymmetry A_{FB} is defined as

$$A_{\mathrm{FB}} = \frac{n_{\mathrm{F}} - n_{\mathrm{B}}}{n_{\mathrm{F}} + n_{\mathrm{B}}} \tag{9.2.27}$$

where $n_{\mathrm{F,B}}$ are the numbers of events in the forward and backward hemisphere respectively. Thus

$$A_{\mathrm{FB}}(\mathscr{P}_e, \mathscr{P}_{\bar{e}}) = \frac{3}{4}\left\{\frac{\mathscr{A}_f\,[\mathscr{P}_{\bar{e}} - \mathscr{P}_e + \mathscr{A}_e(1 - \mathscr{P}_e\mathscr{P}_{\bar{e}})]}{1 - \mathscr{P}_e\mathscr{P}_{\bar{e}} + \mathscr{A}_e(\mathscr{P}_{\bar{e}} - \mathscr{P}_e)}\right\}. \tag{9.2.28}$$

This is a fundamental result and will be used to illustrate the power of utilising polarized beams. For the unpolarized asymmetry we have

$$A_{\mathrm{FB}} = \tfrac{3}{4}\mathscr{A}_f\mathscr{A}_e. \tag{9.2.29}$$

Now recall that from (9.1.6) the vector coupling of the leptons is very small, so that the $\mathscr{A}_l$ are also very small. Then for a given experimental error δA_{FB} we will have for the error on, say, $\mathscr{A}_f$

$$\delta\mathscr{A}_f \approx \frac{1}{\mathscr{A}_e}\delta A_{\mathrm{FB}} \gg \delta A_{\mathrm{FB}} \tag{9.2.30}$$

so that we cannot obtain a sufficiently accurate measurement of $\sin^2\theta_{\mathrm{W}}$.

On the contrary if we have, say, $\mathscr{P}_{\bar{e}} = 0$ but $\mathscr{P}_e$ sizeable then

$$A_{\mathrm{FB}}(\mathscr{P}_e) = -\frac{3}{4}\left\{\frac{\mathscr{A}_f\,(\mathscr{P}_e - \mathscr{A}_e)}{1 + \mathscr{P}_e\mathscr{A}_e}\right\} \simeq -\frac{3}{4}\mathscr{A}_f\mathscr{P}_e. \tag{9.2.31}$$

In this case the error in $\delta\mathscr{A}_f$ will be comparable to that in A_{FB}:

$$\delta\mathscr{A}_f \approx \delta A_{\text{FB}}. \tag{9.2.32}$$

Again, the most advanced studies to date have been carried out at the Stanford SLC. The forward–backward asymmetry has been measured for $e^+e^- \to e^+e^-, \mu^+\mu^-, \tau^+\tau^-$, leading to (see Prepost, 1996)

$$\begin{gathered} \mathscr{A}_e = 0.148 \pm 0.016 \qquad \mathscr{A}_\mu = 0.102 \pm 0.033 \\ \mathscr{A}_\tau = 0.190 \pm 0.034, \end{gathered} \tag{9.2.33}$$

which are compatible with lepton universality.

When combined with the result (9.2.25) for A_{LR} these yield

$$\sin^2\theta_{\text{W}}^{\text{eff}} = 0.23061 \pm 0.00047. \tag{9.2.34}$$

(iii) *Polarization of final state fermion for unpolarized e^-e^+*

If we are interested in the longitudinal polarization of the final state fermion f or in the final state correlations with a polarized initial state, we require the following additional CM reaction parameters:

- *the final fermion longitudinal polarizing power,*

$$\mathscr{P}_f(\theta) \equiv (00|Z0) = -\frac{\mathscr{A}_f(1+\cos^2\theta) + \mathscr{A}_e(2\cos\theta)}{1+\cos^2\theta + \mathscr{A}_e\mathscr{A}_f(2\cos\theta)} \tag{9.2.35}$$

$$= -\mathscr{P}_{\bar{f}}(\theta). \tag{9.2.36}$$

- *the final state correlation parameter,*

$$C_{ZZ} \equiv (00|ZZ) = -1, \tag{9.2.37}$$

  again, a consequence of opposite helicities in the $f\bar{f}$ production;
- *the electron longitudinal depolarization parameter,*

$$\begin{aligned} D_{ZZ}(\theta) &\equiv D_{ZZ}(e^-) \equiv (Z0|Z0) \\ &= \frac{\mathscr{A}_e\mathscr{A}_f(1+\cos^2\theta) + 2\cos\theta}{1+\cos^2\theta + \mathscr{A}_e\mathscr{A}_f 2\cos\theta} \qquad (9.2.38) \\ &= D_{ZZ}(e^+); \qquad (9.2.39) \end{aligned}$$

- *the electron longitudinal polarization transfer parameter,*

$$\begin{aligned} K_{ZZ}(\theta) &\equiv K_{ZZ}(e^-) \equiv (Z0|0Z) \\ &= -D_{ZZ}(\theta) \qquad (9.2.40) \\ &= K_{ZZ}(e^+); \qquad (9.2.41) \end{aligned}$$

- *the three-spin and four-spin correlation parameters,*

$$(Z0|ZZ) = -(0Z|ZZ) = -A_Z(\theta) \tag{9.2.42}$$

$$(ZZ|Z0) = -(ZZ|0Z) = -\mathscr{P}_f(\theta) \tag{9.2.43}$$

$$(ZZ|ZZ) = 1. \tag{9.2.44}$$

For an *unpolarized* initial state the degree of longitudinal polarization of the final fermion is given by (9.2.35), which in principle allows a determination of $\mathscr{A}_f$ and $\mathscr{A}_e$ if the longitudinal polarization of the final fermion can be measured.

If we assume lepton universality and take $\mathscr{A}_e = \mathscr{A}_f \approx 0.16$, corresponding to $\sin^2\theta_W = 0.23$, then we see that $\mathscr{P}_f(\theta)$ varies from 0 at $\theta = \pi$ to about -30% at $\theta = 0$.

However, the measurement of $\mathscr{P}_f(\theta)$ requires an analysis of the angular distribution of the decay products of f (as discussed in subsection 8.2.1), which is a non-trivial matter.

It may therefore be better, from the point of view of statistics, to deal with an integrated quantity. Thus we define

$$\overline{\mathscr{P}}_f \equiv \frac{\int \mathscr{P}_f(\theta) d\sigma/d\Omega}{\int d\sigma/d\Omega}. \tag{9.2.45}$$

From the definition of $\mathscr{P}_f(\theta)$ in terms of relative numbers of right- or left-handed f particles produced at angle θ, it is clear that

$$\overline{\mathscr{P}}_f = \frac{\sigma(f_R) - \sigma(f_L)}{\sigma(f_R) + \sigma(f_L)} \tag{9.2.46}$$

where $\sigma(f_{R,L})$ are the total cross-sections to produce right- or left-handed f particles.

Using (9.2.35) and (9.2.12) in (9.2.45) we see that

$$\overline{\mathscr{P}}_f = -\mathscr{A}_f, \tag{9.2.47}$$

a beautiful and simple result.

In practice it appears that the most accurate results will come from $e^-e^+ \to \tau^-\tau^+$; the τ polarization can be studied via various decays, e.g. $\tau \to \pi\nu$, $\tau \to \mu\bar{\nu}_\mu\nu_\tau$, $\tau \to \rho\nu$, $\tau \to a_1\nu$.

(iv) *Measurement of the τ polarization*

We consider how the τ polarization can be measured. We work within the Standard Model where the τ is produced with longitudinal polarization and the neutrinos are purely left-handed ($\lambda_\nu = -1/2$). All the following results emerge as a straightforward application of the discussion of the decay of unstable particles given in subsection 8.2.1, which should be consulted for notational conventions about angles etc.

(a) $\tau^- \to \pi + \nu_\tau$
Because $\lambda_\nu = -1/2$ and the π is spinless there is only one reduced helicity amplitude, (8.2.1). The decay is trivially magic and the normalized decay distribution of the π in the helicity rest frame of τ^- is given by (8.2.20) and (8.2.31):

$$W(\theta_\pi, \phi_\pi) = \frac{1}{\sqrt{4\pi}}\left[\frac{1}{\sqrt{4\pi}} + t^{1\,*}_m Y_{1m}(\theta_\pi, \phi_\pi)\right]$$

so that, using (3.1.35),

$$W(\theta_\pi) = \tfrac{1}{2}\left(1 + \mathscr{P}_\tau \cos\theta_\pi\right). \tag{9.2.48}$$

(b) $\tau^- \to V + \nu_\tau$, *where* V *is a* spin-*1 meson* $(\rho$ *or* $a_1)$
There are now two independent reduced helicity amplitudes:

$$M(\lambda_V, \lambda_\nu)$$

with $\lambda_\nu = -1/2$ and $\lambda_V = 0$ or -1. (The transition to $\lambda_V = +1$ is impossible by conservation of angular momentum.) Let us label these $M(0)$ and $M(-1)$ respectively. We can identify them by calculating the relevant helicity amplitudes $H_{\lambda_V\lambda_\nu;\lambda_\tau}$ with arbitrary choice of λ_τ and then using (8.2.1).

In the Standard Model the Feynman amplitude is given by

$$\frac{G}{2}\langle V;\lambda_V|h^\mu|0\rangle\left[\bar{u}_{\lambda_\nu}\gamma_\mu(1-\gamma_5)u_{\lambda_\tau}\right] \tag{9.2.49}$$

where h^μ is the hadronic weak current and G is the Fermi coupling constant. (See, for example, Leader and Predazzi (1996), Chapter 1.) In (9.2.49) we have justifiably neglected the effects of the W propagator. We cannot, of course, calculate the hadronic matrix element $\langle V;\lambda_V|h^\mu|0\rangle$, but in any field theory it has to be proportional to the polarization vector $\epsilon^{\mu^*}(\lambda_V)$ of the spin-1 particle. Moreover the proportionality function is just a constant, since in the decay the momentum of V is fixed.

Using the fact that

$$\bar{u}_{-1/2}(\nu)(1+\gamma_5) = 2\bar{u}_{-1/2}(\nu)$$

(see subsection 4.6.3), we can write

$$H_{\lambda_V\lambda_\nu;\lambda_\tau} = C\epsilon^*_\mu(\lambda_V)\bar{u}_{\lambda_\nu}\gamma^\mu u_{\lambda_\tau} \equiv C\epsilon^*_\mu(\lambda_V)V^\mu_{\lambda_\nu\lambda_\tau} \tag{9.2.50}$$

in the notation of subsection 4.6.2.

We can directly use the results (4.6.36)–(4.6.39) together with (4.6.30) and (4.6.28) to evaluate the amplitudes, taking for convenience the produced vector meson to have polar angles θ_V, $\phi_V = 0$ in the helicity rest frame of the τ. The only subtlety is to remember that the neutrino then has polar angles $\theta_\nu = \pi - \theta_V$, $\phi_\nu = \pi$.

For the polarization vector of the V meson, from (3.4.25), (3.4.24), (1.2.23) and (3.1.80), we have

$$\epsilon^{\mu^*}(\pm 1) = \frac{1}{\sqrt{2}}(0, \mp \cos\theta_V, i, \pm \sin\theta_V) \tag{9.2.51}$$

and

$$\epsilon^{\mu^*}(0) = \frac{1}{m_V}(p_V, E_V \hat{\mathbf{p}}_V), \tag{9.2.52}$$

where $\hat{\mathbf{p}}_V = (\sin\theta_V, 0, \cos\theta_V)$,

$$p_V = \frac{m_\tau^2 - m_V^2}{2m_\tau} \qquad \text{and} \qquad E_V = \frac{m_\tau^2 + m_V^2}{2m_\tau}. \tag{9.2.53}$$

After a little algebra involving the Pauli matrices in (4.6.30), one finds, up to a common constant,

$$\begin{aligned} H_{-1-1/2;1/2} &= -\sqrt{2}\sin\theta_V \\ H_{0-1/2;1/2} &= \frac{m_\tau}{m_V}\cos\theta_V. \end{aligned} \tag{9.2.54}$$

Comparing with (8.2.1), using

$$d^{1/2}_{1/2,-1/2}(\theta) = -\sin\theta/2$$

and

$$d^{1/2}_{1/2,1/2}(\theta) = \cos\theta/2$$

(see Appendix 1), we see that the correctly normalized reduced amplitudes are

$$M(0) = \frac{m_\tau}{\sqrt{m_\tau^2 + 2m_V^2}} \qquad M(-1) = \frac{-\sqrt{2}m_V}{\sqrt{m_\tau^2 + 2m_V^2}}. \tag{9.2.55}$$

From (8.2.20), upon using (8.2.12) and (9.2.55) we find for the angular distribution of the vector meson

$$W(\theta_V) = \frac{1}{2}\left[1 + \left(\frac{m_\tau^2 - 2m_V^2}{m_\tau^2 + 2m_V^2}\right)\mathscr{P}_\tau \cos\theta_V\right]. \tag{9.2.56}$$

In practice, in order to use (9.2.48) or (9.2.56) to measure $\mathscr{P}_\tau$ we do not measure the angles θ_h ($h = \pi, \rho, a_1$) but convert the distribution into distributions in the Lab fractional energy $x_h \equiv E_h/E_\tau$ of the decay hadron, using

$$\cos\theta_h = \frac{2x_h - 1 - m_h^2/m_\tau^2}{1 - m_h^2/m_\tau^2}. \tag{9.2.57}$$

An example of the results from a measurement of $\mathscr{P}_\tau(\theta)$ by the L3 Collaboration at CERN (Acciarri *et al.*, 1994) is shown in Fig. 9.1. The curves correspond to fitting $\mathscr{P}_\tau(\theta)$ either with $\mathscr{A}_\tau$ and $\mathscr{A}_e$ as independent parameters (no universality) or enforcing $\mathscr{A}_\tau = \mathscr{A}_e$ (universality). Excellent agreement with the Standard Model is obtained for a value $\sin^2\theta_{\mathrm{W}}^{\mathrm{eff}} = 0.2309 \pm 0.0016$, nicely compatible with (9.2.26). For further experimental studies see: Delphi Collaboration, Abreu *et al.* (1995a, b); Aleph Collaboration, Buskulic *et al.* (1996) and OPAL Collaboration; Alexander *et al.* (1996).

(c) $\tau^- \to \rho^- + \nu_\tau$ *with analysis of* $\rho^- \to \pi^-\pi^0$

Additional information can be obtained by studying the angular distribution of say, π^-, in the $\rho^- \to \pi^-\pi^0$ decay. The theoretical analysis is a very nice example of the power of the methods discussed in subsection 8.2.1.

From (8.2.20) and (8.2.17) the angular decay distribution of the π^-, produced at an angle θ_π to the ρ's direction of flight in the ρ helicity rest

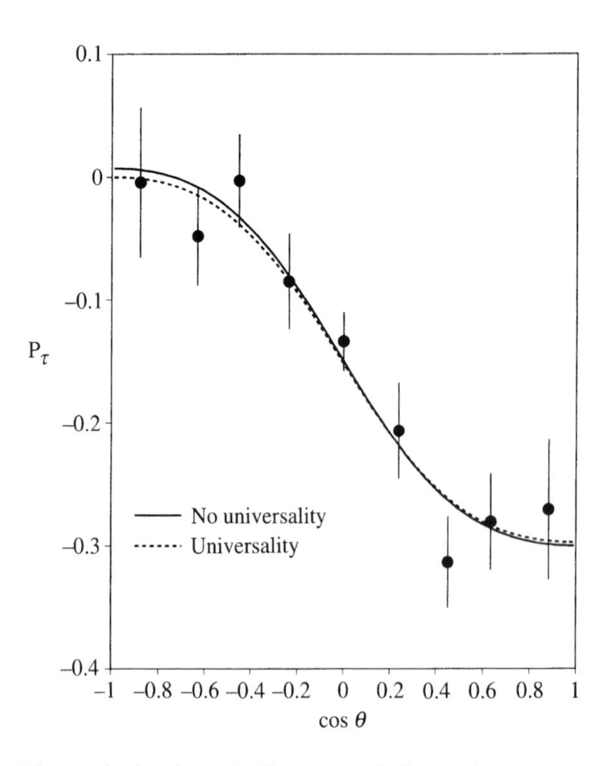

Fig. 9.1 The polarization $\mathscr{P}_\tau(\theta)$ vs. $\cos\theta$ for τ leptons produced in $e^-e^+ \to \tau^-\tau^+$ (from the L3 Collaboration, Acciarri *et al.* 1994). For a discussion of the two curves see the text.

frame is, after integration over the azimuthal angle ϕ_π,

$$W(\theta_\pi) = \tfrac{1}{2}\left[1 - \sqrt{\tfrac{5}{2}}\, t_0^2(\rho, \theta_\rho)(3\cos^2\theta_\pi - 1)\right] \tag{9.2.58}$$

where $t_0^2(\rho, \theta_\rho)$ is the multipole parameter of the ρ, which was produced at angle θ_ρ in the τ helicity rest frame.

The experimental analysis of the π^- angular distribution is done in the ρ helicity rest frame reached from the Lab, where the π^-, π^0 are detected. Thus the Lab multipole parameters $t_0^2|_{S_L}$ needed in (9.2.58) will differ by a Wick rotation from the $t_0^2|_{S_\tau}$ of the ρ in the τ helicity rest frame.

From (3.2.9) and (2.2.5) the connection is

$$t_0^2(\rho, \theta_\rho)\Big|_{S_L} = d^2_{M0}(\theta_{\text{Wick}})\, t_M^2(\rho, \theta_\rho)\Big|_{S_\tau} \tag{9.2.59}$$

in which, from (2.2.6), we have

$$\cos\theta_{\text{Wick}} = \gamma_\tau \frac{p_\rho + \beta_\tau E_\rho \cos\theta_\rho}{p_L^\rho} \tag{9.2.60}$$

$$\sin\theta_{\text{Wick}} = \gamma_\tau\beta_\tau \frac{m_\rho \sin\theta_\rho}{p_L^\rho} \tag{9.2.61}$$

where p_ρ and E_ρ are found from (9.2.53), p_L^ρ is the ρ Lab momentum and γ_τ, β_τ refer to the τ's motion in the Lab.

To compute the $t_m^l(\rho)$ for the decay $\tau^- \to \rho^- + \nu_\tau$, in the τ helicity rest frame, we use (8.2.12), (8.2.14) and (9.2.55). It is easy to see that $t_2^2|_{S\tau} = 0$. For the others we find

$$W(\theta_\rho) t_1^2(\rho, \theta_\rho)|_{S\tau} = \tfrac{1}{2}\sqrt{\tfrac{3}{5}} \frac{m_\tau m_\rho}{m_\tau^2 + 2m_\rho^2} \mathscr{P}_\tau \sin\theta_\rho \tag{9.2.62}$$

$$\begin{aligned} W(\theta_\rho) t_0^2(\rho, \theta_\rho)|_{S\tau} = & -\tfrac{1}{2}\sqrt{\tfrac{2}{5}} \frac{1}{m_\tau^2 + 2m_\rho^2} \\ & \times \left[m_\tau^2 - m_\rho^2 + (m_\tau^2 + m_\rho^2)\mathscr{P}_\tau \cos\theta_\rho\right] \end{aligned} \tag{9.2.63}$$

where $W(\theta_\rho)$ is found from (9.2.56).

Finally putting together (9.2.58)–(9.2.63), using the expressions for d^2_{M0} given in Appendix 1, and eqn (9.2.56), we get for the normalized joint distribution for production of a ρ^- at angle θ_ρ, in the τ helicity rest frame, that then decays into a π^- at angle θ^π, in the ρ helicity rest frame reached from the Lab,

$$W(\theta_\rho|\theta_\pi) = \tfrac{1}{4}\left[f(\theta_\pi) + \mathscr{P}_\tau g(\theta_\rho, \theta_\pi)\right]. \tag{9.2.64}$$

Here

$$f(\theta_\pi) = 1 + \frac{m_\tau^2 - m_\rho^2}{m_\tau^2 + 2m_\rho^2}(3\cos^2\theta_\pi - 1) \tag{9.2.65}$$

and

$$
\begin{aligned}
g(\theta_\rho, \theta_\pi) \\
&= (m_\tau^2 + 2m_\rho^2)^{-1} \\
&\times \Big\{ \Big[m_\tau^2 - 2m_\rho^2 + \tfrac{1}{2}(m_\tau^2 + 2m_\rho^2)(3\cos^2\theta_\pi - 1)(3\cos^2\theta_{\text{Wick}} - 1) \Big] \cos\theta_\rho \\
&\quad + 3m_\tau m_\rho (3\cos^2\theta_\pi - 1)\sin\theta_{\text{Wick}}\cos\theta_{\text{Wick}}\sin\theta_\rho \Big\}
\end{aligned}
\tag{9.2.66}
$$

where it must be remembered that θ_{Wick} is a function of θ_ρ. Thus $\mathscr{P}_\tau$ can also be determined from an analysis of the two-dimensional distribution (9.2.64). Important comments about optimizing the statistical analysis of multi-dimensional distributions are given in Davier *et al.* (1993).

(d) $\tau^- \to a_1^- + \nu_\tau$, *with analysis of* $a_1^- \to 3\pi$

Further information is obtained if one studies the angular distribution of the normal $\mathbf{n}$ (polar angles θ_n, ϕ_n) to the 3π decay plane in the a_1 helicity rest frame. (See subsection 8.2.2 for the three-body decays of an unstable particle.)

In general, as can be seen from (8.2.61) or from (8.2.78), for $a_1 \to$ three spin-0 particles, the angular distribution of the normal depends upon the *unknown* dynamical parameter $2R_{11} - 1$, which multiplies the $t_0^1(a_1)$ multipole parameter of the a_1. However, for $a_1 \to 3\pi$, because of either identical-particle or isotopic-spin symmetry, the correctly symmetrized version of R_{11} vanishes after integration over the Dalitz plot; see (8.2.67).

Hence starting with (8.2.61) and using (8.2.62) we find the following result: after integration over the Dalitz plot, and after integrating over the azimuthal angle ϕ_n of the normal to the decay plane, the normalized angular distribution in θ_n is given by

$$
\overline{W}(\theta_n) = \tfrac{1}{2}\left[1 + \tfrac{1}{2}\sqrt{\tfrac{5}{2}}\, t_0^2(a_1)(3\cos^2\theta_n - 1)\right], \tag{9.2.67}
$$

very similar in form to (9.2.58).

Analogously to the case of $\rho \to 2\pi$, the decay pions in $a_1 \to 3\pi$ detected in the Lab will yield the distribution of $\mathbf{n}$ in the helicity rest frame S_{L} reached from the Lab, so that a Wick rotation (9.2.59) must be carried out.

The result can be read off from the $\rho \to 2\pi$ case. One obtains

$$
\overline{W}(\theta_{a_1}|\theta_n) = \tfrac{1}{4}\left[f'(\theta_n) + \mathscr{P}_\tau g'(\theta_{a_1}, \theta_n)\right] \tag{9.2.68}
$$

where f' and g' are obtained from f and g in (9.2.65) and (9.2.66) by the substitutions

$$\begin{aligned} 3\cos^2\theta_\pi - 1 \quad &\rightarrow \quad \tfrac{1}{2}(1 - 3\cos^2\theta_n) \\ \theta_\rho \rightarrow \theta_{a_1} \qquad & \qquad m_\rho \rightarrow m_{a_1} \end{aligned} \tag{9.2.69}$$

(this must be done also inside θ_{Wick}; see (9.2.60), (9.2.61)).

(e) *Correlation in $\tau^-\tau^+$ production*
The $\tau^-\tau^+$ are created in a correlated state in the $e^-e^+ \rightarrow \tau^-\tau^+$ reaction, so that further information can be obtained by studying the correlated decays of the τ^- and τ^+.

There are two different approaches possible. We could write down from (8.2.22) the most general form for the joint angular distribution of the decay products, measure various correlation coefficients and then test whether their values correspond to the predictions of the Standard Model. We shall carry out, however, the somewhat simpler analysis of *assuming* the structure of the Standard Model and using the correlation analysis to measure the vector and axial-vector couplings, in effect, therefore, measuring $\sin^2\theta_W$ by yet another method.

Because we are dealing with spin-1/2 resonances it is simpler to use the Cartesian spin formalism rather the multipole parameter language.

Firstly, for the production reaction, it is easy to see from (5.6.3), upon using the amplitudes given in (9.2.11), that the only non-zero expectation values are

$$\langle \sigma_Z(\tau^-) \rangle = \mathscr{P}_\tau = -\langle \sigma_Z(\tau^+) \rangle = -\mathscr{P}_{\bar{\tau}} \tag{9.2.70}$$

with $\mathscr{P}_\tau$ as given in (9.2.35),

$$\langle \sigma_Z(\tau^-)\sigma_Z(\tau^+) \rangle = -1 \tag{9.2.71}$$

and

$$\begin{aligned} \langle \sigma_X(\tau^-)\sigma_X(\tau^+) \rangle &= \langle \sigma_Y(\tau^-)\sigma_Y(\tau^+) \rangle \\ &= \left(\frac{v_\tau^2 - a_\tau^2}{v_\tau^2 + a_\tau^2} \right) \frac{\sin^2\theta}{1 + \cos^2\theta + 2\mathscr{A}_e\mathscr{A}_\tau \cos\theta}. \end{aligned} \tag{9.2.72}$$

(Recall that the τ^- is produced at an angle θ to the e^- direction in the e^-e^+ CM; see (9.2.8).)

Secondly, we utilize (8.2.22); on the basis of (5.6.1) and (9.2.71), we substitute

$$\begin{aligned} t^{11}_{00}(\tau^-,\tau^+) &= \tfrac{1}{3}\langle \sigma_Z(\tau^-)\sigma_Z(\tau^+) \rangle = -\tfrac{1}{3} \\ t^{11}_{10} = t^{11}_{01} &= t^{11}_{11} = 0 \\ t^{11}_{1-1} = t^{11}_{-11} &= -\tfrac{1}{3}\langle \sigma_X(\tau^-)\sigma_X(\tau^+) \rangle . \end{aligned}$$

We can deal with all the negative decays $\tau^- \to \pi^- \nu_\tau, \rho^- \nu_\tau, a_1^- \nu_\tau$ and the positive decays $\tau^+ \to \pi^+ \bar{\nu}_\tau, \rho^+ \bar{\nu}_\tau, a_1^+ \bar{\nu}_\tau$ by writing the generic forms

$$\begin{aligned} W(\theta_-, \phi_-) &= \tfrac{1}{2}(1 + \alpha_- \mathscr{P}_\tau \cos\theta_-) \\ W(\theta_+, \phi_+) &= \tfrac{1}{2}(1 + \alpha_+ \mathscr{P}_{\bar{\tau}} \cos\theta_+) \end{aligned} \tag{9.2.73}$$

for their decay distributions with, from (9.2.48) and (9.2.56), and their analogues for τ^+,

$$\alpha_\pm(\pi) = \mp 1 \qquad \alpha_\pm(V) = \mp \frac{m_\tau^2 - 2m_V^2}{m_\tau^2 + 2m_V^2} \tag{9.2.74}$$

where $V = \rho, a_1$.

For the normalized joint distribution we end up with

$$\begin{aligned} &W(\theta_-, \phi_-; \theta_+, \phi_+) \\ &= \frac{1}{16\pi^2}\left[1 + \alpha_- \mathscr{P}_\tau(\theta)\cos\theta_- + \alpha_+ \mathscr{P}_{\bar{\tau}}(\theta)\cos\theta_+ - \alpha_-\alpha_+ \cos\theta_- \cos\theta_+ \right. \\ &\qquad \left. -2\alpha_-\alpha_+ \left(\frac{a_\tau^2 - v_\tau^2}{a_\tau^2 + v_\tau^2}\right) \frac{\sin^2\theta \sin\theta_- \sin\theta_+ \cos(\phi_- - \phi_+)}{1 + \cos^2\theta + 2\mathscr{A}_e \mathscr{A}_\tau \cos\theta}\right] \end{aligned} \tag{9.2.75}$$

where the $\mp$ angles are defined in the τ^-, τ^+ helicity rest frames respectively.

Clearly, the azimuthal dependence can be used to measure the parameter $(a_\tau^2 - v_\tau^2)/(a_\tau^2 + v_\tau^2)$, called C_{TT} in some of the experimental literature.

For experimental data see Abreu *et al.* (1997), where a value of $0.87 \pm 0.20 \pm 0.11$ was obtained, compatible with the value 0.978 corresponding to $\sin^2\theta_W = 0.2236$.

(v) Polarization of final state fermion with polarized e^-e^+

As an example of the benefits of having polarized $e^\pm$ beams, let us study the longitudinal polarization of the final state fermion f when the electron and positron have longitudinal polarizations $\mathscr{P}_e$, $\mathscr{P}_{\bar{e}}$ respectively.

From (5.6.3) and (5.6.5)

$$\begin{aligned} \mathscr{P}_f \left.\frac{d\sigma}{d\Omega}\right|_{\mathscr{P}_e, \mathscr{P}_{\bar{e}}} &= \frac{d\sigma}{d\Omega}\left[(00|Z0) + \mathscr{P}_e(Z0|Z0) + \mathscr{P}_{\bar{e}}(0Z|Z0) + \mathscr{P}_e\mathscr{P}_{\bar{e}}(ZZ|Z0)\right] \\ &= \frac{d\sigma}{d\Omega}\left[(1 - \mathscr{P}_e\mathscr{P}_{\bar{e}})\mathscr{P}_f(\theta) + (\mathscr{P}_e - \mathscr{P}_{\bar{e}}) D_{ZZ}(\theta)\right] \end{aligned} \tag{9.2.76}$$

where $\mathscr{P}_f(\theta)$ is given by (9.2.35) and $D_{ZZ}(\theta)$ by (9.2.38). Substituting for these yields

$$\mathscr{P}_f(\theta;\mathscr{P}_e,\mathscr{P}_{\bar{e}}) = -\frac{\mathscr{A}_f(1+\gamma_1)(1+\cos^2\theta)+(\mathscr{A}_e+\gamma_2)2\cos\theta}{(1+\gamma_1)(1+\cos^2\theta)+\mathscr{A}_f(\mathscr{A}_e+\gamma_2)2\cos\theta} \tag{9.2.77}$$

where $\gamma_{1,2}$ are given in (9.2.20).

The advantages of (9.2.77) are twofold. Firstly, by an appropriate choice of $\mathscr{P}_e$, $\mathscr{P}_{\bar{e}}$ we can obtain a much larger polarization. Figure 9.2 compares $\mathscr{P}_f$ for unpolarized $e^\pm$ with the case $\mathscr{P}_e = -\mathscr{P}_{\bar{e}} = 50\%$. This will enhance the asymmetry in the decay of, say, the τ. Secondly, it is useful from the point of view of statistics since one can use (9.2.77) to extract information from all f decays, no matter what the initial $e^\pm$ polarizations are.

9.2.2 The reaction $e^-e^+ \to$ quark–antiquark pair

All the asymmetry measurements discussed in the previous section for $e^-e^+ \to$ lepton–antilepton pair can, in principle, be carried out for $e^-e^+ \to$

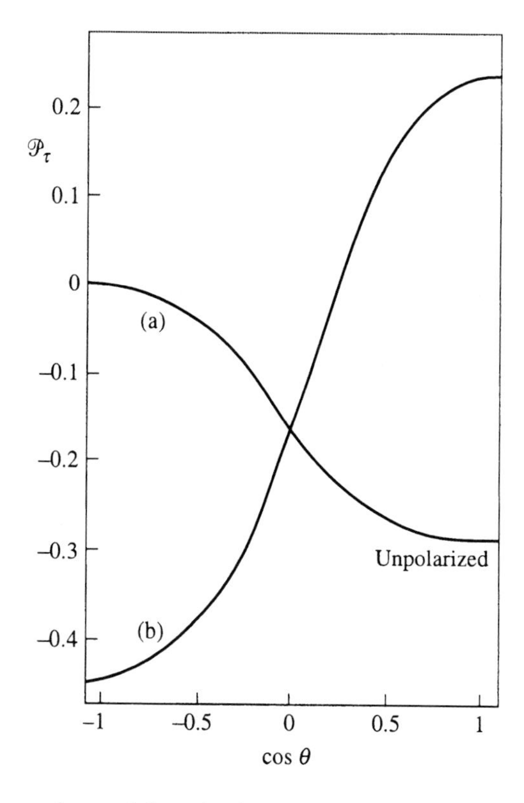

Fig. 9.2 Comparison of Standard Model τ polarization for e^-e^+ collisions: (a) unpolarized; (b) $\mathscr{P}_e = -\mathscr{P}_{\bar{e}} = 50\%$.

$q\bar{q}$. Thereby one is measuring the various vector and axial vector couplings v_f and a_f of (9.1.5) for the quarks.

(i) Production of heavy quarks

Experimentally the identification of a given quark and the determination of its direction of motion is much more complicated than for a lepton, and is probably only feasible for the b and c quarks. All sorts of tagging techniques are required as well as a determination of the thrust axis of the jet produced when the quark hadronizes.

We shall not attempt to cover this subject, but the reader is referred to Prepost (1996) for access to the literature.

The experimental complications are somewhat compensated by the large magnitude of the asymmetries compared with the lepton case. For the key parameters $\mathscr{A}_i$ defined in (9.2.13) we have the following values for $\sin^2\theta_W = 0.23$:

$$\mathscr{A}_e = \mathscr{A}_\mu = \mathscr{A}_\tau = 0.16 \qquad \mathscr{A}_u = \mathscr{A}_c = \mathscr{A}_t = 0.67$$
$$\mathscr{A}_d = \mathscr{A}_s = \mathscr{A}_b = 0.94. \tag{9.2.78}$$

As a consequence the unpolarized forward–backward asymmetry (9.2.29) will be far larger for b and c quarks than for leptons.

For the polarized forward–backward asymmetry, with $\mathscr{P}_{\bar{e}} = 0$ and $\mathscr{P}_e \approx 75\%$, (9.2.31) implies that $|A^b_{\text{FB}}| \approx 0.52$, a huge asymmetry.

Results for the parameters $\mathscr{A}_b$ and $\mathscr{A}_c$ vary somewhat according to the method used for tagging the quark but are essentially compatible with each other and with the Standard Model values. The world averages, given at the 1997 Lepton–Photon Conference (Timmermans, 1998) were

$$\mathscr{A}_b = 0.898 \pm 0.050 \qquad \mathscr{A}_c = 0.649 \pm 0.058, \tag{9.2.79}$$

to be compared with the precise Standard Model predictions

$$\mathscr{A}_b^{\text{SM}} = 0.935 \qquad \mathscr{A}_c^{\text{SM}} = 0.667. \tag{9.2.80}$$

Also, the polarization of the produced quark in an unpolarized e^+e^- collision will be very large and roughly independent of the production angle, as can be seen from (9.2.35):

$$\mathscr{P}_q(\theta) \approx -\mathscr{A}_q. \tag{9.2.81}$$

In principle the state of polarization of the heavy quark could be determined from the lepton energy spectrum in the semileptonic decay $b \to c + l + \bar{\nu}_l$. If we pretend that the quarks are free particles then this is analogous to the determination of the muon polarization from $\mu \to e\nu\bar{\nu}$ discussed in subsection 8.1.1(v). In reality one has to worry about the strong interaction effects, which lead to the hadronization of the quarks (see Mele, 1994). Interestingly, however, this will not be a problem for top decay, if it were ever possible to produce $t\bar{t}$ pairs in lepton–antilepton collisions, since the decay is so rapid that there is no time for strong interaction effects to act.

(ii) Production of light quarks

By eqns (9.2.81) and (9.2.78) the u, d and s quarks are all produced with a high degree of polarization, but there is no sense in considering them as free particles and one is forced to take into account the process of *hadronization*, whereby the quark materializes as a physical particle. Since this is a non-perturbative strong interaction process we are unable to calculate it. The dynamics of the hadronization thus has to be studied experimentally. The focus, therefore, is not so much upon testing the electroweak theory as upon trusting the Standard Model to tell us about the state of the produced quark and then, by measuring the properties of the final state particles, to learn about the process of hadronization.

9.3 Summary

In summary, the measurement of spin-dependent observables has been and will continue to be a very powerful tool in testing the Standard Model to fantastic levels of precision. It is quite remarkable that the SLD measurement of A_{LR} using some 93 000 Z^0 events has achieved the same accuracy in the value of $\sin^2\theta_{\mathrm{W}}$ as all the LEP experiments put together, involving several million Z^0 events!

10
Quantum chromodynamics: spin in the world of massless partons

Quantum chromodynamics (QCD) is the beautiful theoretical structure believed to control the strong interactions of elementary particles. On the one hand, being a theory of strong interactions it is surprising that one can attack certain problems by perturbative methods, and where this has been done the agreement between theory and experiment is generally impressive. On the other hand a number of non-perturbative problems, which used to seem intractable, are now being attacked by lattice methods, but it is too early to say how significant the results are *vis-à-vis* experiment.

Because the theory deals with partons (quarks and gluons), whereas experiments are performed with hadrons, there is always some uncalculable piece in any theoretical treatment of a reaction. Consequently there is, to date, no single crucial experiment, which, analogous to the Lamb shift in QED, could be said to prove or disprove the validity of QCD. It is thus important to test the theory in as many ways as possible.

Historically, spin-dependent experiments have played a seminal rôle in verifying or falsifying theories. QCD has a very simple and clear-cut spin structure, so that the study of spin-dependent reactions should provide an excellent way to probe and test the theory further. In fact, as we shall see in Section 14.3 there is apparently serious disagreement between theory and experiment in several reactions, but it is now believed that this is a result of the naivety of the calculations. The situation is nonetheless tantalizing and should be resolved when results from the giant *pp* collider RHIC at Brookhaven, with polarized proton beams, start to emerge in a year or two.

10.1 A brief introduction to QCD

QCD is a non-abelian gauge theory describing the interaction of massless spin-1/2 objects, the 'quarks', which possess an internal degree of freedom called colour, and a set of massless gauge bosons (vector mesons), the 'gluons', which mediate the force between quarks in much the same way that photons do in QED. Loosely speaking, the quarks come in three colours and the gluons in eight. More precisely, if $q^a(x), a = 1,2,3$ and $A^b_\mu(x), b = 1,\ldots,8$, are the quark and gluon fields respectively then, under an $SU(3)$ transformation acting on the colour indices, q and A are defined to transform as the fundamental ($\underline{3}$) and the adjoint ($\underline{8}$) representations of $SU(3)$ respectively. These $SU(3)$ transformations, acting solely on the colour indices, have nothing at all to do with the usual $SU(3)$ that acts on the quark flavour labels; in what follows it must be understood that these flavour labels play no rôle in QCD since the gluons are taken to be flavourless, i.e. to be singlets under $SU(3)_f$, and electrically neutral, so they will not be displayed unless specifically needed.

The theory is *known* to possess the remarkable property of 'asymptotic freedom' and is *supposed* to possess the property of 'colour confinement'. The former implies that for interactions between quarks at very short distances, i.e. for large momentum transfers, the theory looks more and more like a free-field theory, without interactions. This, ultimately, is the justification for the parton model and for the use of perturbative methods for large momentum reactions. The latter means that only 'colourless' objects, that are colour singlets, can be found existing as real physical particles. In other words the forces between two coloured objects grow stronger with distance, so that they can never be separated. This property of confinement is also referred to as 'infrared slavery'. The proof of confinement is still lacking and remains one of the most burning theoretical questions.

The $SU(3)$ non-abelian, gauge-invariant theory for an octet of massless vector gluons interacting with a triplet of massless spin-1/2 quarks involves:

(1) *generalized field tensors* (i.e. non-abelian generalizations of the electromagnetic $F_{\mu\nu}$)

$$G^a_{\mu\nu} = \partial_\mu A^a_\nu - \partial_\nu A^a_\mu + g f_{abc} A^b_\mu A^c_\nu, \tag{10.1.1}$$

where A^a_μ is the gluon vector potential, the label $a = 1,\ldots,8$ being the octet colour label, and where f_{abc} are the structure constants for $SU(3)$; the group generators obey

$$[T_a, T_b] = i f_{abc} T_c. \tag{10.1.2}$$

Note that colour indices will, for convenience, sometimes be written as subscripts, sometimes as superscripts – there is no difference in meaning;

(2) *quark spinor fields* ψ_j, where $j = 1, 2, 3$ labels the quark colour. There will be a set of ψ_j for each flavour, but we leave out the flavour label to simplify the notation;

(3) *a covariant derivative operator*: symbolically one has the operator

$$\hat{D}_\mu \equiv \partial_\mu - igT_a A^a_\mu. \tag{10.1.3}$$

When acting on some given field that transforms according to a particular representation of the group, one replaces the T_a by the relevant representation matrices. Thus when acting on quark fields $\hat{D}_\mu$ is represented by

$$(D_\mu)_{ij} = \delta_{ij}\partial_\mu - igt^a_{ij}A^a_\mu \tag{10.1.4}$$

where the $t^a, a = 1, \ldots, 8$ are 3×3 hermitian matrices that, for the triplet representation of $SU(3)$, are just one half the Gell–Mann matrices λ^a.

Acting on the gluon fields the T_a are represented by the structure constants $(T_a)_{bc} \to -if_{abc}$, so that $\hat{D}_\mu$ is represented by

$$(D_\mu)_{bc} = \delta_{bc}\partial_\mu - gA^a_\mu f_{abc}. \tag{10.1.5}$$

The gauge-invariant interactions are described by the lagrangian

$$\mathscr{L} = -\tfrac{1}{4}G^a_{\mu\nu}G^{a\mu\nu} + i\bar{\psi}_i\gamma_\mu (D^\mu)_{ij}\,\psi_j, \tag{10.1.6}$$

where the last term is in fact a sum of identical terms, one for each flavour, and where we have assumed massless quarks.

It is usually assumed that there are no quark mass terms in the original QCD lagrangian, so that it is perfectly flavour symmetric and chirally symmetric. The flavour symmetry is presumably spontaneously broken, the quarks acquiring masses from the electroweak Higgs mechanism and/or from non-perturbative spontaneous chiral-symmetry-breaking effects caused by non-zero vacuum expectation values of $\langle 0|\bar{\psi}_f\psi_f|0\rangle$, where f is some fixed flavour. (For an introductory discussion, see Leader and Predazzi, 1996.)

Since quarks are supposed not to exist as free physical particles their masses are not masses in the usual sense. The quark mass should be thought of simply as a parameter in the lagrangian, to be determined in principle from experiment. However, in perturbation theory, a quark propagator has a pole at $p^2 = m^2$, whereas in the exact theory it presumably has no pole at all. So perturbative calculations are only considered reliable in kinematics regions where the momentum transfers, energies etc., are all large compared with m, which can then be neglected. Thus, determination of quark masses must come from non-perturbative studies such as current algebra or QCD sum rules. (A comprehensive review is given in Gasser and Leutwyler, 1982.) One finds that u and d have masses of a few MeV only ($m_u \sim 4$ MeV$/c^2, m_d \sim 7$ MeV$/c^2$) and that $m_s \sim 125$–150 MeV$/c^2$; these are referred to as 'current quark masses' and should not be confused with the 'constituent quark masses' that are used in the non-relativistic treatment of hadron spectroscopy.

The field theory, not surprisingly, is riddled with infinities and has to be renormalized. In the renormalization the bare coupling constant g in the lagrangian, hidden in the $(D^\mu)_{ij}$ of (10.1.6), becomes replaced by the renormalized coupling, which has to be measured by comparing theory and experiment.

It turns out that there is a certain freedom in carrying out the renormalization, but physical quantities must be invariant under changes of the renormalization scheme. This leads to the concept of the *renormalization group*, under whose transformations the physics is invariant. (See, for example, Chapter 20 of Leader and Predazzi, 1996.) The main consequence for our discussion is that one can 'renormalization-group-improve' a perturbative calculation by replacing the strong coupling $\alpha_s \equiv g^2/4\pi$ by an *effective* or *running* coupling $\alpha_s(Q^2)$, where Q is some characteristic energy or momentum scale of the process one is studying. The variation of $\alpha_s(Q^2)$ with Q^2 is determined by the QCD renormalization group, and to lowest order

$$\alpha_s(Q^2) = \frac{12\pi}{(33 - 2N_f)\ln(Q^2/\Lambda^2)} \tag{10.1.7}$$

where N_f is the number of quark flavours and Λ (often written Λ_{QCD}) has to be determined by experiment (strictly speaking it should be called $\Lambda^{(0)}$ because (10.1.7) is only a lowest-order result) and one has $\Lambda \approx 200$ MeV.

In higher orders $\alpha_s(Q^2)$ and therefore Λ become scheme dependent (see Section 11.7) and require a label to indicate the scheme. And N_f is, strictly, not the total number of flavours but the *effective* number that is relevant, i.e. the number playing a rôle at the scale Q.

The power of using $\alpha_s(Q^2)$ is that $\alpha_s(Q^2) \to 0$ as $Q^2 \to \infty$ (asymptotic freedom) so that for reactions at a large scale Q the effective coupling is small and one can justify a perturbative approach.

When a reaction contains several widely disparate scales $Q_1, Q_2, \ldots$ the above argument becomes ambiguous and there is no obvious rule about what value of Q^2 to use in $\alpha_s(Q^2)$. However, there are many important reactions where one large scale does exist, e.g. deep inelastic lepton–hadron scattering at large momentum transfer (Chapter 11), hadron–hadron scattering at large momentum transfer (Chapter 13), the Drell–Yan process

$$\text{hadron} + \text{hadron} \to \left[(l^+l^-), Z^0, W\right] + X$$

at large transverse momentum (Chapter 12) and $e^+e^- \rightarrow$ hadrons at high energies (see Leader and Predazzi, 1996, Section 22.1), so there is a host of experimental data against which the theory can be tested.

In summary we can apply perturbative QCD to hard processes where there is one energy or momentum scale sufficiently large to make $\alpha_s(Q^2) \ll 1$. At these scales we can ignore m_u, m_d and m_s and it is adequate to utilize the massless lagrangian (10.1.6). For the 'heavy' quarks $t(!), b$ and perhaps c, one should modify $\mathscr{L}$ to include quark mass terms, but we shall not have space to discuss this.

10.2 Local gauge invariance in QCD

The QCD lagrangian is invariant under local $SU(3)$ transformations. However, in order to do a concrete calculation one has to choose a definite gauge in which to work. In QED one often uses the covariant Lorentz gauge $\partial_\mu A^\mu(x) = 0$. In QCD covariant gauges are more complicated and it is necessary to include a *ghost* propagator in diagrams involving closed loops. (The Feynman rules are given in Appendix 11.) The reason for this difference is linked to the question whether one may replace a polarization vector $\epsilon_\mu(k)$ by $\epsilon_\mu(k) + ck_\mu$, c arbitrary, in the expression for a Feynman diagram involving external photons or gluons.

In both QED and QCD the total amplitude for a reaction involving any number of external photons or gluons respectively has the structure

$$\begin{aligned} A = {} & \epsilon^*_{\mu_1}(k'_1)\ldots\epsilon^*_{\mu_n}(k'_n)M(k'_1\ldots k'_n;k_1\ldots k_m)^{\mu_1\ldots\mu_n;\nu_1\ldots\nu_m} \\ & \times \epsilon_{\nu_1}(k_1)\ldots\epsilon_{\nu_m}(k_m) \end{aligned} \tag{10.2.1}$$

where in *this* expression all 4-vectors k_i, k'_j are on the mass shell, i.e. $k_i^2 = (k'_j)^2 = 0$. (In QCD M would also have colour labels.)

In QED, either for the whole amplitude or for the amplitude arising from any local-gauge-invariant subset of Feynman diagrams, one has the remarkably powerful property that, for any of the momenta,

$$\begin{aligned} (k'_j)_{\mu_j} M(k'_1\ldots k'_n;k_1\ldots k_m)^{\mu_1\ldots\mu_j\ldots\mu_n;\nu_1\ldots\nu_m} &= 0 \\ M(k'_1\ldots k'_n;k_1\ldots k_m)^{\mu_1\ldots\mu_n;\nu_1\ldots\nu_i\ldots\nu_m}(k_i)_{\nu_i} &= 0 \end{aligned} \tag{10.2.2}$$

irrespective of whether the ks in M are on or off the mass shell.

Clearly, then, in QED one is free to replace any $\epsilon_\mu(k)$ by $\epsilon_\mu(k) + ck_\mu$ as long as one is working with either *all* the diagrams of a given order or some local-gauge-invariant subset of them. (Any single Feynman diagram is usually not invariant!) This, as will be seen, allows huge simplifications in the calculations.

In QCD there is nothing like (10.2.2) involving just M itself. Instead, one gets the following rule.

• *QCD local-gauge-invariance rule.* In (10.2.1) we get zero if we replace one or more $\epsilon_{\mu_j}(k_j)$ by $(k_j)_{\mu_j}$, provided that amongst these ks *at most one* is off shell, i.e. does not satisfy $k^2 = 0$. (All the other ks in (10.2.1) are on mass shell, as previously stated.)

Although much weaker than (10.2.2) this rule is sufficient to permit one to replace any $\epsilon_\mu(k)$ by $\epsilon_\mu(k) + ck_\mu$ in the expression for the amplitude arising from any set of local-gauge-invariant diagrams in QCD. A detailed derivation of these results is given in Section 21.3 of Leader and Predazzi (1996).

The identification of local-gauge-invariant subsets of Feynman diagrams is relatively simple in QED. For the photon of interest, for which one wishes to replace $\epsilon_\mu(k)$ by $\epsilon_\mu(k) + ck_\mu$, one takes the set of diagrams in which this photon is attached to a fermion line in all possible ways. For instance, in lowest-order Compton scattering (see Fig. 10.7) neither diagram is local gauge invariant, but their sum is.

In QCD the identification is much more subtle and was solved in a classic paper by Cvitanović, Lauwers and Scharbach (1981). In any reaction involving gluons and quarks the amplitude will be labelled by a colour label for each external parton, $A(a, b, \ldots; i, j, \ldots)$. In colour space there are invariant tensors $F_r(a, b, \ldots; i, j, \ldots), r = 1, 2, \ldots$, for example t^a_{ij}, $t^a_{ij}t^b_{lm}$, f_{abc} etc., which will emerge from any calculation of any individual Feynman diagram. These tensors are generally not independent and may be related through the fundamental structure relations of the Lie group, for example,

$$\left[t^a, t^b\right] = if_{abc}t^c \tag{10.2.3}$$

or the so-called Jacobi identity

$$f_{abe}f_{ecd} + f_{cbe}f_{aed} + f_{dbc}f_{ace} = 0. \tag{10.2.4}$$

By repeated use of these, one can eliminate various F_r until one is left with a *linearly independent* set of tensors T_r. Note that several different Feynman diagrams could give contributions proportional to some given T_r. This set of tensors is called a *colour basis* for the given reaction.

After grouping together all terms proportional to a given T_r the amplitude will take the form

$$A(a, b, \ldots; i, j, \ldots) = \sum_r T_r(a, b, \ldots; i, j, \ldots)\mathscr{A}_r \tag{10.2.5}$$

where the $\mathscr{A}_r$ are functions of the momenta and helicities of the external partons. Since the QCD local-gauge-invariance rule applies to A, and since the terms in (10.2.5) are linearly independent, the rule must apply separately to each $\mathscr{A}_r$. Examples will be given in Section 10.11.

10.3 Feynman rules for massless particles

Since perturbation methods can only be applied to 'hard' processes, in which energies and momenta are large compared with the scale of a typical nucleon mass, the quark-partons of QCD may in many cases be taken as massless. It then turns out that one can reformulate the rules so that calculating the helicity amplitudes from a Feynman diagram becomes much simpler than in the traditional approach. In fact for low-order diagrams these methods remain efficient even when generalized to allow for non-zero-mass quarks. In addition the methods are especially suitable for numerical computation.

The existence of such methods is important because in a high energy collision of hadrons, final states with many jets or hadrons occur and these arise from partonic reactions involving a large number of partons. The number of Feynman diagrams for this kind of process, even in lowest order (known as Born or tree-level), is horrendous. For example for $GG \to 6G$ there are 34 300 diagrams!

Although it is not easy to imagine studying such reactions in order to test QCD, it often happens that one is trying to look for 'new physics' reactions, involving, for example, a sequential decay of some new heavy particle and giving rise to a multijet, multiparticle final state. The identification of a new reaction is impossible without any accurate knowledge of the standard QCD background.

The pioneering steps in this field were taken by De Causmaecker, Gastmans, Troos and Wu (De Causmaecker *et al.*, 1981), and Farrar and Neri (1983), and there followed many calculations in QED by what became known as the CALKUL collaboration. Berends and Giele (1987) approached the massless spinor problem using the dotted and undotted spinors of Weyl and van der Warden and calculated the cross-section for $2G \to 4G$. A further advance was due to Xu, Zhang and Chang (Xu *et al.*, 1987), who simplified the form of the polarization vectors for gluons. Interesting applications have been made by Kleiss and Stirling (1985) to $\bar{p}p \to W/Z + \text{jets}$, by Mangano, Parke and Xu (1987) to multigluon scattering and by Kleiss (1986) to $e^+e^- \to e^+e^-\gamma$ and $e^+e^- \to f\bar{f}\gamma$ (where f is a fermion). For a review of the subject see Mangano and Parke (1991) and for access to the latest literature see Mahlon and Parke (1997) and Benn *et al.* (1997). The reader is warned that in some of these papers the phase conventions do not correspond to the helicity convention utilized in this book and used widely in the literature. Also, in the CALKUL papers what is labelled as helicity ± 1 corresponds to what is normally called helicity ∓ 1 respectively. However, since these papers calculate only cross-sections, i.e. sum over helicities, this does not affect their results. But there could be confusion regarding signs of polarizations etc.

In the following we reformulate the approach with due care for the phase conventions and in such a fashion that it generalizes to the case of massive particles.

10.3.1 The calculus of massless spinors

The properties of massless spinors are derived in Appendix 12. Here we recall the most important results and introduce a new notation that takes advantage of these properties. As discussed in subsection 4.6.3, in the limit $m \to 0$ the helicity states become states of definitive chirality (R or L) which we shall henceforth designate by + or −. We have then

$$\gamma_5 u_\pm = \pm u_\pm \qquad \gamma_5 v_\pm = \mp v_\pm \tag{10.3.1}$$

and eqns (A12.8) and (A12.29) become, for $p^2 = 0$,

$$\bar{u}_\pm(\mathbf{p})u_\pm(\mathbf{p}) = \bar{v}_\pm(\mathbf{p})v_\pm(\mathbf{p}) = 0 \tag{10.3.2}$$

$$\left.\begin{aligned} u_+(\mathbf{p})\bar{u}_+(\mathbf{p}) + u_-(\mathbf{p})\bar{u}_-(\mathbf{p}) &= \not{p} \\ v_+(\mathbf{p})\bar{v}_+(\mathbf{p}) + v_-(\mathbf{p})\bar{v}_-(\mathbf{p}) &= \not{p} \end{aligned}\right\} \tag{10.3.3}$$

Also, from (A12.53) we have

$$\tfrac{1}{2}(1 \pm \gamma_5)\not{p} = u_\pm(\mathbf{p})\bar{u}_\pm(\mathbf{p}) = v_\mp(\mathbf{p})\bar{v}_\mp(\mathbf{p}). \tag{10.3.4}$$

In the Weyl representation (A12.43) we have a simple form for the spinors

$$\begin{aligned} u_+(\mathbf{p}) = v_-(\mathbf{p}) &= \sqrt{2p^0}\begin{pmatrix} \chi_+(\mathbf{p}) \\ 0 \end{pmatrix} \\ u_-(\mathbf{p}) = v_+(\mathbf{p}) &= \sqrt{2p^0}\begin{pmatrix} 0 \\ \chi_-(\mathbf{p}) \end{pmatrix} \end{aligned} \tag{10.3.5}$$

where the two-component spinors $\chi_\pm(\mathbf{p})$ are given in eqn (4.6.28).

We take advantage of the above by introducing the following notation (*only* when $p^2 = 0$)

$$|\mathbf{p}_\pm\rangle \equiv u_\pm(\mathbf{p}) \qquad \langle\mathbf{p}_\pm| \equiv \bar{u}_\pm(\mathbf{p}) \tag{10.3.6}$$

so that if the 4-vectors p, q are such that $p^2 = q^2 = 0$, $p_0 > 0$, $q_0 > 0$, we have the *spinor product*

$$\langle q_{\lambda'}|p_\lambda\rangle = \bar{u}_{\lambda'}(\mathbf{q})u_\lambda(\mathbf{p}), \tag{10.3.7}$$

where throughout this chapter $\lambda = \pm 1$ is the chirality.

From (A12.47) we have for $\lambda' = \lambda$

$$\langle q_\lambda|p_\lambda\rangle = 0. \tag{10.3.8}$$

The symmetry properties of the spinor product are very simple, and all spinor products can be expressed in terms of a basic one, say $\langle q_-|p_+\rangle$.

If the vectors $\mathbf{q}$, $\mathbf{p}$ have polar angles θ, ϕ and θ', ϕ' respectively, then one finds explicitly

$$\langle q_-|p_+\rangle = 2\sqrt{p_0 q_0}\left[\cos\left(\frac{\phi'-\phi}{2}\right)\sin\left(\frac{\theta'-\theta}{2}\right) + i\sin\left(\frac{\phi'-\phi}{2}\right)\sin\left(\frac{\theta'+\theta}{2}\right)\right] \quad (10.3.9)$$

$$= \sqrt{2p\cdot q}\,e^{i\Phi_{qp}}. \quad (10.3.10)$$

The phase Φ_{qp} is given by

$$\tan\Phi_{qp} = \tan\left(\frac{\phi'-\phi}{2}\right)\sin\left(\frac{\theta'+\theta}{2}\right)\Big/\sin\left(\frac{\theta'-\theta}{2}\right) \quad (10.3.11)$$

and its quadrant is fixed by demanding that

$$\text{sign}\left[\sin\Phi_{qp}\right] = \text{sign}\left[\sin\left(\frac{\phi'-\phi}{2}\right)\sin\left(\frac{\theta'+\theta}{2}\right)\right]. \quad (10.3.12)$$

It is important to remember that spinors are multivalued functions of the components of a vector, so care must be taken to specify polar angles in a consistent fashion.

It is easy to demonstrate the following elegant properties of the spinor product.

(1) Reversal of chiralities:

$$\langle q_+|p_-\rangle = -\langle q_-|p_+\rangle^*. \quad (10.3.13)$$

(2) Interchange of vectors:

$$\langle p_-|q_+\rangle = -\langle q_-|p_+\rangle. \quad (10.3.14)$$

(3) Interchange of initial and final state:

$$\langle p_+|q_-\rangle = \langle q_-|p_+\rangle^*. \quad (10.3.15)$$

Most importantly one finds that

$$|\langle q_-|p_+\rangle|^2 = 2p\cdot q. \quad (10.3.16)$$

It follows that if q is a multiple of p, $q = Cp$, then $\langle Cp_-|p_+\rangle = 0$. Of particular importance is the case $C = 1$:

$$\langle p_-|p_+\rangle = 0. \quad (10.3.17)$$

Let $p^\mu = (p, \mathbf{p})$ be a null vector with polar coordinates $\mathbf{p} = (p, \theta, \phi)$. We define the *conjugate* four vector $\tilde{p}^\mu$ by

$$\tilde{p}^\mu \equiv (p, -\mathbf{p}), \quad (10.3.18)$$

where, in accordance with subsection 1.2.2, the polar coordinates of $-\mathbf{p}$ are given by

$$(-\mathbf{p}) \equiv (p, \pi - \theta, \phi + \pi). \tag{10.3.19}$$

Then from (10.3.9) we find

$$\langle \tilde{p}_- | p_+ \rangle = -2ip = -i\sqrt{2p \cdot \tilde{p}}. \tag{10.3.20}$$

Also if p^μ, q^μ are any two null vectors then one finds that

$$\langle \tilde{q}_- | \tilde{p}_+ \rangle = - \langle q_- | p_+ \rangle^* = \langle q_+ | p_- \rangle . \tag{10.3.21}$$

Furthermore one may interchange the conjugacy:

$$\langle q_- | \tilde{p}_+ \rangle = \langle \tilde{q}_- | p_+ \rangle^* = - \langle \tilde{q}_+ | p_- \rangle . \tag{10.3.22}$$

One should beware of the fact that although $\tilde{\tilde{p}}^\mu = p^\mu$ the polar coordinates of $\tilde{\tilde{p}}$ are $p, \theta, \phi + 2\pi$, so that

$$|\tilde{\tilde{p}}_\pm\rangle = -|p_\pm\rangle \tag{10.3.23}$$

When dealing with Feynman diagrams it will turn out that the vertices give rise to matrix elements of the form $\langle q_\lambda | \gamma^\mu | p_\lambda \rangle$. It is easy to demonstrate the following useful properties.

(1) Reversal of chiralities:

$$\langle q_- | \gamma^\mu | p_- \rangle = \langle q_+ | \gamma^\mu | p_+ \rangle^*. \tag{10.3.24}$$

(2) Interchange of initial and final states:

$$\langle p_+ | \gamma^\mu | q_+ \rangle = \langle q_+ | \gamma^\mu | p_+ \rangle^*. \tag{10.3.25}$$

Combining these we have

$$\langle q_- | \gamma^\mu | p_- \rangle = \langle p_+ | \gamma^\mu | q_+ \rangle. \tag{10.3.26}$$

In the expression for the amplitude of a Feynman diagram the γ^μ in a vertex either will be contracted with the polarization vector of an external vector meson or will be linked via a vector meson propagator to some other vertex. We can choose from the outset to work in the Feynman gauge (see Appendix 11), so that the propagator contains only the term $g_{\mu\nu}$ and we end up with contractions of the form

$$\langle a_+ | \gamma^\mu | b_+ \rangle \langle c_- | \gamma_\mu | d_- \rangle \qquad \text{or} \qquad \langle a_+ | \gamma^\mu | b_+ \rangle \langle c_+ | \gamma_\mu | d_+ \rangle .$$

To evaluate these we use (A12.56):

$$2|b_+\rangle\langle a_+| = \langle a_+ | \gamma_\mu | b_+ \rangle \, \gamma^\mu \tfrac{1}{2}(1 - \gamma_5). \tag{10.3.27}$$

Multiplying on the left by $\langle c_- |$ and on the right by $| d_- \rangle$ yields

$$\langle a_+ | \gamma_\mu | b_+ \rangle \langle c_- | \gamma^\mu | d_- \rangle = 2 \langle a_+ | d_- \rangle \langle c_- | b_+ \rangle . \tag{10.3.28}$$

For the other possibility we use (10.3.26):

$$\begin{aligned}\langle a_+|\gamma_\mu|b_+\rangle\,\langle c_+|\gamma^\mu|d_+\rangle &= \langle a_+|\gamma_\mu|b_+\rangle\,\langle d_-|\gamma^\mu|c_-\rangle\\ &= 2\,\langle a_+|c_-\rangle\,\langle d_-|b_+\rangle\,. \end{aligned} \tag{10.3.29}$$

These are really just special cases of the Fierz rearrangement theorem (Appendix 12).

There are some further useful rearrangement-type results.

Let $|b_+\rangle, |c_+\rangle$ be independent in the sense that the scalar product $b{\cdot}c \neq 0$. Then since the massless spinors are in essence two-component objects it must be possible to expand any $|a_+\rangle$ in terms of $|b_+\rangle$ and $|c_+\rangle$:

$$|a_+\rangle = B|b_+\rangle + C|c_+\rangle.$$

Taking the spinor product with $\langle b_-|, \langle c_-|$ yields

$$\langle c_-|a_+\rangle = B\,\langle c_-|b_+\rangle \qquad \langle b_-|a_+\rangle = C\,\langle b_-|c_+\rangle\,..$$

Therefore

$$|a_+\rangle = \frac{\langle c_-|a_+\rangle}{\langle c_-|b_+\rangle}|b_+\rangle + \frac{\langle b_-|a_+\rangle}{\langle b_-|c_+\rangle}|c_+\rangle. \tag{10.3.30}$$

An analogous expansion holds for $|a_-\rangle$ with all chiralities reversed.

Multiplying on the left by some $\langle d_-|$, using (10.3.14) and relabelling into alphabetical order we get

$$\langle a_-|b_+\rangle\,\langle c_-|d_+\rangle = \langle a_-|d_+\rangle\,\langle c_-|b_+\rangle + \langle a_-|c_+\rangle\,\langle b_-|d_+\rangle\,. \tag{10.3.31}$$

Finally, for *any* 4-vector P^μ we introduce the notation

$$\not{P}_\pm \equiv \tfrac{1}{2}(1 \pm \gamma_5)\not{P}. \tag{10.3.32}$$

If $p^2 = 0$, $p^0 > 0$ then from (10.3.4) we have that

$$\not{p} = \not{p}_+ + \not{p}_- \qquad \not{p}_\pm = |p_\pm\rangle\langle p_\pm|. \tag{10.3.33}$$

10.4 The helicity theorem for massless fermions

Because our primary interest is in QCD, QED and the V–A electroweak theory we consider massless fermions coupled to vector bosons ($\gamma, Z^0, W^\pm, G$) via γ^μ or $\gamma^\mu\gamma_5$ vertices only. There follows a remarkable and powerful result. Consider *any* Feynman diagram, no matter how complicated, in which a fermion line enters in the initial state, continues through the diagram and emerges in the final state, as shown in Fig. 10.1.

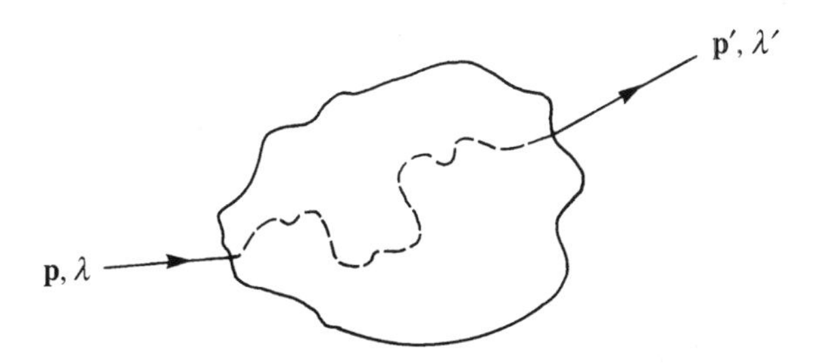

Fig. 10.1 Arbitrary Feynman diagram with fermion line connecting initial and final states.

Label the chiralities of the initial and final fermion f by λ and λ' and call the amplitude $A_{\lambda'\lambda}$. We shall prove that

$$A_{\lambda'\lambda} = 0 \qquad \text{if } \lambda' \neq \lambda. \tag{10.4.1}$$

Focus on the vertices attached to the fermion line under consideration, as shown in Fig. 10.2, where Γ^μ is either γ^μ or $\gamma^\mu\gamma_5$.

Ignoring the denominators of the propagators, the fermion line has associated with it the expression

$$L_{\lambda'\lambda} \equiv \bar{u}_{\lambda'}(\mathbf{p}')\Gamma^{\mu_{n+1}} \not{p}_n \Gamma^{\mu_n} \cdots \not{p}_2 \Gamma^{\mu_2} \not{p}_1 \Gamma^{\mu_1} u_\lambda(\mathbf{p}). \tag{10.4.2}$$

Now replace u and $\bar{u}$ using the fact that for chirality, see (4.6.52),

$$\begin{aligned} u_\lambda(\mathbf{p}) &= \eta_\lambda \gamma_5 u_\lambda(p) \\ \bar{u}_{\lambda'}(\mathbf{p}') &= -\eta_{\lambda'} \bar{u}_{\lambda'}(\mathbf{p}')\gamma_5 \end{aligned} \qquad \eta_\pm = \pm 1 \tag{10.4.3}$$

and commute the rightmost γ_5 through all the Γ^{μ_j} and $\not{p}_j$ until it hits the leftmost γ_5, yielding $\gamma_5^2 = 1$. Now, γ_5 anticommutes with both $\not{p}_j$ and γ^μ or $\gamma^\mu\gamma_5$ so we end up with

$$L_{\lambda'\lambda} = (-1)^N(-\eta_{\lambda'}\eta_\lambda)L_{\lambda'\lambda} \tag{10.4.4}$$

where N is the number of commutations involved. It is easy to see that

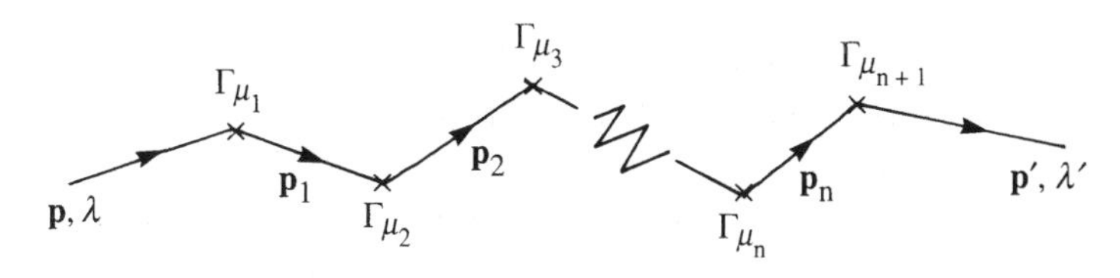

Fig. 10.2. Vertex structure along the fermion line in Fig. 10.1.

N is always an odd number, so that

$$L_{\lambda'\lambda} = \eta_{\lambda'}\eta_{\lambda}L_{\lambda'\lambda} \tag{10.4.5}$$

implying $\eta_{\lambda'}\eta_{\lambda} = +1$ if $L_{\lambda'\lambda} \neq 0$.

Thus we can only have

$$\lambda' = \lambda. \tag{10.4.6}$$

The same result holds for an antifermion passing through the diagram.

For a fermion line that begins and ends in the initial state (i.e. $f\bar{f}$ annihilation) or in the final state (i.e. $f\bar{f}$ production), one finds that the amplitude is zero unless

$$\lambda' = -\lambda. \tag{10.4.7}$$

The conditions for non-zero amplitude are summarized in the diagram in Fig. 10.3.

10.5 Spin structure from a fermion line

Consider the massless fermion line discussed in the previous section (Fig. 10.2) but with all vertices Γ^μ representing γ^μ only. We define the *spin string* associated with it as the ordered product of spinors, propagator factors $\not{p}_j$ and vertices, leaving out all denominators and factors of i. We indicate such a string by the initial and final spinor involved, with a long dash between them. Thus for $\lambda = +1$

$$\begin{aligned}\bar{u}_+(\mathbf{p}')\text{—}u_+(\mathbf{p}) &= \bar{u}_+(\mathbf{p}')\gamma^{\mu_{n+1}}\not{p}_n\cdots\not{p}_1\gamma^{\mu_1}u_+(\mathbf{p})\\ &= \bar{u}_+(\mathbf{p}')\gamma^{\mu_{n+1}}\left(\frac{1+\gamma_5}{2}\not{p}_n\right)\cdots\left(\frac{1+\gamma_5}{2}\not{p}_1\right)\gamma^{\mu_1}u_+(\mathbf{p})\end{aligned} \tag{10.5.1}$$

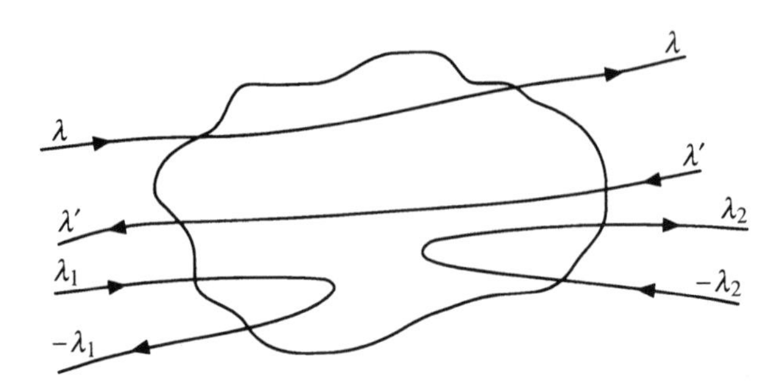

Fig. 10.3. Helicity rules for fermion lines in an arbitrary Feynman diagram.

where we have used the fact that

$$u_+(\mathbf{p}) = \tfrac{1}{2}(1+\gamma_5)u_+(\mathbf{p}) \qquad \text{and} \qquad \left[\tfrac{1}{2}(1+\gamma_5)\right]^2 = \tfrac{1}{2}(1+\gamma_5). \quad (10.5.2)$$

Thus

$$\bar{u}_+(\mathbf{p}')—u_+(\mathbf{p}) = \langle p'_+|\gamma^{\mu_{n+1}}\not{p}^n_+ \cdots \not{p}^1_+\gamma^{\mu_1}|p_+\rangle. \quad (10.5.3)$$

We note that each internal p_j^μ will generally not be a null vector. However, it will always be expressible as a sum of null vectors (in a trivial way in a tree diagram). So each $\not{p}_j$ will give rise, via (10.3.33), to terms of the form $|q_+\rangle\langle q_+|$ with $q^2 = 0$ and the spin string (10.5.3) will be made up of a sum of factors all of the form

$$\langle p'_+|\gamma^{\mu_{n+1}}|q_+\rangle\langle q_+|\gamma^{\mu_n}|r_+\rangle \cdots \langle t_+|\gamma^{\mu_1}|p_+\rangle \quad (10.5.4)$$

with, of course, $q^2 = r^2 = \cdots = t^2 = 0$.

For a string with $\lambda = -1$ we have an analogous expression, except that every factor $\langle r_+|\gamma^\mu|s_+\rangle$ is replaced by $\langle r_-|\gamma^\mu|s_-\rangle$.

For an antifermion line, if the internal momentum labels refer to the flow of physical momentum and are thus directed opposite to the flow of fermion number, the spin string in Fig. 10.4 will be

$$\bar{v}_\lambda(\mathbf{p})—v_{\lambda'}(\mathbf{p}') = \bar{v}_\lambda(\mathbf{p})\gamma^{\mu_1}(-\not{p}_1)\gamma^{\mu_2}(-\not{p}_2)\cdots(-\not{p}_n)\gamma^{\mu_{n+1}}v_{\lambda'}(\mathbf{p}'). \quad (10.5.5)$$

Using (10.3.5) and (10.3.25) one finds

$$\begin{aligned}\bar{v}_\lambda(\mathbf{p})—v_{\lambda'}(\mathbf{p}') &= (-i)^n\left[\bar{u}_{-\lambda'}(\mathbf{p}')—u_{-\lambda}(\mathbf{p})\right]^* \\ &= (-i)^n\left[\bar{u}_{\lambda'}(\mathbf{p}')—u_\lambda(\mathbf{p})\right], \end{aligned} \quad (10.5.6)$$

by (10.3.21).

If some of the vertices, say m of them, are axial-vector, i.e. $\gamma^\mu\gamma_5$, then in (10.5.6) there is an additional factor $(-1)^m$.

The above is sufficient to deal with all processes not involving external vector mesons. We shall illustrate the simplicity of the approach by an example.

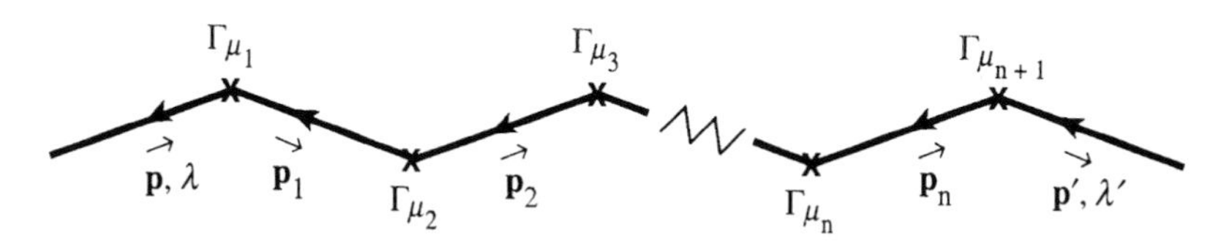

Fig. 10.4. Antifermion line giving rise to the spin string of (10.5.5).

10.6 Example: high energy $e^- + \mu^- \to e^- + \mu^-$

We work in the CM and ignore the lepton masses since it is assumed that $E \gg m$ for all of them. The momentum vectors must be specified with care (see Section 4.1). For the initial particles we take $p^\mu_{\text{electron}} \equiv p^\mu = (p, \mathbf{p})$ with polar coordinates $\mathbf{p} = (p, 0, 0)$, and $p^\mu_{\text{muon}} = (p, -\mathbf{p}) = \tilde{p}^\mu$, as defined in (10.3.18) and (10.3.19). For the final particles, $p'^\mu_{\text{electron}} \equiv p'^\mu = (p, \mathbf{p}')$ with $\mathbf{p} = (p', \theta, \phi)$ and $p'^\mu_{\text{muon}} = (p, -\mathbf{p}') = \tilde{p}'^\mu$.

The Feynman diagram is shown in Fig. 10.5.

Firstly we take out a factor F coming from couplings, propagator denominators etc., using the standard Feynman rules for QED:

$$F = (-ie)^2 \left(-\frac{i}{k^2}\right) = \frac{ie^2}{k^2}. \tag{10.6.1}$$

The Feynman amplitudes are then as follows:

$$\begin{aligned} M_{++;++} &= F\langle p'_+|\gamma^\mu|p_+\rangle\langle\tilde{p}'_+|\gamma_\mu|\tilde{p}_+\rangle \\ &= 2F\,\langle p'_+|\tilde{p}'_-\rangle\,\langle\tilde{p}_-|p_+\rangle \qquad \text{by (10.3.29)} \\ &= 2F(-2ip)^*(-2ip) \qquad \text{by (10.3.15) and (10.3.20)} \\ &= \frac{8ie^2p^2}{k^2} = 2ie^2\frac{s}{t} \end{aligned} \tag{10.6.2}$$

where, as usual,

$$s = (p + \tilde{p})^2 = 4p^2 \qquad t = (p - p')^2 = k^2; \tag{10.6.3}$$

$$\begin{aligned} M_{+-;+-} &= F\langle p'_+|\gamma^\mu|p_+\rangle\langle\tilde{p}'_-|\gamma_\mu|\tilde{p}_-\rangle \\ &= 2F\,\langle p'_+|\tilde{p}_-\rangle\,\langle\tilde{p}'_-|p_+\rangle \qquad \text{by (10.3.28)} \\ &= -\,2F\,\langle\tilde{p}'_-|p_+\rangle^2 \qquad \text{by (10.3.22) and (10.3.13)} \\ &= -\,2F(2ip\cos\theta/2)^2 \qquad \text{by (10.3.9)} \\ &= \frac{8ie^2p^2\cos^2\theta/2}{k^2} = -2ie^2\frac{u}{t} \end{aligned} \tag{10.6.4}$$

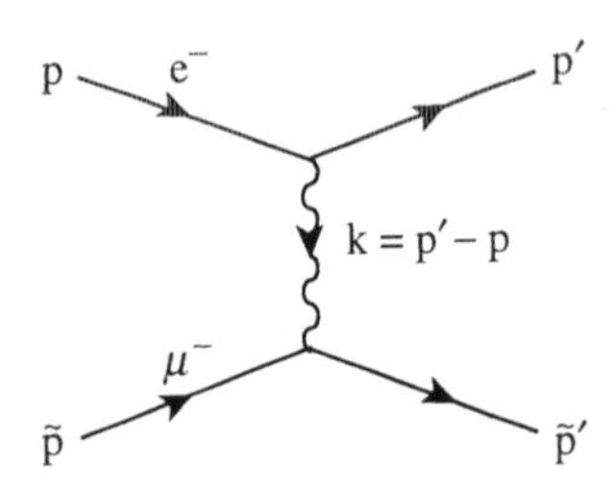

Fig. 10.5. Feynman diagram for $e^- + \mu^- \to e^- + \mu^-$.

where

$$u = (p - \tilde{p}')^2 = -2p^2(1 + \cos\theta). \tag{10.6.5}$$

The remaining non-zero amplitudes are, by parity invariance (see subsection 4.2.1),

$$M_{--;--} = M_{++;++} \quad \text{and} \quad M_{-+;-+} = M_{+-;+-}. \tag{10.6.6}$$

It is then a trivial matter to compute cross-sections, polarizations, spin correlations etc. using (5.6.3) and (5.6.4). Note that if we are only interested in the cross-sections we can immediately use results such as $|\langle\tilde{p}'_-|p_+\rangle|^2 = 2p \cdot \tilde{p}'$, etc.

Let us now consider reactions with external photons or gluons. To begin with, we return to massive spinors and relate them to massless ones.

10.7 Massive spinors

Let $P^\mu = (E, \mathbf{p})$ be a time-like 4-vector with $P^2 = m^2 \neq 0$. With P^μ we associate two null vectors

$$p^\mu \equiv (p, \mathbf{p}) \tag{10.7.1}$$

and its conjugate

$$\tilde{p}^\mu \equiv (p, -\mathbf{p}), \tag{10.7.2}$$

with polar angles as in (10.3.19).

Then from (A12.44), in the Weyl representation,

$$u\left(P, \frac{\lambda}{2}\right) = \frac{1}{\sqrt{2(E+m)}} \begin{pmatrix} E + m + p\lambda \\ E + m - p\lambda \end{pmatrix} \chi_{\lambda/2}(\hat{\mathbf{p}}); \tag{10.7.3}$$

here $\lambda = \pm 1$ corresponds to *helicity* $\pm 1/2$.

Now, for massless spinors

$$\begin{aligned} |p_+\rangle &= \sqrt{2p}\begin{pmatrix} 1 \\ 0 \end{pmatrix} \chi_{1/2}(\hat{\mathbf{p}}) \\ |\tilde{p}_-\rangle &= \sqrt{2p}\begin{pmatrix} 0 \\ 1 \end{pmatrix} \chi_{-1/2}(-\hat{\mathbf{p}}), \end{aligned} \tag{10.7.4}$$

but from (A12.41) one finds

$$\chi_{\lambda/2}(-\hat{\mathbf{p}}) = i\chi_{-\lambda/2}(\hat{\mathbf{p}}). \tag{10.7.5}$$

Thus we can write

$$u_{1/2}(P) = \alpha|p_+\rangle - i\beta|\tilde{p}_-\rangle \tag{10.7.6}$$

where

$$\begin{aligned}\alpha(p) &\equiv \frac{E+m+p}{2\sqrt{p(E+m)}}\\ \beta(p) &\equiv \frac{E+m-p}{2\sqrt{p(E+m)}}.\end{aligned} \tag{10.7.7}$$

Similarly one finds

$$\begin{aligned}u_{-1/2}(P) &= \alpha|p_-\rangle - i\beta|\tilde{p}_+\rangle\\ v_{1/2}(P) &= \alpha|p_-\rangle + i\beta|\tilde{p}_+\rangle\\ v_{-1/2}(P) &= \alpha|p_+\rangle + i\beta|\tilde{p}_-\rangle.\end{aligned} \tag{10.7.8}$$

Note that for $E \gg m$ one has, to leading order in m/E,

$$\left.\begin{aligned}u_{\lambda/2}(P) &= |p_\lambda\rangle - \frac{im}{2p}|\tilde{p}_{-\lambda}\rangle\\ v_{\lambda/2}(P) &= |p_{-\lambda}\rangle + \frac{im}{2p}|\tilde{p}_\lambda\rangle\end{aligned}\right\} \quad (p \gg m). \tag{10.7.9}$$

Using these results it is clear that the amplitude for any Feynman diagram can be expressed as a combination of amplitudes with massless external fermions.

For present-day applications we are mainly interested in high energy collisions so that all external fermions can usually be taken to be massless. But care must be exercised in deciding whether the mass term in an internal fermion propagator is important. For example in the diagram in Fig. 10.6 for $e^-e^+ \to 2\gamma$, for *small momentum transfer* one should keep the full numerator $\not{p} - \not{k} + m$ even at high energies. The term m will induce a non-zero amplitude for annihilation from states of equal helicity or chirality.

10.8 Polarization vectors

Consider a massive vector meson with 4-momentum $K^\mu = (\omega, \mathbf{k}), K^2 = m^2$. As discussed in Section 3.4 the standard polarization vectors for helicity

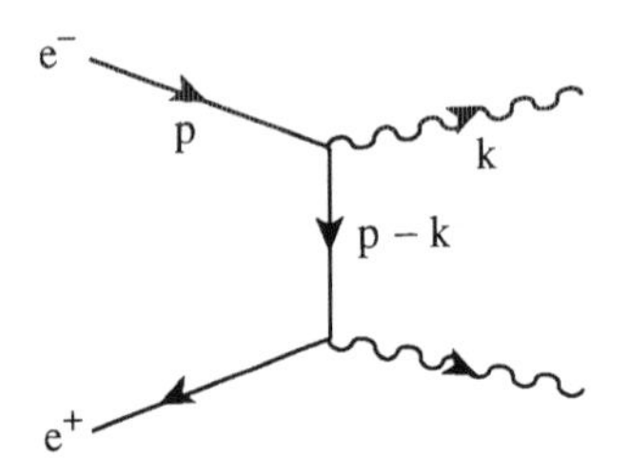

Fig. 10.6. Feynman diagram for $e^-e^+ \to 2\gamma$.

$\lambda = \pm 1, 0$ are

$$\epsilon^{\mu}_{\pm}(K) = \tfrac{1}{\sqrt{2}}(0, \mp \cos\theta\cos\phi + i\sin\phi,$$
$$\mp\cos\theta\sin\phi - i\cos\phi, \pm\sin\theta) \tag{10.8.1}$$
$$\epsilon^{\mu}_{0}(K) = \frac{1}{m}(k, \omega\hat{\mathbf{k}}) \tag{10.8.2}$$

when $\mathbf{k}$ has polar angles θ, ϕ. These are the polarization vectors associated with an *incoming* vector meson. For *outgoing* mesons one uses ϵ^{*}_{λ}. By going to the rest frame, where $K^{\mu} = \overset{\circ}{K}{}^{\mu} = (m, 0, 0, 0)$, one can see how to write the matrices $\not\epsilon(\overset{\circ}{K})$ in terms of products of the rest spinors $u(\overset{\circ}{K}), v(\overset{\circ}{K})$ etc. Then applying the helicity boost $D[h(\mathbf{k})]$, see (A12.24), one eventually obtains an expression for $\not\epsilon(K)$ in terms of massive spinors $u(K), v(K)$ etc. One finds after some labour

$$\not\epsilon_{\lambda=\pm1}(K) = \frac{1}{\sqrt{2}m}\left\{u_{\lambda/2}(K)\bar{v}_{\lambda/2}(K) - v_{-\lambda/2}(K)\bar{u}_{-\lambda/2}(K)\right\} \tag{10.8.3}$$
$$\not\epsilon_{0}(K) = \frac{1}{2m}\left\{u_{1/2}(K)\bar{v}_{-1/2}(K) + u_{-1/2}(K)\bar{v}_{1/2}(K)\right.$$
$$\left. + v_{1/2}(K)\bar{u}_{-1/2}(K) + v_{-1/2}(K)\bar{u}_{1/2}(K)\right\}. \tag{10.8.4}$$

Introducing as before the null vectors

$$k^{\mu} = (k, \mathbf{k}) \qquad \tilde{k}^{\mu} = (k, -\mathbf{k}) \tag{10.8.5}$$

and utilizing (10.7.6) and (10.7.8) one eventually finds the very simple result

$$\not\epsilon_{\lambda=\pm1}(K) = \frac{-i}{\sqrt{2}k}\left\{|k_{\lambda}\rangle\langle\tilde{k}_{\lambda}| + |\tilde{k}_{-\lambda}\rangle\langle k_{-\lambda}|\right\}. \tag{10.8.6}$$

For $\not\epsilon_{0}(K)$ it is simpler to write

$$\epsilon^{\mu}_{0}(K) = \frac{1}{2mk}\left\{(\omega + k)k^{\mu} - (\omega - k)\tilde{k}^{\mu}\right\} \tag{10.8.7}$$

so that

$$\not\epsilon_{0}(K) = \frac{1}{2mk}\left\{(\omega + k)\not k - (\omega - k)\not{\tilde{k}}\right\}. \tag{10.8.8}$$

It is important to note that the above forms for $\not\epsilon_{\lambda}$ correspond to the expressions (10.8.1) and (10.8.2) for the standard ϵ^{μ}_{λ}.

For the *massless* case, i.e. for a photon or gluon of 4-momentum $k^{\mu} = (k, \mathbf{k})$, there is of course no helicity-zero state but (10.8.6) continues to hold for helicities ± 1.

In this case, however, it is possible to make use of gauge invariance to choose other forms for ϵ^{μ} that simplify the calculations.

Let q^μ and q'^μ be any two null vectors such that $q \cdot k \neq 0$, $q' \cdot k \neq 0$. Then from (10.3.30) we can write

$$|\tilde{k}_{-\lambda}\rangle = \frac{\langle q_\lambda|\tilde{k}_{-\lambda}\rangle}{\langle q_\lambda|k_{-\lambda}\rangle}|k_{-\lambda}\rangle + \frac{\langle k_\lambda|\tilde{k}_{-\lambda}\rangle}{\langle k_\lambda|q_{-\lambda}\rangle}|q_{-\lambda}\rangle \tag{10.8.9}$$

$$\langle\tilde{k}_\lambda| = \langle k_\lambda|\frac{\langle q'_\lambda|\tilde{k}_{-\lambda}\rangle}{\langle q'_\lambda|k_{-\lambda}\rangle} + \langle q'_\lambda|\frac{\langle k_\lambda|\tilde{k}_{-\lambda}\rangle}{\langle k_\lambda|q'_{-\lambda}\rangle}; \tag{10.8.10}$$

substituting in (10.8.6) and using

$$\langle k_\lambda|\tilde{k}_{-\lambda}\rangle = 2ik \tag{10.8.11}$$

we get, for $\lambda = \pm 1$,

$$\begin{aligned}\not{\epsilon}_\lambda(k) = \sqrt{2}&\left(\frac{|k_\lambda\rangle\langle q'_\lambda|}{\langle k_\lambda|q'_{-\lambda}\rangle} + \frac{|q_{-\lambda}\rangle\langle k_{-\lambda}|}{\langle k_\lambda|q_{-\lambda}\rangle}\right)\\ &- \frac{i}{\sqrt{2}k}\left(\frac{\langle q'_\lambda|\tilde{k}_{-\lambda}\rangle}{\langle q'_\lambda|k_{-\lambda}\rangle}|k_\lambda\rangle\langle k_\lambda| + \frac{\langle q_\lambda|\tilde{k}_{-\lambda}\rangle}{\langle q_\lambda|k_{-\lambda}\rangle}|k_{-\lambda}\rangle\langle k_{-\lambda}|\right).\end{aligned} \tag{10.8.12}$$

In (10.8.12) let us first choose $q'^\mu = q^\mu$; then the second group of terms becomes

$$-\frac{i}{\sqrt{2}k}\frac{\langle q_\lambda|\tilde{k}_{-\lambda}\rangle}{\langle q_\lambda|k_{-\lambda}\rangle}\left(|k_\lambda\rangle\langle k_\lambda| + |k_{-\lambda}\rangle\langle k_{-\lambda}|\right) = -\frac{i\langle q_\lambda|\tilde{k}_{-\lambda}\rangle}{\sqrt{2}k\,\langle q_\lambda|k_{-\lambda}\rangle}\not{k} \tag{10.8.13}$$

by (10.3.33) and (10.4.1). It thus corresponds to a term proportional to k^μ and may be discarded if gauge invariance allows the substitution $\epsilon^\mu \to \epsilon^\mu + ck^\mu$. When this is so we may therefore simply use the expression

$$\not{\epsilon}_\lambda(k;q) \equiv \frac{\sqrt{2}}{\langle k_\lambda|q_{-\lambda}\rangle}\left(|k_\lambda\rangle\langle q_\lambda| + |q_{-\lambda}\rangle\langle k_{-\lambda}|\right) \tag{10.8.14}$$

where q^μ is *any* null vector for which $q \cdot k \neq 0$. In this expression q should be thought of as a reference vector specifying a family of equivalent polarization vectors.

The polarization vector that corresponds to the expression (10.8.14) is clearly

$$\epsilon^\mu_\lambda(k;q) = \epsilon^\mu_\lambda(k) + \frac{i\langle q_\lambda|\tilde{k}_{-\lambda}\rangle}{\sqrt{2}k\,\langle q_\lambda|k_{-\lambda}\rangle}k^\mu. \tag{10.8.15}$$

There is a very useful form for the $\epsilon^\mu_\lambda(k;q)$, which can be obtained via the relation

$$\epsilon^\mu_\lambda(k;q) = \tfrac{1}{4}\,\mathrm{Tr}\,\left[\gamma^\mu\not{\epsilon}_\lambda(k;q)\right].$$

Using (10.8.14) and (10.3.26) one finds

$$\epsilon^\mu_\lambda(k;q) = \frac{\langle q_\lambda|\gamma^\mu|k_\lambda\rangle}{\sqrt{2}\,\langle k_\lambda|q_{-\lambda}\rangle}. \tag{10.8.16}$$

This is particularly helpful when evaluating scalar products of ϵ^μ_λ with some other 4-vector.

It follows that polarization vectors specified by different reference vectors q differ only by a term proportional to the 4-vector k.

It can be checked via (10.3.9) that

$$q_\mu \epsilon^\mu_\lambda(k;q) = 0. \tag{10.8.17}$$

This implies that using the form (10.8.14) for $\not\epsilon_\lambda(k;q)$ is similar to working in an axial gauge $A^a_\mu q^\mu = 0$, which is convenient for ladder-type diagrams. Other useful properties are

$$\epsilon_\lambda(k_1;q)\cdot\epsilon_\lambda(k_2;q) = 0 \tag{10.8.18}$$

$$\epsilon^*_\lambda(k_1;k_2)\cdot\epsilon_\lambda(k_2;q) = 0 \qquad \text{or} \qquad \epsilon_{-\lambda}(k_1;k_2)\cdot\epsilon_\lambda(k_2;q) = 0. \tag{10.8.19}$$

It is crucial, in a non-abelian gauge theory, where the gauge mesons couple to themselves, to remember that (10.8.15) or (10.8.16) *must* be used in conjunction with (10.8.14), if one wishes to work with the vector ϵ^μ_λ itself.

It is easily checked that, as usual,[1]

$$\epsilon^{\mu *}_\lambda(k;q) = -\epsilon^\mu_{-\lambda}(k;q). \tag{10.8.20}$$

We see now that the standard polarization vectors given in (10.8.6) just correspond to the 4-vector choice $q = \tilde{k}$ in (10.8.14).

The standard form (10.8.6) is adequate for all $2 \to 2$ reactions. For multiparticle production a judicious choice of the reference vector q may simplify the calculation.

Let us return now to the more general expression (10.8.12) in the case where $q'^\mu \neq q^\mu$. We cannot, in general, discard the second group of terms, since, via (10.3.33) and (10.3.32), it contains both $\not k$ and $\gamma_5 \not k$ and thus does not correspond to adding a vector ck^μ to ϵ^μ.

However, in massless QED, in any gauge-invariant subset of diagrams a given photon is attached to one single fermion line. In that case ϵ^μ_λ enters only in the form $\not\epsilon_\lambda$ and the γ_5 is innocuous since it will act on a massless

[1] Note that contrary to all other textbooks on field theory, Mangano and Parke (1991) uses $\epsilon_\lambda(k)$ for *outgoing* photons and gluons. Moreover, in their phase convention (10.8.20) holds with a plus sign on the right-hand side.

spinor and convert itself into ± 1. Thus in massless QED one can use a two-parameter family of polarization vectors

$$\not\epsilon_\lambda(k;q,q') = \sqrt{2}\left(\frac{|k_\lambda\rangle\langle q'_\lambda|}{\langle k_\lambda|q'_{-\lambda}\rangle} + \frac{|q_{-\lambda}\rangle\langle k_{-\lambda}|}{\langle k_\lambda|q_{-\lambda}\rangle}\right). \tag{10.8.21}$$

Spectacular simplifications ensue from a judicious choice of the 4-vectors q, q', usually from choosing q, q' equal to the initial and final momenta of the fermion line to which the photon is attached.

Before looking at some examples we shall introduce a shorthand notation for the spinor products.

10.9 Shorthand notation for spinor products

To simplify the expressions that occur in the calculation of the amplitudes we introduce, for positive-energy null vectors $a, b, c, \ldots$,

$$\langle ab\rangle \equiv \langle a_-|b_+\rangle \tag{10.9.1}$$

$$[ab] \equiv \langle a_+|b_-\rangle\,. \tag{10.9.2}$$

Equations (10.3.13) to (10.3.22) and (10.3.31) then become

$$\langle ab\rangle = -\langle ba\rangle \qquad [ab] = -[ba] \tag{10.9.3}$$

$$[ab] = \langle ba\rangle^* = -\langle ab\rangle^* \tag{10.9.4}$$

$$\langle ab\rangle\,[ba] = 2a\cdot b \tag{10.9.5}$$

$$\langle \tilde{a}a\rangle = -2ia_0 \tag{10.9.6}$$

$$\left\langle \tilde{a}\tilde{b}\right\rangle = [ab] \tag{10.9.7}$$

$$\left\langle a\tilde{b}\right\rangle = -[\tilde{a}b] \tag{10.9.8}$$

$$\langle ab\rangle\,\langle cd\rangle = \langle ad\rangle\,\langle cb\rangle + \langle ac\rangle\,\langle bd\rangle\,. \tag{10.9.9}$$

For the polarization vectors (10.8.16) one has

$$\epsilon_+^\mu(a;b) = \frac{\langle b_+|\gamma^\mu|a_+\rangle}{\sqrt{2}[ab]} \tag{10.9.10}$$

$$\epsilon_-^\mu(a;b) = \frac{\langle b_-|\gamma^\mu|a_-\rangle}{\sqrt{2}\,\langle ab\rangle} \tag{10.9.11}$$

and, for scalar products involving polarization vectors,

$$\epsilon_+(a;b)\cdot\epsilon_+(c;d) = \frac{\langle ac\rangle\,[db]}{[cd]\,[ab]} \tag{10.9.12}$$

$$\epsilon_+(a;b)\cdot\epsilon_-(c;d) = \frac{\langle ad\rangle\,[cb]}{\langle cd\rangle\,[ab]} \tag{10.9.13}$$

$$\epsilon_+(a;b)\cdot k = \frac{\langle ak\rangle\,[kb]}{\sqrt{2}[ab]} \tag{10.9.14}$$

$$\epsilon_-(a;b)\cdot k = \frac{[ak]\,\langle kb\rangle}{\sqrt{2}\,\langle ab\rangle}. \tag{10.9.15}$$

We shall illustrate these techniques by two examples.

10.10 QED: high energy Compton scattering

The lowest-order diagrams are shown in Fig. 10.7. We work in the CM, hence $|\mathbf{k}| = |\mathbf{p}| \equiv k$ and we have the 4-vector association $k = \tilde{p}$, $k' = \tilde{p}'$. We take

$$p^\mu = (p, 0, 0, p) \qquad p'^\mu = (p, \mathbf{p}) \quad \text{with} \quad \mathbf{p} = (p, \theta, 0).$$

While neither diagram is gauge invariant, their sum is, so we may make use of our freedom in choosing the form of the polarization vectors. We may utilize the very general form (10.8.21), in which, with an eye to the structure of the diagrams, we take $q'^\mu = p^\mu$, $q^\mu = p'^\mu$.

For the incoming photon we then have

$$\begin{aligned}\not{\epsilon}_\lambda(k) &\equiv \not{\epsilon}_\lambda(k; p', p)\\ &= \sqrt{2}\left(\frac{|k_\lambda\rangle\langle p_\lambda|}{\langle k_\lambda|p_{-\lambda}\rangle} + \frac{|p'_{-\lambda}\rangle\langle k_{-\lambda}|}{\langle k_\lambda|p'_{-\lambda}\rangle}\right).\end{aligned} \tag{10.10.1}$$

For the outgoing photon we must use $\epsilon_\lambda^{\mu *}(k')$. We shall denote $\gamma_\mu \epsilon_\lambda^{\mu *}(k')$ by $\not{\epsilon}^*_\lambda(k')$. Then via (10.8.20)

$$\begin{aligned}\not{\epsilon}^*_\lambda(k') &= -\not{\epsilon}_{-\lambda}(k')\\ &= -\sqrt{2}\left(\frac{|k'_{-\lambda}\rangle\langle p_{-\lambda}|}{\langle k'_{-\lambda}|p_\lambda\rangle} + \frac{|p'_\lambda\rangle\langle k'_\lambda|}{\langle k'_{-\lambda}|p'_\lambda\rangle}\right).\end{aligned} \tag{10.10.2}$$

We may start with helicity $+1/2$ for the initial fermion, and we know from (10.4.6) that the final helicity must then be $+1/2$ also.

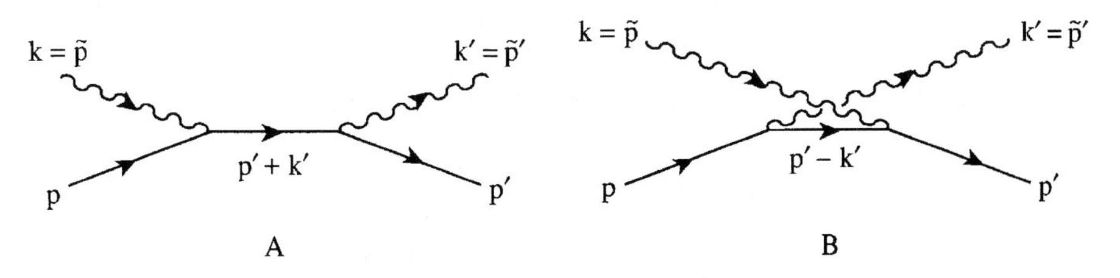

Fig. 10.7. Feynman diagrams for Compton scattering in QED.

From diagram A we have

$$M^{(A)}_{\lambda'+;\lambda+} = (-ie)^2 \frac{i}{2p\cdot k} \langle p'_+|\not\epsilon^*_{\lambda'}(k')(\not p' + \not k')\not\epsilon_\lambda(k)|p_+\rangle. \tag{10.10.3}$$

By (10.3.8) and (10.3.17)

$$\not\epsilon_-(k)|p_+\rangle = 0 \qquad \langle p'_+|\not\epsilon^*_-(k') = 0. \tag{10.10.4}$$

Thus the only independent non-zero amplitude from diagram A is

$$\begin{aligned} M^{(A)}_{1+;1+} &= -\frac{(-ie)^2 2i}{2p\cdot k} \frac{[p'k']\,\langle pk'\rangle\,[k'p']\,\langle kp\rangle}{\langle k'p\rangle\,[kp']} \\ &= (-ie)^2 \frac{i}{p\cdot k} \frac{[p'k'][k'p']\,\langle kp\rangle}{[kp']} \qquad \text{by (10.9.3)} \\ &= -\frac{ie^2}{2p^2} \frac{[p'\tilde p'][\tilde p'p']\,\langle \tilde p p\rangle}{[kp']} \\ &= -\frac{ie^2}{2p^2} \frac{(2ip)(-2ip)(-2ip)}{(-2ip\cos\theta/2)} \end{aligned}$$

by (10.9.3), (10.9.6), (10.9.8) and (10.3.9). Thus

$$M^{(A)}_{1+;1+} = \frac{-2ie^2}{\cos\theta/2} = -2ie^2\sqrt{\frac{s}{-u}}. \tag{10.10.5}$$

(The singularity at $\theta = \pi$ is, of course, an artifact of our having neglected the fermion mass in the Feynman denominators.)

From diagram B we have

$$M^{(B)}_{\lambda'+;\lambda+} = (-ie)^2 \frac{i}{-2p\cdot k'} \langle p'_+|\epsilon_\lambda(k)(\not p - k')\not\epsilon^*_{\lambda'}(k')|p_+\rangle \tag{10.10.6}$$

and now we see that

$$\not\epsilon^*_+(k')|p_+\rangle = 0 \qquad \langle p_+|\not\epsilon_+(k) = 0. \tag{10.10.7}$$

So diagram B does not contribute to $M_{1+;1+}$; for B the only independent non-zero amplitude is

$$\begin{aligned} M^{(B)}_{-1+;-1+} &= \frac{2ie^2}{2p\cdot k'} \frac{[p'k]\,\langle pk'\rangle\,[k'p']\,\langle k'p\rangle}{\langle kp\rangle\,[k'p']} \\ &= \frac{ie^2}{p\cdot k'} \frac{[p'\tilde p]\,\langle p\tilde p'\rangle\,\langle \tilde p'p\rangle}{(-2ip)} \\ &= \frac{-e^2}{2pp\cdot k'} \langle p\tilde p'\rangle^3 = 2ie^2\cos\frac{\theta}{2} \\ &= 2ie^2\sqrt{\frac{-u}{s}}. \end{aligned} \tag{10.10.8}$$

The amplitudes for negative helicity fermions can be obtained by the parity rules in subsection 4.2.1.

The above approach is much simpler and shorter than the conventional one, both for the cross-section and for spin-dependent observables. It is typical of the method that some helicity amplitudes receive contributions from one diagram only.

10.11 QCD: gluon Compton scattering

An important process is

$$G + q \rightarrow G + q,$$

which is the QCD analogue of electromagnetic Compton scattering. There is now an extra diagram in lowest order arising from the triple gluon coupling, as shown in Fig. 10.8, where i, j and a, b are colour labels.

The kinematic structure of Fig. 10.8, parts A and B, is exactly the same as in the QED diagrams A and B in Fig. 10.7. From Appendix 11 we see that

$$M^{(A)}_{\text{QCD}} = (t^b t^a)_{ji} \tilde{M}^{(A)} \qquad (10.11.1)$$

$$M^{(B)}_{\text{QCD}} = (t^a t^b)_{ji} \tilde{M}^{(B)} \qquad (10.11.2)$$

where

$$\tilde{M}^{(A)} = M^{(A)}_{\text{QED}}(e^2 \rightarrow g^2) \qquad \tilde{M}^{(B)} = M^{(B)}_{\text{QED}}(e^2 \rightarrow g^2). \qquad (10.11.3)$$

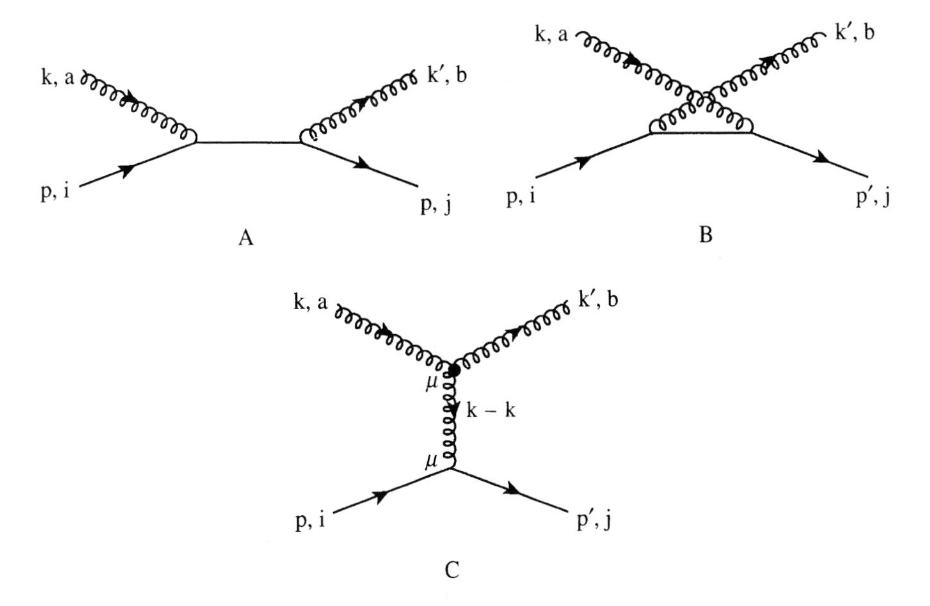

Fig. 10.8. Feynman diagrams for gluon Compton scattering.

In QED, when $\epsilon^\mu(k)$ is replaced by k^μ the gauge invariance is achieved by a cancellation between $M^{(\mathrm{A})}_{\mathrm{QED}}$ and $M^{(\mathrm{B})}_{\mathrm{QED}}$. Clearly, in the QCD case, because of the different colour structure this can no longer happen and gauge invariance is reinstated only when diagram C in Fig. 10.8 is included.

The colour dependent factor in diagram C is $f_{cba}(t^c)_{ji}$. But the fundamental Lie-algebra commutation relation is

$$[t^a, t^b] = if_{abc}t^c \tag{10.11.4}$$

so that we can make the replacement

$$f_{cba}(t^c)_{ji} = i(t^a t^b)_{ji} - i(t^b t^a)_{ji}. \tag{10.11.5}$$

The linearly independent invariant tensors in colour space may thus be taken as $(t^a t^b)_{ji}$ and $(t^b t^a)_{ji}$; diagram C contributes to both. Writing

$$M^{(\mathrm{C})} = f_{cba}(t^c)_{ji}\tilde{M}^{(\mathrm{C})} \tag{10.11.6}$$

we have, for the total amplitude,

$$M = (t^b t^a)_{ji}\left[\tilde{M}^{(\mathrm{A})} - i\tilde{M}^{(\mathrm{C})}\right] + (t^a t^b)_{ji}\left[\tilde{M}^{(\mathrm{B})} + i\tilde{M}^{(\mathrm{C})}\right] \tag{10.11.7}$$

and we may use different choices of polarization vectors in evaluating the combinations in the first and second pairs of square brackets.

For the first term in (10.11.7) we use (10.8.14), with $q^\mu = p^\mu$ for the incoming gluon and $q^\mu = p'^\mu$ for the outgoing gluon. Let us call this gauge 1. Taking the quark helicity to be $+1/2$, by methods similar to the above one finds that diagram A only contributes to $M_{1+;1+}$, and $\tilde{M}^{(\mathrm{A})}_{1+;1+} = -2ig^2\cos\theta/2$.

For the contribution of the three-gluon vertex in diagram C one has, from Appendix 11, aside from the factor gf_{cba},

$$\begin{aligned} V^\mu_{\lambda'\lambda} &= 2k'\cdot\epsilon_\lambda(k)\epsilon^{\mu\,*}_{\lambda'}(k') - (k+k')^\mu\epsilon^*_{\lambda'}(k')\cdot\epsilon_\lambda(k) \\ &\quad + 2k\cdot\epsilon^*_{\lambda'}(k')\epsilon^\mu_\lambda(k) \end{aligned} \tag{10.11.8}$$

leading to

$$\tilde{M}^{(\mathrm{C})}_{\lambda'+;\lambda+} = -\frac{g^2}{2k\cdot k'}\langle p'_+|\not{V}_{\lambda'\lambda}|p_+\rangle, \tag{10.11.9}$$

which, after some straightforward algebra, yields

$$\begin{aligned} &\tilde{M}^{(\mathrm{C})}_{\lambda'+;\lambda+}\Big|_{\mathrm{gauge1}} \\ &= \frac{4g^2 p}{k\cdot k'}\left\{\frac{\sin\theta/2}{\sqrt{2}}\left[\delta_{\lambda'1}k'\cdot\epsilon_\lambda(k;p) - \delta_{\lambda 1}k\cdot\epsilon^*_{\lambda'}(k';p')\right] + p(\cos\theta/2)\epsilon^*_{\lambda'}\cdot\epsilon_\lambda\right\} \\ &= -2g^2(\cos\theta/2)\left[\lambda\delta_{\lambda'1} + \lambda'\delta_{\lambda 1} + \delta_{\lambda'-\lambda} + (\cot^2\theta/2)\delta_{\lambda\lambda'}\right] \\ &= -2g^2(\cos\theta/2)\left[2\delta_{\lambda 1} + \cot^2\theta/2\right]\delta_{\lambda\lambda'}; \end{aligned} \tag{10.11.10}$$

to obtain the penultimate expression we have used (10.8.1) for the polarization vectors.

Thus there is no gluon helicity-flip. For the $1+ \to 1+$ amplitude from diagram C we have

$$\tilde{M}^{(C)}_{1+;1+}\Big|_{\text{gauge1}} = -2g^2(\cos\theta/2)\left(2+\cot^2\theta/2\right). \tag{10.11.11}$$

and the first term in (10.11.7) becomes

$$(t^b t^a)_{ji} 2ig^2 \frac{\cot\theta/2}{\sin\theta/2}. \tag{10.11.12}$$

For the second term in (10.11.7) we choose polarization vectors with reference vectors $q = p'$ in $\epsilon_\lambda(k;q)$ and $q = p$ in $\epsilon_{\lambda'}(k';q)$. Call this gauge 2. Then diagram B does not contribute to $1+ \to 1+$. For diagram C we now find

$$\begin{aligned}\tilde{M}^{(C)}_{\lambda'+;\lambda+}\Big|_{\text{gauge2}} = \frac{4pg^2}{k\cdot k'}\bigg\{ &\frac{\sin\theta/2}{\sqrt{2}}\left[\delta_{\lambda',-1}k'\cdot\epsilon_\lambda(k;p') - \delta_{\lambda,-1}k\cdot\epsilon^*_{\lambda'}(k';p)\right] \\ &+p(\cos\theta/2)\epsilon^*_{\lambda'}\cdot\epsilon_\lambda\bigg\}\end{aligned} \tag{10.11.13}$$

where, via (10.8.15) we find that the polarization vectors in (10.11.13) are related to those in (10.11.10) as follows:

$$\begin{aligned}\epsilon^\mu_\lambda(k;p') &= \epsilon^\mu_\lambda(k;p) - \frac{\lambda}{\sqrt{2}k}(\tan\theta/2)k^\mu \\ \epsilon^\mu_{\lambda'}(k';p) &= \epsilon^\mu_{\lambda'}(k';p') + \frac{\lambda'}{\sqrt{2}k}(\tan\theta/2)k'^\mu.\end{aligned} \tag{10.11.14}$$

Hence

$$\begin{aligned}\tilde{M}^{(C)}_{1+;1+}\Big|_{\text{gauge2}} &= \frac{2g^2\cos\theta/2}{\sin^2\theta/2}\epsilon^*_+(k';p)\cdot\epsilon_+(k;p') \\ &= -\frac{2g^2}{\cos\theta/2\sin^2\theta/2}.\end{aligned} \tag{10.11.15}$$

Thus, for the contribution to the amplitude $1+ \to 1+$, the second term in (10.11.7) yields

$$-(t^a t^b)_{ji}\frac{2ig^2}{\cos\theta/2\sin^2\theta/2}. \tag{10.11.16}$$

The sum of (10.11.12) and (10.11.16) then gives the complete amplitude for $1+ \to 1+$.

It is an interesting exercise to calculate $\tilde{M}^{(A)}$ in gauge 2. One finds

$$\tilde{M}^{(A)}_{1+;1+}\Big|_{\text{gauge2}} = -\frac{2ig^2}{\cos\theta/2}. \tag{10.11.17}$$

Using (10.11.16) we can now calculate the first term of (10.11.7) in gauge 2, finding for $1+ \to 1+$

$$(t^b t^a)_{ji} \left[\frac{-2ig^2}{\cos\theta/2} - i\frac{-2g^2}{\cos\theta/2 \sin^2\theta/2} \right] = (t^b t^a)_{ji} 2ig^2 \frac{\cot\theta/2}{\sin\theta/2} \tag{10.11.18}$$

exactly as in (10.11.12).

It is a straightforward matter to calculate the other independent amplitude $M_{-1+;-1+}$ in a similar fashion. Note that there is no gluon helicity-flip in any of these amplitudes.

Of course, choosing two different gauges for the above is a somewhat sledgehammer approach in such a simple problem. But in more complicated, higher-order, diagrams great simplification can be achieved.

As emphasized by Cvitanović, Lauwers and Scharbach (1981), certain properties of the gauge-invariant subsets of diagrams become clearer if linear combinations of the invariant colour tensors are used that transform simply under permutations of the symmetric group. For example, in gluon Compton scattering we could utilize

$$\begin{aligned} \left(T_+^{ba}\right)_{ji} &\equiv \tfrac{1}{2}(t^b t^a + t^a t^b)_{ji} \\ \left(T_-^{ba}\right)_{ji} &\equiv \tfrac{1}{2}(t^b t^a - t^a t^b)_{ji} \end{aligned} \tag{10.11.19}$$

in which case (10.11.7) becomes

$$M = \left(T_+^{ba}\right)_{ji} \left[\tilde{M}^{(A)} + \tilde{M}^{(B)}\right] + \left(T_-^{ba}\right)_{ji} \left[\tilde{M}^{(A)} - \tilde{M}^{(B)} - 2i\tilde{M}^{(C)}\right]. \tag{10.11.20}$$

We see that the first term contains only the abelian QED amplitudes. This is a general result. For any number of partons the totally symmetric colour tensor singles out the QED-like contributions to the amplitude and the non-abelian effects are contained in the other-gauge invariant subsets.

10.12 QCD: Multigluon amplitudes

In dealing with purely gluonic reactions it is simpler to deal with the symmetric situations where all the gluons are *incoming*. Let the n gluons labelled $1, 2, \ldots, n$ have colours $a_1, \ldots, a_n$, helicities $\lambda_1, \ldots, \lambda_n$ and momenta $k_1, \ldots, k_n$, respectively. We shall abbreviate the amplitude by

$$M(1, 2, \ldots, n) \equiv M(k_1, \lambda_1, a_1; \ldots; k_n, \lambda_n, a_n). \tag{10.12.1}$$

The contribution to M from each Feynman diagram will be of the form

$$F(a_1, \ldots, a_n)\epsilon_{\mu_1}(k_1; \lambda_1) \ldots \epsilon_{\mu_n}(k_n; \lambda_n) M^{\mu_1 \ldots \mu_n}(k_1, \ldots, k_n) \tag{10.12.2}$$

where $F(a_1, \ldots, a_n)$ is a colour factor.

For the purpose of understanding the helicity structure we may ignore energy and momentum conservation and pretend that all incoming gluons have positive energy in (10.12.2).

The amplitude for a reaction with some outgoing gluons is obtained as follows. Let the jth gluon be outgoing with momentum $\bar{k}_j$ and helicity $\bar{\lambda}_j$. Then the amplitude for

$$G_1 + G_2 + \cdots + G_{j-1} + G_{j+1} + \cdots + G_n \to G_j$$

is given by (see (10.8.20)):

(1) replacing $\epsilon_{\mu_j}(k_j, \lambda_j)$ by

$$\epsilon^*_{\mu_j}(\bar{k}_j, \bar{\lambda}_j) = -\epsilon_{\mu_j}(\bar{k}_j, -\bar{\lambda}_j); \tag{10.12.3}$$

(2) putting $k_j = -\bar{k}_j$ in

$$M^{\mu_1 \ldots \mu_n}(k_1, \ldots, k_n). \tag{10.12.4}$$

Thus as far as helicity structure is concerned:

an ingoing helicity λ is equivalent to an outgoing helicity $-\lambda$. (10.12.5)

10.12.1 The colour structure

Now the colour factors, whether due to three-gluon or four-gluon vertices, always contain typical products like $f_{abe}f_{ecd}$. From (10.11.4) and the fact that

$$\text{Tr}\,(t^a t^b) = \tfrac{1}{2}\delta_{ab} \tag{10.12.6}$$

one has that

$$\text{Tr}\,(t^e[t^c, t^d]) = \frac{i}{2} f_{ecd} \tag{10.12.7}$$

and therefore

$$\begin{aligned} f_{abe}f_{ecd} &= -2i\,\text{Tr}\,(f_{abe}t^e[t^c, t^d]) \\ &= -2i\,\text{Tr}\,([t^a, t^b][t^c, t^d]). \end{aligned} \tag{10.12.8}$$

Ultimately one ends up with traces of products, in all possible permutations of all the t^{a_j}. Since the trace is invariant under cyclic permutations the set of independent colour tensors for tree diagrams is just the set of *non-cyclic* permutations of the trace $\text{Tr}\,(t^{a_1} t^{a_2} \cdots t^{a_n})$. The total Feynman amplitude, as will be seen, then has the structure

$$M = \frac{2}{i^n} \sum_{\text{perm}(23\ldots n)} \text{Tr}\,(t^{a_1} t^{a_2} \cdots t^{a_n}) \times \tilde{M}(k_1, \lambda_1; \ldots; k_n, \lambda_n) \tag{10.12.9}$$

where the momentum- and helicity-dependent amplitudes $\tilde{M}$ are gauge invariant. Each $\tilde{M}$, which is defined by the order of the labels in it, will contain contributions from several Feynman diagrams, as was the case for the $\tilde{M}$ in Section 10.10. The $\tilde{M}$ are calculated from Feynman diagrams in which the colour factor f_{abc} is simply left out at each trilinear gluon vertex. For the quadrilinear gluon vertices, one starts from the modified form

$$\frac{2}{i^2}(-ig^2) \sum_{\text{perm}(234)} \text{Tr}\ (t^{a_1}t^{a_2}t^{a_3}t^{a_4}) \\ \times (2g_{\mu_1\mu_3}g_{\mu_2\mu_4} - g_{\mu_1\mu_4}g_{\mu_2\mu_3} - g_{\mu_1\mu_2}g_{\mu_3\mu_4}), \quad (10.12.10)$$

which, as can be checked, coincides with the usual expression given in Appendix 11. So, in calculating contributions to $\tilde{M}$ from a Feynman diagram involving a quadrilinear vertex one must use, for the cyclic order (1234),

$$-ig^2\left[2g_{\mu_1\mu_3}g_{\mu_2\mu_4} - g_{\mu_1\mu_4}g_{\mu_2\mu_3} - g_{\mu_1\mu_2}g_{\mu_3\mu_4}\right]. \quad (10.12.11)$$

Since the $\tilde{M}(1,2,\ldots,n)$ are gauge invariant, the reference vectors q_j used in specifying the polarization vectors $\epsilon(k_j;q_j)$ can be chosen differently for the calculation of each $\tilde{M}$.

Each q_j will always be equal to one of the k_i, say $k_{f(j)}$, i.e. the polarization vectors will have the form $\epsilon(k_j;k_{f(j)})$.

Let the mapping $i \to P_i$ be a permutation of $i = 1,2,\ldots,n$. If, when we calculate $\tilde{M}(P_1,P_2,\ldots,P_n)$, we utilize the set of polarization vectors $\epsilon(k_{P_j};k_{P_{f(j)}})$ then it is clear that we can evaluate $\tilde{M}(P_1,P_2,\ldots,P_n)$ from the result for $\tilde{M}(1,2,\ldots,n)$ by simply carrying out the permutation $i \to P_i$ in the result.

In subsection 10.12.3 we shall illustrate these rather abstract arguments with a concrete example, the four-gluon amplitude, and in subsection 10.12.6 we give some general properties of n-gluon amplitudes. First, however, it will be helpful to deduce two very powerful rules for the helicity structure of gluon amplitudes.

10.12.2 Helicity structure of the *n*-gluon amplitude

Consider the amplitude for the n-gluon reaction. For a tree diagram consisting solely of trilinear couplings it is easy to see that the number of trilinear vertices, N_V, is related to n by

$$N_V = n - 2. \quad (10.12.12)$$

If there are quadrilinear vertices present in the diagram then for the trilinear vertices one will have

$$N_V < n - 2. \quad (10.12.13)$$

Now, as can be seen from (10.11.8) each trilinear vertex has mass dimension 1, i.e. $[m]$, whereas the quadrilinear vertices, (10.12.11), are dimensionless. Hence the mass dimension of the *numerator* of any Feynman tree diagram for the n-gluon reaction is d_n, where

$$d_n \leq n - 2. \tag{10.12.14}$$

However, the numerator of the amplitude is linear in the n polarization vectors ϵ of the gluons, which can only occur in combinations of the type $\epsilon \cdot p$ of dimension $[m]$, where p is some momentum, or $\epsilon_i \cdot \epsilon_j$, which is dimensionless. It follows from (10.12.14) that the number of factors of the type $\epsilon \cdot p$ must be $\leq n-2$ and therefore *at least one factor of the type $\epsilon_i \cdot \epsilon_j$ must occur*. This simple result has powerful consequences, as follows:

(1) Consider the amplitude where *all gluon helicities are equal*. Now choose the same reference vector q in (10.8.15) to define the polarization vectors for all the gluons. Then by (10.8.18) every scalar product $\epsilon_\lambda(k_i;q) \cdot \epsilon_\lambda(k_j;q) = 0$ and thus the entire amplitude vanishes.
    For a physical scattering reaction, for example for $2G \to nG$, this implies, via (10.12.5), that the amplitude for

$$G_1(-\lambda) + G_2(-\lambda) \to G_3(\lambda) + G_4(\lambda) + \ldots + G_{n+2}(\lambda),$$

   i.e. for a maximum change of helicity, is zero.

(2) Suppose now that one gluon, say the lth, has helicity opposite to all the rest, i.e. $\lambda_j = \lambda$ for all $j \neq l$, $\lambda_l = -\lambda$. Now choose the reference vector $q = k_l$ for all gluons except the lth. Then the only possibility for a non-vanishing amplitude must come from scalar products involving $\epsilon_{-\lambda}(k_l;q')$. But these will be of the form $\epsilon_\lambda(k_j;k_l) \cdot \epsilon_{-\lambda}(k_l;q')$, which vanishes by (10.8.19).

For the physical process $2G \to nG$ this implies, via (10.12.5), that

$$A[G_1(-\lambda) + G_2(-\lambda) \to G_3(\lambda) + \cdots + G_l(-\lambda) + \ldots + G_{n+2}(\lambda)] = 0 \tag{10.12.15}$$

$$A[G_1(\lambda) + G_2(-\lambda) \to G_3(\lambda) + \cdots + G_l(\lambda) + \ldots + G_{n+2}(\lambda)] = 0 \tag{10.12.16}$$

etc.

For the important reaction

$$G + G \to G + G$$

we immediately see that there are two non-zero independent amplitudes, for example $M_{11;11}$ and $M_{1-1;1-1}$. The other non-zero amplitudes are obtained via parity invariance or symmetry arguments.

10.12.3 The amplitude for $G + G \to G + G$

To make full use of the symmetry, let us suppose that all the gluons are incoming. The Feynman diagrams are shown in Fig. 10.9.

Using the result (10.12.8), diagrams A, B and C have the form

$$\begin{aligned} &-2\,\mathrm{Tr}\,(1234 + 4321 - 1243 - 3421)\,\tilde{m}^{(\mathrm{A})} \\ &-2\,\mathrm{Tr}\,(1324 + 4231 - 1243 - 3421)\,\tilde{m}^{(\mathrm{B})} \\ &-2\,\mathrm{Tr}\,(1234 + 4321 - 1423 - 3241)\,\tilde{m}^{(\mathrm{C})} \end{aligned} \tag{10.12.17}$$

where the shorthand notation

$$\mathrm{Tr}\,(t^{a_1}t^{a_2}t^{a_3}t^{a_4}) = \mathrm{Tr}\,(1234) \tag{10.12.18}$$

is used, and the $\tilde{m}^{(\mathrm{A})}$ are the momentum- and helicity-dependent amplitudes calculated without the colour factors f_{abc}.

Now, if we write

$$\tilde{m}^{(\mathrm{A})} \equiv \tilde{m}^{(\mathrm{A})}(1,2,3,4) \tag{10.12.19}$$

etc. then it is clear that

$$\tilde{m}^{(\mathrm{B})} = \tilde{m}^{(\mathrm{A})}(1,3,2,4), \qquad \tilde{m}^{(\mathrm{C})} = \tilde{m}^{(\mathrm{A})}(4,1,2,3). \tag{10.12.20}$$

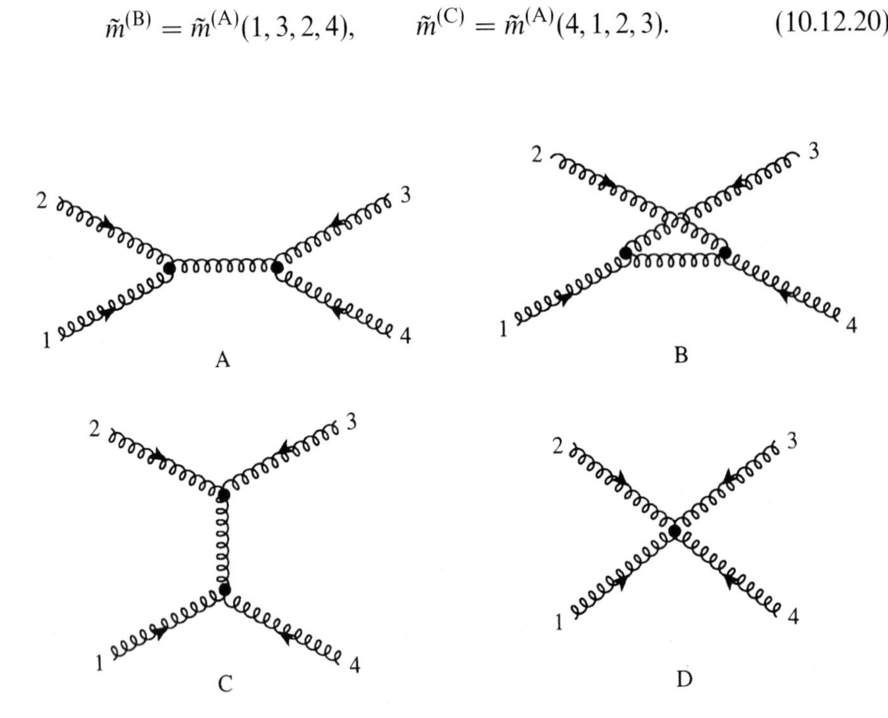

Fig. 10.9. Lowest-order Feynman diagrams for $G + G \to G + G$.

For diagram D, using (10.12.10) we have

$$-2 \sum_{\text{perm }(234)} \text{Tr }(1234)\, \tilde{m}^{(\text{D})}(1,2,3,4) \tag{10.12.21}$$

where $\tilde{m}^{(\text{D})}$ is calculated using (10.12.11).

Now (10.12.17) can be re-arranged in the form

$$\begin{aligned}
&-2 \text{ Tr }(1234+4321)\left[\tilde{m}^{(\text{A})}(1,2,3,4)+\tilde{m}^{(\text{A})}(4,1,2,3)\right] \\
&-2 \text{ Tr }(1324+4231)\left[\tilde{m}^{(\text{A})}(1,3,2,4)-\tilde{m}^{(\text{A})}(4,1,2,3)\right] \\
&+2 \text{ Tr }(1243+3421)\left[\tilde{m}^{(\text{A})}(1,2,3,4)+\tilde{m}^{(\text{A})}(1,3,2,4)\right].
\end{aligned} \tag{10.12.22}$$

We shall soon see explicitly that

$$\tilde{m}^{(\text{A})}(1,2,3,4)=\tilde{m}^{(\text{A})}(2,1,4,3)=\tilde{m}^{(\text{A})}(4,3,2,1) \tag{10.12.23}$$

$$= -\,\tilde{m}^{(\text{A})}(2,1,3,4)=-\tilde{m}^{(\text{A})}(1,2,4,3) \tag{10.12.24}$$

so that (10.12.22) becomes

$$\begin{aligned}
&-2 \text{ Tr }(1234+4321)\left[\tilde{m}^{(\text{A})}(1,2,3,4)+\tilde{m}^{(\text{A})}(4,1,2,3)\right] \\
&-2 \text{ Tr }(1324+4231)\left[\tilde{m}^{(\text{A})}(1,3,2,4)+\tilde{m}^{(\text{A})}(4,1,3,2)\right] \\
&-2 \text{ Tr }(1243+3421)\left[\tilde{m}^{(\text{A})}(1,2,4,3)+\tilde{m}^{(\text{A})}(3,1,2,4)\right]
\end{aligned} \tag{10.12.25}$$

Using the second of eqns (10.12.23) we have finally for diagrams A + B + C + D

$$M = -2 \sum_{\text{perm }(234)} \text{Tr }(t^{a_1}t^{a_2}t^{a_3}t^{a_4})\, \tilde{M}(1,2,3,4) \tag{10.12.26}$$

where

$$\tilde{M}(1,2,3,4)=\tilde{m}^{(\text{A})}(1,2,3,4)+\tilde{m}^{(\text{A})}(4,1,2,3)+\tilde{m}^{(\text{D})}(1,2,3,4). \tag{10.12.27}$$

Equation (10.12.26) is precisely in the form (10.12.9). In Section 10.12.4 we shall explain how the amplitudes $\tilde{M}$ are constructed in the general n-gluon case. Here we look in more detail into the four-gluon amplitude.

Using the form of the three-gluon vertex given in Appendix 11, without the colour factor f_{abc}, one finds after some algebra that

$$\begin{aligned}\tilde{m}^{(A)}(1,2,3,4) = &-\frac{ig^2}{(k_1+k_2)^2}\Big((\epsilon_1\cdot\epsilon_2)(\epsilon_3\cdot\epsilon_4)(k_1-k_2)\cdot(k_3-k_4)\\ &+4\{(\epsilon_1\cdot\epsilon_2)[(\epsilon_3\cdot k_1)(\epsilon_4\cdot k_2)-(\epsilon_3\cdot k_2)(\epsilon_4\cdot k_1)]\\ &\quad+(\epsilon_3\cdot\epsilon_4)[(\epsilon_1\cdot k_3)(\epsilon_2\cdot k_4)-(\epsilon_1\cdot k_4)(\epsilon_2\cdot k_3)]\\ &\quad+(\epsilon_1\cdot\epsilon_3)(\epsilon_2\cdot k_1)(\epsilon_4\cdot k_3)+(\epsilon_2\cdot\epsilon_4)(\epsilon_1\cdot k_2)(\epsilon_3\cdot k_4)\\ &\quad-(\epsilon_1\cdot\epsilon_4)(\epsilon_2\cdot k_1)(\epsilon_3\cdot k_4)\\ &\quad-(\epsilon_2\cdot\epsilon_3)(\epsilon_1\cdot k_2)(\epsilon_4\cdot k_3)\}\Big)\end{aligned} \tag{10.12.28}$$

from which the properties (10.12.23) and (10.12.24) can be read off.

As we know from subsection 10.12.2 we can take as independent non-zero amplitudes just $M(++;--)$ and $M(+-;+-)$, corresponding to $\lambda_1 = \lambda_2 = +1$, $\lambda_3 = \lambda_4 = -1$ and $\lambda_1 = \lambda_3 = +1$, $\lambda_2 = \lambda_4 = -1$ respectively.

Consider first, in an obvious notation, the amplitude

$$\tilde{M}(1^+, 2^+, 3^-, 4^-).$$

Choose the reference q_j in such a way that the polarization vectors are

$$\epsilon_+(k_1;k_3) \quad \epsilon_+(k_2;k_3) \quad \epsilon_-(k_3;k_2) \quad \epsilon_-(k_4;k_2). \tag{10.12.29}$$

Then, via (10.8.18) and (10.8.19), the only non-zero scalar product of two polarization vectors is $\epsilon_+(k_1;k_3)\cdot\epsilon_-(k_4;k_2)$. Hence

$$\tilde{m}^{(D)}(1^+, 2^+, 3^-, 4^-) = 0 \tag{10.12.30}$$

and

$$\tilde{m}^{(A)}(1^+, 2^+, 3^-, 4^-) = \frac{4ig^2}{(k_1+k_2)^2}(\epsilon_1^+\cdot\epsilon_4^-)(\epsilon_2^+\cdot k_1)(\epsilon_3^-\cdot k_4). \tag{10.12.31}$$

Further, from (10.12.28),

$$\tilde{m}^{(A)}(4^-, 1^+, 2^+, 3^-) \propto (\epsilon_4\cdot\epsilon_1)[(\epsilon_2\cdot k_4)(\epsilon_3\cdot k_1)-(\epsilon_2\cdot k_1)(\epsilon_3\cdot k_4)],$$

which, using (10.8.17), will vanish when energy–momentum conservation is enforced. Thus we end up with the remarkably simple result

$$\tilde{M}(k_1^+, k_2^+, k_3^-, k_4^-) = \frac{4ig^2}{(k_1+k_2)^2}(\epsilon_1^+\cdot\epsilon_4^-)(\epsilon_2^+\cdot k_1)(\epsilon_3^-\cdot k_4), \tag{10.12.32}$$

Consider now the amplitude $\tilde{M}(1^+, 2^-, 3^+, 4^-)$. Choose reference momenta such that the polarization vectors are:

$$\epsilon_+(k_1;k_4) \quad \epsilon_+(k_3;k_4) \quad \epsilon_-(k_2;k_1) \quad \epsilon_-(k_4;k_1). \tag{10.12.33}$$

and the only non-zero scalar product is $\epsilon_+(k_3;k_4)\cdot\epsilon_-(k_2;k_1)$. Hence

$$\tilde{m}^{(\mathrm{D})}(1^+,2^-,3^+,4^-) = 0 \qquad (10.12.34)$$

$$\tilde{m}^{(\mathrm{A})}(1^+,2^-,3^+,4^-) = \frac{4ig^2}{(k_1+k_2)^2}(\epsilon_2^- \cdot \epsilon_3^+)(\epsilon_1^+ \cdot k_2)(\epsilon_4^- \cdot k_3) \qquad (10.12.35)$$

and, as before,

$$\tilde{m}^{(\mathrm{A})}(4^-,1^+,2^-,3^+) = 0.$$

Hence

$$\tilde{M}(k_1^+,k_2^-,k_3^+,k_4^-) = \frac{4ig^2}{(k_1+k_2)^2}(\epsilon_2^- \cdot \epsilon_3^+)(\epsilon_1^+ \cdot k_2)(\epsilon_4^- \cdot k_3). \qquad (10.12.36)$$

The physical amplitudes $\tilde{M}_{\lambda_1'\lambda_2';\lambda_1\lambda_2}$ for the reaction

$$(k_1,\lambda_1) + (k_2,\lambda_2) \to (k_1',\lambda_1') + (k_2',\lambda_2')$$

are then, via (10.12.3), (10.12.4) and (10.12.17),

$$M_{11;11} = -2 \sum_{\text{perm (234)}} \mathrm{Tr}\,(t^{a_1}t^{a_2}t^{a_3}t^{a_4})\,(-2ig^2)$$
$$\times \left\{ \frac{1}{k_1 \cdot k_2} \left[\epsilon_1^+(k_1;-k_3) \cdot \epsilon_4^-(-k_4;k_2)\right] \left[\epsilon_2^+(k_2;-k_3) \cdot k_1\right] \right.$$
$$\left. \times \left[\epsilon_3^+(-k_3;k_2) \cdot (-k_4)\right] \right\}, \qquad (10.12.37)$$

where

$$k_3 \equiv -k_1' \qquad k_4 \equiv -k_2', \qquad (10.12.38)$$

and

$$M_{1-1;1-1} = -2 \sum_{\text{perm (234)}} \mathrm{Tr}\,(t^{a_1}t^{a_2}t^{a_3}t^{a_4})\,(-2ig^2)$$
$$\times \left\{ \frac{1}{k_1 \cdot k_2} \left[\epsilon_2^-(k_2;k_1) \cdot \epsilon_3^+(-k_3;-k_4)\right] \left[\epsilon_1^+(k_1;-k_4) \cdot k_2\right] \right.$$
$$\left. \times \left[\epsilon_4^-(-k_4;k_1) \cdot (-k_3)\right] \right\} \qquad (10.12.39)$$

Using eqns (10.9.1)–(10.9.11) we get for the parts of (10.12.37) and (10.12.39) within the braces

$$-\frac{\langle k_1k_2\rangle^2\,[k_1'k_2']^2}{4(k_1 \cdot k_2)(k_2 \cdot k_3)} \qquad \text{and} \qquad -\frac{\langle k_1k_1'\rangle^2\,[k_2'k_2]^2}{4(k_1 \cdot k_2)(k_2 \cdot k_3)} \qquad (10.12.40)$$

respectively; we have used the fact that $k_1 \cdot k_4 = k_2 \cdot k_3$.

It is clear that the structure of the numerators just corresponds to a pairing of the momenta of the gluons with the same helicity label, $(\pm)$, *after* each ϵ^*_λ is replaced by $-\epsilon_{-\lambda}$, with the correspondence

$$(+) \to \langle \quad \rangle$$
$$(-) \to [\quad].$$

Thus, since permutation does not alter which gluon has which helicity, one has

$$M_{11;11} = -ig^2 \langle k_1 k_2 \rangle^2 [k'_1 k'_2]^2 \times \sum_{\text{perm }(234)} \frac{\text{Tr}\,(t^{a_1} t^{a_2} t^{a_3} t^{a_4})}{(k_1 \cdot k_2)(k_2 \cdot k_3)} \tag{10.12.41}$$

$$M_{1-1;1-1} = -ig^2 \langle k_1 k'_1 \rangle^2 [k'_2 k_2]^2 \times \sum_{\text{perm }(234)} \frac{\text{Tr}\,(t^{a_1} t^{a_2} t^{a_3} t^{a_4})}{(k_1 \cdot k_2)(k_2 \cdot k_3)}. \tag{10.12.42}$$

Note that since $k_1 + k_2 + k_3 + k_4 = 0$,

$$\sum_{\text{perm }(234)} \frac{1}{(k_1 \cdot k_2)(k_2 \cdot k_3)} = 0 \tag{10.12.43}$$

which is actually a reflection of a general property referred to as a *dual Ward identity* (see subsection 10.12.6 below).

We turn now to consider the colour structure.

10.12.4 Colour sums for gluon reactions

All physical observables are bilinear in the helicity amplitudes and one almost always wishes to carry out a sum over the colours of the gluons. One thus has to carry out colour sums of the type

$$S = \sum_{\text{all } a_j} [\,\text{Tr}\,(t^{a_1} t^{a_2} \cdots t^{a_n})] \left[\,\text{Tr}\,(t^{a_1} t^{b_2} \cdots t^{b_n})\right]^* \tag{10.12.44}$$

where $(b_2 \dots b_n)$ is some permutation of $(a_2 \dots a_n)$. Because the t^a are hermitian one has

$$S = \sum_{\text{all } a_j} \text{Tr}\,(t^{b_n} \cdots t^{b_2} t^{a_1})\,\text{Tr}\,(t^{a_1} t^{a_2} \cdots t^{a_n}). \tag{10.12.45}$$

The sum can be carried out step by step using the relations, valid for $SU(N)$,

$$\sum_a t^a_{ij} t^a_{kl} = \frac{1}{2}\left[\delta_{il}\delta_{jk} - \frac{1}{N}\delta_{ij}\delta_{kl}\right], \tag{10.12.46}$$

$$\sum_a (t^a t^a)_{ij} = \frac{N^2 - 1}{2N}\delta_{ij}. \tag{10.12.47}$$

We briefly indicate how this works. We have

$$S = \sum_{\text{all } a_j} (t^{b_n} \cdots t^{b_2} t^{a_1})_{ji} t^{a_1}_{ij} t^{a_1}_{kl} (t^{a_2} \cdots t^{a_n})_{lk}$$
$$= \frac{1}{2} \sum_{a_2 \ldots a_n} \Big[\text{Tr}\,(t^{b_n} \cdots t^{b_2} t^{a_2} \cdots t^{a_n})$$
$$- \frac{1}{N} \text{Tr}\,(t^{b_n} \cdots t^{b_2})\, \text{Tr}\,(t^{a_2} \cdots t^{a_n}) \Big]. \qquad (10.12.48)$$

To reduce further the second term one uses invariance under cyclic permutations to put it in the form

$$\text{Tr}\,(t^{b_n} \cdots t^{b_2})\, \text{Tr}\,(t^{b_2} t^{c_3} \cdots t^{c_n}),$$

where $(c_3 c_4 \ldots c_n)$ is a permutation of $(b_3 b_4 \ldots b_n)$, and then repeats the process used in (10.12.48).

For the first term there are two possibilities. If $b_2 = a_2$ we have

$$\sum_{a_2} \text{Tr}\,(t^{b_n} \cdots t^{b_3} t^{a_2} t^{a_2} t^{a_3} \cdots t^{a_n})$$
$$= \frac{N^2 - 1}{N} \text{Tr}\,(t^{b_n} \cdots t^{b_3} t^{a_3} \cdots t^{a_n}). \qquad (10.12.49)$$

If $b_2 \neq a_2$ then by cyclic permutation the first term can be put in the form $\text{Tr}\,(\Lambda_1 t^{a_2} \Lambda_2 t^{a_2})$, where $\Lambda_{1,2}$ are products of t^{a_j}, $j = 3, \ldots, n$. Then by (10.12.46)

$$\sum_{a_2} \text{Tr}\,(\Lambda_1 t^{a_2} \Lambda_2 t^{a_2}) = \frac{1}{2} \left[\text{Tr}\,(\Lambda_1)\, \text{Tr}\,(\Lambda_2) - \frac{1}{N} \text{Tr}\,(\Lambda_1 \Lambda_2) \right]. \qquad (10.12.50)$$

Some useful results that hold when the colour group is $SU(3)$ are given below. For brevity we write

$$(a_1 a_2 a_3 a_4) \equiv \text{Tr}\,(t^{a_1} t^{a_2} t^{a_3} t^{a_4}); \qquad (10.12.51)$$

then

$$\sum_{a_i} (a_1 a_2) = 4 \qquad (10.12.52)$$

$$\sum_{a_i} (a_1 a_2 a_1 a_2) = -2/3 \qquad (10.12.53)$$

$$\sum_{a_i} (a_1 a_2 a_2 a_1) = 16/3 \qquad (10.12.54)$$

$$\sum_{a_i} (a_1 a_2)(a_1 a_2) = 2. \qquad (10.12.55)$$

There are two independent products of traces of three ts; the rest can be obtained by cyclic permutations.

$$\sum_{a_i} (a_3 a_2 a_1)(a_1 a_2 a_3) = 7/3 \tag{10.12.56}$$

$$\sum_{a_i} (a_2 a_3 a_1)(a_1 a_2 a_3) = -2/3. \tag{10.12.57}$$

There are six independent products of traces of four ts; the rest can be obtained by cyclic permutations.

$$\sum_{a_i} (a_4 a_3 a_2 a_1)(a_1 a_2 a_3 a_4) = 19/6 \tag{10.12.58}$$

$$\sum_{a_i} (a_1 a_2 a_3 a_4)(a_1 a_2 a_3 a_4) = 2/3 \tag{10.12.59}$$

$$\sum_{a_i} (a_3 a_4 a_2 a_1)(a_1 a_2 a_3 a_4) = -1/3 \tag{10.12.60}$$

$$\sum_{a_i} (a_1 a_2 a_4 a_3)(a_1 a_2 a_3 a_4) = -1/3 \tag{10.12.61}$$

$$\sum_{a_i} (a_4 a_2 a_3 a_1)(a_1 a_2 a_3 a_4) = -1/3 \tag{10.12.62}$$

$$\sum_{a_i} (a_1 a_3 a_2 a_4)(a_1 a_2 a_3 a_4) = -1/3. \tag{10.12.63}$$

10.12.5 Colour sum for $GG \to GG$

Let us now apply these results to gluon–gluon scattering using the amplitudes (10.12.41), (10.12.42). Suppose we are interested in calculating the cross-section. In that case we need for example

$$\begin{aligned}\sum_{\text{colours}} |M_{11;11}|^2 = g^4 |\langle k_1 k_2 \rangle|^4 |[k_1' k_2']|^4 \\ \times \sum_{\text{colours}} \left[\sum_{\text{perm}(234)} \frac{(a_1 a_2 a_3 a_4)}{(k_1 \cdot k_2)(k_2 \cdot k_3)} \right] \\ \times \left[\sum_{\text{perm}(234)} \frac{(a_1 a_2 a_3 a_4)}{(k_1 \cdot k_2)(k_2 \cdot k_3)} \right]^* .\end{aligned} \tag{10.12.64}$$

It is clear from (10.12.25) that actually the trace $(a_1 a_2 a_3 a_4)$ and the trace $(a_1 a_4 a_3 a_2) = (a_4 a_3 a_2 a_1)$, obtained by the reflection permutation $(1234) \to (4321)$, should be multiplied by the same kinematic amplitude. This is seen to be true since

$$(k_1 \cdot k_2)(k_2 \cdot k_3) = (k_1 \cdot k_4)(k_4 \cdot k_3).$$

Thus we can replace $(a_1 a_2 a_3 a_4)$ by

$$[a_1 a_2 a_3 a_4] \equiv (a_1 a_2 a_3 a_4) + (a_4 a_3 a_2 a_1) \tag{10.12.65}$$

and sum only over permutations of (234) that are not reflections (NR) of each other, i.e. NR permutations of (234) are (234), (243), (324).

Furthermore the two separate sets of permutations can be rearranged so that

$$\sum_{\text{colours}} |M_{11;11}|^2 = g^4 |\langle k_1 k_2 \rangle|^4 |[k_1' k_2']|^4$$

$$\times \sum_{\text{colours}} \sum_{\substack{\text{NRperm} \\ (234)}} \frac{[a_1 a_2 a_3 a_4]}{(k_1 \cdot k_2)(k_2 \cdot k_3)}$$

$$\times \left\{ \frac{[a_1 a_2 a_3 a_4]^*}{(k_1 \cdot k_2)(k_2 \cdot k_3)} + \frac{[a_1 a_2 a_4 a_3]^*}{(k_1 \cdot k_2)(k_2 \cdot k_4)} + \frac{[a_1 a_3 a_2 a_4]^*}{(k_1 \cdot k_3)(k_3 \cdot k_2)} \right\}. \tag{10.12.66}$$

Using eqns (10.12.58)–(10.12.63) one obtains

$$\sum_{\text{colours}} [a_1 a_2 a_3 a_4][a_1 a_2 a_3 a_4]^* = 23/3 \tag{10.12.67}$$

$$\sum_{\text{colours}} [a_1 a_2 a_3 a_4][a_1 a_2 a_4 a_3]^* = -4/3 \tag{10.12.68}$$

$$\sum_{\text{colours}} [a_1 a_2 a_3 a_4][a_1 a_3 a_2 a_4]^* = -4/3 \tag{10.12.69}$$

Writing $23/3 = 9 - 4/3$, the terms in multiplying $-4/3$ in (10.12.66), vanish by (10.12.43). So we are left with

$$\sum_{\text{colours}} |M_{11;11}|^2 = 9g^4 |\langle k_1 k_2 \rangle|^4 |[k_1' k_2']|^4$$

$$\times \sum_{\substack{\text{NRperm} \\ (234)}} \left[\frac{1}{(k_1 \cdot k_2)(k_2 \cdot k_3)} \right]^2. \tag{10.12.70}$$

Writing this in terms of the usual Mandelstam variables,

$$\sum_{\text{colours}} |M_{11;11}|^2 = 144 s^4 \left(\frac{1}{s^2 t^2} + \frac{1}{s^2 u^2} + \frac{1}{t^2 u^2} \right). \tag{10.12.71}$$

Summing over helicities and using symmetry arguments to evaluate $M_{1-1;1-1}$ and parity invariance for the other Ms,

$$\sum_{\substack{\text{helicities} \\ \text{colours}}} |M^2| = 288(s^4 + t^4 + u^4) \left(\frac{1}{s^2 t^2} + \frac{1}{s^2 u^2} + \frac{1}{t^2 u^2} \right), \tag{10.12.72}$$

which can be simplified using $s + t + u = 0$. Dividing by 4×64 to obtain an average over initial spins and colours gives

$$\overline{|M|^2} = \frac{9}{2} g^4 \left(3 - \frac{ut}{s^2} - \frac{us}{t^2} - \frac{st}{u^2} \right) \tag{10.12.73}$$

which is a well-known result.

10.12.6 Some properties of *n*-gluon amplitudes

The set of all Feynman tree diagrams, for a given n, consists of certain 'basic' structures from which the rest can be generated by permuting the gluon labels in all possible ways, subject to the restriction that the diagrams thus generated are topologically independent. For example, for the Feynman diagrams for four gluons shown in Fig. 10.9, diagrams (B) and (C) were obtained from (A) by permuting gluon labels; see (10.12.20).

As a consequence it turns out that the kinematic amplitudes $\tilde{M}$ in (10.12.9) are invariant under *cyclic* permutations of the gluon momentum and helicity labels. Note, however, that the application of a permutation to the result for an amplitude $\tilde{M}$ can be rather subtle. For example, in (10.12.41) the factors $\langle k_1 k_2 \rangle$ and $[k_1' k_2']$ refer to the momenta of gluons *with particular helicities* and this does not change under a permutation, i.e. if gluon k_1 has helicity $+1$ it still does so after permuting the arguments of the function.

For a given n the Feynman tree diagrams can be grouped into subsets, each subset J being represented by one characteristic diagram, D_J, from which the other members of the Jth subset can be generated by permutation of the gluons.

The coefficient of a particular trace, say $\mathrm{Tr}\,(t^{a_1} t^{a_3} \cdots t^{a_j})$, is

$$\tilde{M}(k_1, \lambda_1; k_3, \lambda_3; \ldots; k_j, \lambda_j),$$

where the kinematic amplitude $\tilde{M}$ is a sum of contributions labelled by the characteristic diagram D_j, with the kinematic variables in the same order as the gluon labels inside the trace,

$$\tilde{M}(1, 3, \ldots, j) = \sum_{D_j} \tilde{M}_{D_j}(1, 3, \ldots, j) \tag{10.12.74}$$

and for each characteristic diagram D_j, $\tilde{M}_{D_j}$ is a sum of the amplitude $\tilde{m}_{D_j}$ arising from diagram D_j plus those cyclic permutations of it that correspond to topologically independent diagrams:

$$\tilde{M}_{D_j}(1, 3, \ldots, j) = \sideset{}{'}\sum_{\substack{\text{cyclic} \\ \text{perms}}} \tilde{m}_{D_j}(1, 3, \ldots, j). \tag{10.12.75}$$

In the review of Mangano and Parke (1991), it is shown that the kinematic amplitudes $\tilde{M}$ possess remarkable general properties, being essentially so-called *dual amplitudes* and related to string amplitudes. Two important properties, which were shown, in subsection 10.12.4, to be true for the four-gluon amplitude, are the *dual Ward identity*

$$\tilde{M}(1,2,3,\ldots,n) + \tilde{M}(2,1,3,\ldots,n) + \cdots + \tilde{M}(2,3,4,\ldots,1,n) = 0 \tag{10.12.76}$$

and the symmetry under reversal of the order of the labels,

$$\tilde{M}(n,n-1,\ldots,2,1) = (-1)^n \tilde{M}(1,2,\ldots,n-1,n).$$

Moreover, it is shown how supersymmetry can be used to relate amplitudes for pure gluonic reactions to those where a pair of gluons is replaced by a quark–antiquark pair. For these general developments and many results for specific amplitudes the reader is referred to the above review, but care must be taken since the same symbols have been given differing normalizations in the review and in the earlier papers of the same authors.

11
The spin of the nucleon: polarized deep inelastic scattering

11.1 Introduction

The static or low energy properties of the 'lighter' baryons (the nucleons and the hyperons and their resonances) are quite well explained in the *constituent quark model*, in which a baryon is visualized as made up of three *constituent quarks* (up, down and strange) with masses typically of about one third of the nucleon mass (see, for example, Close, 1979). In this picture the properties of the baryons are calculated using a non-relativistic Schrödinger equation or some more sophisticated version thereof. But the essential point for our discussion is that an unexcited baryon corresponds to the ground state of the three-particle system in which all the quarks are in relative s-states with zero orbital angular momentum and with no explicit rôle being played by any gluonic degrees of freedom.

As a consequence the spin of the baryon is equal to the sum of the spins of its constituent quarks. For a baryon moving along the Z-axis with helicity $\Lambda = 1/2$ one would thus expect to have

$$\sum S_z^{\text{quarks}} = \tfrac{1}{2}, \qquad (11.1.1)$$

where the sum is over the flavours of quark present in the baryon.

At the other end of the scale, for high energy interactions involving large momentum transfers, a baryon is visualized as made up of point-like constituents, *partons*, consisting of quarks and gluons. The partonic quarks have the same internal quantum numbers as the constituent quarks, but very different effective masses ($m_u \approx 4$ MeV$/c^2$, $m_d \approx 7$ MeV$/c^2$, $m_s \approx 125$–150 MeV$/c^2$). The gluons are massless and mediate the strong force between the quarks. Originally this was a purely phenomenological picture but was later subsumed into the beautiful gauge theory of strong interactions, QCD (see, for example, Leader and Predazzi, 1996).

The precise relation between constituent quarks and partonic quarks is not known and, as we shall see, experiments on polarized deep inelastic scattering (DIS) raise intriguing questions as to how the spin of a baryon is related to the spins of its partonic quarks. In particular the analogue of (11.1.1) is now known to be significantly violated.

This experimental discovery of the European Muon Collaboration (EMC) at CERN in 1988 (Ashman *et al.*, 1988, 1989) came as a great surprise and catalysed a major programme of experimental investigation and a host of papers on the theoretical aspects of the problem, with some results as surprising as the experimental one had been.

Since high energy large-momentum-transfer interactions between baryons are determined by the QCD-controlled interaction of their constituents it is clearly of the greatest importance to know what the structure of a baryon is in term of its partonic quarks and gluons. This structure cannot be studied easily by theoretical methods since it involves the non-perturbative regime of a strong interaction field theory. Consequently one relies on information gleaned from experimental studies in which the hadron is probed by hard photons or W and Z bosons. The prime example is DIS, to which we now turn.

11.2 Deep inelastic scattering

Deep inelastic lepton–hadron scattering has played a seminal rôle in the development of our present understanding of the substructure of elementary particles. The discovery of Bjorken scaling in the late 1960s provided the critical impetus for the idea that elementary particles contain almost point-like constituents and for the subsequent invention of the parton model, in which the reaction (Fig. 11.1)

$$l + N \to l + X \tag{11.2.1}$$

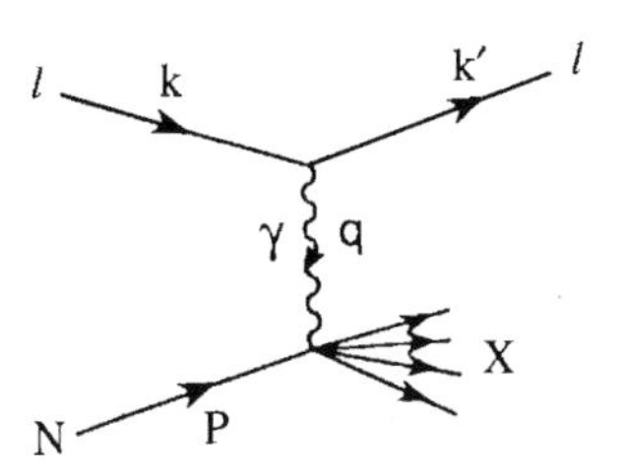

Fig. 11.1. Feynman diagram for inelastic lepton–nucleon scattering $lN \to lX$.

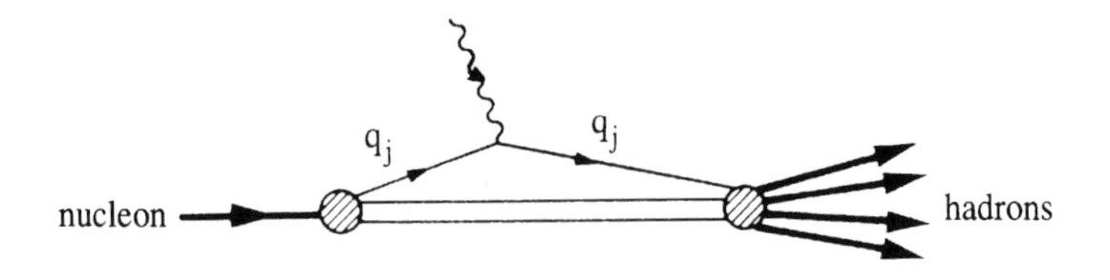

Fig. 11.2. Partonic interpretation of the lower part of the diagram in Fig. 11.1.

is viewed, as shown in Fig. 11.2, as the interaction of a hard virtual photon with a parton constituent of the nucleon. ('Hard' will mean $Q^2 \equiv -q^2 \gg M^2$, where M is the nucleon mass.)

DIS continued to play an essential rôle in the period of consolidation that followed, in the gradual linking of partons and quarks, in the discovery of the existence of missing constituents, later identified as gluons, and in the wonderful confluence of all the different parts of the picture into a coherent dynamical theory of quarks and gluons–quantum chromodynamics (QCD).

In more recent times the emphasis has shifted to the detailed study of the x-dependence of the parton distribution functions and to the study of their Q^2-evolution, probably the most direct test of the perturbative aspects of QCD.

Polarized DIS, involving the collision of a longitudinally polarized lepton beam on a polarized target (either longitudinally or transversely polarized) provides a different, complimentary and equally important insight into the structure of the nucleon. Whereas ordinary DIS probes simply the number density of partons with a fraction x of the momentum of the parent hadron, polarized DIS can partly answer the more sophisticated question about the number density of partons with given x and given spin polarization, in a hadron of definite polarization.

But what is quite extraordinary and unexpected *ab initio* is the richness and subtlety of the dynamical effects associated with the polarized case. Whereas the unpolarized scaling functions $F_{1,2}(x)$ have a simple interpretation in the naive parton model (where the nucleon is considered as an ensemble of essentially free massless partons) and a straightforward generalization in the framework of perturbative QCD, the spin-dependent scaling functions $g_{1,2}(x)$ are much more subtle, each fascinating in its own way. The function $g_1(x)$, which, at first sight, seems trivial to deal with in the naive parton model, turns out to have an anomalous gluon contribution associated with it, within perturbative QCD. In addition the first moment of $g_1(x)$ turns out to be connected with an essentially non-perturbative aspect of QCD, the axial ghost that is invoked to resolve the

$U(1)$ problem of the mass of the η'. And $g_2(x)$ turns out not to have any interpretation at all in purely partonic language.

What is also fascinating is the extraordinary interplay of theory and experiment in the study of $g_1(x)$. For a long time the theory of $g_1(x)$ remained comfortably at the level of the naive parton model. Then, in 1988, came the disturbing result of the EMC, which differed significantly from the naive theoretical predictions. These results could be argued to imply that the expectation value of the sum of the spins carried by the quarks in a proton, $\langle S_z^{\text{quarks}} \rangle$, was consistent with zero rather than with 1/2 as given in the quark model, suggesting a 'spin crisis in the parton model' (Leader and Anselmino, 1988). This led to an intense scrutiny of the basis of the theoretical calculation of $g_1(x)$ and the discovery of the anomalous gluon contribution (Efremov and Teryaev, 1988). (As often happens in theoretical physics it turns out that such an effect had already been studied to some extent in a largely overlooked paper of 1982 (Lam and Li, 1982).) So surprising was this discovery that the calculation was immediately checked by three different groups (Altarelli and Ross, 1988; Carlitz, Collins and Mueller, 1988; Anselmino and Leader, 1988), who all arrived at the same result. (Somewhat fortuitously, as it turns out, as was demonstrated in Carlitz *et al.*)

In the modified theoretical picture, the quantity $\Delta\Sigma \equiv 2\langle S_z^{\text{quark}} \rangle$, whose value had to be consistent with zero as a consequence of the EMC experiment, is replaced by the linear combination (for three flavours) $\Delta\Sigma - (3\alpha_s/2\pi)\Delta g$, where $\Delta g = \int_0^1 dx \Delta g(x)$ and $\Delta g(x)$ is the polarized gluon number density.

It has to be stressed that as a consequence of QCD a measurement of the first moment of $g_1(x)$ *does not* measure the z-component of the sum of the quark spins. It measures only the superposition $(1/9)\,[\Delta\Sigma - 3\alpha_s \Delta g/2\pi]$ and this linear combination can be made small by a cancellation between quark and gluon contributions. Thus the EMC results *cease to imply that* $\Delta\Sigma$ *is small.*

The function $g_2(x)$, however, does not have any simple interpretation in the naive parton model and it is a triumph of perturbative QCD that one can derive a sensible, gauge-invariant, result for it in the QCD field-theoretic model (Efremov and Teryaev, 1984).

In the following we shall concentrate almost exclusively on the polarized case. A good survey of the unpolarized case can be found in the review by Altarelli (1982) or, at a more introductory level, in Leader and Predazzi (1996).

For lack of space we shall also restrict our discussion almost entirely to *neutral current* interactions of the type (11.2.1), where the lepton is an electron or muon and the particle exchanged between the lepton and the hadron is a virtual photon. For very large Q^2, Z^0-exchange must also be taken into account, but for the spin-dependent case of interest the present generation of experiments does not require that. Also of importance are the *charged current* reactions of the type.[1]

$$l^- + N \to \nu_l + X, \tag{11.2.2}$$

which will be studied by the HERMES group at HERA. For a detailed discussion of these matters and the extension to nuclear targets, the reader should consult the review article of Anselmino, Efremov and Leader (1995).

Finally, a word about notation. In this chapter, in order to follow the convention in the literature, the covariant spin vector $\mathscr{S}_\mu$ (Section 3.4) for a spin-1/2 particle will be normalized in such a way that

$$\mathscr{S}^2 = -1. \tag{11.2.3}$$

Then for a fermion or antifermion of mass m and given $\mathscr{S}$ one has from subsections 4.6.2 and 4.6.3 the following useful results:

$$\begin{aligned}\bar{u}(p,\mathscr{S})\gamma^\mu\gamma_5 u(p,\mathscr{S}) &= -\bar{v}(p,\mathscr{S})\gamma^\mu\gamma_5 v(p,\mathscr{S})\\ &= 2m\mathscr{S}^\mu\end{aligned} \tag{11.2.4}$$

and for a massless fermion of helicity $\lambda = \pm 1/2$

$$\begin{aligned}\bar{u}(p,\lambda)\gamma^\mu\gamma_5 u(p,\lambda) &= -\bar{v}(p,\lambda)\gamma^\mu\gamma_5 v(p,\lambda)\\ &= 4\lambda p^\mu.\end{aligned} \tag{11.2.5}$$

11.3 General formalism and structure functions

Consider the reaction Fig. 11.1 in the Lab frame, where the proton is at rest. For the initial and final lepton momenta we write

$$k^\mu = (E, \mathbf{k}) \qquad k'^\mu = (E', \mathbf{k}') \tag{11.3.1}$$

and for the initial nucleon momentum

$$P^\mu = (M, \mathbf{O}). \tag{11.3.2}$$

Then the differential cross-section to find the scattered lepton in solid angle $d\Omega$ with energy in the range $(E', E' + dE')$ can be written (see, for example, Leader and Predazzi, 1996)

$$\frac{d^2\sigma}{d\Omega dE'} = \frac{\alpha^2}{2Mq^4}\frac{E'}{E} L_{\mu\nu} W^{\mu\nu}, \tag{11.3.3}$$

[1] For spin-dependent measurements it has not been possible up to now to contemplate using a neutrino *beam*, because of the impossibility of polarizing the huge target needed. See, however, Section 11.10.

where $q = k - k'$ and α is the fine structure constant. In (11.3.3) the leptonic tensor $L_{\mu\nu}$ is given by QED as

$$L_{\mu\nu}(k,s;k',s') = \left[\bar{u}(k',s')\gamma_\mu u(k,s)\right]^* \left[\bar{u}(k',s')\gamma_\nu u(k,s)\right], \tag{11.3.4}$$

where s is the lepton covariant spin vector. It can be split into symmetric (S) and antisymmetric (A) parts under μ, ν interchange and, after summing over the spin of the final lepton, takes the form

$$L_{\mu\nu}(k,s;k') = 2\left[L^{(S)}_{\mu\nu}(k;k') + iL^{(A)}_{\mu\nu}(k,s;k')\right] \tag{11.3.5}$$

where

$$L^{(S)}_{\mu\nu} = k_\mu k'_\nu + k'_\mu k_\nu - g_{\mu\nu}(k \cdot k' - m^2) \tag{11.3.6}$$

$$L^{(A)}_{\mu\nu}(k,s;k') = m\epsilon_{\mu\nu\alpha\beta}s^\alpha(k-k')^\beta. \tag{11.3.7}$$

The hadronic tensor $W_{\mu\nu}$, which contains the strong interaction dynamics, can be written in terms of four scalar *inelastic form factors*, $W_{1,2}$ and $G_{1,2}$, functions at most of q^2 and $P \cdot q$:

$$W_{\mu\nu}(q;P,\mathscr{S}) = W^{(S)}_{\mu\nu}(q;P) + iW^{(A)}_{\mu\nu}(q;P,\mathscr{S}) \tag{11.3.8}$$

with

$$\begin{aligned}\frac{1}{2M}W^{(S)}_{\mu\nu}(q;P) = {} & \left(-g_{\mu\nu} + \frac{q_\mu q_\nu}{q^2}\right) W_1(P\cdot q, q^2) \\ & + \left[\left(P_\mu - \frac{P\cdot q}{q^2}q_\mu\right)\left(P_\nu - \frac{P\cdot q}{q^2}q_\nu\right)\right]\frac{W_2(P\cdot q,q^2)}{M^2}\end{aligned} \tag{11.3.9}$$

$$\begin{aligned}\frac{1}{2M}W^{(A)}_{\mu\nu}(q;P,\mathscr{S}) = \epsilon_{\mu\nu\alpha\beta}q^\alpha \Bigg\{ & M\mathscr{S}^\beta G_1(P\cdot q,q^2) \\ & + \left[(P\cdot q)\mathscr{S}^\beta - (\mathscr{S}\cdot q)P^\beta\right]\frac{G_2(P\cdot q,q^2)}{M}\Bigg\}\end{aligned} \tag{11.3.10}$$

Putting eqns (11.3.8) and (11.3.5) into (11.3.3) one finds

$$\frac{d^2\sigma}{d\Omega dE'} = \frac{\alpha^2}{Mq^4}\frac{E'}{E}\left[L^{(S)}_{\mu\nu}W^{\mu\nu(S)} - L^{(A)}_{\mu\nu}W^{\mu\nu(A)}\right]. \tag{11.3.11}$$

Note that only the antisymmetric part $W^{(A)}_{\mu\nu}$ depends on the nucleonic spin and that the cross-section (11.3.11) is independent of the nucleon spin if the lepton is unpolarized.

The spin-independent inelastic form factors $W_{1,2}$ and the spin-dependent ones $G_{1,2}$, which can be measured experimentally, are written in terms of scaling functions $F_{1,2}$ and $g_{1,2}$ as follows:

$$MW_1(P \cdot q, Q^2) \equiv F_1 \qquad \nu W_2(P \cdot q, Q^2) \equiv F_2 \tag{11.3.12}$$

$$\frac{(P \cdot q)^2}{\nu} G_1(P \cdot q, Q^2) \equiv g_1 \qquad \nu(P \cdot q) G_2(P \cdot q, Q^2) \equiv g_2, \tag{11.3.13}$$

where ν is the energy of the virtual photon in the Lab,

$$\nu = E - E'. \tag{11.3.14}$$

The structure functions $F_{1,2}$ have played a seminal rôle in the history of elementary particle physics. They are, in principle, functions of the two variables, $P \cdot q$ and Q^2 or, equivalently, of Q^2 and 'Bjorken-x' (we shall sometimes write x_{Bj} for clarity),

$$x \equiv \frac{Q^2}{2P \cdot q} = \frac{Q^2}{2M\nu}, \tag{11.3.15}$$

and were expected to decrease rapidly as Q^2 increases at fixed x as a consequence of the inability of an extended object – the proton – to absorb very large momentum transfers. The discovery of 'Bjorken scaling', i.e. that $F_{1,2}(Q^2, x)$ are almost independent of Q^2 in the *Bjorken limit* $Q^2 \to \infty$, x fixed, catalysed the invention of partons, hard point-like constituents within the proton, and led eventually to the invention of QCD.

The spin-dependent structure functions $g_{1,2}$ are much more difficult to measure, requiring polarized beams and targets, but, as mentioned in Section 11.2, tremendous progress has been made in the last decade, largely as a consequence of the stimulus provided by the remarkable results of the EMC collaboration at CERN.

The most direct way to measure $g_{1,2}$ is to utilize a longitudinally polarized beam and a nucleon target polarized either along the direction of the lepton beam or transversely to it and in the scattering plane. In each case one measures the cross-section difference upon reversal of the nucleon spin. Indicating lepton and nucleon spin directions by $\rightarrow$ and $\Rightarrow$ respectively, one has

$$\frac{d^2\sigma^{\overset{\rightarrow}{\Rightarrow}}}{d\Omega dE'} - \frac{d^2\sigma^{\overset{\rightarrow}{\Leftarrow}}}{d\Omega dE'} = -\frac{4\alpha^2 E'}{Q^2 E M \nu} \left[(E + E' \cos\theta) g_1 - 2xMg_2\right] \tag{11.3.16}$$

$$\frac{d^2\sigma^{\rightarrow\Uparrow}}{d\Omega dE'} - \frac{d^2\sigma^{\rightarrow\Downarrow}}{d\Omega dE'} = -\frac{8\alpha^2 (E')^2}{Q^2 M \nu^2} \left(\frac{\nu}{2E} g_1 + g_2\right) \sin\theta \cos\phi \tag{11.3.17}$$

where θ is the Lab scattering angle of the lepton. Since θ is typically a few milliradians, it is much more difficult to make an accurate determination of the left-hand side of (11.3.17) than that of (11.3.16). The final-state-lepton azimuthal angle ϕ is defined in Fig. 11.18.

Because of the relative magnitudes of the coefficients of g_1 and g_2 in (11.3.16) and (11.3.17) it is usually assumed that the left-hand side of (11.3.16) is essentially a measurement of g_1 whereas the left-hand side of (11.3.17) largely determines g_2. Only in the past year or so has it become possible to extract g_1 and g_2 from measurements of both these types of cross-section difference at the Stanford linear collider (see Abe *et al.*, 1997b, c; Anthony *et al.*, 1999).

In the following we shall first consider what is known theoretically about $g_{1,2}$ and then turn to consider the experimental situation. Note, incidentally, that (11.3.16) and (11.3.17) are not the only possible measurable quantities. For a more general approach, see Anselmino (1979).

11.4 The simple parton model

We shall sketch briefly how one derives parton-model expressions for $g_{1,2}$. For a very detailed and historical treatment see Leader and Predazzi (1996).

In an *infinite momentum frame* S^∞, i.e. a Lorentz frame where the nucleon is moving very fast, the latter is visualized as made up of fast-moving constituents (partons) and the collision of the projectile with a constituent is treated in an impulse approximation, as if the constituent were a free particle.

If in this reference frame we imagine taking a snapshot of the target as seen by the projectile we may see a set of constituents of mass m_j and momenta p_j and we may ask for how long this fluctuation or virtual state will exist. Its lifetime, τ_V, by the uncertainty principle is likely to be of order $(\Delta E)^{-1}$,

$$\tau_V \simeq \frac{1}{\Delta E} = \frac{1}{E_V - E_N}, \tag{11.4.1}$$

where E_V is the energy of the virtual state and E_N the energy of the nucleon, in the given reference frame.

The impulse approximation will be valid when:

(1) the time of interaction τ_{int} between the projectile and the constituent is much smaller than τ_V, so that the constituent is basically free during the period of its interaction with the projectile, and
(2) the impulse given to the constituent is so large that after interaction its energy is much larger than the binding energy, and so it continues to behave as a free particle.

The second condition is automatically satisfied in the deep inelastic regime since a very large momentum is imported to the struck constituent. The first can be shown to be satisfied if the energy of internal motion of the constituents is limited, say $O(M)$, where M is the mass of the nucleon, and one avoids events where x is very small.

In the impulse approximation the interaction of the projectile with the nucleon is simply an incoherent sum over its interactions with the individual constituents.

Let $n_q(\mathbf{p}, s; \mathscr{S})d^3\mathbf{p}$ be the number of quark-partons q, charge e_q, with momentum in the range $\mathbf{p} \to \mathbf{p} + d\mathbf{p}$ and with covariant spin vector s^μ inside a nucleon of four-momentum P^μ with covariant spin vector $\mathscr{S}^\mu$, as seen in the frame S^∞.

Then one finds the rather intuitive result[1]

$$W_{\mu\nu}^{(A)}(q; P, \mathscr{S}) = \sum_{q,s} e_q^2 \int d^3\mathbf{p} \left(\frac{p_0}{E_q}\right) \delta(2p \cdot q - Q^2)$$
$$\times n_q(\mathbf{p}, s; \mathscr{S}) \tilde{w}_{\mu\nu}^{(A)}(q; p, s) \tag{11.4.2}$$

where $e_q^2 \tilde{w}_{\mu\nu} \delta(2p \cdot q - Q^2)$ is the analogue of $W_{\mu\nu}$ for the interaction of a hard photon with a 'free' quark-parton, as shown in Fig. 11.3. The factor p_0/E_q, where E_q is the energy of the struck quark, arises from the relativistic normalization of the states.

Since the parton in Fig. 11.3 is treated as an elementary, point-like charged fermion we can calculate $\tilde{w}_{\mu\nu}$ using QED, and the strong interaction dynamics is then hidden in the parton number-density or distribution functions in (11.4.2).

Note that since the whole approach only makes sense if the partons can be considered as essentially free, any result that turns out to depend critically on the parton mass must be treated with suspicion, because the mass of a constituent reflects its binding energy. Indeed, we shall see in a moment that for this very reason g_2 cannot be calculated reliably in the parton model (Anselmino, Efremov and Leader, 1995).

Now we consider the calculation of $\tilde{w}_{\mu\nu}^{(A)}$, describing as mentioned above the interaction of the hard photon with a quark of given flavour depicted in Fig. 11.3. The final state quark is a 'free' quark and is on the mass shell: $(p')^2 = m_q^2$. In the impulse approximation also the initial quark is

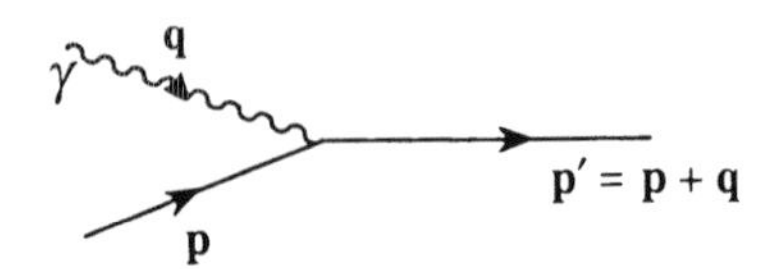

Fig. 11.3. Point-like interaction of a hard photon with a quark.

[1] An analogous result holds for the symmetric, spin-independent, part.

considered to be free and to have the same mass. But to see the potential danger of this assertion, let us put $p^2 = m^2$, where for the moment we allow $m^2 \neq m_q^2$ in order to represent the fact that the initial quark is really a bound quark. Aside from this we treat the incoming quark as free, i.e. its wave function is taken as the usual free-particle Dirac spinor $u(p, s)$ for a particle of mass m.

One then finds

$$\tilde{w}_{\mu\nu}(q; p, s) = \tfrac{1}{2} \text{ Tr } \left[(1 + \gamma_5 \not{s})(\not{p} + m)\gamma_\mu(\not{p} + \not{q} + m_q)\gamma_\nu\right] \tag{11.4.3}$$

from which one obtains

$$\tilde{w}^{(A)}_{\mu\nu}(q; p, s) = (2\epsilon_{\mu\nu\alpha\beta})(m_q s^\alpha) \left[\left(1 - \frac{m}{m_q}\right) p^\beta - \frac{m}{m_q} q^\beta\right]. \tag{11.4.4}$$

Equation (11.4.4) is extremely revealing. We see immediately that for a general s^μ the result is not gauge invariant ($q^\mu \tilde{w}^{(A)}_{\mu\nu} \neq 0$) unless $m = m_q$. Moreover, the offending term, when $m \neq m_q$, is not small in an infinite-momentum frame (where the impulse approximation is supposed to be most justifiable) even if $m - m_q$ is small.

However, in the special case of longitudinal (L) polarization, if the quark has high momentum so that $m_q/p_z \ll 1$, the product $m_q s_L^\beta \to \pm p^\beta$ (see Section 3.4) and the non-gauge-invariant term vanishes because of the antisymmetric $\epsilon_{\mu\nu\alpha\beta}$ in (11.4.4). Let us therefore choose $\mathcal{S}^\beta$ in (11.4.2) to correspond to a nucleon of definite helicity and s^α in (11.4.4) to correspond to a quark of definite helicity.

Then, putting (11.4.4) into (11.4.2) and integrating over the assumed-negligible quark transverse momentum, one finds, on comparing (11.4.2) and (11.3.10), that for a proton

$$g_1(x, Q^2) = \tfrac{1}{2} \sum_q e_q^2 \left[\Delta q(x) + \Delta\bar{q}(x)\right] \tag{11.4.5}$$

where

$$\Delta q(x) \equiv q_+(x) - q_-(x) \tag{11.4.6}$$

and $q_\pm$ are the number densities of quarks with momentum fraction x and with helicity $\lambda = \pm 1/2$ respectively inside a *proton* of momentum P with helicity $\Lambda = +1/2$.

In terms of the original parton densities,

$$q_\lambda(x) \equiv P \int d^2\mathbf{p}_\perp n_q(\mathbf{p}, \lambda; \Lambda = 1/2) \tag{11.4.7}$$

and the usual, unpolarized, number density is then

$$q(x) = q_+(x) + q_-(x). \tag{11.4.8}$$

Note that in (11.4.5) the sum is over all quarks and antiquarks. For antiquarks one commonly uses the notation $\Delta\bar{q}(x)$.

The result for g_1 seems to be unambiguous – it is not sensitive to the value of the quark mass. Yet, as we shall see later, (11.4.5) is not the full story because of the axial anomaly.

Quite the opposite happens for g_2 where the *transverse* spin is relevant. There is an extreme sensitivity to whether m equals m_q and one cannot expect to make a reliable calculation of $g_2(x)$ in the parton model.

One can put the case even more forcefully. The whole point of quarks is that, in their point-like interaction with a hard photon, they produce the large momentum-transfer reactions that we are trying to generate for the photon–hadron interactions. But even if we *define* the model by insisting that $m = m_q$, comparing the expression (11.4.4) with the general structure of $W_{\mu\nu}^{(A)}$ for a spin-1/2 particle (11.3.10), we see that

$$g_2(x)|_{\text{quark}} = 0. \tag{11.4.9}$$

Thus the hard-photon–free-quark interaction does not possess the cross-section asymmetry that we are seeking to explain in the photon–hadron interaction. It is clearly unrealistic therefore to try to produce such an asymmetry from quark-partons.

Of course the parton model predates QCD and, as treated above, is rather simplistic. Its value lies in the intuitive nature of its expressions in terms of parton number densities. When interactions come into play Bjorken scaling is broken and the main effect is that gluons become important and the parton densities get replaced by Q^2-dependent densities $\Delta q(x, Q^2)$, whose Q^2-dependence or evolution can be handled perturbatively.

We turn now to a more serious approach to the subject in the framework of field theory.

11.5 Field-theoretic generalization of the parton model

Historically there have been two approaches to DIS in QCD. The earlier one, the *operator product expansion* (OPE), concentrated on current commutators and their behaviour on the light-cone; the newer deals directly with Feynman diagrams. The latter is the more general and reproduces the OPE results wherever they are supposed to be valid. Neither is a complete scheme; in each, one has to make certain reasonable-sounding assumptions about the behaviour of non-calculable hadronic matrix elements of operators.

Because of its greater generality we shall base our treatment on the Feynman diagram field-theoretic approach.

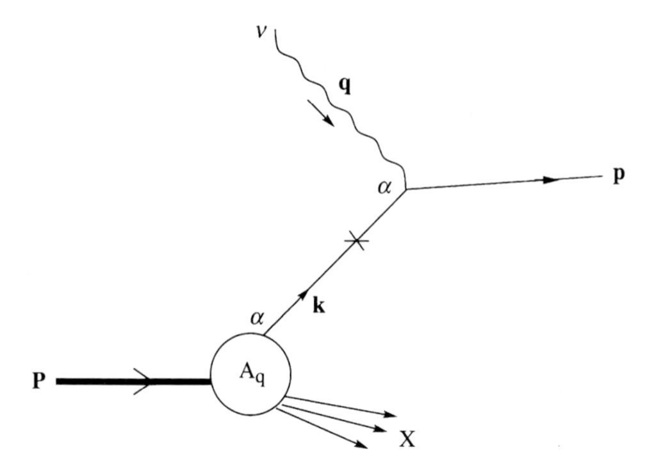

Fig. 11.4 Feynman diagram for $\gamma^* p \to$ anything. The virtual photon has Lorentz index ν and momentum q; the proton has momentum P.

The parton model, viewed as an impulse approximation, long predates QCD. Once QCD is accepted as a theory of the strong interactions it is clearly important to reformulate and extend the parton model using a field-theoretic framework.

We shall illustrate how this is carried out for the simplest case of deep inelastic scattering. For this purpose the rôle of the leptons is superfluous so we consider the reaction

$$\gamma^*(q) + p(P) \to \text{ anything}$$

where the virtual photon has Lorentz index ν. The Feynman amplitude for this is shown in Fig. 11.4, where the amplitude has been split into a soft part controlling the emission of a quark from the proton and a hard part where the hard photon interacts with the quark. Note this is a *Feynman* diagram, not a probabilistic parton diagram. For simplicity we ignore flavour and colour; they may be dealt with trivially.

The soft vertex for the emission of a quark with Dirac spinor index α is called $A_{q\alpha}(k, X; P)$ and is defined to include the propagator for the quark of momentum k. The symbol $\times$ on the quark line indicates that no propagator is to be inserted in the expression for the Feynman amplitude.

It will help to think of $A_q(k, X; P)$ as a column spinor. The Feynman amplitude is then

$$M^\nu(p, X; P) = \bar{u}_\lambda(p)(i\gamma^\nu)A_q(p - Q, X; P), \tag{11.5.1}$$

where, since it is inessential for our discussion, we have put the quark charge equal to unity.

A cross-section would involve the modulus squared of M^μ summed over all states X, and over the helicity λ of the final state quark and integrated over the momentum $\mathbf{p}$. However, because our virtual γ is actually attached to a lepton, we must sum over the Lorentz index ν. Hence we need to consider

$$w^{\mu\nu} \equiv \sum_\lambda \sum_X \int \frac{d^3\mathbf{p}}{2E(2\pi)^3} M^{\mu *} M^\nu (2\pi)^4 \delta^4(p + p_X - q - P). \tag{11.5.2}$$

Since M^μ is a number we can replace the complex conjugate by the hermitian conjugate †, so that

$$\begin{aligned} w^{\mu\nu} &\equiv \sum_\lambda \sum_X \int \frac{d^3\mathbf{p}}{2E} A_q^\dagger (-i\gamma^{\mu\dagger})\gamma^0 u_\lambda(p)\bar{u}_\lambda(p)(i\gamma^\nu) A_q \\ &\quad \times 2\pi\delta^4(p + p_X - q - P) \\ &= 2\pi \sum_X \int \frac{d^3\mathbf{p}}{2E} A_q^\dagger \gamma^0 (-i\gamma^\mu) \left\{ \sum_\lambda u_\lambda(p)\bar{u}_\lambda(p) \right\} (i\gamma^\nu) A_q \\ &\quad \times \delta^4(p + p_X - q - P) \end{aligned} \tag{11.5.3}$$

where we have used $\gamma^0\gamma^{\mu\dagger}\gamma^0 = \gamma^\mu$. Carrying out the sum on λ, we can now write (11.5.2) in the form

$$w^{\mu\nu} = 2\pi \int \frac{d^3\mathbf{p}}{2E} \left\{ \sum_X \bar{A}_{q\beta} A_{q\alpha} \delta^4(p + p_X - q - P) \right\} \mathscr{E}^{\mu\nu}_{\beta\alpha} \tag{11.5.4}$$

where, obviously, $\bar{A}_q = A_q^\dagger \gamma^0$ and the hard, *short-distance*, piece is

$$\mathscr{E}^{\mu\nu}_{\beta\alpha} = [(-i\gamma^\mu)(\not{p} + m_q)(i\gamma^\nu)]_{\beta\alpha}. \tag{11.5.5}$$

The structure involving the soft vertices is written conventionally[1] as

$$\Phi_{\alpha\beta}(k = p - Q; P) = \sum_X \bar{A}_{q\beta} A_{q\alpha} \delta^4(p + p_X - q - P). \tag{11.5.6}$$

Note that the back-to-front convention for the Dirac indices α, β allows (11.5.6) to be written in matrix form

$$\begin{aligned} w^{\mu\nu} &= 2\pi \int \frac{d^3\mathbf{p}}{2E} \Phi_{\alpha\beta} \mathscr{E}^{\mu\nu}_{\beta\alpha} \\ &= 2\pi \int \frac{d^3\mathbf{p}}{2E} \, \mathrm{Tr}\,(\Phi \mathscr{E}^{\mu\nu}) \end{aligned} \tag{11.5.7}$$

where Φ and $\mathscr{E}^{\mu\nu}$ are 4×4 matrices in Dirac spinor space.

[1] Sometimes in the literature Φ is defined with a factor $(2\pi)^4$ on the right-hand side of (11.5.6).

Finally we convert to $\int d^4p$ by using

$$\int \frac{d^3\mathbf{p}}{2E} = \int d^4p\,\delta(p^2 - m_q^2)\theta(E - m_q) \tag{11.5.8}$$

so that (11.5.7) can be written

$$w^{\mu\nu} = \int d^4p \ \mathrm{Tr}\ (\Phi E^{\mu\nu}) \tag{11.5.9}$$

where, restoring the charge ee_f of the quark of flavour f,

$$\begin{aligned} E^{\mu\nu} &= (ee_f)^2 2\pi\delta(p^2 - m_q^2)\theta(E - m_q)\mathscr{E}^{\mu\nu} \\ &= -(i\gamma^\mu ee_f)\left[(\not p + m_q)2\pi\delta(p^2 - m_q^2)\theta(E - m_q)\right](i\gamma^\nu ee_f). \end{aligned} \tag{11.5.10}$$

We shall now see that $E^{\mu\nu}$ is the discontinuity of the Feynman amplitude in Fig. 11.5, with external spinors and polarization vectors removed. For the latter amplitude has the form

$$M^{\mu\nu} = (i\gamma^\mu ee_f)\left[\frac{i(\not p + m_q)}{p^2 - m_q^2 + i\epsilon}\right](i\gamma^\nu ee_f) \tag{11.5.11}$$

and using

$$\frac{1}{p^2 - m_q^2 + i\epsilon} = \mathrm{P}\left(\frac{1}{p^2 - m_q^2}\right) - i\pi\delta(p^2 - m_q^2) \tag{11.5.12}$$

one will obtain a result for $E^{\mu\nu}$, if one makes the following replacement in the propagator in $M^{\mu\nu}$:

$$\begin{aligned} \frac{1}{p^2 - m_q^2 + i\epsilon} &\to \left(\frac{1}{p^2 - m_q^2 - i\epsilon} - \frac{1}{p^2 - m_q^2 + i\epsilon}\right)\theta(E - m_q) \\ &= 2i\pi\delta(p^2 - m_q^2)\theta(E - m_q). \end{aligned} \tag{11.5.13}$$

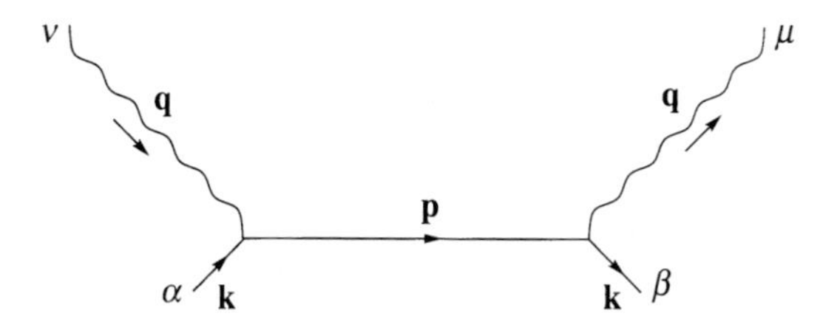

Fig. 11.5. Amputated Feynman diagram for $\gamma^* q \to \gamma^* q$. The virtual photon has Lorentz index ν and momentum q; the quark has momentum k.

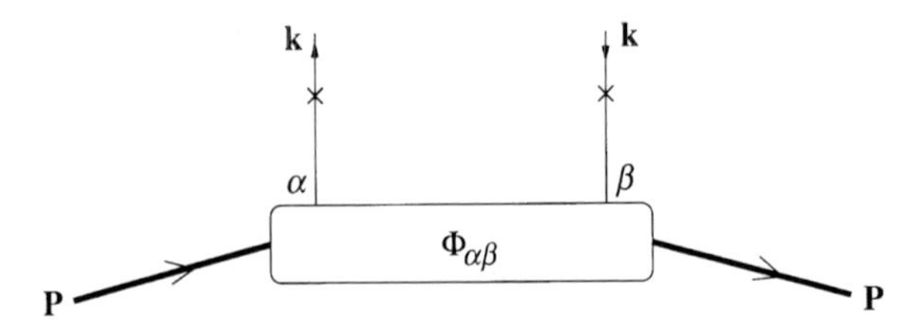

Fig. 11.6. Graphical representation of Φ.

Thus if we depict Φ graphically as in Fig. 11.6 then the entire result (11.5.9) corresponds to the Feynman diagram of Fig. 11.7, often called the 'handbag diagram', where, as before the symbol $\times$ indicates no propagator and the diagonal slash means the replacement (11.5.13) should be made in the propagator. Conventionally one changes integration variable from p^μ to k^μ so that the loop integration becomes $\int d^4k$ (note this differs from the conventional $\int d^4k/(2\pi)^4$ used in Feynman diagrams).

Finally, then, (11.5.9) becomes

$$w^{\mu\nu} = \int d^4k \; \mathrm{Tr}\; (\Phi E^{\mu\nu}) \tag{11.5.14}$$

Figure 11.7 is the basis for the field-theoretic generalization of the parton model for deep inelastic scattering.

Let us turn now to the structure of Φ in (11.5.6). One can show that[1]

$$A_{q\alpha}(k, X; P) = \langle X|\Psi_\alpha(0)|P\rangle \tag{11.5.15}$$

where $\Psi_\alpha(z)$ is the usual quark field operator with spinor index α. We have, up to now, glossed over the colour structure, but actually both $A_{q\alpha}$

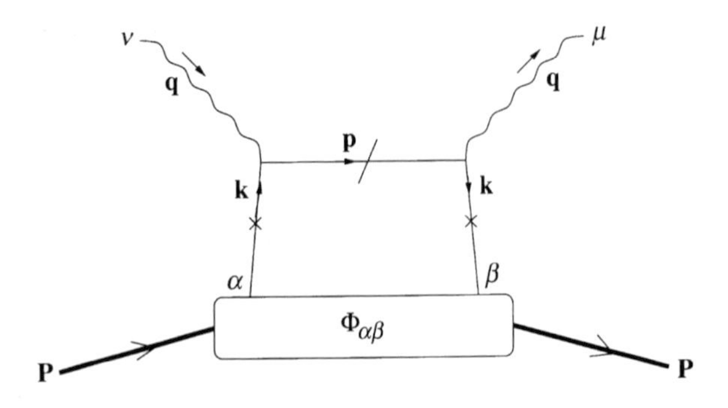

Fig. 11.7. Field-theoretic Feynman diagram for deep inelastic scattering.

[1] Strictly speaking $|X\rangle$ is an 'out' state.

and $\Psi_\alpha(0)$ should be considered as column vectors in colour space. The combination in (11.5.6) is a colour singlet. Then

$$\begin{aligned}\Phi_{\alpha\beta} &= \sum_X \delta^4(p - p_X - q - P)\langle P|\bar{\Psi}_\beta(0)|X\rangle\langle X|\Psi_\alpha(0)|P\rangle \\ &= \sum_X \int \frac{d^4z}{(2\pi)^4} e^{ix\cdot(p+p_X-q-P)} \\ &\quad \times \langle P|\bar{\Psi}_\beta(0)|X\rangle\langle X|\Psi_\alpha(0)|P\rangle. \end{aligned} \tag{11.5.16}$$

Using translational invariance this becomes

$$\Phi_{\alpha\beta} = \int \frac{d^4z}{(2\pi)^4} e^{iz\cdot(p-q)} \sum_X \langle P|\bar{\Psi}_\beta(0)|X\rangle\langle X|\Psi_\alpha(z)|P\rangle$$

so that, using completeness, the sum over X can be carried out, yielding the final result

$$\Phi_{\alpha\beta}(P, \mathscr{S}, K) = \int \frac{d^4z}{(2\pi)^4} e^{ik\cdot z} \langle P, \mathscr{S}|\bar{\Psi}_\beta(0)\Psi_\alpha(z)|P, \mathscr{S}\rangle \tag{11.5.17}$$

where we have specified the proton state more precisely by including its covariant polarization vector $\mathscr{S}^\mu$, and we have used $k = p - q$.

Thus Φ is expressed as a matrix element of a bilocal operator. It contains all the non-perturbative information about the state of a quark inside a proton in a given spin state. Φ is, at this stage, much more general than in the parton model, and if we expanded it in terms of a set of specific Dirac matrices, it would be hopeless to try to learn about the coefficient functions from experiment. To reach a manageable structure one has to make the key assumption that Φ decreases rapidly with increasing virtuality of the quark, i.e. as $|k^2|$ increases, and also decreases rapidly as $|P \cdot k|$ increases. These conditions guarantee that the dominant part of the k-integration region corresponds to the collinear situation $k^\mu \propto P^\mu$. We shall see later how these conditions are utilized to recover a structure recognizable as the parton model.

Φ is sometimes called the quark–quark correlation function. In reality it is the unnormalized density matrix of a virtual quark inside a proton. This can be seen by considering the expression for the cross-section of the process $\gamma^* + q(k) \to \gamma^* + q(k)$, with the initial virtual quark in an arbitrary state of polarization. One obtains the expression (11.5.14), of course without the d^4k integration, with Φ replaced by the density matrix of the initial quark.

The quark density matrix Φ was introduced in a seminal paper by Ralston and Soper (1979) and was generalized and much utilized by Efremov, Teryaev and collaborators (e.g. Efremov and Teryaev, 1984) and more recently by Mulders and collaborators (e.g. Mulders, 1997). The

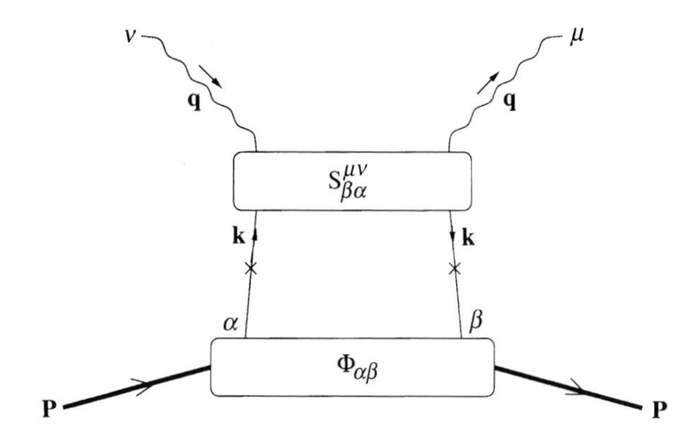

Fig. 11.8. Generalized version of Fig. 11.7.

most complete discussion of Φ and the analogous function for gluons can be found in the Ph.D. thesis of Boer (1998). See also Anselmino, Efremov and Leader (1995), Section 10 and Appendices B and C.

In the above discussion we used, for pedagogical purposes, the simplest possible diagram (Figs. 11.4 and 11.5) for the hard part of the scattering $\gamma^* + q \to q$. If one allows a more general perturbative QCD amplitude for the hard scattering, then the result (11.5.14) corresponding to Fig. 11.7 generalizes to

$$w^{\mu\nu} = \int d^4k \ \mathrm{Tr}\ (\Phi S^{\mu\nu}) \tag{11.5.18}$$

corresponding to Fig. 11.8, in which $S^{\mu\nu}_{\beta\alpha}$ is the Feynman amplitude for $\gamma^{*\nu} + q \to \gamma^{*\mu} + q$, with external spinors and polarization vectors removed and with the replacement (11.5.13) made in the propagators.

We turn now to the question of reducing the general field-theoretic form (11.5.9) to the standard parton-model picture for polarized DIS.

Firstly, the hadronic tensor $W^{\mu\nu}$ is defined in such a way that for a quark of flavour f whose charge, in units of e is e_f,

$$W^{\mu\nu} = \frac{1}{2\pi e^2} w^{\mu\nu} = \frac{1}{2\pi e^2} \int d^4k \ \mathrm{Tr}\ (\Phi S^{\mu\nu}) \tag{11.5.19}$$

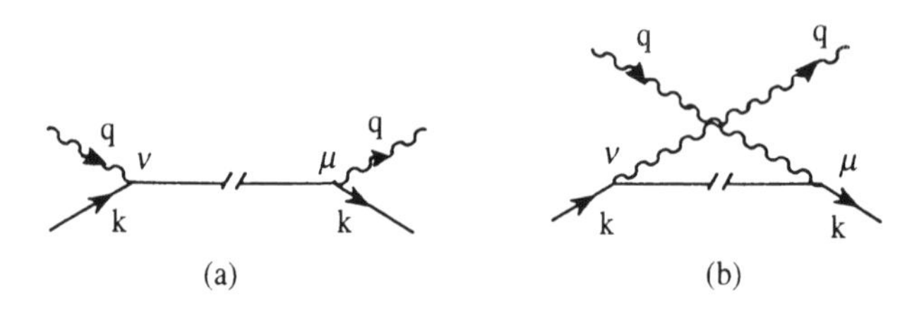

Fig. 11.9. Born diagrams for 'hard' $\gamma^* q$ interaction.

The parton model usually emerges upon making the following approximations.

(1) The $\gamma^* q$ interaction is given its simplest form, as shown in Fig. 11.9. In the following we calculate only with the uncrossed Born diagram (a). The result for the crossed diagram is obtained at the end by the replacement $x_{\mathrm{Bj}} \to -x_{\mathrm{Bj}}$ in the hadronic matrix elements connected with $\Phi_{\alpha\beta}$ and is simply to be added to the uncrossed result. In order to isolate the antisymmetric part of $W^{\mu\nu}$ we make the replacements in (11.5.10)

$$\gamma^\nu \gamma^\rho \gamma^\mu \to \tfrac{1}{2}\left(\gamma^\nu \gamma^\rho \gamma^\mu - \gamma^\mu \gamma^\rho \gamma^\nu\right) = -i\epsilon^{\mu\nu\rho\sigma}\gamma_\sigma \gamma_5 \tag{11.5.20}$$

$$\gamma^\nu \gamma^\mu \to \tfrac{1}{2}\left(\gamma^\nu \gamma^\mu - \gamma^\mu \gamma^\nu\right) = i\sigma^{\mu\nu} \tag{11.5.21}$$

and recalling (11.3.8) we find

$$W^{(\mathrm{A})}_{\mu\nu} = \frac{e_f^2}{2} \int d^4k\, \delta\left[(k+q)^2 - m_q^2\right] \left[\epsilon_{\mu\nu\rho\sigma}(q^\rho + k^\rho)\ \mathrm{Tr}\ (\gamma^\sigma \gamma_5 \Phi) \right.$$
$$\left. - m_q\ \mathrm{Tr}\ (\sigma_{\mu\nu}\Phi)\right]. \tag{11.5.22}$$

(2) One assumes that the soft matrix element cuts off rapidly for k^2 off the mass shell $k^2 = m_q^2$, and for k^μ non-collinear with respect to the hadron momentum P^μ. This is implemented as follows. Consider a reference frame where the hadron is moving at high momentum along OZ, so that

$$P^\mu = (E, 0, 0, P) \qquad \text{with} \quad E \approx P \tag{11.5.23}$$

is a 'large' 4-vector. We introduce a 'small' null vector

$$n^\mu = \left(\frac{1}{P+E},\ 0,\ 0,\ -\frac{1}{P+E}\right) \tag{11.5.24}$$

such that

$$n \cdot P = 1 \qquad n^2 = 0. \tag{11.5.25}$$

One can then write for k^μ

$$k^\mu = (k \cdot n) P^\mu + \frac{1}{2}\left[\frac{k^2 + \mathbf{k}_T^2}{(k \cdot n)} - M^2 (k \cdot n)\right] n^\mu + k_T^\mu \tag{11.5.26}$$

where

$$k_T^\mu = (0, \mathbf{k}_T, 0). \tag{11.5.27}$$

In view of the assumption about Φ we can say that

$$k^\mu \approx (k \cdot n) P^\mu. \tag{11.5.28}$$

It should be noted that some care is necessary in deciding whether the approximation (11.5.28) is adequate. We shall see that this depends crucially upon whether we are considering a nucleon with longitudinal (L) or with transverse (T) polarization.

11.5.1 Longitudinal polarization: the quark contribution to $g_1(x)$

For the study of g_1 we consider a nucleon with helicity $\lambda = \pm 1/2$ and it is sufficient to approximate (11.5.22) by putting

$$(q+k)^\rho \approx q^\rho + (k\cdot n)P^\rho \tag{11.5.29}$$

and dropping the term proportional to the quark mass m_q. Then writing

$$q^\rho + (k\cdot n)P^\rho = \int dx\, \delta(x - k\cdot n)(q+xP)^\rho \tag{11.5.30}$$

we can take the integration over d^4k in (11.5.22) through to obtain the antisymmetric component

$$W_{\mu\nu}^{(A)} = \frac{e_f^2}{2}\epsilon_{\mu\nu\rho\sigma}\int dx \frac{\delta(x - x_{Bj})}{2P\cdot q}(q+xP)^\rho A^\sigma(x), \tag{11.5.31}$$

where (using $\mathscr{S}_L$ to denote longitudinal spin)

$$\begin{aligned} A^\sigma(x) &\equiv \int \frac{d^4k}{(2\pi)^4} d^4z\, \delta(x - k\cdot n)e^{ik\cdot z}\langle P, \mathscr{S}_L|\bar{\psi}(0)\gamma^\sigma\gamma_5\psi(z)|P,\mathscr{S}_L\rangle \\ &= \int \frac{d\lambda}{2\pi} e^{i\lambda x}\langle P, \mathscr{S}_L|\bar{\psi}(0)\gamma^\sigma\gamma_5\psi(\lambda n)|P,\mathscr{S}_L\rangle \end{aligned} \tag{11.5.32}$$

is a pseudovector which can depend only upon the vectors P^μ, n^μ and $v^\mu \equiv \epsilon^{\mu\alpha\beta\gamma}\mathscr{S}_\alpha P_\beta n_\gamma$ and the pseudovector $\mathscr{S}^\mu$ and which must be linear in $\mathscr{S}^\mu$. Given that $\mathscr{S}\cdot P = 0$ the only possibilities are $\mathscr{S}^\sigma$ and $(n\cdot\mathscr{S})P^\sigma$. Note that with the normalization

$$\langle \mathbf{P}|\mathbf{P}'\rangle = (2\pi)^3 2E\delta^3(\mathbf{P}-\mathbf{P}') \tag{11.5.33}$$

$A^\sigma(x)$ has dimensions $[M]$.

Recall that for a nucleon with 4-momentum given by (11.5.23)

$$\mathscr{S}^\mu(\lambda) = \frac{2\lambda}{M}(P,0,0,E) \qquad \mathscr{S}^2 = -1, \tag{11.5.34}$$

where λ is the helicity ($\lambda = \pm 1/2$), so that

$$\mathscr{S}^\mu(\lambda) = \frac{2\lambda}{M}(P^\mu - M^2 n^\mu) \tag{11.5.35}$$

and we may take

$$\mathscr{S}^\mu(\lambda) \approx \frac{2\lambda}{M}P^\mu, \tag{11.5.36}$$

i.e. $M\mathscr{S}^\mu(\lambda)$ is a 'large' vector. In view of (11.5.36) the structures $(n \cdot \mathscr{S})P^\sigma$ and $\mathscr{S}^\sigma(\lambda)$ are equivalent in leading order and the only possibility is then (the factor 4 is for later convenience)

$$A^\sigma(x, \lambda) = 4Mh_L(x)\mathscr{S}^\sigma(\lambda) \qquad (11.5.37)$$

where the dimensionless *longitudinal distribution function* is given by

$$4h_L(x) = \frac{n_\sigma A^\sigma(x)}{2\lambda} = \int \frac{d\tau}{2\pi} e^{i\tau x} \tilde{h}_L(\tau). \qquad (11.5.38)$$

Here

$$\tilde{h}_L(\tau) = \frac{1}{Mn \cdot \mathscr{S}_L} \langle P, \mathscr{S}_L | \bar{\psi}(0) \not{n} \gamma_5 \psi(\tau n) | P, \mathscr{S}_L \rangle. \qquad (11.5.39)$$

Substituting into (11.5.31) and adding the contribution from the crossed Born diagram, Fig. 11.9(b), yields

$$W_{\mu\nu}^{(A)}(L) = e_f^2 \frac{M}{P \cdot q} \left[h_L(x_{Bj}) + h_L(-x_{Bj}) \right] \epsilon_{\mu\nu\rho\sigma} q^\rho \mathscr{S}^\sigma(\lambda). \qquad (11.5.40)$$

Note that the term xP^ρ in (11.5.31) does not contribute on account of (11.5.36). Consequently (11.5.40) is gauge invariant, $q^\mu W_{\mu\nu}^{(A)} = 0$. Note that (11.5.36), which holds only for longitudinal spin, is crucial for the gauge invariance.

Comparing with (11.3.10) in the approximation $\mathscr{S}^\mu \propto P^\mu$ we obtain, for the contribution of a quark of flavour f,

$$g_1(x) = \tfrac{1}{2} e_f^2 \left[h_L^f(x) + h_L^f(-x) \right]. \qquad (11.5.41)$$

If one treats the quark fields in (11.5.39) as free fields and regards the nucleon as an assemblage of free partons one finds

$$h_L^f(x) = \Delta q_f(x) \qquad h_L^f(-x) = \Delta \bar{q}_f(x) \qquad (11.5.42)$$

so that (11.5.41) reproduces the simple parton-model result for $g_1(x)$ in (11.4.5). Equation (11.5.41) provides a field-theoretic generalization of the parton-model result.

In the above we have been a little careless in not mentioning that most of the operators that appear require to be renormalized. That involves choosing a renormalization scale μ, and the matrix elements of the operators then depend upon μ. Physical, measurable quantities, of course, must not depend upon μ.

Although we shall be interested mainly in g_1, it is instructive to compare the case above with the case of transverse polarization, which involves g_2.

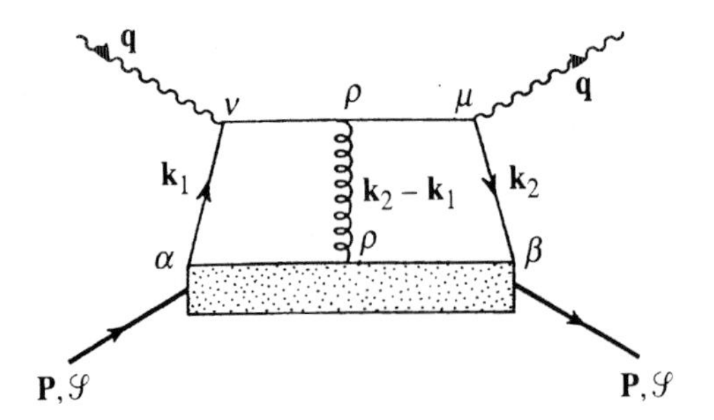

Fig. 11.10 Contribution to $\gamma^* p \to \gamma^* p$ involving quark–quark–gluon correlations.

11.5.2 *Transverse polarization:* $g_2(x)$

In order to see the essential difference between the longitudinal and transverse spin case, firstly consider again the result (11.5.40). In the CM of the γ^*–nucleon collision, for the longitudinal case we have, as far as magnitudes are concerned,

$$|q_\sigma| \sim |M\mathscr{S}_\sigma(\lambda)| = O(\nu) \qquad P \cdot q = M\nu \tag{11.5.43}$$

so that for the large components of $W^{(A)}_{\mu\nu}(L)$,

$$|W^{(A)}_{12}(L)| = O(\nu/M) \tag{11.5.44}$$

assuming that $|h_L(x)| = O(1)$. In the transverse spin case the analogue of (11.5.32) can only be proportional to $\mathscr{S}^\sigma_T$, since $n \cdot \mathscr{S}_T = 0$, and will produce a result like (11.5.40) with $\mathscr{S}(\lambda) \to \mathscr{S}_T$. Given that $|\mathscr{S}_T| = O(1)$ one has, for the 'large' components, only

$$|W^{(A)}_{\mu\nu}(L)| = O(1). \tag{11.5.45}$$

This immediately suggests that care must be exercised in neglecting non-leading terms, e.g. in (11.5.26).

Secondly, note that in (11.5.29) the term $(k \cdot n)P^\mu$ of (11.5.26) did not contribute because of the fact that, in leading order, $P^\mu \propto \mathscr{S}^\mu(\lambda)$. In the transverse case this will not happen and the analogue of (11.5.40) will contain a term $\epsilon_{\mu\nu\rho\sigma}P^\sigma\mathscr{S}^\sigma_T$ in $W^{(A)}_{\mu\nu}(T)$, which (analogously to the parton model case (11.4.4) when $m \neq m_q$) is not gauge invariant.

We must therefore return to (11.5.10) and improve upon approximation (11.5.28). However, it will then turn out that a more complicated, non-parton-model, diagram involving gluon exchange, Fig. 11.10, contributes to the same order.

Amazingly, as was shown by Efremov and Terayev (1984) this diagram just cancels the unwanted contribution from the $(k \cdot n)\not{P}$ and mass terms of the handbag diagram and the final result is gauge invariant. Essential in this proof is the use of the equations of motion for the quark field.

The analysis to show the cancellation is rather complicated and is carried out, for example, in Appendix C of Anselmino, Efremov and Leader (1995). It is helpful in this to utilize a definite QCD gauge $A^a_\mu(x) \cdot n^\mu = 0$, where $A^a_\mu(x)$ is the gluon vector potential of colour a. We shall only state the result here. It is the exact analogue of (11.5.40), namely, including the contribution of the crossed Born diagram in Fig. 11.9:

$$W^{(A)}_{\mu\nu}(T) = e_f^2 \frac{M}{P \cdot q} \left[f_T(x_{Bj}) + f_T(-x_{Bj}) \right] \epsilon_{\mu\nu\rho\sigma} q^\rho \mathcal{S}_T^\sigma \tag{11.5.46}$$

where the analogue of (11.5.37) is

$$A^\sigma(x, T) = 4M f_T(x) \mathcal{S}_T^\sigma \tag{11.5.47}$$

with

$$4 f_T(x) = \int \frac{d\tau}{2\pi} e^{i\tau x} \tilde{f}_T(\tau). \tag{11.5.48}$$

Here

$$\tilde{f}_T(\tau) = \frac{1}{M} \langle P, \mathcal{S}_T | \bar{\psi}(0) \gamma_5 \, \not{\mathcal{S}}_T \psi(\tau n) | P, \mathcal{S}_T \rangle. \tag{11.5.49}$$

Comparing (11.5.46) with (11.3.10), for the case of transverse polarization we obtain for the contribution of a quark of flavour f,

$$g_1(x) + g_2(x) = \frac{e_f^2}{2} \left[f_T^f(x) + f_T^f(-x) \right]. \tag{11.5.50}$$

Although the surviving contribution comes from the 'handbag' diagram, it does not, in fact, have any simple parton interpretation. This looks mysterious given that (11.5.49) only involves *quark* fields. The subtle point is that (11.5.49) vanishes if the fields are treated as free fields and the nucleon as an assemblage of parallel moving quarks.

It is possible to obtain a non-zero result for $g_1(x) + g_2(x)$ if the partons are allowed to have transverse momentum $\mathbf{k}_T$, but the result then depends upon the specific assumption about the $\mathbf{k}_T$ behaviour and is outside the traditional parton-model form.

11.6 Moments of the structure functions, sum rules and the spin crisis

Because of their relationship to the absorptive part of the scattering amplitudes for virtual Compton scattering one knows that $g_{1,2}(x) = 0$ for $|x| > 1$ and that $g_{1,2}(-x) = g_{1,2}(x)$. Consequently if we define the nth moment of $g_{1,2}(x)$ by $\int_0^1 dx\, x^{n-1} g_{1,2}(x) dx$ and we substitute the field-theoretic expressions from (11.5.40), (11.5.38), (11.5.46) and (11.5.48), we can extend the integration region to $-\infty \le x \le \infty$ so that integration over x in (11.5.38) and (11.5.48) will produce a δ-function $\delta(\tau)$. Subsequently performing the integral over τ will turn the bilocal operator products in (11.5.39) and (11.5.49) into a product of *local* operators. (Note that we are interchanging orders of integration so some care may be needed regarding convergence questions.) Powers of x can be turned into derivatives with respect to τ. Proceeding in this way one ends up with expressions for the odd moments of $g_{1,2}(x)$ and the even moments of what can be considered the *valence* parts of $g_{1,2}(x)$, in terms of hadronic matrix elements of local operators (Efremov, Leader and Teryaev, 1997).

The most important result of the latter type is the so-called ELT sum rule, involving the valence parts (V) of g_1 and g_2:

$$\int_0^1 \left[g_1^{\mathrm{V}}(x) + 2g_2^{\mathrm{V}}(x)\right] x dx = 0 \qquad (11.6.1)$$

This is unusual in that, like the Bjorken sum rule to be discussed below, it is a rigorous result. Testing the sum rule requires data on both protons and neutrons. Unfortunately the data on $g_2(x)$ are not yet accurate enough for a significant test.

In the operator product approach one begins with an expression for $W_{\mu\nu}$ in terms of the commutator of electromagnetic currents, which can be deduced from the Feynman diagram Fig. 11.1 to be

$$W_{\mu\nu}(q; P, \mathscr{S}) = \frac{1}{2\pi} \int d^4x\, e^{iq\cdot x} \langle P, \mathscr{S}|[J_\mu(x), J_\nu(0)]|P, \mathscr{S}\rangle \qquad (11.6.2)$$

and which implies that it is, up to a numerical factor, the imaginary part of the tensor $T_{\mu\nu}$ that appears in the expression $\epsilon^{\mu *} T_{\mu\nu} \epsilon^\nu$ for the forward virtual-photon Compton scattering amplitude.

The behaviour of $W_{\mu\nu}$ in the deep inelastic limit is controlled by the behaviour of the product of currents near the light cone $x^2 = 0$ and can be derived from Wilson's operator product expansion. A lengthy analysis is needed involving the use of dispersion relations for forward virtual Compton scattering and leads to expressions for the odd moments only.

In either approach, keeping only the leading *twist*[1] operators, which give the dominant contribution in the Bjorken limit, the final result for the moments has the form

$$\int_0^1 dx\, x^{n-1} g_1(x, Q^2) = \tfrac{1}{2} \sum_i \delta_i a_n^i E_{1,i}^n(Q^2, g) \qquad n = 1, 3, 5, \ldots \qquad (11.6.3)$$

and

$$\int_0^1 dx\, x^{n-1} g_2(x, Q^2) = \frac{1-n}{2n} \sum_i \delta_i \left[a_n^i E_{1,i}^n(Q^2, g) - d_n^i E_{2,i}^n(Q^2, g)\right]$$
$$n = 3, 5, 7, \ldots \qquad (11.6.4)$$

[1] Twist is defined as the mass dimension of an operator minus its spin.

where the δ_i are numerical coefficients, the $E_i^n(Q^2, g)$ are *coefficient functions* that can be calculated perturbatively in the strong coupling constant g and the a_n^i and d_n^i are related to the hadronic matrix elements of the local operators. The label i indicates what kind of operator is contributing: for flavour-non-singlet operators, only quark fields and their covariant derivatives occur and $i = 1, \ldots, 8$ corresponds to the components of an $SU(3)$ octet of flavours; for the flavour-singlet case, $i = \Psi$ or G corresponds to flavour-singlet combinations of quark operators or gluon operators respectively (and their covariant derivatives).

For details and for a discussion of the tantalizing question whether it is permissible to put $n = 1$ in (11.6.4), thereby obtaining the Burkhardt–Cottingham sum rule

$$\int_0^1 dx\, g_2(x, Q^2) = 0 \tag{11.6.5}$$

the reader may consult Anselmino, Efremov and Leader (1995). The data on $g_2(x, Q^2)$ are not yet accurate enough for a significant test of (11.6.5).

Here we shall concentrate on the very interesting question of the first moment of $g_1(x, Q^2)$, because it is related to the puzzle about the spin content of the nucleon. In this case (11.6.3) can be written, for the proton, as

$$\Gamma_1^p(Q^2) \equiv \int_0^1 dx g_1^p(x, Q^2) \tag{11.6.6}$$

$$= \frac{1}{12}\left[\left(a_3 + \frac{a_8}{\sqrt{3}}\right) E_{\rm NS}(Q^2) + \frac{4}{3} a_0(Q^2) E_{\rm S}(Q^2)\right] \tag{11.6.7}$$

where the non-singlet and singlet coefficient functions have the expansion[1]

$$E_{\rm NS}(Q^2) = 1 - \frac{\alpha_s}{\pi} - \begin{pmatrix} 3.58 \\ 3.25 \end{pmatrix} \left(\frac{\alpha_s}{\pi}\right)^2 \cdots \tag{11.6.8}$$

$$E_{\rm S}(Q^2) = 1 - \frac{\alpha_s}{\pi} - \begin{pmatrix} 1.10 \\ -0.07 \end{pmatrix} \left(\frac{\alpha_s}{\pi}\right)^2 \cdots \tag{11.6.9}$$

where $\alpha_s = \alpha_s(Q^2)$ is the running QCD coupling and the upper and lower numbers correspond to taking either three flavours of quark or four flavours if one includes the charm quark.

In the above, a_3 and a_8 are measures of the proton matrix elements of an $SU(3)$ flavour octet of quark axial-vector currents:

$$\langle P, \mathcal{S} | J_{5\mu}^j | P, \mathcal{S} \rangle = M a_j \mathcal{S}_\mu \qquad j = 1, \ldots, 8 \tag{11.6.10}$$

[1] These coefficients are 'scheme dependent' (see Section 11.7) and the result quoted corresponds to the $\overline{\rm MS}$ scheme.

where

$$J^j_{5\mu} = \bar{\psi}\gamma_\mu\gamma_5 \left(\frac{\lambda_j}{2}\right) \psi. \tag{11.6.11}$$

Here the λ_j are the usual Gell-Mann matrices and ψ is a column vector in flavour space,

$$\psi = \begin{pmatrix} \psi_u \\ \psi_d \\ \psi_s \end{pmatrix}. \tag{11.6.12}$$

The coefficient a_0 in (11.6.7) is a measure of the flavour-singlet operator. Now in (11.6.3), for $n \geq 2$ there are both gluonic and quarkonic flavour-singlet operators but for $n = 1$, the case we are now considering, the OPE has only one operator, the quark flavour-singlet current

$$J^0_{5\mu} = \bar{\psi}\gamma_\mu\gamma_5\psi \tag{11.6.13}$$

and a_0 is defined by

$$\langle P, \mathscr{S}|J^0_{5\mu}|P, \mathscr{S}\rangle = 2Ma_0\mathscr{S}_\mu. \tag{11.6.14}$$

The absence of a gluonic operator in the first moment of g_1 will turn out to be a non-trivial issue.

To the extent that flavour $SU(3)$ is a global symmetry of the strong interactions the non-singlet octet of currents will be conserved currents, and this will lead to the a_j ($j = 1, \ldots, 8$) being independent of Q^2. The singlet current is not conserved, as a consequence of the axial anomaly (Adler, 1969; Bell and Jackiw, 1969), so that a_0 depends on Q^2. (This will be discussed in subsection 11.6.2.)

Now what is remarkable is that the octet of axial-vector currents (11.6.11) is precisely the set of currents that controls the weak β-decays of the neutron and of the spin-1/2 hyperons. Consequently a_3 and a_8 can be expressed in terms of two parameters F and D *measured* in hyperon β-decay (see, for example, Chapter 4 of Bailin, 1982):

$$a_3 = F + D \equiv g_A = 1.2573 \pm 0.0028 \tag{11.6.15}$$

$$\frac{1}{\sqrt{3}}a_8 = \tfrac{1}{3}(3F - D) = 0.193 \pm 0.008. \tag{11.6.16}$$

It follows that the measurement of $\Gamma_1^p(Q^2)$ in polarized DIS can be interpreted, via (11.6.7), as a measurement of $a_0(Q^2)$.

The determination of $\Gamma_1^p(Q^2)$ is not entirely straightforward, firstly since extrapolations of the data on $g_1(x, Q^2)$ have to be made to the regions $x = 0$ and $x = 1$ in calculating the integral in (11.6.6) and secondly because the data at different x-values correspond to different ranges of Q^2. At present the value of Γ_1^p at $Q^2 = 10$ (GeV/c)^2 is believed to lie in the range

$$0.130 \leq \Gamma_1^p(Q^2 = 10\ (\mathrm{GeV}/c)^2) \leq 0.142 \tag{11.6.17}$$

which leads to

$$0.22 \leq a_0(Q^2 = 10\ (\text{GeV}/c)^2) \leq 0.34. \tag{11.6.18}$$

In the famous EMC experiment the measured value of a_0 was consistent with zero and led to a 'crisis in the parton model' (Leader and Anselmino, 1988). The present value measured of a_0 is still disturbingly small.

Before turning to discuss this intriguing question, note that in going from the case of a proton to a neutron, a_8 and a_0 in (11.6.7) remain unchanged whereas a_3 reverses its sign. Consequently one has the *Bjorken sum rule*:

$$\int_0^1 dx\, \left[g_1^p(x, Q^2) - g_1^n(x, Q^2)\right] = \frac{g_A}{6} E_{\text{NS}}(Q^2). \tag{11.6.19}$$

This is considered to be a very fundamental result. Moreover the right-hand side is known to great accuracy (see (11.6.15)) and much effort has gone into trying to test (11.6.19). Up to the present (11.6.19) seems to be well satisfied by the data, as will be discussed in Section 11.8.

11.6.1 A spin crisis in the parton model

In the naive parton model the nucleon is simply an ensemble of free, parallel-moving quarks. The picture can be recovered by putting the QCD coupling g equal to zero. In that case the quark fields in $J^0_{5\mu}$ in (11.6.13) and (11.6.11) become free fields, and treating the nucleon state in (11.6.14) as a superposition of free-quark states one easily finds that

$$a_0 = \Delta\Sigma \equiv \int_0^1 \Delta\Sigma(x) dx \tag{11.6.20}$$

where

$$\Delta\Sigma(x) \equiv \Delta u(x) + \Delta\bar{u}(x) + \Delta d(x) + \Delta\bar{d}(x) + \Delta s(x) + \Delta\bar{s}(x). \tag{11.6.21}$$

Now given the physical significance of the number densities $q_\pm(x)$ discussed in Section 11.4 it is clear that the integral in (11.6.20) is just twice the expectation value of the sum of the z-components of the quark and antiquark spins, i.e.

$$a_0 = \Delta\Sigma = 2\left\langle S_z^{\text{quarks}}\right\rangle, \tag{11.6.22}$$

which implies, if we adopt (11.1.1) uncritically, that we expect $a_0 \approx 1$.

As mentioned, the EMC experiment found a_0 compatible with zero provoking a 'crisis in the parton model' and the present value, given in (11.6.18), is still small compared with naive expectations. It is not clear how to quantify the extent to which we expect a_0 to differ from its naive value, but it is generally felt that the small value in (11.6.18) is not in accord with intuition.

We shall see in the next section that (11.6.20) and hence (11.6.22) are, surprisingly, not correct in the interacting theory.

11.6.2 The gluon anomaly

Consider the axial current

$$J^f_{5\mu} = \bar{\psi}_f(x)\gamma_\mu\gamma_5\psi_f(x) \tag{11.6.23}$$

made up of quark operators of definitive flavour f (an implicit colour sum is always implied). From the free Dirac equation of motion one finds that

$$\partial^\mu J^f_{5\mu} = 2im_q\bar{\psi}_f(x)\gamma_5\psi_f(x) \tag{11.6.24}$$

where m_q is the mass of the quark of flavour f. In the chiral limit $m_q \to 0$ (11.6.24) appears to imply that $J^f_{5\mu}$ is conserved. If this were really true then there would be a symmetry between left- and right-handed quarks, leading to a parity degeneracy of the hadron spectrum, e.g. there would exist two protons of opposite parity. However, the formal argument from the free equations of motion is not reliable and there is an anomalous contribution from the triangle diagram shown in Fig. 11.11, where two gluons couple to the current of (11.6.23).

As a consequence the axial current is not conserved when $m_q = 0$. One has instead, for the QCD case,

$$\partial^\mu J^f_{5\mu} = \frac{\alpha_s}{4\pi} G^a_{\mu\nu}\tilde{G}^{\mu\nu}_a \tag{11.6.25}$$

where $\tilde{G}^{\mu\nu}_a$ is the dual field tensor

$$\tilde{G}^{\mu\nu}_a \equiv \tfrac{1}{2}\epsilon^{\mu\nu\rho\sigma}G^a_{\rho\sigma}. \tag{11.6.26}$$

The result (11.6.25), which emerges from a calculation of the triangle diagram using $m_q = 0$ and the gluon virtuality $k^2 \neq 0$, is really a particular

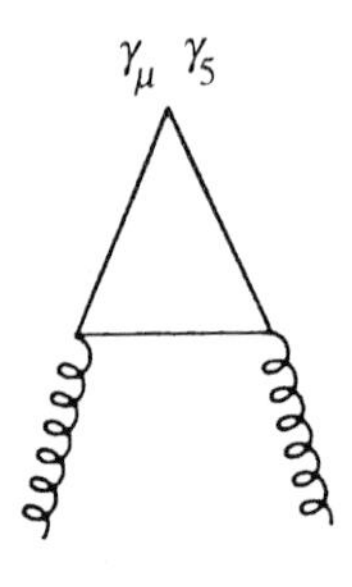

Fig. 11.10. Triangle diagram giving rise to the gluon anomaly.

limit of a highly non-uniform function. If we take $m_q \neq 0$, $k^2 \neq 0$ the right-hand side of (11.6.25) is multiplied by

$$T\left(m_q^2/k^2\right) = 1 - \frac{2m_q^2/k^2}{\sqrt{1+4m_q^2/k^2}} \ln\left(\frac{\sqrt{1+4m_q^2/k^2}+1}{\sqrt{1+4m_q^2/k^2}-1}\right). \qquad (11.6.27)$$

We see that on the one hand the gluon anomaly corresponds to $T \to 1$ for $m_q^2/k^2 \to 0$. On the other hand, for on-shell gluons $k^2 = 0$ and $m_q \neq 0$, i.e. in the limit $m_q^2/k^2 \to \infty$ the terms cancel, $T \to 0$ and there is no anomaly. In our case the gluons are strictly speaking, bound inside the proton so they are off shell and $k^2 \neq 0$ is the relevant case.

The anomaly induces an effectively point-like interaction between $J^0_{\mu 5}$ and gluons. Using the expression given in Adler (1969), generalized to QCD, the matrix element of $J^0_{\mu 5}$ between almost free gluons is

$$\langle k;\lambda|J^f_{5\mu}|k;\lambda\rangle = \frac{i\alpha_s}{2\pi}\epsilon_{\mu\nu\rho\sigma}k^\nu \epsilon^{*\rho}(\lambda)\epsilon^\sigma(\lambda)T(m_q^2/k^2);$$

this, via (3.4.28) and (3.1.80), becomes

$$\langle k;\lambda|J^f_{5\mu}|k;\lambda\rangle = -\frac{\alpha_s}{2\pi}\mathcal{S}^{\text{gluons}}_\mu(k,\lambda)T(m_q^2/k^2), \qquad (11.6.28)$$

where λ is the gluon helicity and we may take

$$\mathcal{S}^{\text{gluons}}_\mu(k,\lambda) \approx \lambda k_\mu$$

as the covariant spin vector for almost massless gluons.

The component of the proton wave function containing almost free gluons then yields a gluonic contribution to a_0

$$a_0^{\text{gluons}}(Q^2) = -3\frac{\alpha_s}{2\pi}\Delta G(Q^2) \qquad (11.6.29)$$

where

$$\Delta G(Q^2) \equiv \int_0^1 \Delta G(x,Q^2)dx. \qquad (11.6.30)$$

Here $\Delta G(x)$ is the analogue for gluons of $\Delta q(x)$. The concept of Q^2-dependent parton distributions such as $\Delta G(x,Q^2)$ is explained in Section 11.7. The factor 3 in (11.6.29) arises from taking $m_u, m_d, m_s \ll k^2$, $m_c, m_b, m_t \gg k^2$ so that $T = 1$ for m_u, m_d, m_s and $T = 0$ for m_c, m_b, m_t.

Instead of (11.6.20) we now have (Efremov and Teryaev, 1988; Altarelli and Ross, 1988; Carlitz, Collins and Mueller, 1988; Anselmino and Leader, 1988)

$$a_0(Q^2) = \Delta\Sigma - \frac{3\alpha_s(Q^2)}{2\pi}\Delta G(Q^2). \qquad (11.6.31)$$

The result (11.6.31) is remarkable. It shows that there *is* a gluonic contribution to the first moment of g_1. Moreover it is quite anomalous in the sense that it look like a perturbative correction that will disappear at large Q^2, where $\alpha_s(Q^2) \to 0$, but in reality does not do so because $\Delta G(Q^2)$ can be shown to grow in precisely the right way to compensate for the decrease of $\alpha_s(Q^2)$. It also has the fundamental implication that *the small measured value of* a_0 *does not necessarily imply that* $\Delta\Sigma$ *is small.*

So, for example, we could let the quarks carry 60% of the proton's spin at $Q^2 = 10\ (\text{GeV}/c)^2$ and the experimental value (11.6.18) would then imply

$$2.2 \leq \Delta G\left(Q^2 = 10\ (\text{GeV}/c)^2\right) \leq 3.3. \tag{11.6.32}$$

Now similarly to (11.6.22), $\Delta G(Q^2)$ measures the contribution to the proton's spin arising from the spin of the gluon constituents. Bearing in mind (11.1.1) we have the apparently surprising result that

$$\left\langle S_z^{\text{gluons}}\left(Q^2 = 10\ (\text{GeV}/c)^2\right)\right\rangle \approx 2 \sim 3.$$

However, the operator corresponding to the spin of a gluon is not a conserved operator, so its matrix elements depend upon the renormalization scale (this causes the Q^2-dependence) and $\langle S_z^G(Q^2)\rangle$ does not have a simple physical interpretation as a fixed number. Indeed $\langle S_z^G(Q^2)\rangle \to \infty$ as $Q^2 \to \infty$ and this is compensated by the fact that the gluon *orbital* angular momentum grows in the opposite sense:

$$\left\langle L_z^{\text{gluons}}(Q^2)\right\rangle \to -\infty \qquad \text{as} \quad Q^2 \to \infty.$$

Given that gluons play no rôle in the low energy constituent quark model, it is somewhat reassuring that the above value of $\langle S_z^{\text{gluons}}\rangle$ at $Q^2 = 10\ (\text{GeV}/c^2$ leads, via a perturbative calculation, to

$$\left\langle S_z^{\text{gluons}}(Q^2 = 4\Lambda_{\text{QCD}}^2)\right\rangle \lesssim 0.6.$$

Below this value of Q^2 we enter the non-perturbative regime so cannot estimate how $\langle S_z^{\text{gluons}}\rangle$ behaves.

In contrast to this, $\Delta\Sigma$ or $\langle S_z^{\text{quarks}}\rangle$ can be linked to a conserved operator (see Anselmino, Efremov and Leader, 1995, Section 6.3) and so are independent of Q^2. It thus makes sense to expect that $\Delta\Sigma \approx 1$.

11.7 QCD corrections and evolution

The field-theoretic approach of Section 11.5 leads to the simple parton model when the hard scattering amplitude E in Fig. 11.4 is treated in the Born approximation and the quark fields as free fields. When gluonic corrections are included, problems arise from the so-called *mass or collinear singularities* linked to the effective masslessness of the partonic quarks. A subtle process of *factorization at scale* μ (chosen for simplicity to be the same as the renormalization scale) allows the singular terms to be absorbed into the non-calculable parton distribution functions. These then depend on μ, leaving the finite terms as Q^2-dependent correction terms in the expressions for the structure functions, which thus no longer obey exact Bjorken scaling. In fact they develop a slow logarithmic dependence on Q^2.

In the *leading logarithmic approximation* (LLA) one keeps only the most dominant terms, proportional to $\alpha_s \ln(Q^2/\mu^2)$, and finds that the parton-model expressions remain valid provided the replacements

$$q(x) \to q(x, Q^2) \qquad \Delta q(x) \to \Delta q(x, Q^2) \tag{11.7.1}$$

to the Q^2-dependent parton distributions are made.

In this approximation the $q(x, Q^2)$ and $\Delta q(x, Q^2)$ are universal, i.e. they are a property of the nucleon and will appear in any hard reaction involving the nucleon.

The x-dependence of the $\Delta q(x, Q^2)$ cannot be calculated, of course, but the variation with Q^2 is controlled by the Gribov, Lipatov, Altarelli and Parisi *evolution equations* (Gribov and Lipatov, 1972; Altarelli and Parisi, 1977), which have the generic form

$$\frac{d}{d\ln Q^2}\Delta q(x, Q^2) = \frac{\alpha_s(Q^2)}{2\pi}\int_x^1 \frac{dy}{y}\left\{\Delta P_{qq}^{(0)}(x/y)\Delta q(y, Q^2) + \Delta P_{qG}^{(0)}\Delta G(y, Q^2)\right\} \tag{11.7.2}$$

$$\frac{d}{d\ln Q^2}\Delta G(x, Q^2) = \frac{\alpha_s(Q^2)}{2\pi}\int_x^1 \frac{dy}{y}\left\{\Delta P_{Gq}^{(0)}(x/y)\Delta q(y, Q^2) + \Delta P_{GG}^{(0)}(x/y)\Delta G(y, Q^2)\right\} \tag{11.7.3}$$

The $\Delta P_{ij}^{(0)}$ are the lowest-order *polarized splitting functions*, first given for QCD in Altarelli and Parisi (1977).

11.7.1 Beyond leading order; scheme dependence

When the *non-dominant* correction terms are included and when one works to order α_s^2, i.e. to the next-to-leading order (NLO), two new features arise.

(1) The expression for $g_1(x, Q^2)$ in (11.4.5) is modified to

$$g_1(x, Q^2) = \tfrac{1}{2}\sum_q e_q^2 \left\{\Delta q(x, Q^2) + \Delta\bar{q}(x, Q^2) + \right. \\ + \frac{\alpha_s(Q^2)}{2\pi}\int_x^1 \frac{dy}{y}\left[\Delta C_q(x/y)\left[\Delta q(y, Q^2) + \Delta\bar{q}(y, Q^2)\right]\right. \\ \left.\left. + \Delta C_G(x/y)\Delta G(y, Q^2)\right]\right\} \tag{11.7.4}$$

where the sum is over the *flavours* of the quarks and antiquarks and the *coefficient functions* ΔC_q, ΔC_G are related to calculable short-distance, i.e. hard, photon–quark and photon–gluon cross-sections. Figure 11.12 shows the mechanism whereby the photon couples to the gluon. The convolutions in (11.7.4) are often symbolized by $\Delta C_q \otimes \Delta q$, $\Delta C_G \otimes \Delta G$ etc. and have the property that the moment of the convolution is the product of the moments of the functions: $(f \otimes g)^{(n)} = f^{(n)} g^{(n)}$.

At this order the evolution equations have the form of (11.7.2) and (11.7.3) but the splitting functions, now calculated to higher order, have the form

$$\Delta P_{ij} = \Delta P_{ij}^{(0)} + \frac{\alpha_s}{2\pi} \Delta P_{ij}^{(1)}.$$

All these functions are given in a very clear form in Vogelsang (1996).

(2) The massive calculations involved are plagued by ambiguities linked to the renormalization of operators containing γ_5. In *any* theory requiring infinite renormalization the subtraction of the infinite terms is clearly defined, but the handling of associated finite terms is a matter of taste, giving rise to a *renormalization-scheme dependence* of the auxiliary quantities in any calculation. Physically measurable quantities, like g_1 for example, must be independent of the choice of scheme. But in NLO the coefficient functions, and therefore the parton distributions, become scheme dependent and one must specify in what scheme one is working.

For the unpolarized case this is straightforward and there are simple unambiguous ways to define the various schemes in use, DIS, MS (minimal subtraction), $\overline{\text{MS}}$ etc. Moreover the parton distributions in the various schemes differ from each other only by terms of order $\alpha_s(Q^2)$. In the polarized case, because of the complexity of the calculations one often renormalizes using the modern *dimensional regularization* technique (for a simple introduction, see Leader and Predazzi, 1996, Vol. 2) and it then turns out that specifying the finite subtractions as MS or $\overline{\text{MS}}$ is not enough

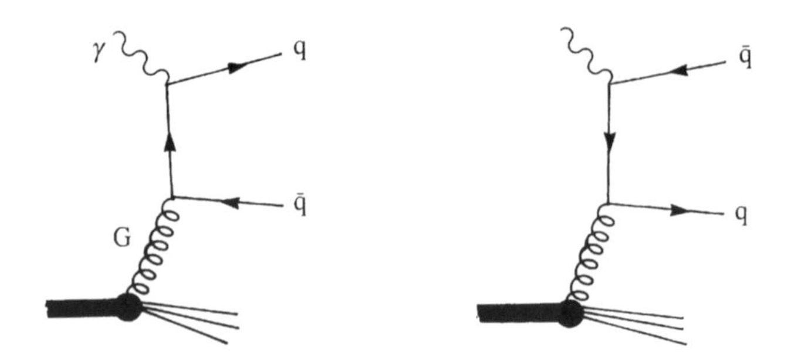

Fig. 11.11 Diagram illustrating how the photon couples to the gluon in the nucleon.

– there remain ambiguities linked to the freedom in defining γ_5 in more than four-dimensions.

The classic method of handling γ_5, due to 't Hooft and Veltman (1972) and to Breitenlohner and Maison (1977), leads to what we shall label the $\overline{\text{MS}}$–HVBM scheme. But this has the undesirable peculiarity that the renormalized isovector current $J^3_{\mu 5}$, see eqn (11.6.11), is *not* conserved.

The main schemes in use are:

(1) a modified $\overline{\text{MS}}$–HVBM scheme, due to Mertig and van Neerven (1996) and to Vogelsang (1996), which does conserve $J^3_{\mu 5}$ and which we shall refer to as the $\overline{\text{MS}}$–MNV *scheme*;
(2) a scheme referred to as the *AB scheme* in Ball, Forte and Ridolfi (1996), which modifies only the first moment of $\Delta\Sigma(x, Q^2)$, so as to make it independent of Q^2;
(3) the more physically motivated scheme advocated by Carlitz, Collins and Mueller (1988), by Anselmino, Efremov and Leader (1995) and, on the basis of more general arguments, by Teryaev and Müller (1997). This has the advantage that the contribution to $g_1(x, Q^2)$ arising from the reaction virtual-photon $+p \to \text{jet}(\mathbf{k}_T) + \text{jet}(-\mathbf{k}_T) + X$ (see Fig. 11.11), for the production of two jets with *large* transverse momentum $\mathbf{k}_T$, is directly given by the gluon term in (11.7.4) with, clearly, the coefficient function appropriate to this scheme, which is given below. We shall label this the *JET scheme*. Of course, this scheme also has the desirable property that the first moment of $\Delta\Sigma(x, Q^2)$ is independent of Q^2.

The most important difference between these schemes shows up in the first moment of g_1. One finds that

$$\int_0^1 dx\, \Delta C_G^{\overline{\text{MS}}-\text{MNV}}(x) = 0 \tag{11.7.5}$$

so that the gluon makes no contribution to the first moment of g_1. Thereby one loses the nice explanation given by (11.6.31) for the smallness of a_0. Moreover the first moment of the quark-singlet combination, $\Delta\Sigma^{\overline{\text{MS}}\text{-MNV}}(Q^2)$, varies with Q^2 so cannot be compared with the constituent quark result (11.1.1).

On the contrary, in the AB and JET schemes one has

$$\int_0^1 dx\, \Delta C_G^{\text{AB}}(x) = \int_0^1 dx\, \Delta C_G^{\text{JET}}(x) = -1 \tag{11.7.6}$$

in exact agreement with the result for a_0 in (11.6.31). Moreover the quark-singlet-contribution first moment, $\Delta\Sigma^{\mathrm{AB}} = \Delta\Sigma^{\mathrm{JET}}$, is independent of Q^2, allowing an intuitive interpretation as the spin carried by the quarks. For reasons explained earlier we feel the JET scheme has a more direct physical interpretation than the AB scheme.

There is another strange feature peculiar to the polarized case. The gluon distributions are the same in the $\overline{\mathrm{MS}}$–MNV, AB and JET schemes, but the first moments of the singlet quark densities are related by

$$\Delta\Sigma^{\mathrm{AB}} = \Delta\Sigma^{\mathrm{JET}} = \Delta\Sigma^{\overline{\mathrm{MS}}-\mathrm{MNV}}(Q^2) + \frac{1}{3}\frac{\alpha_s(Q^2)}{2\pi}\Delta G(Q^2) \tag{11.7.7}$$

so that, as explained after eqn (11.6.31), the difference between them is not really of order $\alpha_s(Q^2)$ and could be quite large. Scheme changes of this type are thus quite anomalous compared with the unpolarized case.

In the JET scheme the coefficient $\Delta C_G(x)$ appearing in the expression (11.7.4) for $g_1(x, Q^2)$ is given by

$$\Delta C_G^{\mathrm{JET}}(x) = (2x-1)\left[\ln\left(\frac{1-x}{x}\right) - 1\right] \tag{11.7.8}$$

and $\Delta C_q(x)$ is the same in both types of scheme.

In NLO the quark non-singlet distributions are actually of two kinds, combinations like $(\Delta u + \Delta\bar{u}) - (\Delta d + \Delta\bar{d})$, which are genuinely *flavour non-singlets*, and combinations like $(\Delta u - \Delta\bar{u}) \equiv \Delta u_V$, which are *valence non-singlets*. These have different evolution properties, as explained in Vogelsang (1996). In comparing the $\overline{\mathrm{MS}}$–MNV, AB and JET schemes, we find that all non-singlet distributions are the same in these schemes. Also the gluon density is the same. Only the singlet quark density $\Delta\Sigma(x, Q^2)$ changes. For *any* scheme change of this type, one has

$$\begin{aligned}\Delta\Sigma(x, Q^2)|_{\mathrm{new}} &= \Delta\Sigma(x, Q^2)|_{\overline{\mathrm{MS}}-\mathrm{MNV}} \\ &\quad + \frac{N_f}{2}\frac{\alpha_s(Q^2)}{2\pi} h_G \otimes \Delta G\end{aligned} \tag{11.7.9}$$

where $h_G(x)$ is a function specifying the change; for the transformation $\overline{\mathrm{MS}}$–MNV $\rightarrow$ JET one has

$$h_G(x) = 4(1-x). \tag{11.7.10}$$

Of course the NLO part of the splitting functions, which control the evolution in (11.7.2) and (11.7.3), is different in the two schemes. The connection is given by

$$\begin{aligned}\left(\Delta P_{qq}^{\mathrm{PS}}\right)_{\mathrm{new}} &= \left(\Delta P_{qq}^{\mathrm{PS}}\right)_{\overline{\mathrm{MS}}-\mathrm{MNV}} \\ &\quad + \frac{N_f}{2}\frac{\alpha_s}{2\pi} h_G \otimes \Delta P_{Gq}^{(0)}\end{aligned} \tag{11.7.11}$$

$$(\Delta P_{qG})_{\text{new}} = (\Delta P_{qG})_{\overline{\text{MS}}-\text{MNV}} + \frac{N_f}{2}\frac{\alpha_s}{2\pi} \times \left[h_G \otimes \left(\Delta P^{(0)}_{GG} - \Delta P^{(0)}_{qq}\right) - \frac{\beta_0}{2} h_G\right] \tag{11.7.12}$$

$$(\Delta P_{Gq})_{\text{new}} = (\Delta P_{Gq})_{\overline{\text{MS}}-\text{MNV}} \tag{11.7.13}$$

$$(\Delta P_{GG})_{\text{new}} = (\Delta P_{GG})_{\overline{\text{MS}}-\text{MNV}} - \frac{N_f}{2}\frac{\alpha_s}{2\pi} h_G \otimes \Delta P^{(\circ)}_{Gq} \tag{11.7.14}$$

where

$$\beta_0 = 11 - 2N_f/3. \tag{11.7.15}$$

For the case $\overline{\text{MS}}$–MNV → JET these simplify to

$$\left(\Delta P^{PS}_{qq}\right)_{\text{JET}} = \left(\Delta P^{PS}_{qq}\right)_{\overline{\text{MS}}-\text{MNV}} - 8\frac{N_f}{3}\frac{\alpha_s}{2\pi}\left[(x+2)\ln x + 3(1-x)\right] \tag{11.7.16}$$

$$(\Delta P_{qG})_{\text{JET}} = (\Delta P_{qG})_{\overline{\text{MS}}-\text{MNV}} + 4\frac{N_f}{3}\frac{\alpha_s}{2\pi} \times \{5(1-x)\left[\ln(1-x) - 7\right] - (11x+16)\ln x\} \tag{11.7.17}$$

$$(\Delta P_{Gq})_{\text{JET}} = (\Delta P_{Gq})_{\overline{\text{MS}}-\text{MNV}} \tag{11.7.18}$$

$$(\Delta P_{GG})_{\text{JET}} = (\Delta P_{GG})_{\overline{\text{MS}}-\text{MNV}} + 8\frac{N_f}{3}\frac{\alpha_s}{2\pi}\left[(x+2)\ln x - 3(1-x)\right] \tag{11.7.19}$$

Note that the connection between the nth moments following from (11.7.9) is

$$\Delta\Sigma^{(n)}(Q^2)|_{\text{JET}} = \Delta\Sigma^{(n)}(Q^2)|_{\overline{\text{MS}}-\text{MNV}} + N_f \frac{\alpha_s(Q^2)}{2\pi}\frac{2}{n(n+1)}\Delta G^{(n)}(Q^2). \tag{11.7.20}$$

We remind the reader that detailed expressions for all the $\overline{\text{MS}}$–MNV functions can be found in Vogelsang (1996). A study of scheme dependence in the analysis of data is given in Leader, Sidorov and Stamenov (1998b).

11.8 Phenomenology: the polarized-parton distributions

Pioneering experiments with polarized electron beams and polarized proton targets at SLAC in the 1970s demonstrated significant spin dependence, but it was not until the surprising results of the EMC experiment in 1988 that the field really took off. A vast programme of experiments has been, and is, under way at CERN (the SMC group), at SLAC (experiments E 142, 143, 154, 155) and at HERA (the HERMES collaboration) and precision data on the polarized structure function g_1 is available for protons, neutrons and deuterons over a reasonable range of x and Q^2. Data on g_2 have recently begun to be published. For access to the experimental literature, see Abe *et al.* (1997a).

Several NLO analyses of the data have been carried out during the last year or two, leading to much improved information on the polarized parton densities (Glück, Reya, Stratmann and Vogelsang, 1996; Ball, Forte and Ridolfi, 1996; Abe *et al.*, 1997b; Altarelli *et al.*, 1997; Leader, Sidorov and Stamenov, 1998a, 1999).

However, it would be wrong to imagine that the polarized densities can now be determined to the same accuracy with which the unpolarized densities are known. This can be understood quite simply. Up to the present the polarized data consist solely of fully inclusive neutral current (in effect, photon-induced) reactions on nucleons, i.e. one has information on two polarized structure functions $g_1^p(x, Q^2)$ and $g_1^n(x, Q^2)$. Even if one makes some simplifying assumptions about the polarized sea, one is still expressing two experimental functions in terms of four densities $\Delta u, \Delta d, \Delta\bar{q}$ and ΔG. What is lacking here is information from charged current reactions, which play an important rôle in pinning down the unpolarized densities. Neutrino experiments on a polarized target have been inconceivable up to now and charged current reactions of the type $ep \to \nu X$ will be extremely difficult. The situation is somewhat alleviated by the constraints coming from the beautiful connection between the first moments of the polarized parton densities and weak interaction physics, as discussed in Section 11.6; see eqns (11.6.15) and (11.6.16). For one has

$$a_3 = \int_0^1 dx \left[\Delta u(x, Q^2) + \Delta\bar{u}(x, Q^2) - \Delta d(x, Q^2) - \Delta\bar{d}(x, Q^2)\right] \qquad (11.8.1)$$

and

$$a_8 = \frac{1}{\sqrt{3}} \int_0^1 dx \left[\Delta u(x, Q^2) + \Delta\bar{u}(x, Q^2) + \Delta d(x, Q^2) \right.$$
$$\left. + \Delta\bar{d}(x, Q^2) - 2\Delta s(x, Q^2) - 2\Delta\bar{s}(x, Q^2)\right]. \qquad (11.8.2)$$

In all analyses one chooses some parametrization for the functional form of the distribution at some initial scale Q_0^2, in terms of a small number of the unknown parameters, and then evolves the distributions up to the values of Q^2 corresponding to the data and determines the parameters by a best fit to the data. A typical parametrization might involve the unpolarized distribution in the generic form

$$\Delta f(x, Q_0^2) = Ax^\alpha(1-x)^\beta f(x, Q_0^2) \qquad (11.8.3)$$

where f is the unpolarized version of Δf and where A, α and β are to be determined from the data. Or, in the approach of Brodsky, Burkhardt and Schmidt (1995), both $f(x, Q_0^2)$ and $\Delta f(x, Q_0^2)$ are parametrized as polynomials in x and a simultaneous fit is made to both the polarized and the unpolarized data.

There seem to be two sensible choices of distribution to parametrize at Q_0^2:

(1) $\Delta u + \Delta \bar{u}$, $\Delta d + \Delta \bar{d}$, Δs and ΔG, or

(2) Δu_V, Δd_V, ΔG, together with some *ansatz* about the sea, e.g. $SU(3)$-symmetric $\Delta \bar{u} = \Delta \bar{d} = \Delta s$ or a weighting in favour of the lightest quarks, e.g. $\Delta \bar{u} = \Delta \bar{d} = \lambda_s \Delta s$, $\lambda_s > 1$. Here $\Delta q_V \equiv \Delta q - \Delta \bar{q}$ are the *valence* parton densities.

In Fig. 11.12 we show recent results on g_1^p from the SMC collaboration. Of great interest is the comparison of their 1996 and 1993 data, especially at small x. We shall discuss this issue in the next section.

In Fig. 11.13 the results on g_1^n from experiments 142 and 154 at SLAC are shown. Here the data are not strictly measured values; they have been evolved to a common value $Q^2 = 5$ $(\text{GeV}/c)^2$, but for these experiments this gives only a very small effect. Again, the behaviour at small x raises interesting questions.

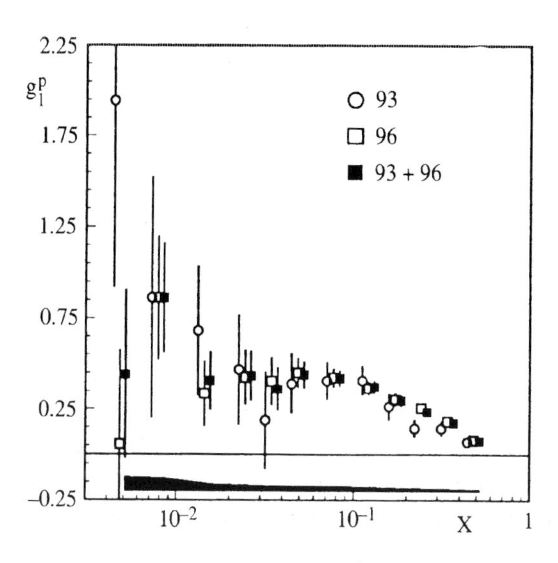

Fig. 11.12 SMC data on $g_1^p(x, Q^2)$ at the measured Q^2 for each x and comparing 1996 and 1993 data. The solid band indicates the systematic uncertainty. '93 + 96' means a weighted average of the two sets of results. (From Adeva *et al.*, 1997.)

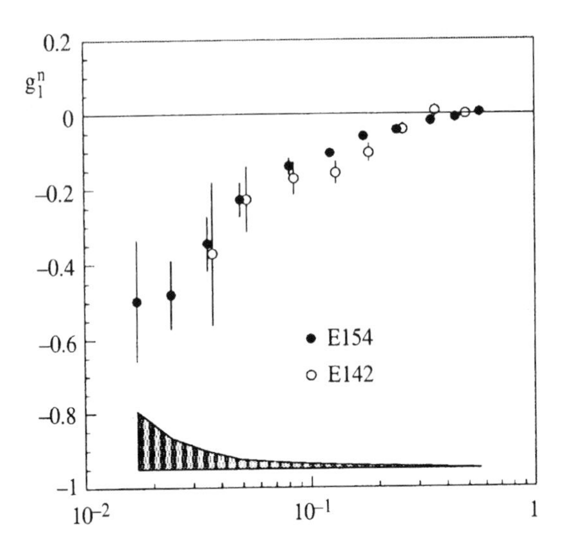

Fig. 11.13 SLAC data on $g_1^n(x, Q^2)$ from experiments E142 and the later E154. The data have been evolved to a common value $Q^2 = 5\ (\mathrm{GeV}/c)^2$ assuming that g_1/F_1 is independent of Q^2 in the range of these measurements. The shaded band indicates the systematic uncertainty. (From Abe *et al.*, 1997c.)

The HERMES group at HERA has recently begun to publish results. An example, comparing their data on g_1^n with the SLAC E142 data, is shown in Fig. 11.15.

In Fig. 11.16 we show typical shapes of the polarized-parton densities at $Q^2 = 4\ (\mathrm{GeV}/c)^2$. In comparing different analyses one finds that on the one hand $\Delta u(x)$ is rather well determined, as is $\Delta d(x)$ for medium

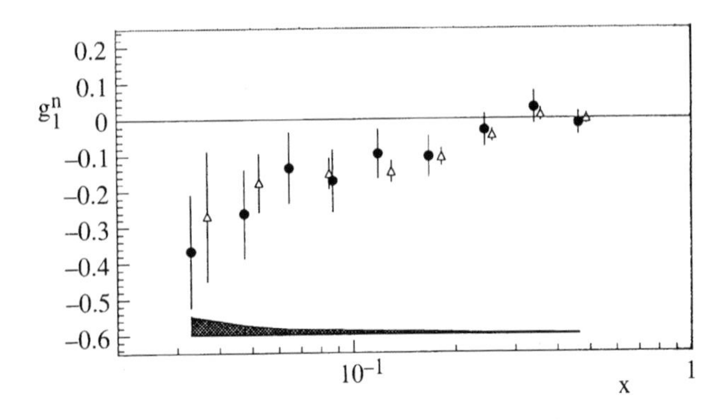

Fig. 11.14 HERMES data on g_1^n, compared with SLAC E142 data: •, HERMES; △, E142. (From Ackerstaff *et al.*, 1997.)

values of x, but the behaviour of $\Delta d(x)$ for small x and for $x \gtrsim 0.35$ is not yet accurately known. On the other hand the sea-quark distribution is still largely undetermined though it is claimed that there is a tendency to favour an $SU(3)$ flavour symmetric sea. This is quite misleading since *in principle*, this cannot be determined form the data (see Leader, Sidorov and Stamenov, 1998a). Unfortunately the most interesting quantity of all, the gluon distribution $\Delta G(x, Q^2)$, is still relatively poorly determined. That this is so can be understood from the facts that its *direct* contribution to $g_1(x, Q^2)$ is of order $\alpha_s(Q^2)$, see (11.7.4) and that its main rôle is in the evolution equations. However, the range of Q^2 thus far measured is too small for the latter to play a definitive rôle.

In order to give greater weight to the process of evolution, Glück, Reya, Stratmann and Vogelsang (1996) chose the very low value $Q_0^2 = 0.34\ (\text{GeV}/c)^2$ at which to parametrize their initial distributions. One may wonder whether it is meaningful to use perturbative methods at such values of Q^2, where $\alpha_s(Q^2)$ is relatively large, but there is no doubt that excellent fits to the data were achieved. The same approach was shown to work for the unpolarized case. Stratmann (see Blumlein *et al.*, 1997) claimed that even in this approach $\Delta G(x, Q^2)$ is virtually undetermined. However, a more recent study, including much new and precise data, was shown by Leader, Sidorov and Stamenov (1999) to constrain $\Delta G(x)$ within reasonable limits, as shown in Fig. 11.17.

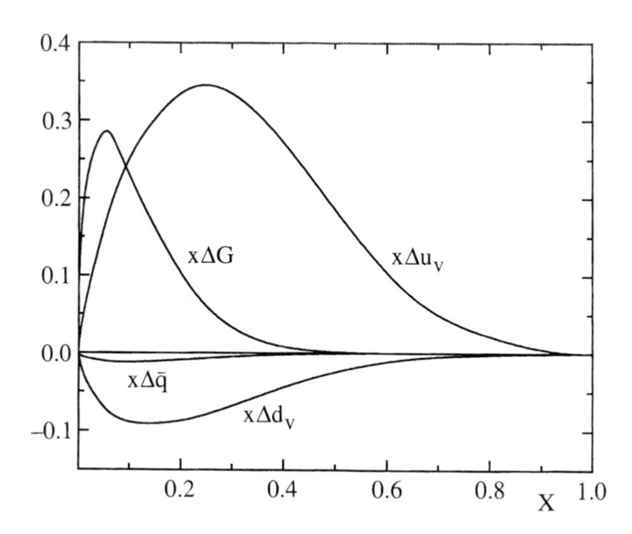

Fig. 11.15 Typical shapes of the polarized-parton densities (multiplied by x) at $Q^2 = 4\ (\text{GeV}/c)^2$, obtained from a NLO analysis assuming an $SU(3)$-symmetric sea. (From Leader, Sidorov and Stamenov, 1998a.)

A knowledge of $\Delta G(x, Q^2)$ is of great importance for understanding the spin structure of the nucleon, but it is clear that polarized DIS is not the best place to seek this information; future experiments, though, in which the proton beam at HERA is polarized, would cover a larger range of Q^2 and thus have better control over the gluon.

The quest for more precise knowledge about $\Delta G(x, Q^2)$ has inspired a whole new series of experiments involving polarized nucleons, which will begin in the very near future. The COMPASS experiment at CERN will study polarized *semi-inclusive* DIS, where, for example, reactions like

$$\mu + p \to \mu + p + \text{jet} \qquad \text{or} \qquad \mu + p + \text{ two jets}$$

are very sensitive to the gluon distribution. The RHIC collider at Brookhaven will have high energy colliding beams of polarized protons, where reactions like Drell–Yan scattering $pp \to l^+l^-X$, for lepton pairs with large transverse momentum, are also sensitive to $\Delta G(x, Q^2)$. Both COMPASS and an upgraded HERMES experiment will look at the semi-inclusive production of charm. For further information about these new experiments the reader should consult the paper 'Towards future measurements of $\Delta G/G$' in the Proceedings of the DESY workshop, *Deep Inelastic Scattering Off Polarized Targets: Theory Meets Experiment* (Blümlein *et al.*, 1997).

11.8.1 Behaviour as $x \to 0$ and $x \to 1$

In order to test sum rules one must integrate the experimentally measured $g_{1,2}(x, Q^2)$ from $x = 0$ to $x = 1$ and this inevitably means making a

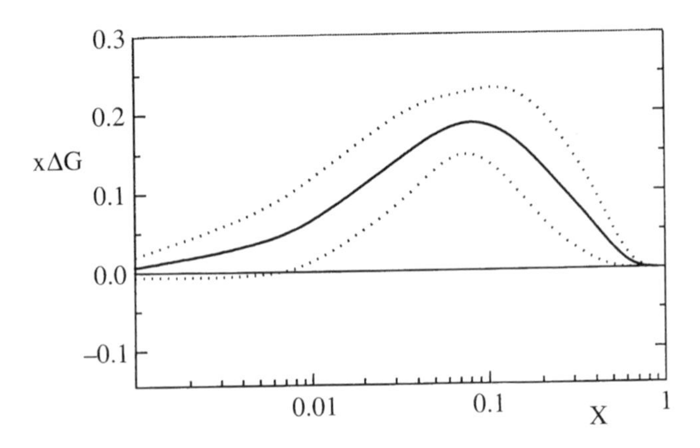

Fig. 11.16 Polarized-gluon density and error band (shown by dotted curves). (From Leader, Sidorov and Stamenov, 1999.)

theoretically biased extrapolation from the region actually covered in the experiment.

In the region $x \to 1$ there are perturbative QCD arguments (Farrar and Jackson, 1975) that suggest that

$$\frac{\Delta q_V(x)}{q_V(x)} \to 1 \qquad \frac{\Delta \bar{q}(x)}{\bar{q}(x)} \to 1 \qquad \frac{\Delta G(x)}{G(x)} \to 1. \tag{11.8.4}$$

Even if detailed fits to the data do not always support this behaviour, the disagreement is innocuous from the point of view of the sum rules, since in all cases the parton densities drop rapidly to zero as $x \to 1$ and the contribution to the sum rule from the large-x region is essentially negligible. Surprisingly, the behaviour as $x \to 1$ turns out to be quite critical in the analysis of $p^{\uparrow}p \to \pi x$ with a transversely polarized proton, is discussed in Section 13.4.

Quite the contrary happens in the region $x \to 0$, where it is not at all clear what is happening, nor what is expected to happen, theoretically. In view of the connection between DIS and the imaginary part of the forward virtual-photon Compton scattering amplitude (Section 11.6) one would expect the structure functions to have a Regge-type behaviour as the energy $\nu \to \infty$ at fixed photon 'mass' Q^2. Regge behaviour (see, for example, Collins and Martin, 1984) describes the highly non-perturbative region of forward scattering, so cannot be derived rigorously in QCD, but there are powerful reasons to believe in its validity. In that case we should have at fixed Q^2

$$g_1 \overset{\nu \to \infty}{\approx} \nu^{-\alpha_1} f(Q^2),$$

a behaviour arising from the trajectories associated with the $a_1(1260)$ and $f_1(1285)$ mesons, for which one expects $\alpha_{f_1}(t) \approx \alpha_{a_1}(t)$ and $\alpha_1 \equiv \alpha_{a_1}(0) = -0.14 \pm 0.20$.

Since, via (11.3.15), $\nu \propto 1/x$ we may deduce that, on the one hand, at fixed Q^2

$$g_1(x, Q^2) \overset{x \to 0}{\approx} x^{-\alpha_1} \times (\text{function of } Q^2), \tag{11.8.5}$$

which would imply a rather flat, almost constant behaviour of g_1 as $x \to 0$. (A detailed analysis of the Regge contributions to DIS is given in Heimann, 1973.)

On the other hand, data on the growth with energy of several exclusive reactions initiated by virtual photons, for example,

$$\text{`}\gamma\text{'} + p \to p + \text{ vector meson},$$

do not seem to follow standard Regge behaviour: they grow much faster with energy ν when Q^2 is significantly different from zero, even though in these reactions $\nu \gg Q$, a condition which used to be thought sufficient to justify Regge behaviour. Moreover the behaviour of the *unpolarized* structure functions $F_{1,2}(x, Q^2)$ does not seem to follow Regge predictions at small x. This is particularly evident in the HERA data (see, for example, Adloff *et al.*, 1997), where there is a very rapid growth of $F_1(x, Q^2)$, faster than x^{-1}, as $x \to 0$. This has consequences for the polarized case since there are arguments relating the small-x behaviour of polarized and unpolarized densities, namely, for $x \to 0$

$$\frac{\Delta q(x)}{q(x)} \propto x \qquad \frac{\Delta G(x)}{G(x)} \propto x$$

which would imply then that $|g_1(x, Q^2)|$ should grow faster than the Regge behaviour (11.8.5) as $x \to 0$.

Attempts have been made to study the small-x behaviour via a selective summation of terms in perturbation theory. Berera (1992) and Ball, Forte and Ridolfi (1995) studied the small-x behaviour of the evolution equations. Very interesting results emerge. If the starting distribution at Q_0^2 is singular enough as $x \to 0$, namely $g_1(x, Q_0^2) \propto x^{-\lambda}$ with $\lambda > 0$, then this behaviour remains unchanged by the evolution. But if one starts with a relatively flat distribution, for example one corresponding to the Regge behaviour (11.8.5), then as Q^2 grows the behaviour near $x = 0$ for $\Delta q_{\mathrm{NS}}(x, Q^2)$, $\Delta\Sigma(x, Q^2)$ and $\Delta G(x, Q^2)$ tends to

$$(\xi\zeta)^{-1/4} \exp\left(2\gamma\sqrt{\xi\zeta} - \delta\zeta\right) \tag{11.8.6}$$

where

$$\xi = \ln\frac{x_0}{x} \qquad \zeta = \ln\left[\frac{\alpha_s(Q_0^2)}{\alpha_s(Q^2)}\right] \tag{11.8.7}$$

and x_0 is a small value of x below which the asymptotic treatment is valid. The coefficients γ and δ depend upon what density one is studying.

For Δq_{NS} one has

$$\begin{aligned} \gamma = \gamma_{\mathrm{NS}} &= \left(\frac{8}{33 - 2N_f}\right)^{1/2} \\ \delta = \delta_{\mathrm{NS}} &= \frac{4}{33 - 2N_f}. \end{aligned} \tag{11.8.8}$$

For both $\Delta\Sigma$ and ΔG,

$$\begin{aligned} \gamma = \gamma_+ &= \left[\frac{8}{33 - 2N_f}\left(5 + 4\sqrt{1 - \frac{3}{32}N_f}\right)\right]^{1/2} \\ \delta = \delta_+ &= \frac{1}{2(33 - 2N_f)}\left[35 + 2N_f + 43\left(\frac{1 - \frac{11}{86}N_f}{\sqrt{1 - \frac{3}{32}N_f}}\right)\right]. \end{aligned} \tag{11.8.9}$$

Moreover one finds the remarkable result that as $x \to 0$

$$\Delta\Sigma(x, Q^2) \to -2\left(1 - \sqrt{1 - \tfrac{3}{32}N_f}\right)\Delta G(x, Q^2). \tag{11.8.10}$$

The implications of these results are fascinating.

(1) All the polarized distributions grow faster than any power of $\ln x_0/x$, and the growth rate increases with Q^2.
(2) Since $\gamma_+ > \gamma_{\rm NS}$, $\Delta\Sigma$ and ΔG dominate in magnitude over $\Delta q_{\rm NS}$.
(3) If $\Delta G(x, Q^2)$ is positive and reasonably large the negative contribution of $\Delta\Sigma$ will then make the sum $g_1^p + g_1^n$, and eventually each of g_1^p and g_1^n, *negative* at small enough x.

Interestingly, a similar analysis for the unpolarized case produces the required growth at small x to account for the HERA data mentioned above provided the starting value Q_0^2 is chosen small enough.

The situation is somewhat muddied by the work of Bartels, Ermolaev and Ryskin (1996a, b) and Manayenkov and Ryskin (1998), who sum 'double logarithmic' terms, of the form $(\alpha_s \ln^2 x)^n$, that are not included in the evolution equations, with the startling results that all the densities diverge as $x^{-\lambda}$, with $\lambda_{\rm NS} \approx 0.5$ and $\lambda_S = \lambda_G \gtrsim 1$. The latter would imply that the first moment of $g_1^{p,n}$ diverges! It is difficult to believe that these results reflect the physical behaviour of the densities. It may be that such selective summations at fixed Q^2 are dangerous and that major cancellations can occur between different subsets of terms.

All the above results, based on a selective summation of terms in perturbation theory, disagree with Regge behaviour. Is this a genuine incompatibility or are the perturbative arguments unreliable at small enough x or small enough Q^2? And if the latter, at what scale should we expect to see Regge behaviour setting in? These are fascinating questions to which we have, as yet, no answer. A good discussion can be found in Altarelli, Ball, Forte and Ridolfi (1997). These authors also show that with a careful extrapolation to small x the Bjorken sum rule (11.6.19) is well satisfied by the present data.

As a final word on the subject of small-x behaviour, note that the *new*, 1996, SMC data on g_1^p (Fig. 11.12), combined with the new E154 data on g_1^n (Fig. 11.13), do suggest that $g_1^p + g_1^n$ might become negative at the smallest measured x-values!

There is a major experimental push towards smaller x. The results are awaited with great interest.

11.9 The general partonic structure of the nucleon

In Section 11.5 we saw how the parton model for DIS can be given a more fundamental field-theoretic formulation. Crucial to that derivation was the separation of the physics into 'hard' and 'soft' parts, exemplified by $E^{\mu\nu}$ and Φ in (11.5.14) respectively. In Section 11.5 we took a specific form for $E^{\mu\nu}$ and concentrated upon its antisymmetic part under $\mu \leftrightarrow \nu$. But the steps taken are actually more general and would apply to any structure of the form

$$W = \int d^4k \ \mathrm{Tr} \ [E(q,k)\Phi(P,\mathcal{S};k)] \tag{11.9.1}$$

so long as $E(q,k)$ represents a hard process whose scale is set by q^2 and provided the approximation (11.5.28) is adequate.

The generalization of (11.5.31) and (11.5.32) is then

$$W = \mathrm{Tr} \ \left[E(q,k)\hat{\Phi}(P,\mathcal{S};n,x)\right] \tag{11.9.2}$$

where

$$\hat{\Phi}_{\alpha\beta} = \int \frac{d\lambda}{2\pi} e^{i\lambda x} \langle P,\mathcal{S}|\bar{\psi}_\beta(0)\psi_\alpha(\lambda n)|P,\mathcal{S}\rangle \tag{11.9.3}$$

and $k = xP$ in $E(q,k)$.[1]

Being a 4×4 matrix in its Dirac labels, $\hat{\Phi}$ can be expanded in the form

$$\hat{\Phi} = sI + v_\mu\gamma^\mu - ip_{\mu\nu}\gamma_5\sigma^{\mu\nu} + a_\mu\gamma^\mu\gamma_5 + ip\gamma_5. \tag{11.9.4}$$

The coefficients are given by traces of the form (Γ is some Dirac matrix)

$$\begin{aligned} \mathrm{Tr} \ \left(\hat{\Phi}\Gamma\right) &= \int \frac{d\lambda}{2\pi} e^{i\lambda x} \langle P,\mathcal{S}|\bar{\psi}(0)\Gamma\psi(\lambda n)|P,\mathcal{S}\rangle \\ &\equiv \langle\Gamma\rangle \end{aligned} \tag{11.9.5}$$

for brevity. One has

$$\left.\begin{aligned} s = \tfrac{1}{4}\langle I\rangle \qquad & p = -\tfrac{1}{4}\langle i\gamma_5\rangle \\ v_\mu = \tfrac{1}{4}\langle\gamma_\mu\rangle \qquad & a_\mu = \tfrac{1}{4}\langle\gamma_5\gamma_\mu\rangle \\ p_{\mu\nu} = \tfrac{1}{4} & \langle i\gamma_5\sigma_{\mu\nu}\rangle, \end{aligned}\right\} \tag{11.9.6}$$

these coefficients being scalar, pseudoscalar, vector, axial-vector and pseudo-tensor respectively. The coefficients can only be constructed from the large vector P_μ, the small vector n_μ and the axial vector $\mathcal{S}_\mu$, and they can at most be linear in $\mathcal{S}_\mu$. In addition the behaviour under hermitian conjugation, and the transformation laws for the fields under space inversion and time reversal (see Section 2.3), lead to the requirements

$$\hat{\Phi}^\dagger(P,\mathcal{S};n) = \gamma_0\hat{\Phi}(P,\mathcal{S};n)\gamma_0 \tag{11.9.7}$$

$$\hat{\Phi}(P,\mathcal{S};n) = \gamma_0\hat{\Phi}(\tilde{P},-\tilde{\mathcal{S}};\tilde{n})\gamma_0, \tag{11.9.8}$$

[1] In fact one can go beyond this approximation by making a Taylor expansion of $E(q,k)$ about the point $k = xP$. For details see Boer (1998).

where for any 4-vector $\tilde{V}^\mu = (V^0, -\mathbf{V})$, and

$$\hat{\Phi}(P, \mathscr{S}; n) = \gamma_5 C^{-1} \hat{\Phi}^T(P, -\mathscr{S}; n) C \gamma_5 \tag{11.9.9}$$

where T indicates the transpose and $C\gamma_\mu C^{-1} = -\gamma_\mu^T$.

It follows that $p = 0$ in (11.9.6). For the other coefficients, several terms are possible. They can be ordered into sets according to magnitude, the largest $O(|P|)$, the next $O(1)$ and so on. When linked to the hard process they give rise to terms of twist 2 and twist 3 respectively. ('Twist' was briefly mentioned in Section 11.7. For a more detailed explanation of this concept see, for example, Section 22.2 of Leader and Predazzi (1996)). For the purpose of this classification it is convenient to split the covariant spin vector into a longitudinal part $\mathscr{S}^\mu(\lambda)$ given by (11.5.35), and therefore 'large', and a transverse part

$$\mathscr{S}^\mu_T = (0, \boldsymbol{S}_T, 0) \tag{11.9.10}$$

of $O(1)$. We have

$$\begin{aligned}\hat{\Phi} = & \frac{\not{P}}{2} [f(x) - 2\lambda h_L(x)\gamma_5 + h_T(x)\gamma_5 \not{\mathscr{S}}_T] \\ & + \frac{M}{2} \{e(x) I + f_T(x)\gamma_5 \not{\mathscr{S}}_T + f_L(x)\lambda\gamma_5[\not{P}, \not{n}]\} \\ & + \cdots \end{aligned} \tag{11.9.11}$$

There is confusion in the literature about the nomenclature of the coefficient functions in (11.9.11). We have essentially followed the logical notation in the ground-breaking paper of Ralston and Soper (1979) and in the later discussion of Cortes, Pire and Ralston (1992).

Unfortunately, an influential paper of Jaffe and Ji (1991) uses a quite different and potentially misleading notation. This labels some of the coefficient functions as g_1, g_2, g_3, thereby confusing experimentally defined and measured quantities with approximate theoretical expressions for them. Up to an overall constant the relations between our coefficients and those of Jaffe and Ji are

$$\begin{aligned} h_L = g_1^{\mathrm{JJ}} \qquad & h_T = h_1^{JJ} \\ f_T = (g_1 + g_2)^{\mathrm{JJ}} \qquad & f = f_1^{JJ} \\ f_L = (h_2 + \tfrac{1}{2}h_1)^{\mathrm{JJ}} & \end{aligned} \tag{11.9.12}$$

In principle a complete knowledge of the partonic structure of the nucleon would require a knowledge of all the coefficient functions in (11.9.11). It is hard to imagine that we will ever possess such detailed knowledge. In (11.9.11) the first three terms correspond to twist 2 and are the parton-model terms that would emerge if the quark fields ψ were treated as free fields. As explained in subsection 11.5.2 the term f_T, which occurs in the expression (11.5.50) for $g_1 + g_2$, is not of twist 2 and does not emerge from the parton model.

All the coefficient functions in (11.9.11) are given by the nucleon matrix elements of bilocal light-cone operators. As such they depend not only upon x but, strictly speaking, upon the renormalization scale μ as well. Since this is, in principle, arbitrary, it is usual to choose it equal to the large scale in the reaction. For example, in DIS one chooses $\mu^2 = Q^2$.

In the shorthand notation of (11.9.5), with the spin state of the nucleon indicated by a subscript λ or $\mathscr{S}_T$ for the longitudinal and transverse cases respectively, one has

$$\begin{aligned} h_L &= -\frac{1}{4\lambda}\langle\gamma_5 \not{n}\rangle_\lambda \\ h_T &= -\frac{1}{2}\langle i\gamma_5\sigma_{\mu\nu}\mathscr{S}_T^\mu n^\nu\rangle_{\mathscr{S}_T} \\ f_T &= -\frac{1}{2M}\langle\gamma_5\not{\mathscr{S}}_T\rangle_{\mathscr{S}_T} \\ f_L &= \frac{1}{4M\lambda}\langle i\gamma_5\sigma_{\mu\nu}n^\mu P^\nu\rangle_\lambda \\ f &= \frac{1}{2}\langle\not{n}\rangle \\ e &= \frac{1}{2M}\langle I\rangle . \end{aligned} \qquad (11.9.13)$$

We mentioned in Section 11.5 that in the free-field or parton model

$$h_L(x) = \Delta q(x) = q_+(x) - q_-(x) \qquad (11.9.14)$$

where $\pm$ refers to the quark helicity inside a nucleon of helicity $+1/2$.

The function $h_T(x)$ is the analogue of this when the nucleon is polarized perpendicular to its momentum, and in the parton model

$$h_T(x) = \Delta_T q(x) = q_\uparrow(x) - q_\downarrow(x) \qquad (11.9.15)$$

where $\uparrow\downarrow$ refer to the quark's transverse covariant spin vector being along or opposite to the spin of the transversely polarized nucleon.

Conventionally the transverse spin direction is chosen as the Y direction for a particle moving along the Z-axis. Then by (1.1.18)

$$\begin{aligned} |\mathbf{p};\uparrow\rangle_y &= \tfrac{1}{\sqrt{2}}\left(|\mathbf{p};1/2\rangle + i|\mathbf{p};-1/2\rangle\right) \\ |\mathbf{p};\downarrow\rangle_y &= \tfrac{1}{\sqrt{2}}\left(|\mathbf{p};1/2\rangle - i|\mathbf{p};-1/2\rangle\right) \end{aligned} \qquad (11.9.16)$$

where $\uparrow\downarrow$ means along or opposite to OY respectively.

Sometimes the X-direction is chosen, for which we have

$$\begin{aligned} |\mathbf{p};\uparrow\rangle_x &= \tfrac{1}{\sqrt{2}}\left(|\mathbf{p};1/2\rangle + |\mathbf{p};-1/2\rangle\right) \\ |\mathbf{p};\downarrow\rangle_x &= -\tfrac{1}{\sqrt{2}}\left(|\mathbf{p};1/2\rangle - |\mathbf{p};-1/2\rangle\right) . \end{aligned} \qquad (11.9.17)$$

In DIS it is easy to see from (11.5.10) that the leading-order piece of $E(q,k)$ in (11.9.1), which describes the 'hard' process, always involves a product of an odd number of γ-matrices. Hence in the trace in (11.9.2) only that part of $\hat{\Phi}$ involving an odd number of γ-matrices will contribute, i.e. only the terms v_μ and a_μ in (11.9.4). The large term $h_T(x)$ is connected to the structure $\gamma_5\sigma_{\mu\nu}$, and hence does not appear in leading order in DIS. It can be measured in polarized Drell–Yan-type reactions, as is discussed in Section 12.4, and possibly also in single-spin asymmetries in semi-inclusive hadron–hadron reactions. (See, however, Section 13.4.)

It should be stressed that the functions $f(x), h_L(x)$ and $h_T(x)$ or, equivalently, $q(x), \Delta q(x)$ and $\Delta_T q(x)$ are on an equal footing and contain the most essential information about the internal partonic-spin structure of the nucleon. There is steady progress in the experimental determination of the $\Delta q(x)$ but, to date, we possess very little experimental information about the transverse densities $\Delta_T q(x)$. (See, however, Section 13.4.) Indeed, the only unambiguous information we have about $\Delta_T q(x)$ is the *Soffer bound* (Soffer, 1995):

$$|\Delta_T q(x)| \leq \tfrac{1}{2}\left[q(x) + \Delta q(x)\right]. \qquad (11.9.18)$$

The importance of $\Delta_T q(x)$ was first stressed in a seminal paper by Ralston and Soper (1979), and the possibility of its measurement was discussed by Artru and Mekhfi (1990), by Jaffe and Ji (1991) and by Cortes, Pire and Ralston (1992). We shall discuss the phenomenological aspects of $\Delta_T q(x)$ in Section 12.4 and Chapter 13.

11.9.1 Evolution for $\Delta_T q(x, Q^2)$

The evolution equations for $\Delta_T q(x, Q^2)$ are similar in form to (11.7.2), but simpler since there is no gluon contribution. The evolution is thus analogous to that of a flavour non-singlet combination of polarized-quark densities.

The transverse-polarization splitting functions, in leading order, $\Delta_T P_{qq}^{(0)}$, were given by Artru and Mekhfi (1990). The next-to-leading-order result can be found in Vogelsang (1998).

At leading order the moments of the transversely polarized quark densities vary is a very simple way with Q^2:

$$\Delta_T q^{(n)}(Q^2) = \left[\frac{\alpha_s(Q^2)}{\alpha_s(Q_0^2)}\right]^{-2\Delta_T P_{qq}^{(0)(n)}/\beta_0} \times \Delta_T q^{(n)}(Q_0^2) \qquad n = 1, 2, \ldots \quad (11.9.19)$$

where $\beta_0 = 11 - 2n_f/3$, n_f being the number of active flavours, and the moments $\Delta_T P_{qq}^{(0)(n)}$ of the splitting functions turn out to be negative for all n. It follows that all $\Delta_T q^{(n)}(Q^2)$ decrease as Q^2 increases. In general one cannot conclude that $\Delta_T q(x, Q^2)$ decreases in magnitude for all x as

Q^2 increases, but one can do so if $\Delta_T q(x, Q_0^2)$ is a monotonic function of x.

Strictly speaking, in NLO the combinations $\Delta_T q_\pm \equiv \Delta_T q \pm \Delta_T \bar{q}$ evolve with different splitting functions $\Delta_T P^{(1)}_{qq\pm}$, but it turns out that the difference between $\Delta_T P^{(1)}_{qq+}$ and $\Delta_T P^{(1)}_{qq-}$ is completely negligible. Hence, in practice $\Delta_T q$ and $\Delta_T \bar{q}$ can be considered to evolve with essentially the same splitting functions.

11.10 The future: neutrino beams?

There has been much discussion recently about the possibility of constructing a *muon collider* involving the collision of two circulating high energy muon beams. A prerequisite for this is a muon storage ring, which, it was suddenly realized, could provide a clean high energy neutrino beam of staggering intensity – 10^3 or 10^4 times more intense than present fluxes and well focussed. In fact the production via, say, $\mu^- \to e^- + \nu_\mu + \bar{\nu}_e$ actually produces, in a well-defined way, a mixture of neutrinos and antineutrinos. But it is a trivial matter to separate the neutrino from the antineutrino charged current reactions in the target by simply identifying the charge of the final state lepton. It is *not* necessary to separate high energy same-sign muons from electrons, which would have been a daunting task.

With this sort of flux it becomes possible to use targets of a few kilograms, rather than kilotonnes and, for the first time ever, to contemplate polarized target experiments with neutrino beams. This would indeed be a dramatic development. Flavour separation, i.e. the separate determination of the parton and antiparton densities of a given flavour, has only been possible for the unpolarized case because of the ability to combine data from neutral current and charged current reactions. With this exciting prospect in view we shall here present a brief outline of how and what could be measured for neutrino and antineutrino CC reactions.

Because of the parity-violating electroweak coupling one no longer has the correspondence that the symmetric part $W^{(S)}_{\mu\nu}$ in (11.3.9) is spin independent and the antisymmetric part $W^{(A)}_{\mu\nu}$ in (11.3.10) is spin dependent. Now the spin-dependent part of $W_{\mu\nu}$ is a superposition of symmetric and antisymmetric pieces and involves five independent structure functions $g_j(x, Q^2)$, which, in the absence of strong interactions, would obey Bjorken scaling, i.e. would be independent of Q^2. There is a plethora of different definitions of the g_j in the literature. We shall follow the definitions used in the recent very important paper of Blümlein and Kochelev (1997).[1]

[1] These g_j are related to the $g_j^{\rm AEL}$ used in the review article by Anselmino, Efremov and Leader (1995) via $g_{1,2} = g_{1,2}^{\rm AEL}$; $g_3 = -g_3^{\rm AEL}$; $g_4 = g_4^{\rm AEL} - g_3^{\rm AEL}$; $g_5 = -g_5^{\rm AEL}$.

Then, using the same kinematic variables as in Section 11.3 and defining

$$\hat{P}_\mu = P_\mu - \frac{P \cdot q}{q^2} q_\mu \tag{11.10.1}$$

$$\hat{\mathscr{S}}_\mu = \mathscr{S}_\mu - \frac{\mathscr{S} \cdot q}{q^2} q_\mu \tag{11.10.2}$$

one has for the spin-dependent part

$$\begin{aligned}
\frac{1}{2M} W_{\mu\nu} = {} & \frac{\epsilon_{\mu\nu\alpha\beta}}{P \cdot q} q^\alpha \left[\mathscr{S}^\beta g_1 + \frac{(P \cdot q \mathscr{S}^\beta - \mathscr{S} \cdot q P^\beta)}{P \cdot q} g_2 \right] \\
& + \left(\frac{\hat{P}_\mu \hat{\mathscr{S}}_\nu + \hat{\mathscr{S}}_\mu \hat{P}_\nu}{2} - \mathscr{S} \cdot q \frac{\hat{P}_\mu \hat{P}_\nu}{P \cdot q} \right) g_3 \\
& + \mathscr{S} \cdot q \left[\frac{\hat{P}_\mu \hat{P}_\nu}{P \cdot q} g_4 + \left(-g_{\mu\nu} + \frac{q_\mu q_\nu}{q^2} \right) g_5 \right]
\end{aligned} \tag{11.10.3}$$

where we have suppressed the neutrino labels ν and $\bar{\nu}$ that should be attached to $W_{\mu\nu}$ and the g_j.

The differential cross-section differences for the longitudinally and transversely polarized target cases are related to the g_j as follows.

For ν or $\bar{\nu}$ beams on a target polarized longitudinally, along ($\Rightarrow$) or opposite ($\Leftarrow$) to the initial lepton beam direction,

$$\begin{aligned}
\frac{d^2\sigma^{\nu,\bar{\nu}}(\Leftarrow)}{dxdy} - \frac{d^2\sigma^{\nu,\bar{\nu}}(\Rightarrow)}{dxdy} = {} & 32\pi s \frac{\alpha^2}{Q^4} \eta_W \\
& \times \left[\pm \left(2 - y - 2xy\frac{M^2}{s} \right) yxg_1 \mp \frac{4M^2}{s} yx^2 g_2 \right. \\
& + \frac{2M^2}{s} \left(1 - y - xy\frac{M^2}{s} \right) xg_3 \\
& - \left(1 + 2x\frac{M^2}{s} \right) \left(1 - y - xy\frac{M^2}{s} \right) g_4 \\
& \left. + \left(1 + 2x\frac{M^2}{s} \right) y^2 x g_5 \right]
\end{aligned} \tag{11.10.4}$$

where $\sqrt{s}$ is the CM energy of the lepton–nucleon collision ($s \approx 2ME$), and

$$\eta_W = \frac{1}{2} \left(\frac{GM_W^2}{4\pi\alpha} \frac{Q^2}{Q^2 + M_W^2} \right)^2 \tag{11.10.5}$$

In (11.10.4) one must use $g_j^{W^-}$ for neutrino beams, and $g_j^{W^+}$ for antineutrino beams.

It is perfectly reasonable to neglect terms of order M^2/s, so that one obtains the simpler result

$$\frac{d^2\sigma^{\nu,\bar{\nu}}(\Leftarrow)}{dxdy} - \frac{d^2\sigma^{\nu,\bar{\nu}}(\Rightarrow)}{dxdy}$$
$$\approx 32\pi s\frac{\alpha^2}{Q^4}\eta_W\left[\pm(2-y)yxg_1^{W\mp} - (1-y)g_4^{W\mp} + y^2xg_5^{W\mp}\right]. \quad (11.10.6)$$

For a transversely polarized target ($\Uparrow$ or $\Downarrow$), with ϕ, the azimuthal angle of the final state lepton, measured with respect to the plane formed by the initial lepton momentum and the nucleon spin direction $\Uparrow$ (see Fig. 11.17), one has

$$\frac{d^3\sigma^{\nu,\bar{\nu}}(\Uparrow)}{dxdyd\phi} - \frac{d^3\sigma^{\nu,\bar{\nu}}(\Downarrow)}{dxdyd\phi}$$
$$= 16M\sqrt{s}\frac{\alpha^2}{Q^4}\eta_W \cos\phi\left[xy\left(1-y-xy\frac{M^2}{s}\right)\right]$$
$$\times\left\{\pm 2yxg_1 \pm 4g_2 - \frac{1}{y}\left(2-y-2xy\frac{M^2}{s}\right)g_3\right.$$
$$\left.+\frac{2}{y}\left(1-y-xy\frac{M^2}{s}\right)g_4 + 2y^2xg_5\right\}. \quad (11.10.7)$$

Just as for the case of g_2 in electromagnetic neutral current reactions, (Section 11.4 and subsection 11.5.2) the structure functions g_2^W and g_3^W cannot be calculated in the simple parton model. For the other g_j^W one finds

$$g_1^{W^-} = \Delta u + \Delta c + \Delta\bar{d} + \Delta\bar{s} \quad (11.10.8)$$
$$g_5^{W^-} = \Delta\bar{d} + \Delta\bar{s} - \Delta u - \Delta c \quad (11.10.9)$$

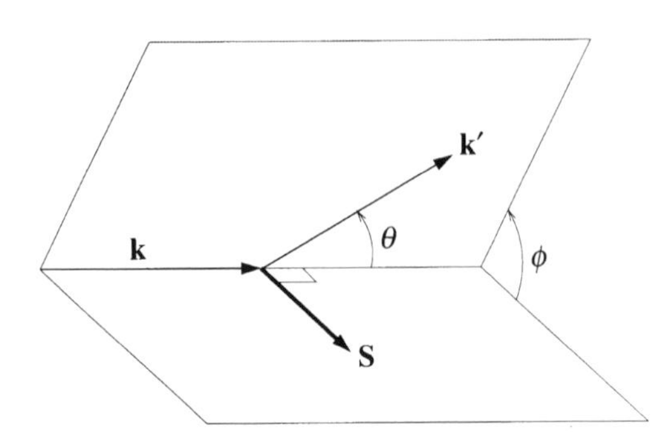

Fig. 11.17. Definition of azimuthal angle ϕ for transverse nucleon polarization. The bold horizontal arrow gives the nucleon spin direction.

and (Dicus, 1972)

$$g_4^{W^\mp} = 2xg_5^{W^\mp} \tag{11.10.10}$$

and

$$g_1^{W^+} = \Delta d + \Delta s + \Delta\bar{u} + \Delta\bar{c} \tag{11.10.11}$$

$$g_5^{W^+} = \Delta d + \Delta s - \Delta\bar{u} - \Delta\bar{c}. \tag{11.10.12}$$

In all these it is probably safe to take $\Delta c = \Delta\bar{c} = 0$.

At the parton level, (11.10.6) simplifies to

$$\frac{d^2\sigma^{\nu,\bar{\nu}}(\Leftarrow)}{dxdy} - \frac{d^2\sigma^{\nu,\bar{\nu}}(\Rightarrow)}{dxdy}$$

$$\approx 32\pi s\frac{\alpha^2}{Q^4}\eta_W \left\{\pm(2-y)yxg_1^{W^\mp} + \left[y^2 - 2(1-y)\right] xg_5^{W^\mp}\right\} \tag{11.10.13}$$

so that it might not be too difficult to determine g_1 and g_5 separately.

In that case one would be able to make a direct measurement of the flavour-singlet combination $\Delta\Sigma$, which plays such a crucial rôle in the 'spin crisis' (Section 11.2 and subsection 11.6.1). For one has

$$g_1^{W^-} + g_1^{W^+} = \Delta u + \Delta\bar{u} + \Delta d + \Delta\bar{d} + \Delta s + \Delta\bar{s} + \Delta c + \Delta\bar{c}. \tag{11.10.14}$$

The advent of neutrino-induced polarized DIS would open up an extremely rich and valuable source of information on the internal structure of the nucleon. It is to be hoped that such experiments do not lie too far into the future.

12

Two-spin and parity-violating single-spin asymmetries at large scale

Besides inclusive deep inelastic lepton–hadron scattering, which was treated in detail in Chapter 11, there is an enormous amount of data on spin dependence in a wide variety of other reactions, inclusive, semi-inclusive and exclusive, and at a range of momentum transfers from zero (forward scattering) to moderately large values $p_T^2 \lesssim 8$ $(\mathrm{GeV}/c)^2$. Also there is the prospect of a huge increase in p_T with the coming into operation of the polarized RHIC collider at Brookhaven. Broadly speaking the large-momentum-transfer reactions or those involving some large scale can be treated using perturbative QCD; however, there is a major distinction between inclusive or semi-inclusive reactions, where the parton densities play a crucial rôle, and exclusive reactions where, in principle, one needs to know the parton *wave function* of the hadron. Our knowledge of the latter is much less secure than that of the parton densities.

As for the reactions at small momentum transfer, with no large mass scale, they fall into the regime of non-perturbative QCD and are presently treated in a less fundamental way. Indeed there is really no satisfactory theoretical explanation of the dramatic spin dependence seen in many of these reactions. This will be discussed in Chapter 14.

There is a further distinction between *single-spin asymmetries*, such as the polarization of a particle produced in an unpolarized collision or the analysing power of a reaction, and *double-spin asymmetries* such as the dependence of a cross-section upon the initial spins in the collision of a polarized beam on a polarized target. There are two types of single-spin asymmetry: helicity or longitudinal spin asymmetries, which are parity violating and occur in electroweak processes such as the production of massive lepton–antilepton pairs involving γ–Z^0 interference, or the production of $W^\pm$; and transverse spin asymmetries, which, as will be seen in Chapter 13, are difficult to generate in perturbative QCD.

The double-spin asymmetries, however, will be seen to emerge in a very natural way from perturbative QCD. Indeed partonic reactions have very large double-spin asymmetries and the relation between hadronic and partonic asymmetries will play a crucial rôle in gaining further insight into the polarized-parton densities discussed in Chapter 11. This subject is about to take a major leap forward with the coming into operation of RHIC, where two beams of 250 GeV polarized protons will collide head on. For the first time ever it will be possible to study proton–proton spin asymmetries at truly high energies and truly large momentum transfers. At the same time COMPASS will begin to produce results on semi-inclusive deep inelastic lepton–hadron scattering, promising much new information about the polarized densities of antiquarks and gluons.

12.1 Inclusive and semi-inclusive reactions: general approach

In the following we concentrate largely on hadron–hadron collisions, but lepton–hadron reactions are mentioned in subsection 12.2.5.

Consider the hadronic collision

$$A + B \to C + X \tag{12.1.1}$$

in which C is produced with large transverse momentum p_T. The product C could be a jet, a specific hadron, a lepton–antilepton pair (the Drell–Yan reaction) or an electroweak gauge boson. In practice A and B will be protons or antiprotons.

For a hadronic final state the reaction is visualized as in Fig. 12.1 and is interpreted in a probabilistic sense. The beam and target are simply sources of partons, the parton number densities being either known from deep inelastic scattering or parametrized in some simple form. Partons from A and B undergo a large p_T collision, the *partonic subprocess*, which is calculated perturbatively, and one of the final state partons may then *hadronize* into C. The hadronization, being non-perturbative, is treated phenomenologically in terms of a *fragmentation* function.

In order to understand the spin structure, consider a collision of longitudinally polarized protons that produces a particle C with large p_T. Ignore all momenta and all integrations and pretend that a quark of only one flavour from each proton participates in the reaction. Symbolically, then, Fig. 12.1 yields

$$d\sigma_{\Lambda\Sigma} \sim \sum_{\lambda,\sigma,\rho} q_\lambda^\Lambda q_\sigma^\Sigma d\hat{\sigma}_{\rho;\lambda\sigma} D_C^\rho \tag{12.1.2}$$

where Λ, Σ are the proton helicities, λ and σ the helicities of the quarks, $d\hat{\sigma}_{\rho;\lambda\sigma}$ the quark–quark differential cross-section for incoming quarks of

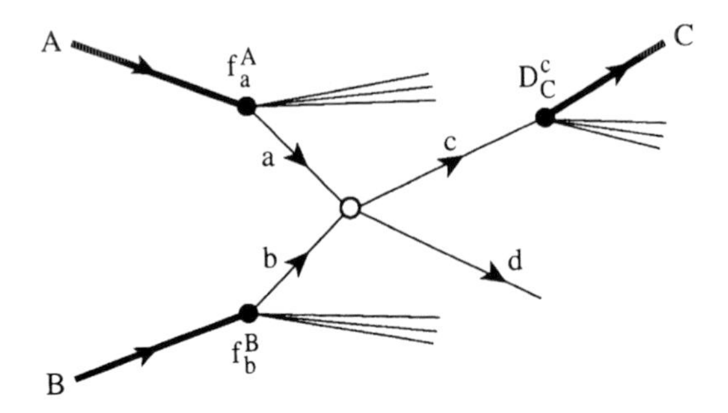

Fig. 12.1 Mechanism for $AB \to CX$ in the parton model. Heavy lines represents hadrons, and light lines represent partons.

helicities λ and σ to produce a quark of helicity ρ, and D_C^ρ is the probability for the quark of helicity ρ to fragment into a C. If the spin of C is not measured the latter cannot, in fact, depend on ρ so can be left out in the following discussion.

We then have symbolically

$$d\sigma_{++} \sim q_+^+ q_+^+ d\hat{\sigma}_{++} + q_-^+ q_+^+ d\hat{\sigma}_{-+} + q_-^+ q_-^+ d\hat{\sigma}_{--} + q_+^+ q_-^+ d\hat{\sigma}_{+-}. \quad (12.1.3)$$

By parity invariance

$$d\hat{\sigma}_{++} = d\hat{\sigma}_{--} \quad \text{and} \quad d\hat{\sigma}_{+-} = d\hat{\sigma}_{-+} \quad (12.1.4)$$

so that (12.1.3) becomes

$$d\sigma_{++} \sim (q_+^+ q_+^+ + q_-^+ q_-^+) d\hat{\sigma}_{++} + (q_-^+ q_+^+ + q_+^+ q_-^+) d\hat{\sigma}_{+-}. \quad (12.1.5)$$

Similarly

$$d\sigma_{+-} \sim (q_+^+ q_+^- + q_-^+ q_-^-) d\hat{\sigma}_{++} + (q_-^+ q_+^- + q_+^+ q_-^-) d\hat{\sigma}_{+-} \quad (12.1.6)$$

but again, by parity invariance,

$$q_-^- = q_+^+ \quad \text{and} \quad q_+^- = q_-^+ \quad (12.1.7)$$

so that (12.1.6) becomes

$$d\sigma_{+-} \sim (q_+^+ q_-^+ + q_-^+ q_+^+) d\hat{\sigma}_{++} + (q_-^+ q_-^+ + q_+^+ q_+^+) d\hat{\sigma}_{+-}. \quad (12.1.8)$$

Finally then

$$\begin{aligned} d\sigma_{++} - d\sigma_{+-} &\sim (q_+^+ \Delta q - q_-^+ \Delta q) d\hat{\sigma}_{++} + (q_-^+ \Delta q - q_+^+ \Delta q) d\hat{\sigma}_{+-} \\ &= \Delta q \Delta q d\hat{\sigma}_{++} - \Delta q \Delta q d\hat{\sigma}_{+-} \\ &= \Delta q \Delta q d\Delta\hat{\sigma} \end{aligned} \quad (12.1.9)$$

where, as in Section 11.3, $\Delta q = q_+^+ - q_-^+$ and $d\Delta\hat{\sigma} = d\hat{\sigma}_{++} - d\hat{\sigma}_{+-}$.

The spin structure in (12.1.9) is quite general and will always emerge if Fig. 12.1 is interpreted probabilistically.

In detail, let $f_a^A(x_a)$ be the number density of partons of type a in hadron A and let $D_C^c(z)$ be the fragmentation function for parton c to fragment into C, with C having a fraction z of the parton's momentum.

Then for the unpolarized differential cross-section for $AB \to CX$, where C has energy E_C and momentum $\mathbf{p}_C = (\mathbf{p}_T, p_z)$,

$$E_C \frac{d^3\sigma}{d^3\mathbf{p}_C} = \frac{1}{\pi} \sum_{a,b,c,d} \int_{\xi_a} dx_a \int_{\xi_b} dx_b f_a^A(x_a) f_b^B(x_b) \times \left[\frac{d\hat{\sigma}}{d\hat{t}}^{ab\to cd}(\hat{s},\hat{t}) \frac{1}{z} D_C^c(z) \right] \tag{12.1.10}$$

where the right-hand side is independent of the azimuthal angle ϕ and where $\hat{s}$, $\hat{t}$ are the Mandelstam variables for the partonic reaction

$$\hat{s} \equiv (p_a + p_b)^2 \simeq 2p_a \cdot p_b \simeq x_a x_b s \tag{12.1.11}$$

$$\hat{t} \equiv (p_a - p_c)^2 \simeq -2p_a \cdot p_c \simeq x_a t/z. \tag{12.1.12}$$

Here, for large momentum transfer,

$$s \equiv (p_A + p_B)^2 \simeq 2p_A \cdot p_B \tag{12.1.13}$$

$$t \equiv (p_A - p_C)^2 \simeq -2p_A \cdot p_C. \tag{12.1.14}$$

In (12.1.10) the value of z is fixed in terms of x_a and x_b, as follows. Let

$$x_T = \frac{2p_T}{\sqrt{s}} \tag{12.1.15}$$

be the ratio of p_T to its maximum allowed value, and define the rapidity by

$$y \equiv -\ln\tan\theta/2 \tag{12.1.16}$$

where θ is the production angle of C in the AB centre of mass frame. Then

$$z = \frac{x_T}{2}\left(\frac{e^y}{x_a} + \frac{e^{-y}}{x_b}\right). \tag{12.1.17}$$

In terms of these variables one has

$$\hat{t} = -\frac{s}{2}\left(\frac{x_a x_T}{z}\right) e^{-y} \tag{12.1.18}$$

$$\hat{u} \equiv (p_b - p_c)^2 = -\frac{s}{2}\left(\frac{x_b x_T}{z}\right) e^y. \tag{12.1.19}$$

The lower limits of integration in (12.1.10) are given by

$$\xi_a = \frac{x_T e^y}{2 - x_T e^{-y}} \qquad \xi_b = \frac{x_a x_T e^{-y}}{2x_a - x_T e^y}. \tag{12.1.20}$$

The cross-section to produce a jet of energy E, momentum $\mathbf{p} = (\mathbf{p}_T, p_z)$ is obtained from (12.1.10) by summing over C, and making the replacement

$$\sum_C D_C^c(z) \to \delta(1-z) \tag{12.1.21}$$

and then carrying out the integration over x_b, which forces the equality

$$x_b = \frac{x_a x_T e^{-y}}{2x_a - x_T e^y}. \tag{12.1.22}$$

For the production of a hard photon,

$$A + B \to \gamma + X,$$

to order $\alpha\alpha_s$ the square bracket in (12.1.10) is replaced by

$$\frac{d\hat{\sigma}^{ab\to\gamma d}}{d\hat{t}}\delta(1-z) \tag{12.1.23}$$

and the relevant partonic subprocesses are $Gq \to \gamma q$ and $q\bar{q} \to \gamma G$. QCD corrections to order $\alpha\alpha_s^2$ involve also final states with three partons, e.g. $GG \to \gamma q\bar{q}$, with integration over the phase space of the additional parton.

As written eqn (12.1.10) is at the level of the simple parton model. Allowing for QCD corrections in the leading logarithmic approximation (the LLA, see Section 11.6) one simply makes the replacements

$$\begin{aligned} f_a^A(x_a) &\to f_a^A(x_a, Q^2) \\ D_C^c(z) &\to D_C^c(z, Q^2) \end{aligned} \tag{12.1.24}$$

where $Q^2 \approx p_T^2$ and the parton densities remain universal, i.e. the same densities appear in DIS and in all large-p_T reactions. When one goes beyond the LLA care must be taken with scheme dependence, and the densities that appear in one reaction will differ by terms of order $\alpha_s(Q^2)$ from the densities occurring in other reactions.

12.2 Longitudinal two-spin asymmetries

For longitudinally polarized beam and target and for a parity-conserving theory one conventionally defines

$$A_{\mathrm{LL}} = \frac{d\sigma_{++} - d\sigma_{+-}}{d\sigma_{++} + d\sigma_{+-}} \equiv \frac{d\Delta\sigma}{2d\sigma} \tag{12.2.1}$$

where + and − refer to the *helicities* and where $d\sigma$ is the unpolarized differential cross-section. According to the discussion leading to (12.1.9), and suppressing the Q^2-dependence (12.1.24),

$$E_C \frac{d^3\sigma}{d^3\mathbf{p}_C} = \frac{1}{\pi} \sum_{a,b,c,d} \int_{\xi_a} dx_a \int_{\xi_b} dx_b \, \Delta f_a^A(x_a) \Delta f_b^B(x_b) \times \left[\frac{d\Delta\hat{\sigma}}{d\hat{t}}^{ab \to cd} (\hat{s}, \hat{t}) \frac{1}{z} D_C^c(z) \right] \tag{12.2.2}$$

where

$$\frac{d\Delta\hat{\sigma}}{d\hat{t}} = \frac{d\hat{\sigma}_{++}}{d\hat{t}} - \frac{d\hat{\sigma}_{+-}}{d\hat{t}} \equiv 2\hat{a}_{LL} \frac{d\hat{\sigma}}{d\hat{t}}. \tag{12.2.3}$$

Note that there is no dependence on the azimuthal angle ϕ in the above. The polarized-parton distributions are the helicity-dependent Δq, ΔG that appear in the spin-dependent structure function g_1 in polarized DIS (Sections 11.3 and 11.7). With the exception of the gluon, they are reasonably well determined; see Fig. 11.15.

The observable A_{LL} is particularly important because at the parton level the asymmetries

$$\hat{a}_{LL} = \frac{d\hat{\sigma}_{++} - d\hat{\sigma}_{+-}}{d\hat{\sigma}_{++} + d\hat{\sigma}_{+-}} \tag{12.2.4}$$

are very large on account of the helicity structure discussed in Section 10.4. The lowest-order expressions for the unpolarized cross-sections, and the expressions for $\hat{a}_{LL}$, due to Babcock, Monsay and Sivers (1979) and to Ranft and Ranft (1978), are listed in Tables 12.1 and 12.2 for all partonic processes, and shown in Fig. 12.2 as a function of the parton scattering angle θ^* in the parton–parton CM.

In Table 12.1, for purely strong interactions the cross-sections in the parton–parton CM are given by

$$\frac{d\hat{\sigma}}{d\hat{t}} = \frac{\pi\alpha_s^2}{\hat{s}^2} \overline{|M|^2}. \tag{12.2.5}$$

For cross-sections involving one photon in the initial or final state, we have

$$\frac{d\hat{\sigma}}{d\hat{t}} = \frac{\pi e_q^2 \alpha\alpha_s}{\hat{s}^2} \overline{|M|^2}, \tag{12.2.6}$$

where e_q is the quark electric charge, and for the process $q\bar{q} \to \gamma \to l\bar{l}$ we have

$$\frac{d\hat{\sigma}}{d\hat{t}} = \frac{\pi e_q^2 \alpha^2}{\hat{s}^2} \overline{|M|^2}. \tag{12.2.7}$$

Of course these large asymmetries get diluted by the rather small

Table 12.1. Partonic cross-sections (for the normalization see eqns (12.2.5)–(12.2.7)). Note that qq' means that the quarks q' and q differ in flavour

Reaction	$\overline{\|M\|^2}$
$qq \to qq$	$\frac{4}{9}\left(\frac{\hat{s}^2+\hat{u}^2}{\hat{t}^2}+\frac{\hat{s}^2+\hat{t}^2}{\hat{u}^2}-\frac{2}{3}\frac{\hat{s}^2}{\hat{u}\hat{t}}\right)$
$q\bar{q} \to q\bar{q}$	$\frac{4}{9}\left(\frac{\hat{s}^2+\hat{u}^2}{\hat{t}^2}+\frac{\hat{u}^2+\hat{t}^2}{\hat{s}^2}-\frac{2}{3}\frac{\hat{u}^2}{\hat{s}\hat{t}}\right)$
$qq' \to qq'$ $q\bar{q}' \to q\bar{q}'$	$\frac{4}{9}\frac{\hat{s}^2+\hat{u}^2}{\hat{t}^2}$
$q\bar{q} \to q'\bar{q}'$	$\frac{4}{9}\frac{\hat{u}^2+\hat{t}^2}{\hat{s}^2}$
$q\bar{q} \to GG$	$\frac{8}{3}\left(\frac{4}{9}\frac{\hat{u}^2+\hat{t}^2}{\hat{u}\hat{t}}-\frac{\hat{u}^2+\hat{t}^2}{\hat{s}^2}\right)$
$GG \to q\bar{q}$	$\frac{9}{64}\left[\frac{8}{3}\left(\frac{4}{9}\frac{\hat{u}^2+\hat{t}^2}{\hat{u}\hat{t}}-\frac{\hat{u}^2+\hat{t}^2}{\hat{s}^2}\right)\right]$
$qG \to qG$ $\bar{q}G \to \bar{q}G$	$(\hat{u}^2+\hat{s}^2)\left(\frac{1}{\hat{t}^2}-\frac{4}{9}\frac{1}{\hat{u}\hat{s}}\right)$
$GG \to GG$	$\frac{9}{2}\left(3-\frac{\hat{u}\hat{t}}{\hat{s}^2}-\frac{\hat{u}\hat{s}}{\hat{t}^2}-\frac{\hat{s}\hat{t}}{\hat{u}^2}\right)$
$qG \to q\gamma$ $\bar{q}G \to \bar{q}\gamma$	$-\frac{1}{3}\left(\frac{\hat{u}}{\hat{s}}+\frac{\hat{s}}{\hat{u}}\right)$
$q\bar{q} \to \gamma G$	$\frac{8}{9}\left(\frac{\hat{u}}{\hat{t}}+\frac{\hat{t}}{\hat{u}}\right)$
$q\bar{q} \to \gamma \to l\bar{l}$	$\frac{4}{3}\frac{\hat{u}^2+\hat{t}^2}{\hat{s}^2}$

polarizations of the partons inside the hadron, so that $|A_{LL}| \ll |\hat{a}_{LL}|$. Nonetheless, measurable asymmetries survive and will be very useful in learning about the polarized gluon density, which, as mentioned, is poorly known from polarized DIS, and in improving the precision of the polarized sea-quark densities. We shall give a few illustrations involving reactions that will be studied by the COMPASS and HERMES groups and at the RHIC collider. We draw the reader's attention to the fact that these

Table 12.2. Partonic longitudinal double-spin asymmetries

Reaction	$\hat{a}_{LL}$
$qq \to qq$	$\dfrac{\hat{u}^2(\hat{s}^2-\hat{u}^2)+\hat{t}^2(\hat{s}^2-\hat{t}^2)-\frac{2}{3}\hat{s}^2\hat{t}\hat{u}}{\hat{u}^2(\hat{s}^2+\hat{u}^2)+\hat{t}^2(\hat{s}^2+\hat{t}^2)-\frac{2}{3}\hat{s}^2\hat{t}\hat{u}}$
$q\bar{q} \to q\bar{q}$	$\dfrac{\hat{s}^2(\hat{s}^2-\hat{u}^2)-\hat{t}^2(\hat{u}^2+\hat{t}^2)+\frac{2}{3}\hat{u}^2\hat{s}\hat{t}}{\hat{s}^2(\hat{s}^2+\hat{u}^2)+\hat{t}^2(\hat{u}^2+\hat{t}^2)-\frac{2}{3}\hat{u}^2\hat{s}\hat{t}}$
$qq' \to qq'$, $q\bar{q}' \to q\bar{q}'$	$\dfrac{\hat{s}^2-\hat{u}^2}{\hat{s}^2+\hat{u}^2}$
$q\bar{q} \to q'\bar{q}'$	-1
$q\bar{q} \to GG$	-1
$GG \to q\bar{q}$	-1
$qG \to qG$, $\bar{q}G \to \bar{q}G$	$\dfrac{\hat{s}^2-\hat{u}^2}{\hat{s}^2+\hat{u}^2}$
$GG \to GG$	$\dfrac{-3+2\hat{s}^2/(\hat{u}\hat{t})+(\hat{u}\hat{t})/\hat{s}^2}{3-(\hat{s}\hat{u})/\hat{t}^2-(\hat{s}\hat{t})/\hat{u}^2-(\hat{u}\hat{t})/\hat{s}^2}$
$qG \to q\gamma$, $\bar{q}G \to \bar{q}\gamma$	$\dfrac{\hat{s}^2-\hat{u}^2}{\hat{s}^2+\hat{u}^2}$
$q\bar{q} \to \gamma G$	-1
$q\bar{q} \to \gamma \to l\bar{l}$	-1

and other spin asymmetries can also be used in testing for non-standard physical effects – supersymmetry, technicolour etc. – which we do not have space to discuss. A comprehensive discussion can be found in the review article by Bourrely, Renard, Soffer and Taxil (1989).

12.2.1 $pp \to \pi^0 X$

In the lower part of Fig. 12.3 we show $A_{LL}^{\pi^0}$ at $y=0$ for the reaction $pp \to \pi^0 X$ as measured by the E581/704 collaboration at Fermilab (Adams *et al.*, 1991) at $p_L = 200$ GeV/c, corresponding to $\sqrt{s} = 20$ GeV.

Before discussing the theoretical curves, two points should be noted.

(1) The range of p_T is small, $1 \leq p_T \leq 3.5$ GeV/c, and it is known that the *unpolarized* cross-section is poorly described by QCD in this region unless the partons are given some intrinsic transverse momentum of order $\langle k_T \rangle \approx 0.45$ GeV/c. Equation (12.2.2) can be generalized to

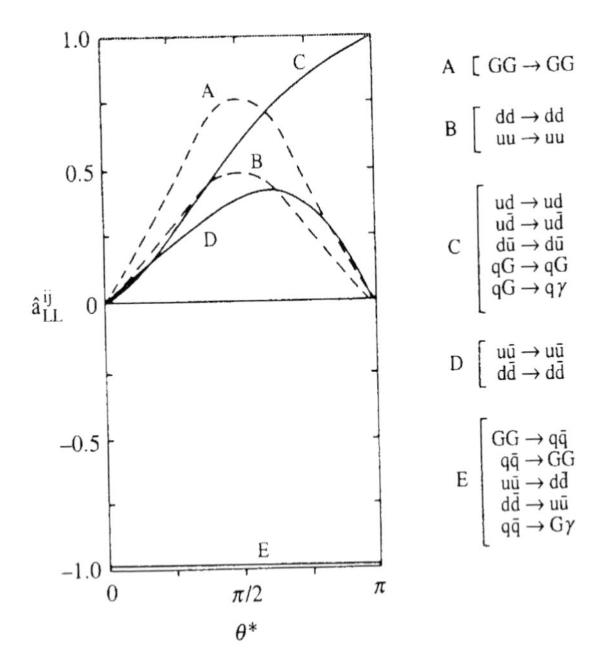

Fig. 12.2 Two-spin longitudinal asymmetry $\hat{a}_{LL}$ for various partonic reactions vs. CM scattering angle θ^*. (From Craigie, Hidaka, Jacob and Renard, 1983.)

allow for intrinsic k_T effects (Vogelsang and Weber, 1992), but this is a somewhat *ad hoc* procedure. Measurements at larger p_T will not be sensitive to k_T.

(2) The very small experimental asymmetries were claimed to prove that the polarized gluon density $\Delta G(x)$ is negligibly small, in contradiction with the use of ΔG to explain the 'spin crisis' discussed in Section 11.5. However, at the relatively small CM energy of 20 GeV, the p_T values involved imply, via (12.1.20), that typical x-values are outside the region where $\Delta G(x)$ is expected to be large (see Fig. 11.15).

The curves in the upper part of Fig. 12.3 correspond to a variety of $\Delta G(x)$, including a very hard version, (b), and a negative version, (c(−)); $-\Delta G(x)$ is shown for the case (c(−)). It is seen that the data have a very poor ability to discriminate amongst these, with the exception of (b), which is clearly ruled out. (A recent analysis of polarized DIS (Leader, Sidorov and Stamenov, 1998a) would rule out (b) as well as (c(±)); see Fig. 11.15.)

The general insensitivity to $\Delta G(x)$ in Fig. 12.3 is due to the low energy of the reaction and the fact that for the present data it is the $qG \to qG$

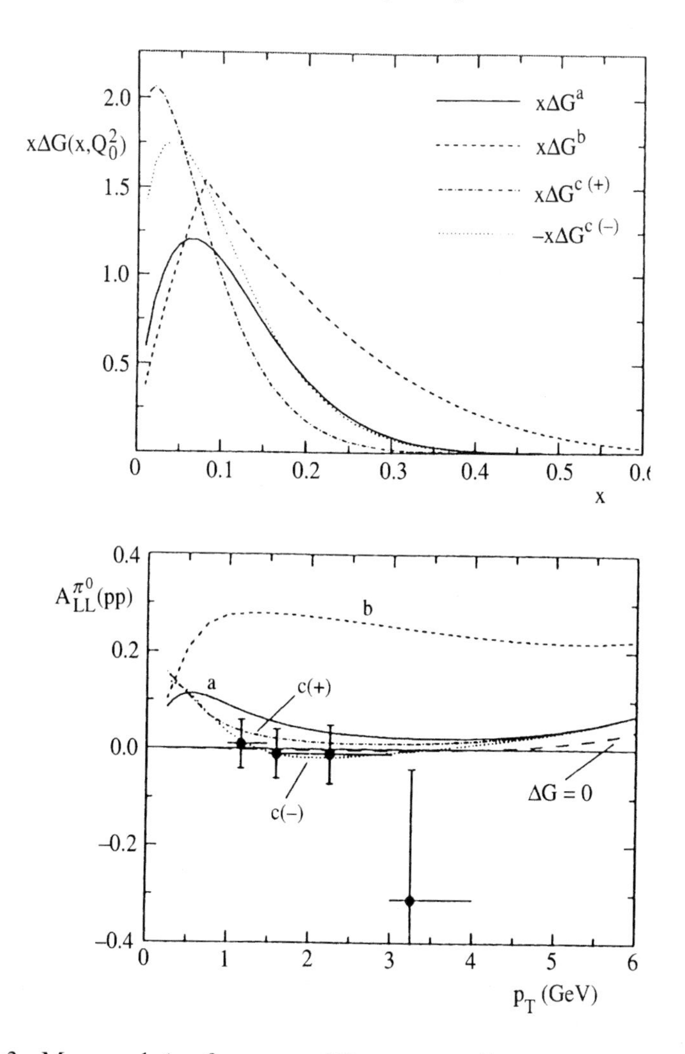

Fig. 12.3 Measured A_{LL} for $pp \to \pi^\circ X$ vs. p_T at $\sqrt{s} = 20$ GeV and $y = 0$ (Adams *et al.*, 1991) compared with theoretical predictions (Vogelsang, 1993) using various models of $\Delta G(x)$. The curves in the lower part of the figure correspond to the $\Delta G(x, Q^2)$ shown in the upper part at $Q^2 = 4(\text{GeV}/c)^2$. See the text for further explanation. (a) Kunszt (1989), (b) Ramsey and Sivers (1991) and (c) Glück, Reya and Vogelsang (1992).

process that dominates. At higher energies the reaction will be more discriminating.

12.2.2 Prompt photon production

It is expected that prompt photon production, $pp \to \gamma X$, will be rather sensitive to $\Delta G(x)$. The partonic process $qG \to q\gamma$ has a significant positive

$\hat{a}_{LL}$ (see Fig. 12.2), while the competing process $q\bar{q} \to G\gamma$, although it has $\hat{a}_{LL} = -1$, is relatively unimportant because of the smallness of the polarized-sea density (see Fig. 11.15). A very detailed study, at next-to-leading order, has been carried out by Gordon and Vogelsang (1993, 1994).

Figure 12.4 shows $A_\gamma \equiv A^\gamma_{LL}$ versus p_T at $\sqrt{s} = 100$ GeV and $y = 0$ for two models of ΔG. Model (a) corresponds closely to the curve (a) in Fig. 12.3. It has a relatively large ΔG and the polarized sea-quark density is taken to be zero below a scale of $Q^2 = 10$ (GeV/c)^2. In model (b) the polarized gluon density is zero below 10 (GeV/c)^2 and the polarized sea is relatively large and negative. It is seen that A_γ is quite sensitive to ΔG both in leading order and in next-to-leading order.

12.2.3 The Drell–Yan reaction $pp \to l^+l^-X$

The reaction, in lowest order, is visualized as in Fig. 12.5, where the lepton–antilepton pair is produced via virtual one-photon exchange $q\bar{q} \to$ 'γ' $\to l^+l^-$ and should thus be sensitive to the polarized antiquark density. For l^+l^- pairs of very large mass M it is also necessary to include Z^0 exchange and one then finds interesting single-spin parity-violating asymmetries as well (see subsection 12.3.3).

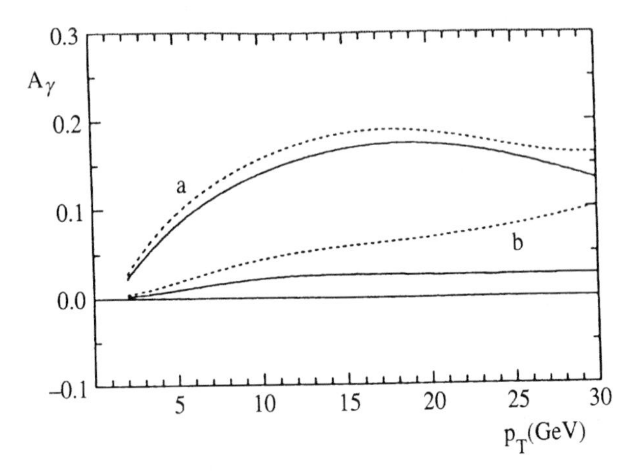

Fig. 12.4 Calculated values of $A_\gamma \equiv A^\gamma_{LL}$ for $pp \to \gamma X$ at $\sqrt{s} = 100$ GeV and $y = 0$ vs. p_T (from Vogelsang, 1993). The solid curves refer to next-to-leading order calculations and the broken curves to leading-order calculations for two models for ΔG (see text).

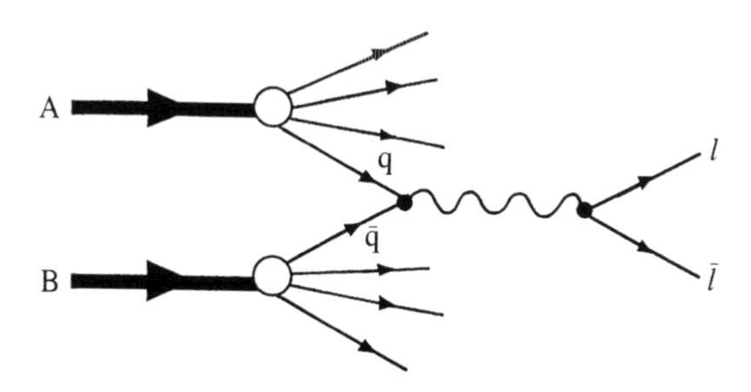

Fig. 12.5. Mechanism for the Drell–Yan reaction $pp \to l^+l^-X$ in lowest order.

Conventionally we define

$$\tau = \frac{M^2}{s} \tag{12.2.8}$$

and 'Feynman x'

$$x_{\mathrm{F}} = \frac{2}{\sqrt{s}} p_z \tag{12.2.9}$$

where p_z is the longitudinal momentum of the *pair*, which, in the lowest-order process, moves essentially along the pp collision axis. We then find that x_a and x_b are fixed once τ and x_{F} are determined:

$$x_a x_b = \tau \tag{12.2.10}$$

$$x_a - x_b = x_{\mathrm{F}}. \tag{12.2.11}$$

Since $x_{a,b}$ are fixed the result (12.2.2) simplifies greatly and one ends up with

$$A_{LL}^{\mathrm{D-Y}} = -\frac{\sum_f e_f^2 \left[\Delta q_f(x_a)\Delta\bar{q}_f(x_b) + \Delta q_f(x_b)\Delta\bar{q}_f(x_a)\right]}{\sum_f e_f^2 \left[q_f(x_a)\bar{q}_f(x_b) + q_f(x_b)\bar{q}_f(x_a)\right]} \tag{12.2.12}$$

where the sum is over the flavours f and where the parton densities should be evaluated at a scale $Q^2 \approx M^2$.

Equation (12.2.12) can be given a simple intuitive interpretation. Since the protons are just 'beams' of partons, (12.2.12) expresses the fact that

$$A_{LL}^{\mathrm{D-Y}} = \langle P_q P_{\bar{q}} \rangle_{x_a, x_b} \hat{a}_{LL}^{q\bar{q}\to l^+l^-} \tag{12.2.13}$$

where $\langle P_q P_{\bar{q}} \rangle$ plays the rôle of the product of quark and antiquark degrees of polarization and where, for the Drell–Yan process, $q\bar{q} \to \gamma \to l^+l^-$, $\hat{a}_{LL} = -1$.

A remarkable simplification can be achieved by a careful choice of the kinematics. Consider only events with $x_{\mathrm{F}} \gtrsim 0.2$. Then $x_a \gtrsim 0.2 + x_b$ and by keeping τ not too small, via (12.2.10) one can avoid very small x_b-values.

For such a range of x_a, only the up quark is important in the denominator of (12.2.12). Moreover $\Delta s(x_a)$ is then negligible, so that

$$A_{LL}^{\text{D-Y}} \simeq -\frac{1}{u(x_a)\bar{u}(x_b)}\left[\Delta u(x_a)\Delta\bar{u}(x_b) + \tfrac{1}{4}\Delta d(x_a)\Delta\bar{d}(x_b)\right]. \qquad (12.2.14)$$

Further, *in this region*, one can take for DIS (see (11.4.5))

$$\begin{aligned} F_1^p(x) &\approx \frac{2}{9}u(x) \\ g_1^p(x) &\approx \frac{1}{18}\left[4\Delta u(x) + \Delta d(x)\right] \qquad (12.2.15) \\ g_1^n(x) &\approx \frac{1}{18}\left[\Delta u(x) + 4\Delta d(x)\right]. \end{aligned}$$

Now $g_1^n(x)$ is almost zero for the region $0.25 \leq x \leq 0.5$ (see Figs. 11.13 and 11.14), suggesting that $\Delta d(x) \approx -\Delta u(x)/4$. Using this, (12.2.13) takes the beautifully simple form

$$A_{LL}^{\text{D-Y}}(x_a, x_b) = -A_1(x_a)\frac{\Delta\bar{u}(x_b)}{\bar{u}(x_b)} \qquad x_a \gtrsim 0.2 + x_b \qquad (12.2.16)$$

where $A_1(x)$ is one of the two photon–proton asymmetries often used in discussing polarized DIS,

$$A_1(x) = \frac{g_1^p(x)}{F_1^p(x)} \qquad (12.2.17)$$

and its value is therefore known from experiment. Equation (12.2.16) thus provides a tight link between D–Y asymmetries and $\Delta\bar{u}$. Note that although radiative corrections are known to have a significant effect upon the lowest-order result for the cross-section, Ratcliffe (1983) showed that the result (12.2.12) is a good approximation. For a more complete study of NLO corrections, see Gehrmann (1997).

It should soon become possible to study the Drell–Yan reaction for an l^+l^- pair, in which the pair has large mass M and may have large p_T. This would involve the additional partonic process

$$q\bar{q} \to (\gamma \text{ or } Z^0) + G \qquad (12.2.18)$$

$$qG \to (\gamma \text{ or } Z^0) + q. \qquad (12.2.19)$$

The basic two-spin asymmetry $A_{LL}^{\text{D-Y}}$ now becomes a function of x_T and the rapidity y of the pair, (see (12.1.15) and (12.1.16), and involves an integration over x_a, x_b subject to (12.1.22), so is not so directly related as before to the parton densities at specific x_a and x_b. However, as shown by Leader and Sridhar (1994), it does now becomes sensitive to the polarized gluon distribution. Moreover, there are other interesting asymmetries accessible if one can measure the distribution in the polar angle θ_l of the lepton l^- in the l^+l^- rest frame, in which the axis OZ is chosen along the direction of the pair in the CM of the reaction.

The cross-section takes the form

$$\frac{d^4\sigma_{++/+-}}{dx_T dy d\tau d\cos\theta_l} = \sum_{j=0}^{2} D^j_{++/+-} Y_{j0}(\theta_l) \tag{12.2.20}$$

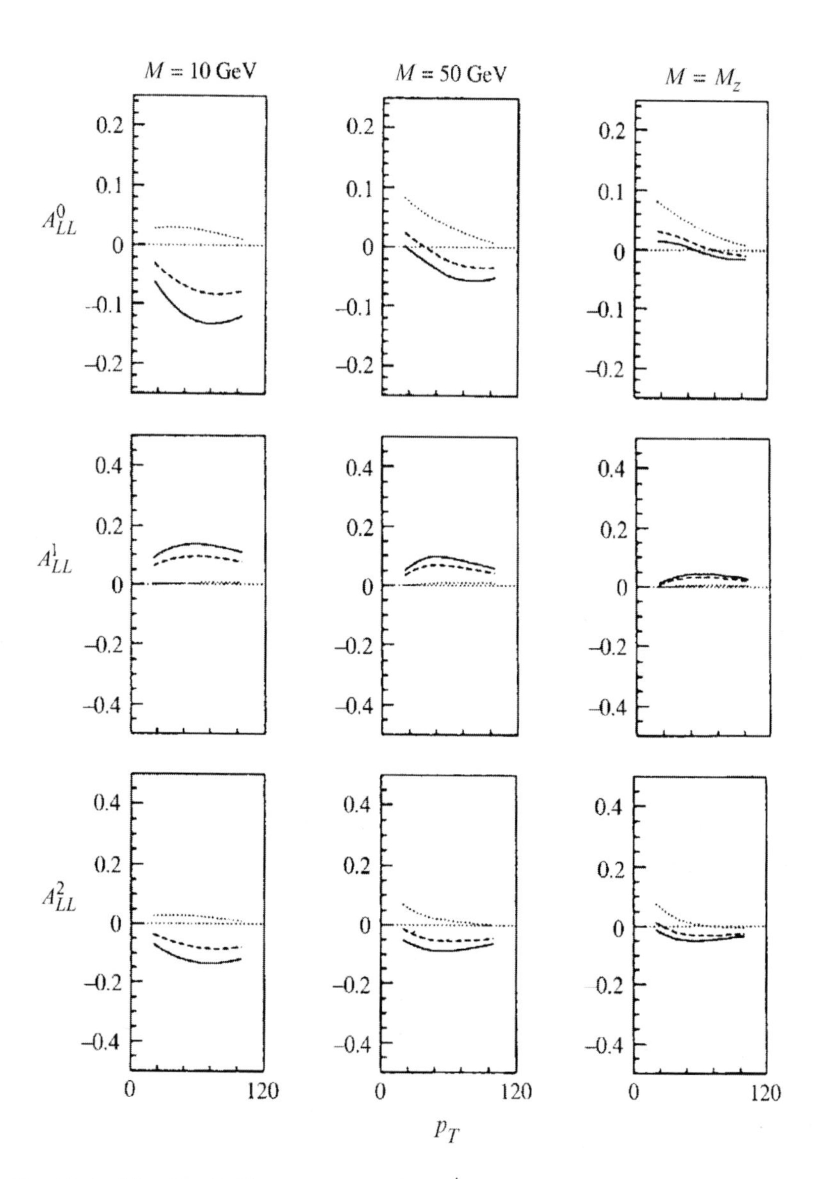

Fig. 12.6 The Drell–Yan asymmetries A^j_{LL} vs. p_T for several values of M and at $y = 0$. Solid curves; set I; broken curves, set II; dotted curves, set III. See the text for further explanation. (From Leader and Sridhar, 1994.)

where the Y_{j0} are spherical harmonics and the coefficients D^j depend on x_T, y and τ and involve an integration over the momentum fractions carried by the partons. Explicit expressions can be found in Leader and Sridhar (1994).

There are then three double spin asymmetries[1] $j = 0, 1, 2$:

$$A^j_{LL} = \frac{D^j_{++} - D^j_{+-}}{D^j_{++} + D^j_{+-}} \tag{12.2.21}$$

of which $A^0_{LL} \equiv A^{\text{D-Y}}_{LL}$.

Some results are shown in Fig. 12.6 as functions of p_T, for various pair masses, and for $y = 0$. The curves correspond to the three kinds of polarized gluon density given in Sridhar and Leader (1992).

$$\begin{aligned}
&\text{set I (solid curves), } \Delta G \text{ large, } \Delta s = 0;\\
&\text{set II (broken curves), } \Delta G \text{ and } \Delta s \text{ both moderately large;}\\
&\text{set III (dotted curves), } \Delta G = 0,\ \Delta s \text{ large.}
\end{aligned} \tag{12.2.22}$$

It is seen that these double-spin asymmetries are reasonably large and offer some hope of discriminating between the various models of ΔG.

Another interesting possibility, suggested by Contogouris and Papadopoulos (1991), is to look for longitudinal spin correlations with only one of the initial hadrons longitudinally polarized and with detection of the longitudinal polarization of one of the produced leptons. Perhaps this is feasible for $\mu^+\mu^-$ production.

The above suggestions look somewhat futuristic, since so far there have been no experiments on polarized Drell–Yan reactions; they will be studied for the first time at the polarized pp collider RHIC, which came into operation early in the year 2000.

12.2.4 Drell–Yan production of J/Ψ and χ_2

We comment briefly upon the possibility of using these reactions to learn about ΔG. The most important partonic processes are shown in Fig. 12.7 and are gluon initiated. The processes shown in Fig. 12.8 are much less important.

Note that in process A in Fig. 12.7

$$x_a x_b = M^2_\chi / s \tag{12.2.23}$$

[1] These are referred to as A^j_{d} (d for double) in the above-cited paper.

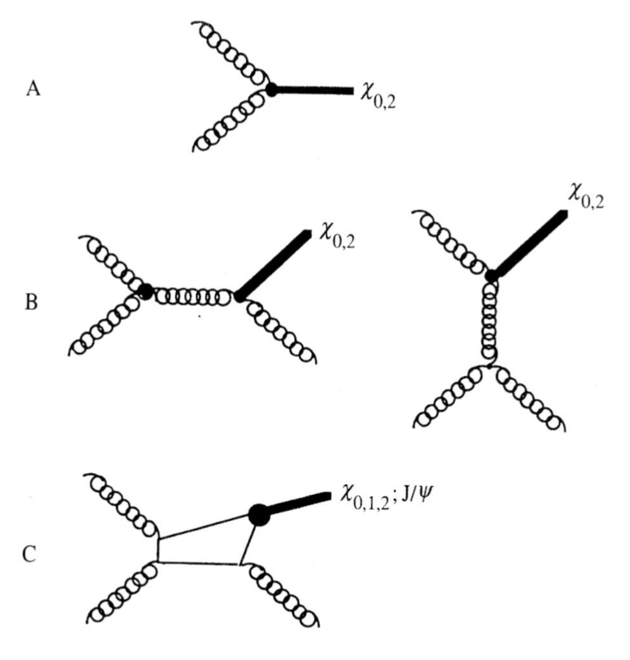

Fig. 12.7. The most important partonic reactions involved in $pp \to J/\Psi$ or $\chi_2 + X$.

whereas in processes B and C

$$1 \geq x_a x_b \geq M_\chi^2/s. \tag{12.2.24}$$

χ_2 production. This is very clean when reaction A of Fig. 12.7 dominates and, as pointed out by Cortes and Pire (1988), should have a large asymmetry since the partonic asymmetry is maximal,

$$\hat{a}_{LL}^{GG\to\chi_2} = -1. \tag{12.2.25}$$

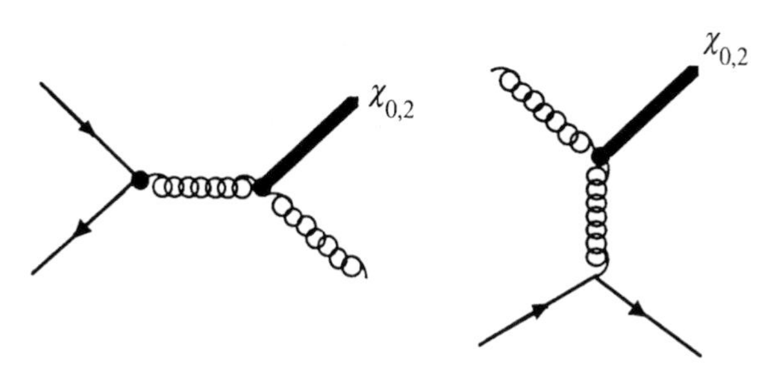

Fig. 12.8. Less important ($q\bar{q}$ initiated) partonic reactions.

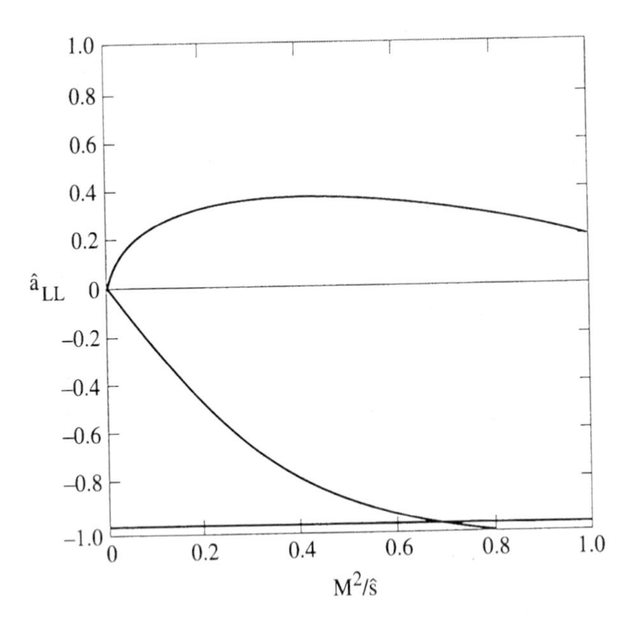

Fig. 12.9 Longitudinal partonic asymmetry for (top curve) $GG \to J/\Psi + G$, (bottom curve) $GG \to \chi_2$ and (middle curve) $GG \to \chi_2 + G$, extracted from the results of Doncheski and Robinett (1994). See the text for an explanation of the curves.

At low energies reaction A should dominate, in which case knowing x_F fixes x_a and x_b. Thus one measures directly (analogously to (12.2.12))

$$A_{LL}^{\chi_2} = -\frac{\Delta G(x_a)}{G(x_a)}\frac{\Delta G(x_b)}{G(x_b)}. \tag{12.2.26}$$

If there is a significant contribution from reaction B in Fig. 12.7 the effective gluon polarization gets smeared out by an integration over, say, x_a, and the partonic asymmetry is no longer maximal. $a_{LL}^{GG \to \chi_2 G}$ is shown in Fig. 12.9 in the limit $\hat{t} \to 0$ as function of $M^2/\hat{s}$. The hadronic asymmetry will thus be smaller than predicted by (12.2.26) but should still yield measurable values. This depends, however, upon ensuring that the collision energy is not too large, i.e. not forcing x_a, x_b to be too small, since one expects $\Delta G(x)/G(x)$ to tend to zero as $x \to 0$.

There is also the problem that the theory is somewhat less reliable when more than one partonic channel contributes. Suppose there are two channels (A) and (B). From the intuitive expression (12.2.13) we will have

$$\begin{aligned} A_{LL} &= \left(\frac{\Delta\sigma_A}{\sigma_A + \sigma_B}\right) + \left(\frac{\Delta\sigma_B}{\sigma_A + \sigma_B}\right) \\ &= \left(\frac{\sigma_A}{\sigma_A + \sigma_B}\right)\hat{a}_{LL}^A + \left(\frac{\sigma_B}{\sigma_A + \sigma_B}\right)\hat{a}_{LL}^B \end{aligned} \tag{12.2.27}$$

and various uncertain factors, such as K-factors, non-relativistic radial wave-function values etc., no longer cancel out in the asymmetry. In practice this may not be a serious problem but must be investigated.

J/Ψ production. If the χ_2 particles cannot be reconstructed adequately one can get less clean information from the J/Ψ asymmetry. All three processes A, B and C in Fig. 12.7 now contribute, since the branching ratio $BR(\chi_2 \to J/\Psi\gamma) \approx 14\%$ is sizeable. The same gluon polarization appears, but the partonic asymmetry is diluted since $\hat{a}_{LL}(GG \to J/\Psi G)$ is generally *positive*.

The integrated partonic cross-section asymmetry is shown versus $M^2/\hat{s}$ in Fig. 12.9. It would seem from this that the hadronic asymmetry ought to be reasonably sensitive to the structure of $\Delta G(x)$, but detailed calculations, using up-to-date parametrizations of $\Delta G(x)$, are needed.

12.2.5 Semi-inclusive lepton–hadron scattering

Consider the deep inelastic scattering of a charged lepton l, momentum k^μ, on a nucleon N, momentum P^μ:

$$l + N \to l + C + X, \tag{12.2.28}$$

where C is some identified hadron, set of hadrons or jet. We assume that lepton and proton are longitudinally polarized. The incoming lepton and the exchanged photon have energies E and ν respectively, in the Lab.

Of particular interest is the case of charm production since it is controlled by the mechanism in Fig. 11.11 and may prove to be one of the best ways to learn about the polarized gluon density. The cross-section for

$$l + p \to l + (c\bar{c}) + X, \tag{12.2.29}$$

based upon the mechanism depicted in Fig. 11.11, can be expressed in terms of various cross-sections $\hat{\sigma}, \hat{\sigma}_0, \Delta\hat{\sigma}$ for the partonic reaction

$$\text{`}\gamma\text{'} + G \to c + \bar{c}. \tag{12.2.30}$$

For real photons the flux factor needed in calculating a cross-section is $K = \nu$. For virtual photons there is no unambiguous definition. It is simply a matter of convention, provided only that $K = \nu$ at $Q^2 = 0$. Various conventions for K exist in the literature: $\sqrt{\nu^2 + Q^2}$, $\nu - Q^2/(2M)$, $\nu/\sqrt{1 + Q^2/\nu^2}$, but clearly the *physical* cross-section for any reaction is independent of the choice of K.

For the reaction (12.2.29) one has

$$\frac{d\sigma^{\stackrel{\rightarrow}{\Leftarrow}}}{dv dQ^2} + \frac{d\sigma^{\stackrel{\rightarrow}{\Rightarrow}}}{dv dQ^2}$$
$$= \frac{\alpha K}{2\pi Q^2 v^2} \int_{x_{\min}}^{x_{\max}} dx G(x) \left\{ \left[1 + (1-y)^2\right] \hat{\sigma}(\hat{s}, Q^2) + 2(1-y)\hat{\sigma}_0(\hat{s}, q^2) \right\} \tag{12.2.31}$$

$$\frac{d\sigma^{\stackrel{\rightarrow}{\Leftarrow}}}{dv dQ^2} - \frac{d\sigma^{\stackrel{\rightarrow}{\Rightarrow}}}{dv dQ^2} = \frac{\alpha K}{2\pi Q^2 v^2} \left[1 - (1-y)^2\right] \int_{x_{\min}}^{x_{\max}} dx \Delta G(x) \Delta\hat{\sigma}(\hat{s}, Q^2) \tag{12.2.32}$$

where

$$y \equiv \frac{P \cdot q}{P \cdot k} = \frac{v}{E} \tag{12.2.33}$$

$$\hat{s} = (q + xP)^2 \approx 2Mvx - Q^2 \tag{12.2.34}$$

and the range of integration variable x corresponds to the requirement on the charm-quark mass

$$4m_c^2 \leq \hat{s} \leq s = 2Mv - Q^2. \tag{12.2.35}$$

Expressed in terms of cross-sections with fixed initial state helicities, the 'γ' $G \to c\bar{c}$ partonic cross-sections are

$$\begin{aligned} \hat{\sigma} &= \hat{\sigma}_{++} + \hat{\sigma}_{-+} \\ \Delta\hat{\sigma} &= \hat{\sigma}_{++} - \hat{\sigma}_{-+} \\ \hat{\sigma}_0 &= \hat{\sigma}_{0+} + \hat{\sigma}_{0-}. \end{aligned} \tag{12.2.36}$$

The results for the partonic cross-sections can be extracted from the calculations of Jones and Wyld (1978), Watson (1982) and Glück and Reya (1988). One finds

$$\hat{\sigma}(\hat{s}, Q^2) = \frac{2\pi\alpha_s \alpha e_c^2}{\hat{s}\left[1 + (Q^2/\hat{s})\right]^3} \left(\frac{v}{K}\right) \left\{ -\beta(2-\beta^2) + \frac{1}{2}(3-\beta^4) \ln\left(\frac{1+\beta}{1-\beta}\right) + 2\frac{Q^2}{\hat{s}}\beta + \frac{Q^4}{\hat{s}^2}\left[\ln\left(\frac{1+\beta}{1-\beta}\right) - \beta\right] \right\} \tag{12.2.37}$$

$$\Delta\hat{\sigma}(\hat{s}, Q^2) = \frac{2\pi\alpha_s \alpha e_c^2}{\hat{s}\left[1 + (Q^2/\hat{s})\right]^3} \left(\frac{v}{K}\right) \left\{ \ln\left(\frac{1+\beta}{1-\beta}\right) - 3\beta - 2\frac{Q^2}{\hat{s}}\beta + \frac{Q^4}{\hat{s}^2}\left[\beta - \ln\left(\frac{1+\beta}{1-\beta}\right)\right] \right\} \tag{12.2.38}$$

$$\hat{\sigma}_0(\hat{s}, Q^2) = \frac{2\pi\alpha_s\alpha e_c^2}{\hat{s}\left[1 + (Q^2/\hat{s})\right]^3}\left(\frac{\nu}{K}\right)\frac{Q^2}{\hat{s}}\left\{2\beta + (\beta^2 - 1)\ln\left(\frac{1+\beta}{1-\beta}\right)\right\} \tag{12.2.39}$$

where

$$\beta = \sqrt{1 - 4m_c^2/\hat{s}} \tag{12.2.40}$$

and $e_c^2 = 4/9$.

In the above we have not shown the scale μ^2 at which $G(x)$ and $\Delta G(x)$ are to be evaluated. It should be clear from Fig. 11.11 that the scale is not Q^2, and it can be argued that the relevant scale is $\mu^2 \approx \hat{s}$.

In practice, because the cross-sections (12.2.31), (12.2.32) drop rapidly with Q^2, the first planned experiments will concentrate on small Q^2, where it is probably safe to put $Q^2 = 0$ in the expressions (12.2.37)–(12.2.39) for the partonic cross-sections. However, this is only justified if $Q^2/\hat{s}$ is negligible, and since $\hat{s}$ can be as small as $4m_c^2$ the approximation really requires $Q^2 \ll 4m_c^2$.

The partonic longitudinal asymmetry

$$\hat{a}_{LL}^{c\bar{c}} \equiv \frac{\Delta\hat{\sigma}(\hat{s}, q^2)}{\hat{\sigma}(\hat{s}, q^2)} \tag{12.2.41}$$

has an interesting shape as a function of β. At threshold, $\beta = 0$ and $\hat{a}_{LL}^{c\bar{c}} = -1$, whereas at infinite energy, $\beta \to 1$ and $\hat{a}_{LL}^{c\bar{c}} \to 1$ and this is true for all Q^2.

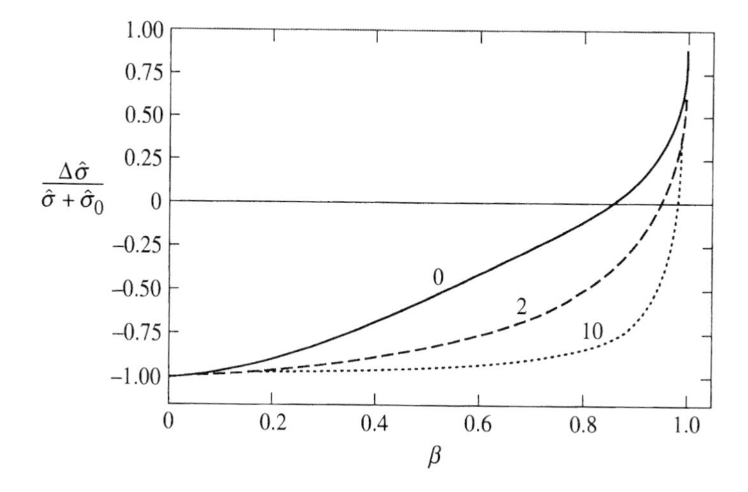

Fig. 12.10 The asymmetry $\Delta\hat{\sigma}/(\hat{\sigma} + \hat{\sigma}_0)$, which is effectively equal to $\hat{a}_{LL}^{c\bar{c}}$, as a function of β for three values of $Q^2/4m_c^2$.

In Fig. 12.10 we show $\Delta\hat{\sigma}/(\hat{\sigma}+\hat{\sigma}_0)$ as a function of β for three values of $Q^2/4m_c^2$. In fact the rôle of $\hat{\sigma}_0$ is completely negligible, for two reasons: firstly $\hat{\sigma}_0/\hat{\sigma}$ is generally very small and secondly $\hat{\sigma}_0/\hat{\sigma}$ reaches its maximum of 10%–20% just where $\Delta\hat{\sigma} \approx 0$.

The $ep \to e + (c\bar{c}) + X$ cross-sections (12.2.31) and (12.2.32) involve an integration over x or, equivalently, over $\hat{s}$ or β (see (12.2.34) and (12.2.40)); there will thus be a diminution of the cross-section difference (12.2.32), and therefore of the asymmetry, owing to a cancellation between the regions where $\hat{a}_{LL}^{c\bar{c}}$ is negative and those where it is positive, if $\Delta G(x)$ is monotonic. This effect grows in importance as ν increases, since larger ν implies larger $\beta_{\max}$ in the integration.

The effect of the cancellation can be combatted by going to larger Q^2, as can be seen from Fig. 12.10, but, as mentioned, one loses event rate.

In the approximation $Q^2 = 0$ one has for the measured asymmetry, upon changing integration variable from x to $\hat{s}$,[1]

$$A_{\parallel}^{c\bar{c}X}(E,\nu,Q^2=0) \equiv \left(\frac{d\sigma^{\overset{\rightarrow}{\Leftarrow}}}{d\nu dQ^2} - \frac{d\sigma^{\overset{\rightarrow}{\Rightarrow}}}{d\nu dQ^2}\right) \Big/ \left(\frac{d\sigma^{\overset{\rightarrow}{\Leftarrow}}}{d\nu dQ^2} + \frac{d\sigma^{\overset{\rightarrow}{\Rightarrow}}}{d\nu dQ^2}\right)$$

$$= \left[\frac{1-(1-y)^2}{1+(1-y)^2}\right] \frac{\int_{4m_c^2}^{2M\nu} d\hat{s}\,\Delta G(x=\hat{s}/(2M\nu),\hat{s})\Delta\hat{\sigma}^{\gamma G}(\hat{s},Q^2=0)}{\int_{4m_c^2}^{2M\nu} d\hat{s}\,G(x=\hat{s}/(2M\nu),\hat{s})\hat{\sigma}^{\gamma G}(\hat{s},Q^2=0)}. \tag{12.2.42}$$

It is unfortunate that in (12.2.42) the parton densities are in principle integrated over both the gluon momentum fraction x and the hard scale $\hat{s}$. It would be much simpler, from a theoretical point of view, if one could measure cross-sections differential in $\hat{s}$.

An idea of the size of the asymmetry at $E = 100$ GeV and $Q^2 = 0$, and its sensitivity to various models for $\Delta G(x)$, due to Gehrmann and Stirling (1996), is shown in Fig. 12.11. The sign, and the rapid change with energy for model C, is a consequence of the change in sign of $\Delta G(x)$ at $x \approx 0.1$.

12.3 Parity-violating longitudinal single-spin asymmetries

In a parity-conserving theory, if only one of the initial hadrons is polarized one has

$$d\sigma_+ = d\sigma_- \tag{12.3.1}$$

i.e. there is no asymmetry under reversal of helicity. However, in electroweak reactions such as the production of $W^\pm$ and Z^0, and in the

[1] Here we use the designation $A_\parallel$ that is conventional in papers on this subject.

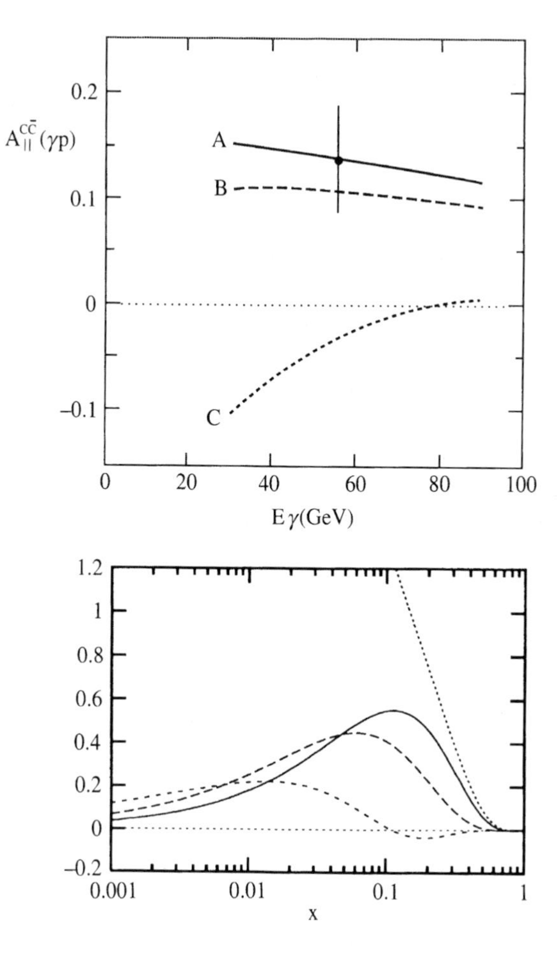

Fig. 12.11 The longitudinal asymmetry $A_{\parallel}^{c\bar{c}}$ vs. E_γ, for $lp \to l(c\bar{c})X$ at $E = 100$ GeV and $Q^2 = 0$. The data point with error bar indicates the expected accuracy in the COMPASS experiment at CERN; the curves correspond to $x\Delta G(x, 4\text{ GeV}^2)$ in *NLO* for the three models of $\Delta G(x)$ shown in the lower figure (courtesy of G. Mallot and T. Gehrmann.) In the latter figure the dotted line gives $xG(x)$.

production of very massive Drell–Yan pairs where Z^0 exchange is important, parity is not conserved and one obtains interesting information from the longitudinal single-spin asymmetries.

12.3.1 Small-p_T single-spin $W^\pm$ and Z^0 production

The production of $W^\pm$ and Z^0 in Drell–Yan-type reactions at *small* p_T should provide a formidable test for our knowledge of the polarized *sea-quark* densities at the huge scale of $Q^2 \approx M_W^2$.

Consider production of W^+ in

$$\vec{p}_A + p_B \to W^+ + X$$

with the axis OZ taken along the momentum of the longitudinally polarized proton, labelled A, which collides with the unpolarized target proton B.

The W^+ is produced with 4-momentum $p^\mu \approx (p^0, 0, 0, p_z)$ and rapidity

$$y = \frac{1}{2} \ln \left(\frac{p^0 + p_z}{p^0 - p_z} \right). \tag{12.3.2}$$

Since only left-handed u quarks and right-handed $\bar{d}$ antiquarks can couple to the W^+ one has

$$\begin{aligned} d\sigma_+ &\propto u_-(x_a)\bar{d}_+(x_b) + \bar{d}_+(x_a)u_-(x_b) \\ &\propto u_-(x_a)\bar{d}(x_b) + \bar{d}_+(x_a)u(x_b) \end{aligned} \tag{12.3.3}$$

since, with B unpolarized, for any parton $q_+(x) = q_-(x) = q(x)/2$.

Similarly

$$d\sigma_- \propto u_+(x_a)\bar{d}(x_b) + \bar{d}_-(x_a)u(x_b) \tag{12.3.4}$$

Thus the parity-violating longitudinal spin asymmetry is[1]

$$A_L^{W^+} = \frac{d\sigma_- - d\sigma_+}{d\sigma_- + d\sigma_+} = \frac{\Delta u(x_a)\bar{d}(x_b) - \Delta\bar{d}(x_a)u(x_b)}{u(x_a)\bar{d}(x_b) + \bar{d}(x_a)u(x_b)} \tag{12.3.5}$$

where

$$x_a = \sqrt{\tau} e^y \qquad x_b = \sqrt{\tau} e^{-y} \tag{12.3.6}$$

with, here,

$$\tau = M_W^2/s \tag{12.3.7}$$

and where the parton densities should be evaluated at a scale $Q^2 \approx M_W^2$.

For the production of W^- one has

$$A_L^{W^-} = \frac{\Delta d(x_a)\bar{u}(x_b) - \Delta\bar{u}(x_a)d(x_b)}{d(x_a)\bar{u}(x_b) + \bar{u}(x_a)d(x_b)}. \tag{12.3.8}$$

As pointed out by Bourrely and Soffer (1993), the expressions (12.3.5) and (12.3.8) take a particularly simple form in certain kinematic regimes.

For $y = 0$ one has $x_a = x_b = \sqrt{\tau}$ and

$$\begin{aligned} A_L^{W^+} &= \frac{1}{2}\left(\frac{\Delta u}{u} - \frac{\Delta\bar{d}}{\bar{d}} \right) \\ A_L^{W^-} &= \frac{1}{2}\left(\frac{\Delta d}{d} - \frac{\Delta\bar{u}}{\bar{u}} \right) \end{aligned} \tag{12.3.9}$$

[1] We here follow the sign convention of Bourrely and Soffer (1993).

For large negative y, x_a is small and x_b large so that, for the antiquark, $\bar{q}(x_b)$ should be negligible. Hence

$$A_L^{W^+} \approx -\frac{\Delta\bar{d}(x_a)}{\bar{d}(x_a)} \qquad A_L^{W^-} \approx -\frac{\Delta\bar{u}(x_a)}{\bar{u}(x_a)}. \tag{12.3.10}$$

Similarly, for y large and positive, x_a is large, so

$$A_L^{W^+} \approx \frac{\Delta u(x_a)}{u(x_a)} \qquad A_L^{W^-} \approx \frac{\Delta d(x_a)}{d(x_a)}. \tag{12.3.11}$$

For the productions of Z^0 at *small* p_T, the argument leading to (12.3.5) is slightly complicated by the fact that both left- and right-handed quarks can couple to Z^0 with a strength that can be read off from (9.1.8). One ends up with

$$A_L^{Z^0} = \frac{\sum_f \mathscr{A}_f \left[\Delta\bar{q}_f(x_a)q_f(x_b) - \Delta q_f(x_a)\bar{q}_f(x_b)\right]}{\sum_f \left[\bar{q}_f(x_a)q_f(x_b) + q_f(x_a)\bar{q}_f(x_b)\right]} \tag{12.3.12}$$

where $f = u, d$ and $\mathscr{A}_f$ is defined in (9.2.13) and (9.1.5).

Some estimates of the single-spin parity-violating asymmetries, as a function of y, at $\sqrt{s} = 500$ GeV, for $W^\pm$ and Z^0 with very small transverse momentum, are shown in Figs. 12.12 and 12.13. The solid curves correspond to a reasonable choice of sea-quark polarization; the broken curves have $\Delta\bar{u} = \Delta\bar{d} = 0$. It is seen that there are regions for W^- and Z^0 production where there is significant sensitivity to the sea-quark polarization. This is not so for W^+ production, since it is dominated by the large positive up-quark polarization.

12.3.2 Larger-p_T single-spin $W^\pm$ production

We consider the production of $W^\pm$ with 4-momentum $q^\mu = (q_0, \mathbf{q}_T, q_z)$, where q_T need not be small, in collisions of a proton (A) with longitudinal polarization $\mathscr{P}$, and an unpolarized proton or antiproton (B):

$$\vec{p}_A(\mathscr{P}) + (p_B \text{ or } \bar{p}_B) \to W^\pm + X. \tag{12.3.13}$$

The Z-axis lies along the momentum of the polarized proton. Several partonic processes contribute to the reaction: (a) Drell–Yan $q\bar{q} \to W$, (b) $q\bar{q} \to WG$ and (c) $qG \to Wq$. However, there are two kinematic regions where (a) and (b) dominate and where, consequently, the results depend principally on the quark and antiquark distributions. They are specified as region 1,

$$q_z > 0 \qquad q_z \gtrsim 2q_T \qquad q_T \ll M_W,$$

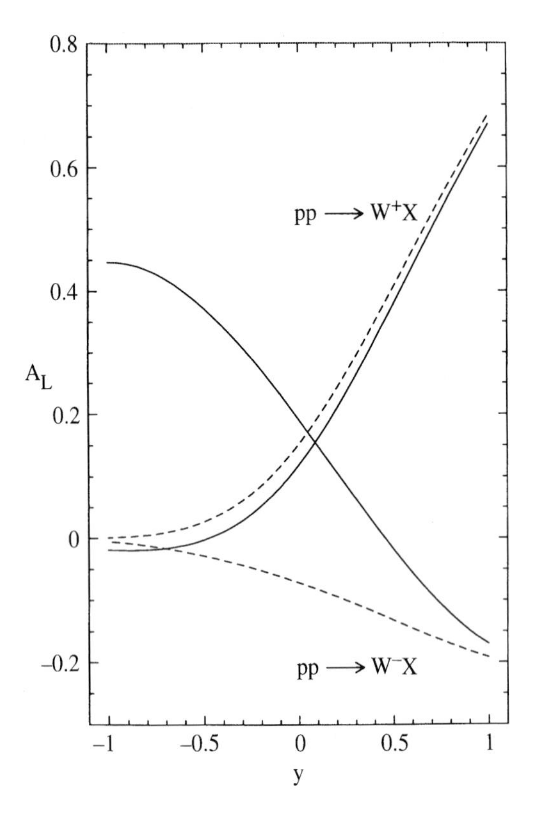

Fig. 12.12 Single-spin parity-violating asymmetry in $\vec{p}p \to W^{\pm}X$ vs. rapidity y at $\sqrt{s} = 500$ GeV for $W^{\pm}$with very small transverse momentum (from Bourrely and Soffer, 1993). For an explanation of the curves see the text.

and region 2,

$$q_z < 0 \qquad |q_z| \gtrsim 2q_T \qquad q_T \leq \bar{q}(s),$$

where $\bar{q}(s)$ is plotted in Fig. 12.14. The results that follow (Leader, 1986) are reliable only in regions 1 and 2.

In these regions the momentum fractions carried by the partons can be approximated by

$$\begin{aligned} x_a \equiv \hat{x}_a &= \frac{1}{\sqrt{s}}(q_0 + q_z + q_T) \\ x_b \equiv \hat{x}_b &= \frac{1}{\sqrt{s}}(q_0 - q_z + q_T) \end{aligned} \tag{12.3.14}$$

with $q^2 \approx M_W^2$, provided that $\hat{x}_{a,b} \ll 1$.

In the specified kinematic regions the normalized decay angular distribution for the $l^{\pm}$, in the rest frame of the $W^{\pm}$, with OZ along the motion

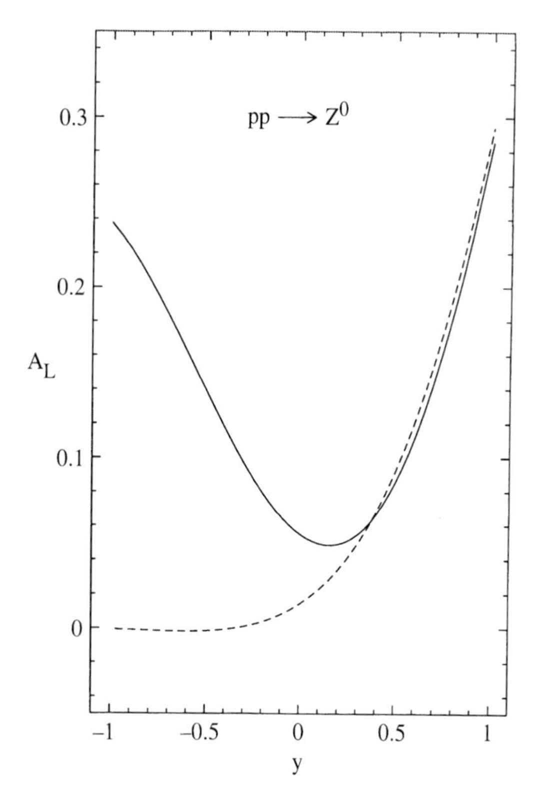

Fig. 12.13 Single-spin parity-violating asymmetry in $\vec{p}p \to Z^0X$ vs. rapidity y at $\sqrt{s} = 500$ GeV for Z^0 with small transverse momentum (from Bourrely and Soffer, 1993). For curves see the text.

of the $W^{\pm}$, may be taken to be[1]

$$W_{\pm}(\theta_l, \phi_l) \approx \frac{3}{16}\left(1 + \cos^2\theta_l \pm 2\mathscr{P}_{\pm}\cos\theta_l\right) \tag{12.3.15}$$

where $\mathscr{P}_{\pm}$ is the helicity polarization of the produced $W^{\pm}$. Experimental study of the decay distribution yields the values of $\mathscr{P}_{\pm}$.

The results are most simply expressed in term of the auxiliary quantities

$$\begin{aligned} \alpha_{\pm}(q_z, q_T; \mathscr{P}) &\equiv \frac{1+\mathscr{P}_{\pm}}{1-\mathscr{P}_{\pm}} \qquad \text{for } q_z > 0 \\ &\equiv \frac{1-\mathscr{P}_{\pm}}{1+\mathscr{P}_{\pm}} \qquad \text{for } q_z < 0 \end{aligned} \tag{12.3.16}$$

[1] We have used eqns (3.1.70) and (8.2.20) and the fact that in the above kinematic regions the Ws are produced predominantly with helicity ± 1.

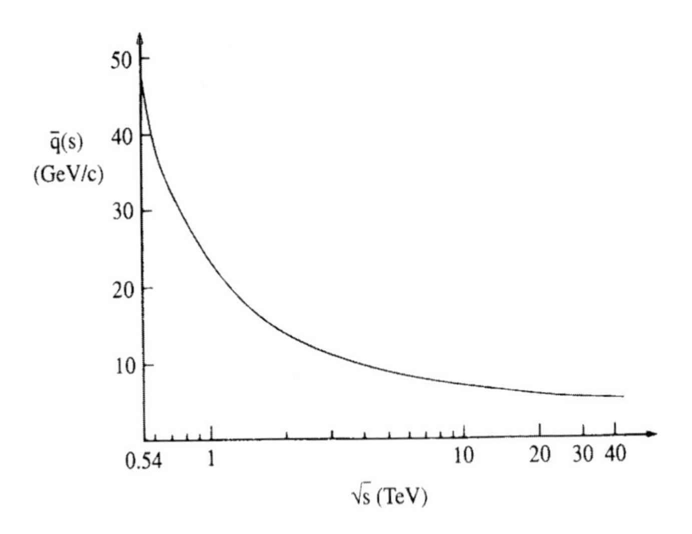

Fig. 12.14. The function $\bar{q}(s)$ vs. $\sqrt{s}$.

and in the comparison of these quantities for polarized and unpolarized protons, i.e. by constructing the ratio

$$r_{\pm}(q_z, q_T; \mathscr{P}) \equiv \frac{\alpha_{\pm}(q_z, q_T; \mathscr{P})}{\alpha_{\pm}(q_z, q_T; 0)}. \tag{12.3.17}$$

Then, for the various reactions one has

$$\text{for} \quad \left.\begin{array}{l} \vec{p} + p \to W^+ + X \\ \vec{p} + \bar{p} \to W^+ + X \end{array}\right\} \quad r_+(q_z, q_T; \mathscr{P}) = \frac{1 + \mathscr{P}\Delta\bar{d}(\hat{x}_a)/\bar{d}(\hat{x}_a)}{1 - \mathscr{P}\Delta u(\hat{x}_a)/u(\hat{x}_a)} \tag{12.3.18}$$

$$\text{for} \quad \left.\begin{array}{l} \vec{p} + p \to W^- + X \\ \vec{p} + \bar{p} \to W^- + X \end{array}\right\} \quad r_-(q_z, q_T; \mathscr{P}) = \frac{1 + \mathscr{P}\Delta\bar{u}(\hat{x}_a)/\bar{u}(\hat{x}_a)}{1 - \mathscr{P}\Delta d(\hat{x}_a)/d(\hat{x}_a)} \tag{12.3.19}$$

where the parton densities are at scale $q^2 = M_W^2$.

It should be noted that, whereas the $r_{\pm}$ are the same for proton–proton and proton–antiproton collisions, the $\alpha_{\pm}$ are quite different (see Leader, 1986) and it may turn out that the $\alpha_{\pm}$ for $p\bar{p}$ collisions are too small for a significant measurement in practice — indeed, early tests of the standard model assumed $\alpha_{\pm} = 0$ for the $p\bar{p}$ collisions at the CERN collider.

Note that if one can observe the process

$$\vec{p}_A + (p_B \text{ or } \bar{p}_B) \to W^{\pm}(\mathbf{q}) + \text{jet}(\mathbf{k}) \tag{12.3.20}$$

then (12.3.14) can be replaced by the exact equations

$$\begin{aligned} x_a &= \frac{1}{\sqrt{s}}(q_0 + k_0 + q_z + k_z) \\ x_b &= \frac{1}{\sqrt{s}}(q_0 + k_0 - q_z - k_z). \end{aligned} \tag{12.3.21}$$

Note also that the validity of the whole approach can be checked by testing whether the measured quantities in (12.3.18) and (12.3.19) are independent of x_b, as they are supposed to be.

Finally, it is a nice feature that the measurement of $r_\pm$ gives information directly about the *polarization* of the quarks and antiquarks, i.e. about $\Delta q(x)/q(x)$, a quantity that is of interest because of theoretical ideas about its behaviour for small and large x (see subsection 11.7.1).

12.3.3 Larger-p_T single-spin massive Drell-Yan production

For a Drell–Yan reaction where the lepton pair is produced with non-negligible transverse momentum, the analogue of (12.3.20) is

$$\frac{d^4\sigma_{+/-}}{dx_T dy d\tau d\cos\theta_l} = \sum_{j=0}^{2} D^j_{+/-} Y_{j0}(\theta_l) \tag{12.3.22}$$

and the single-spin longitudinal asymmetries are

$$A^j_L = \frac{D^j_+ - D^j_-}{D^j_+ + D^j_-}. \tag{12.3.23}$$

Because parity violation only becomes significant when Z^0 exchange starts to be important, one finds that the A^j_L are exceedingly small for $M \ll M_Z$. However, as seen in Fig. 12.15 the asymmetry is fairly large for $M = M_Z$ and discriminates quite well between the polarized gluon densities described in (12.2.22).

12.4 Transverse two-spin asymmetries

We consider a transversely polarized beam and target in a parity conserving theory. Particle A is taken to be incoming along OZ and we fix the X-axis to lie along the polarization of particle A. The spin directions of A and B are denoted by $\uparrow\uparrow$ or $\uparrow\downarrow$, corresponding to the transverse covariant spin vector of B being either along or opposite to that of A. The final state jet or hadron C emerges with azimuthal angle ϕ in the Lab.

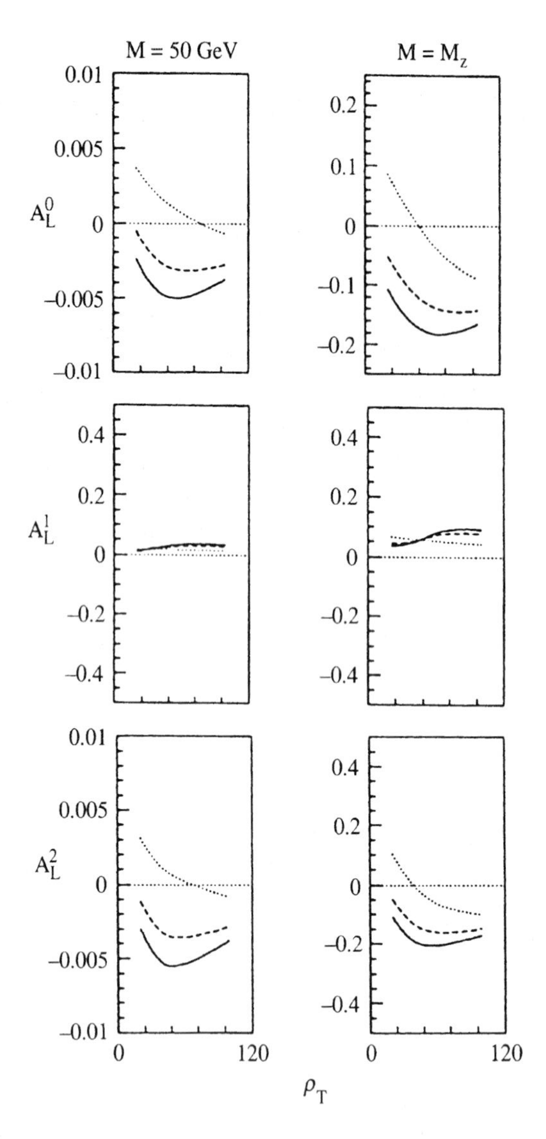

Fig. 12.15 The single-spin parity-violating Drell–Yan longitudinal asymmetries A_L^j vs. p_T for two values of M at $y = 0$. For the curves see the caption to Fig. 12.6. From Leader and Sridhar (1994), where A_L^j is labelled A_s^j ('s' for single).

Because the difference of cross-sections now depends upon ϕ, the expression (12.2.2) is somewhat altered. One defines

$$\frac{d^3\Delta_T\sigma}{d^3\mathbf{p}_C} \equiv \frac{d^3\sigma^{\uparrow\uparrow}}{d^3\mathbf{p}_C} - \frac{d^3\sigma^{\uparrow\downarrow}}{d^3\mathbf{p}_C} \tag{12.4.1}$$

and finds

$$E_C \frac{d^3 \Delta_T \sigma}{d^3 \mathbf{p}_C} = 2 \sum_{a,b,c,d} \int_{\xi_a} dx_a \int_{\xi_b} dx_b \Delta_T f_a^A(x_a) \Delta_T f_b^B(x_b) \times \frac{d^2 \Delta_T \hat{\sigma}^{ab \to cd}}{d\hat{t} d\phi}(\hat{s}, \hat{t}, \phi) \frac{1}{z} D_C^c(z) \tag{12.4.2}$$

where

$$\frac{d^2 \Delta_T \hat{\sigma}^{ab \to cd}}{d\hat{t} d\phi} \equiv \frac{d^2 \hat{\sigma}^{\uparrow\uparrow}}{d\hat{t} d\phi} - \frac{d^2 \hat{\sigma}^{\uparrow\downarrow}}{d\hat{t} d\phi}. \tag{12.4.3}$$

The transversely polarized parton densities $\Delta_T f$ that occur in (12.4.2) were introduced in Section 11.9; see eqn (11.9.15). Expressions for the integration limits and relations between x_a, x_b and $\hat{s}$ and $\hat{t}$ and definitions of other kinematic variables were given in Section 12.1. Because the partonic single spin asymmetries are zero in leading order, it follows from (5.6.12) that one can write

$$\frac{d^2 \hat{\sigma}^{\uparrow\uparrow/\uparrow\downarrow}}{d\hat{t} d\phi} = \frac{1}{2\pi} \frac{d\hat{\sigma}}{d\hat{t}} \left[1 \pm \hat{a}_{TT}(\hat{s}, \hat{t}) \cos 2\phi \right] \tag{12.4.4}$$

so that

$$\frac{d^2 \Delta_T \hat{\sigma}^{ab \to cd}}{d\hat{t} d\phi} = \frac{1}{\pi} \frac{d\hat{\sigma}}{d\hat{t}} \hat{a}_{TT} \cos 2\phi. \tag{12.4.5}$$

The double-spin partonic transverse asymmetry parameters $\hat{a}_{TT}$ are given in Table 12.3 for various reactions and shown as a function of the CM scattering angle θ^* in Fig. 12.16. Some of these results are due to Ranft and Ranft (1978), to Hidaka, Monsay and Sivers (1979) and to

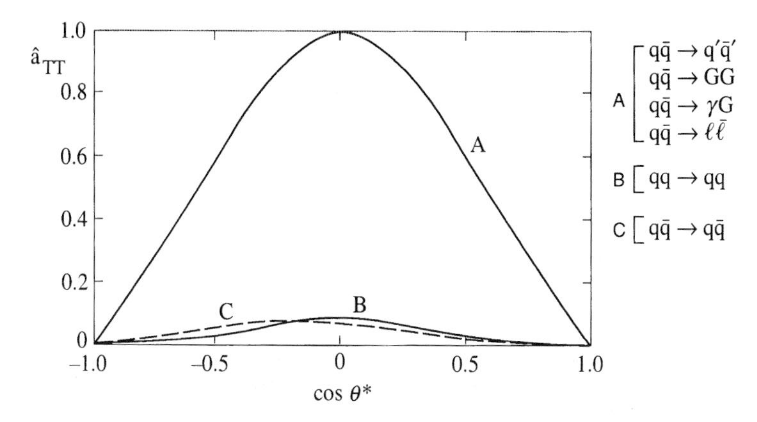

Fig. 12.16 Double-spin partonic transverse polarization asymmetry parameter vs. $\cos\theta^*$ for various partonic processes, in lowest order.

Table 12.3. Partonic transverse double spin asymmetry parameter. A prime on q means that q' has different flavour from q.

Reaction	$\hat{a}_{TT}$
$qq \to qq$	$\dfrac{2\hat{t}^2\hat{u}^2/3}{\hat{u}^2(\hat{s}^2+\hat{u}^2)+\hat{t}^2(\hat{s}^2+\hat{t}^2)-(2\hat{s}^2\hat{t}\hat{u}/3)}$
$q\bar{q} \to q\bar{q}$	$\dfrac{2\hat{u}\hat{t}^2(\hat{t}-\hat{s}/3)}{\hat{s}^2(\hat{s}^2+\hat{u}^2)+\hat{t}^2(\hat{u}^2+\hat{t}^2)-(2\hat{u}^2\hat{s}\hat{t}/3)}$
$qq' \to qq'$ $q\bar{q}' \to q\bar{q}'$	0
$q\bar{q} \to q'\bar{q}'$ $q\bar{q} \to GG$	$\dfrac{2\hat{u}\hat{t}}{\hat{t}^2+\hat{u}^2}$
$q\bar{q} \to \gamma G$	$\dfrac{2\hat{u}\hat{t}}{\hat{u}^2+\hat{t}^2}$
$q\bar{q} \to \gamma \to l\bar{l}$	$\dfrac{2\hat{u}\hat{t}}{\hat{t}^2+\hat{u}^2}$

Jaffe and Saito (1996).[1] Others were calculated in an interesting paper of Artru and Mekhfi (1990).

It is seen that large asymmetries occur for $q\bar{q} \to q'\bar{q}', GG, \gamma G$ and $l\bar{l}$, the rest being an order of magnitude smaller. However, it is not at all clear how big the asymmetries will be at the hadronic level, since we have little knowledge about the transversely polarized parton densities which have never been measured directly; see, however, Section 13.3.

Various models for the $\Delta_T q(x)$ have been constructed, but none of them is very convincing. As mentioned in subsection 11.9.1, the one reliable result we have to guide us is the Soffer bound (Soffer, 1995):

$$|\Delta_T q(x)| \leq \tfrac{1}{2}\left[q(x) + \Delta q(x)\right]. \qquad (12.4.6)$$

Since for d quarks $\Delta d(x)$ is negative and fairly large whereas $\Delta u(x)$ is large and positive, we might expect to have

$$|\Delta_T d(x)| \ll |\Delta_T u(x)|, \qquad (12.4.7)$$

but this does not strictly follow from (12.4.6). Interestingly, however, an attempt to estimate $\Delta_T u(x)$ and $\Delta_T d(x)$ using QCD sum rules (Ioffe and

[1] Note that there are errors in these three papers.

Khodjamirian, 1995), while only reliable in the range $0.3 \lesssim x \lesssim 0.5$ does suggest a large $\Delta_T u(x)$, with $|\Delta_T u(x)| > |\Delta u(x)|$, and $|\Delta_T d(x)| \ll |\Delta_T u(x)|$.

On the basis of non-relativistic models it has sometimes been supposed that $\Delta_T q(x) = \Delta q(x)$, but it is clear that such an assumption may violate (12.4.6), especially if Δq is negative.

Given the magnitudes of the partonic $\hat{a}_{TT}$, the best way to study the $\Delta_T q(x)$ would be via the small-p_T Drell-Yan reaction $p\bar{p} \to l\bar{l} + X$, with both beams transversely polarized, but there is no prospect of such a measurement in the foreseeable future. In reactions initiated by proton–proton collisions the hadronic asymmetry will be diminished by those partonic reactions with large cross-sections but small $\hat{a}_{TT}$, or worse, gluon–gluon reactions that give no contribution to the asymmetry. Drell-Yan reactions *at small* p_T at least do not suffer from the latter problem, but in pp collisions the relevant parton densities that enter are of the form $\Delta_T q(x_a)\Delta_T \bar{q}(x_b)$, and it may well be that the transversely polarized antiquark densities are very small.

Drell–Yan reactions at large p_T have been studied by Vogelsang and Weber (1993) and, more recently, by Martin, Schäfer, Stratman and Vogelsang (1997). In the earlier work, $\Delta_T q_V = \Delta q_V$ was assumed for the valence densities and two models for the antiquark $\Delta_T \bar{q}$ were studied. One, labelled (a), is small and negative; the other, (b), large and negative. The double-spin transverse asymmetry

$$A_{TT} = \frac{d\sigma^{\uparrow\uparrow} - d\sigma^{\uparrow\downarrow}}{d\sigma^{\uparrow\uparrow} + d\sigma^{\uparrow\downarrow}} \tag{12.4.8}$$

was analysed for various choices of cross-section, for example differential in the mass of the lepton pair or in the p_T of the photon (i.e. of the lepton pair) etc. Typically, at $\sqrt{s} = 100$ GeV, $M^2 = 49\ (\text{GeV}/c^2)^2$ and $\cos 2\phi = 1, A_{TT}(p_T)$ is very small for $p_T \gtrsim 2$ GeV/c, of the order of a few per cent. Of course, in this region the fundamental lowest-order process, which has a large asymmetry, does not contribute. Its contribution in the integrated cross-section, defined as $d^2\sigma/dMd\phi$, is considerable, and $A_{TT}(M)$ is much larger. In fact the integrated A_{TT} is roughly a scaling function of $\tau = M^2/s$, apart from scale breaking in the parton densities, and is shown in Fig. 12.17 as a function of τ. Also shown is the difference between the leading-order, $O(\alpha^2)$, and the next-to-leading-order $O(\alpha^2\alpha_s)$, results. The parton densities used were at scale $M^2 = 49\ (\text{GeV}/c^2)^2$.

In the more recent study cited above an attempt is made to estimate the maximum possible A_{TT} by suitably saturating the Soffer bound, i.e. by taking

$$\Delta_T(x, Q_0^2) = \tfrac{1}{2}\left[q(x, Q_0^2) + \Delta q(x, Q_0^2)\right]. \tag{12.4.9}$$

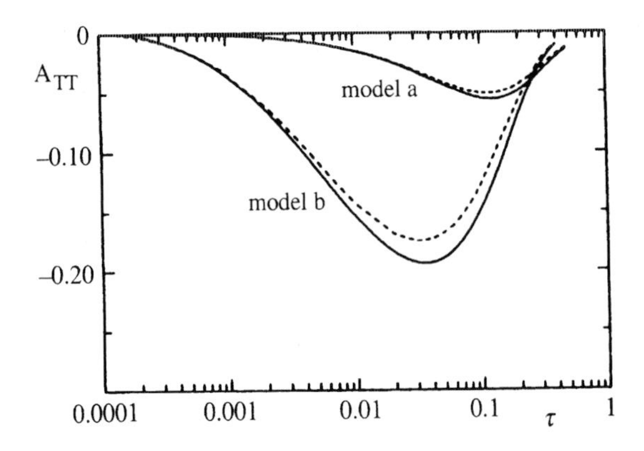

Fig. 12.17 The double-spin transverse asymmetry for the integrated Drell–Yan cross sections for $pp \to l\bar{l} + X$ vs. $\tau = M^2/s$, for two models of $\Delta_T\bar{q}(x)$; see text. Solid lines, $O(\alpha^2\alpha_s)$ calculation; broken lines, $O(\alpha^2)$ calculation. (From Vogelsang, 1993.)

The tricky question is the choice of Q_0^2 at which (12.4.9) is imposed. Imposing it at a low value of Q_0 is acceptable, because the evolution to larger Q^2 continues to respect (12.4.6), as demonstrated by Bourrely, Soffer and Teryaev (1998), whereas imposing (12.4.9) at high Q_0^2 will lead to a violation of (12.4.6) at $Q^2 < Q_0^2$.

In fact a very low scale, $Q_0 = 0.6$ GeV$/c$, is chosen, corresponding to using the parton densities of the so-called 'radiative parton model' of Glück, Reya, Stratmann and Vogelsang (1996).

The results of this study are shown in Figs. 12.18 and 12.19, which show both the asymmetry A_{TT} as function of the lepton pair mass M and the cross-section difference

$$\frac{d\Delta_T\sigma}{dM} = \left(\int_{-\pi/4}^{\pi/4} d\phi + \int_{3\pi/4}^{5\pi/4} d\phi - \int_{\pi/4}^{3\pi/4} d\phi - \int_{5\pi/4}^{7\pi/4} d\phi\right) \frac{d^2\Delta_T\sigma}{dM d\phi} \tag{12.4.10}$$

at two different values $\sqrt{s} = 150$ GeV and $\sqrt{s} = 500$ GeV, corresponding to the lower and upper regions of the RHIC energy range. At the higher energy, where much larger massses M can be produced, both photon and Z^0 exchange have been included, and this leads to an interesting M-dependence of A_{TT}.

It is seen that the cross-section differences are small and that A_{TT}, which, it should not be forgotten, has been maximized, is at the level of 1%–4%. The 'error bars' shown are an estimate of the expected statistical errors assuming 70% beam polarization and integrated luminosities of 240 pb^{-1} and 800 pb^{-1} at $\sqrt{s} = 150$ GeV and $\sqrt{s} = 500$ GeV respectively.

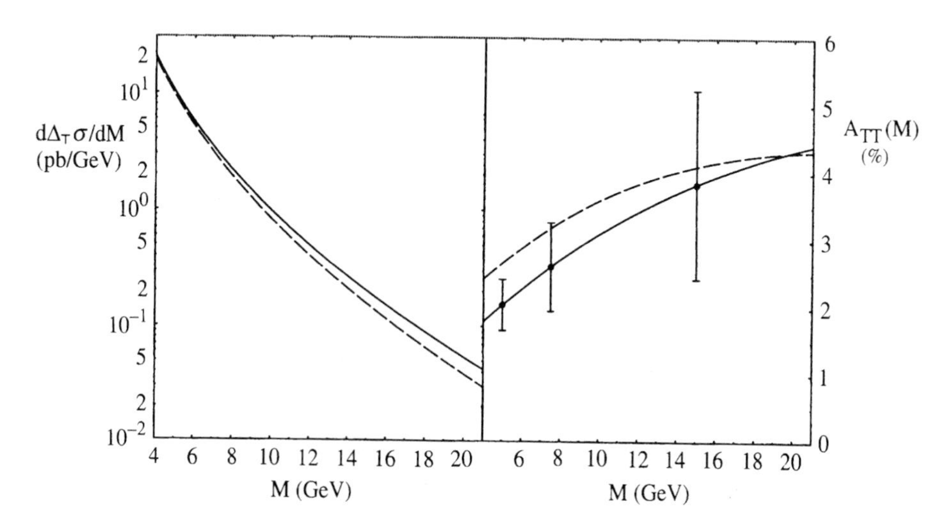

Fig. 12.18 $d\Delta_T\sigma/dM$ and A_{TT} vs. M for $\sqrt{s} = 150$ GeV. The solid lines are NLO, the broken line LO; see text. (Adapted from Martin, Schäfer, Stratman and Vogelsang, 1997.)

Of course, the asymmetries chosen by Nature could be considerably smaller than shown. It is thus not going to be easy to learn about the transverse densities $\Delta_T q(x)$, but with luck our first information should soon emerge from the RHIC programme. Another interesting source of information on $\Delta_T q(x)$ is discussed in Section 13.3.

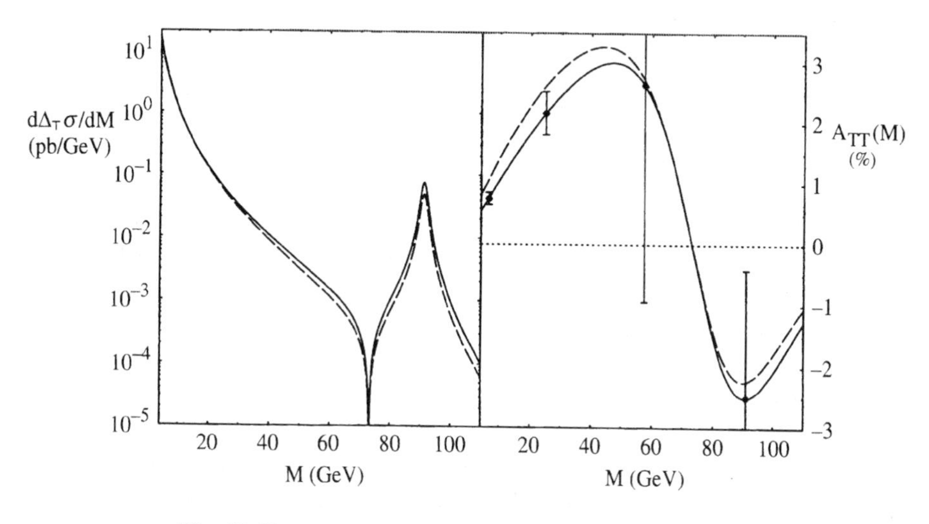

Fig. 12.19. As for Fig. 12.18 but at $\sqrt{s} = 500$ GeV

13
One-particle inclusive transverse single-spin asymmetries

One of the most interesting and challenging issues at the moment concerns the question of asymmetries involving either an initial transversely polarized hadron, in which case we consider the analysing power A_N of the reaction, or the production of a transversely polarized final state hadron in an unpolarized collision, in which case we consider the polarizing power P of the reaction.

The problem is that the *lowest-order* QCD partonic cross-sections yield $A_N = P = 0$, whereas experimentally there is a mass of data showing large asymmetries or large polarizations, both in elastic and semi-inclusive reactions.

The treatment of elastic reactions is very different from that of the semi-inclusive case, requiring consideration of hadronic *wave functions* rather than parton densities. We shall therefore deal with the elastic case separately in Chapter 14.

The most dramatic examples in the one-particle inclusive case are the transverse asymmetries A_N in proton–proton and in antiproton–proton collisions ($pp^\uparrow \to \pi^\pm X$ and $\bar{p}^\uparrow p \to \pi^\pm X$) and the hyperon polarization in $pp, p+$ nucleus and $Kp \to$ hyperon $+X$.

Broadly speaking the effects have the following characteristics.

(1) They increase linearly with p_T out to $p_T \approx 2$–2.5 GeV/c (see Figs. 13.1, 13.2) though there is a hint of a flattening out beyond $p_T \approx 1$ GeV/c in the lower energy data (see Fig. 13.3).
(2) They increase linearly with Feynman x_F ($x_F > 0$) out to the maximum $x_\mathrm{F}^\mathrm{max} \approx 0.9$ (see Figs. 13.4 and 13.5).
(3) They seem to be roughly energy-independent (see Fig. 13.2).
(4) There are interesting dependences on the charge and strangeness of the final state hadron (see Fig. 13.6) and upon the correlation between the quark contents of the initial and produced hadrons.

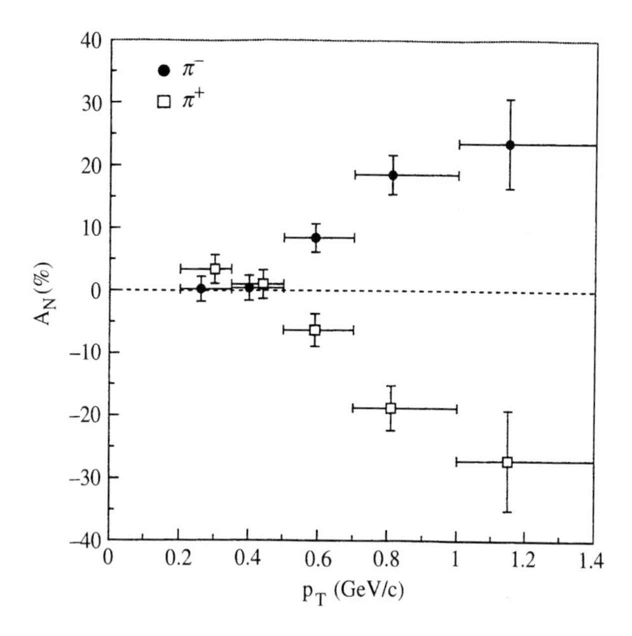

Fig. 13.1 Analysing power A_N vs. p_T measured at FERMILAB for 200 GeV/c polarized antiprotons in the reaction $\bar{p}^{\uparrow}p \to \pi^{\pm}X$. (From Bravar *et al.*, 1996.)

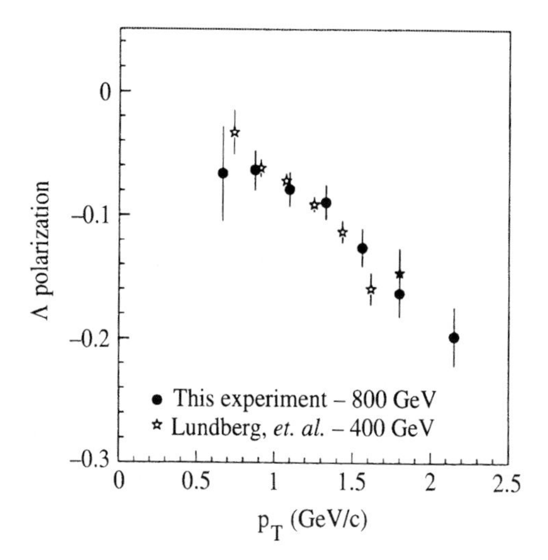

Fig. 13.2 The average Λ polarizations vs. p_T measured at FERMILAB for 800 GeV/c protons in the reaction $pBe \to \Lambda X$ (from Ramberg *et al.*, 1994). Also shown are the 400 GeV/c data of Lundberg *et al.*, 1989.

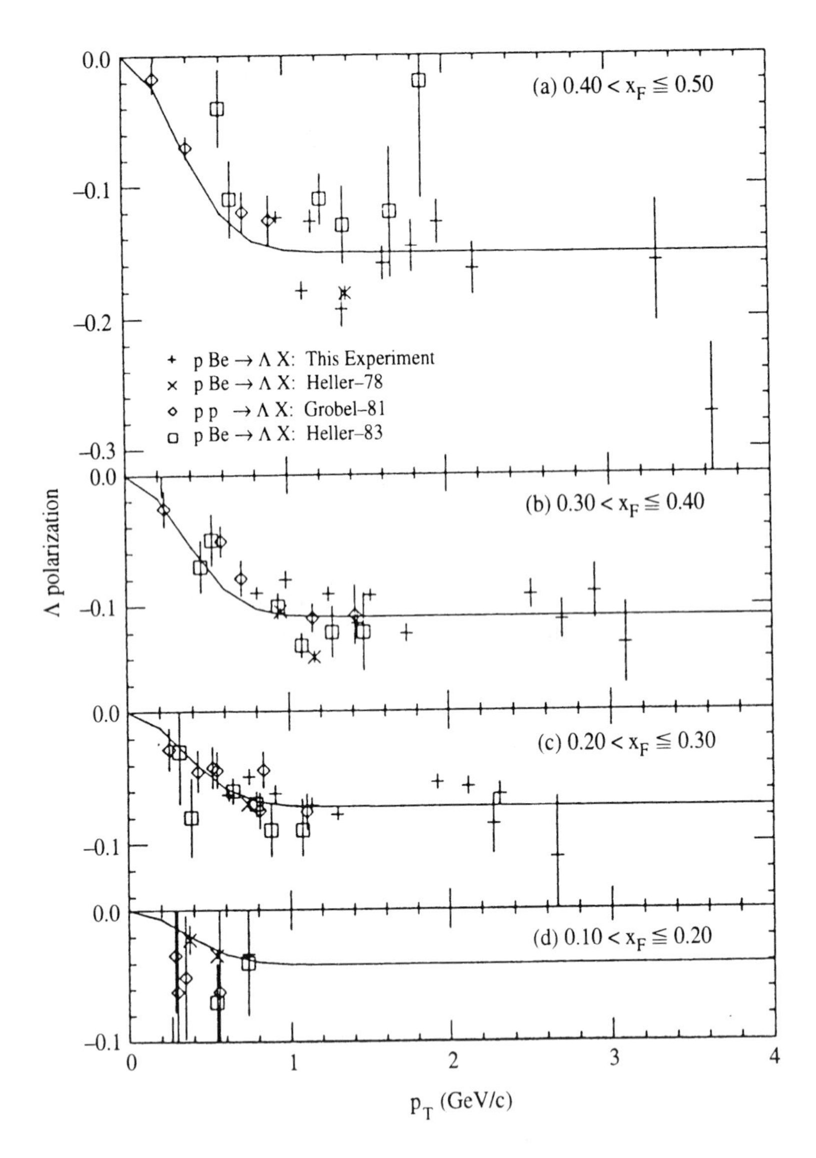

Fig. 13.3 The Λ polarization vs. p_T in bins of x_F measured at FERMILAB for 400 GeV$/c$ protons in the reaction $pBe \to \Lambda X$. (From Lundberg *et al.*, 1989.)

13.1 Theoretical approaches

If one takes the same approach theoretically as was done for the asymmetries discussed in the previous chapter, i.e. a simple parton approach refined by QCD corrections with a perturbative QCD amplitude for the hard scattering, one can generate transverse asymmetries only by going

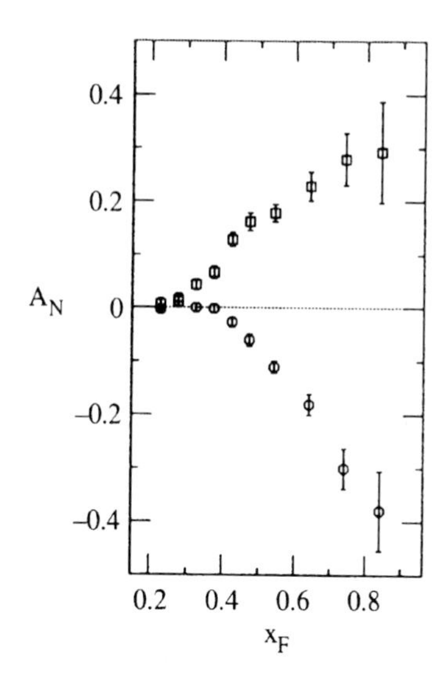

Fig. 13.4 Analysing power A_N vs. x_F measured at FERMILAB for 200 GeV/c polarized protons in the reaction $p^{\uparrow}p \to \pi^{\pm}X$: □, π^+; ○, π^-. (From Adams *et al.*, 1991b.)

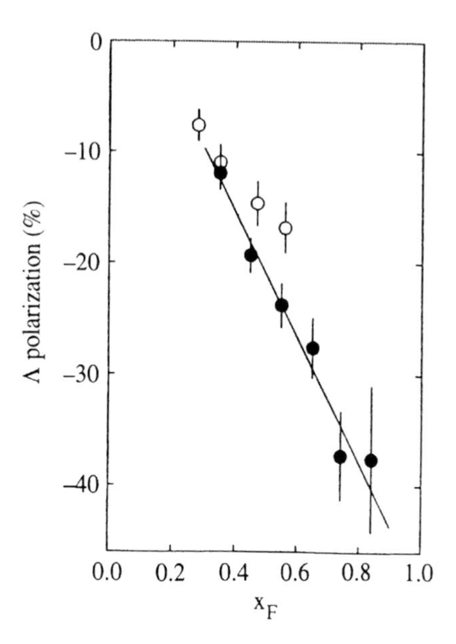

Fig. 13.5 The Λ polarization for $p_T \geq 0.96$ GeV/c vs. x_F measured at CERN for the reaction $pp \to \Lambda X$ at $\sqrt{s} = 62$ GeV (solid circles) (from Smith *et al.*, 1987). Also shown (open circles) are the FERMILAB data at $\sqrt{s} = 27$ GeV of Lundberg *et al.*, 1989.

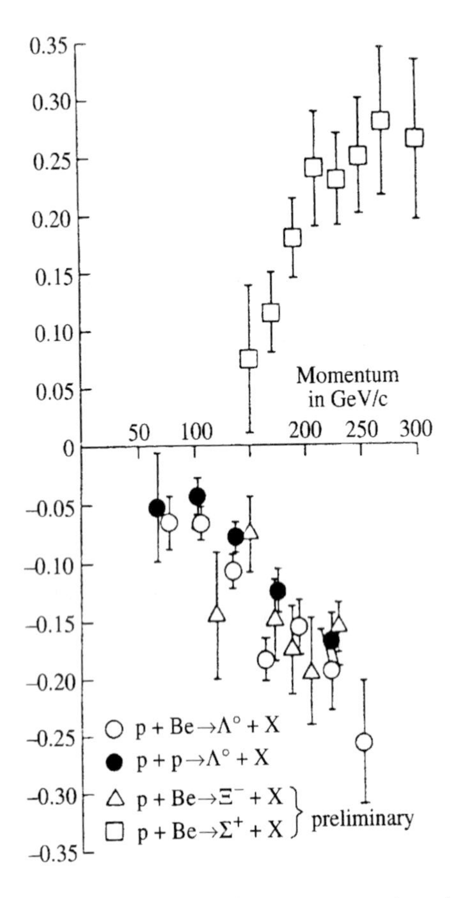

Fig. 13.6 Polarization of various hyperons produced by 400 GeV/c protons at FERMILAB at fixed Lab angle of 5 mrad. (From Heller, 1981.)

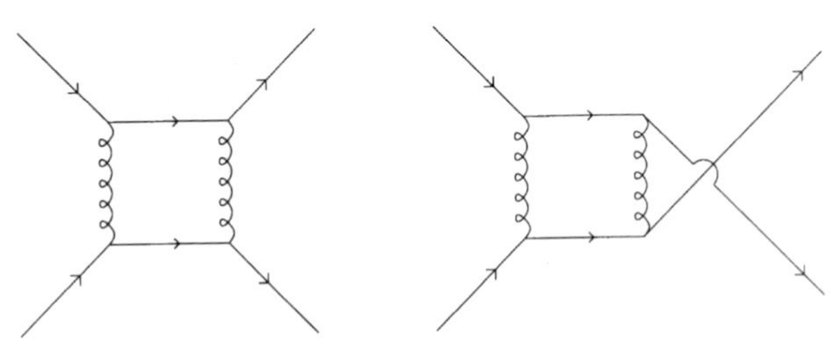

Fig. 13.7. Feynman diagram for quark–quark scattering amplitude at order α_s^2.

beyond the Born approximation in the partonic amplitude. Thus at one-loop level, as shown for example in Fig. 13.7 for $qq \to qq$, one finds a non-zero value for $\hat{a}_N$, the partonic analogue of A_N, but it is exceedingly small, much too small to explain the data and moreover is proportional to the quark mass. It has the typical form

$$\hat{a}_N = \alpha_s \frac{m_q}{\sqrt{s}} f(\theta^*). \tag{13.1.1}$$

This is similar to the problem of $g_2(x)$ in polarized deep inelastic scattering (see Section 11.4) and really signals the failure of the model to produce the asymmetry. In our discussion of $g_{1,2}(x)$ we followed the traditional approach using the hadronic and partonic tensors $W_{\mu\nu}$ and $w_{\mu\nu}$; however, we could have treated the asymmetries in terms of cross-section asymmetries in the $eq \to eq$ partonic collision and would have found that the partonic asymmetry is zero for transverse polarization of the quark in the transversely polarized nucleon. Put another way, (13.1.1) indicates that the asymmetry is not a leading twist effect. Most interestingly, we shall see that the field-theoretic mechanism needed to discuss $g_2(x)$ also provides a mechanism for a non-zero $\hat{a}_N$.

There are two diverse attitudes to the above situation. One point of view is that p_T is simply too small to justify the use of perturbative QCD, so one should try to construct phenomenological models for the non-perturbative aspects of the problem. We shall briefly examine this approach in Section 13.5. The alternative, which we shall follow, is to remain within the framework of the standard QCD-parton model but either to adopt a more sophisticated approach to the non-perturbative elements, i.e. the parton densities and the process of parton fragmentation (Sections 13.2, 13.3), or to generalize the partonic reactions beyond the usual $2 \to 2$ Born amplitudes (Section 13.4).

Before discussing this it is important to note a major distinction between semi-inclusive lepton–hadron deep inelastic scattering and one particle inclusive hadron–hadron scattering, concerning the measurement of the analysing power A_N of the reaction.

In a hadron–hadron collision $AB^\uparrow \to CX$, with OZ along the collision axis in the CM, we know from Sections 5.4 and 5.8 that the differential cross-section depends upon the azimuthal angle ϕ of C, measured with respect to the quantization plane containing OZ and the transverse spin polarization $\mathcal{P}_T$. A_N can then be measured either by studying the azimuthal dependence with fixed $\mathcal{P}_T$ or by studying the asymmetry when $\mathcal{P}_T$ is reversed. The key point is that for an *unpolarized* collision there cannot be any dependence on an azimuthal angle about OZ – there simply is no physically defined plane from which to measure the angle.

The latter is not true for a reaction like $ep \to e'CX$. Even with unpolarized initial particles there can, in principle, be an azimuthal dependence about the 'γ'–proton CM collision axis, since in this frame $\mathbf{p}(e)$ and $\mathbf{p}(e')$ define a plane from which the angle can be measured. This cannot occur in the simple parton model, where partons are *collinear* with the momenta of their parent hadrons and partons fragment collinearly into hadrons. But it can happen, for example, if one allows partons to have an intrinsic $\mathbf{p}_T$ (Cahn, 1978). The azimuthal dependence arises in the following way. Consider the partonic electron–quark reaction $eq(\mathbf{p}) \to e'q'(\mathbf{p}')$ in the 'γ'–proton CM, where γ has momentum $\mathbf{q}$ along OZ. Since $\mathbf{p}' = \mathbf{p} + \mathbf{q}$ we have that $\mathbf{p}'_T = \mathbf{p}_T$. Moreover the Mandelstam variables $\hat{s}, \hat{t}, \hat{u}$ of the partonic reaction depend upon ϕ, the azimuthal angle of $\mathbf{p}_T$, and hence the cross-section depends on the azimuthal angle of the final parton and therefore of C. Such azimuthal dependence is indeed seen experimentally (Arneodo *et al.*, 1987) and implies that A_N can only be measured by measuring an asymmetry under reversal of $\mathcal{P}_T$.

One final general comment is necessary. *All* the mechanisms we shall discuss are able to produce asymmetries that increase with transverse momentum, but they are able to do this only out to values $\approx$ 1–2 GeV/c. Beyond this the asymmetries decrease and eventually vanish. If, therefore, the experimental asymmetries continue to grow with p_T we shall find ourselves in a critical state of ignorance.

13.2 Standard QCD-parton model with soft-physics asymmetries

In this section we discuss a standard parton-model approach, with the hard scattering controlled by a $2 \to 2$ partonic reaction but with allowance for transverse momentum of the partons. It will be seen that a transverse single-spin asymmetry could arise from possible asymmetries in the soft-physics aspects, i.e. in the parton number densities and fragmentation functions.

Consider the reaction

$$A^{\uparrow} + B \to C + X \tag{13.2.1}$$

where the momentum of A lies along OZ in the CM of the reaction; A is polarized transversely with spin along ($\uparrow$) or opposite ($\downarrow$) to OY.

Let us consider the cross-section for (13.2.1) in the spirit of the simplified analysis in Section 12.1. Since B is unpolarized here, it plays a passive rôle in the spin dependence, so we show only the rôle played by the partons in A and C. Then symbolically

$$d\sigma^{\uparrow} \sim f_{\uparrow}^{\uparrow}\hat{\sigma}_{\uparrow}D(\mathcal{P}_c) + f_{\downarrow}^{\uparrow}\hat{\sigma}_{\downarrow}D(-\mathcal{P}_c) \tag{13.2.2}$$

where $f_{\uparrow}^{\uparrow}$, $f_{\downarrow}^{\uparrow}$ are the number densities of quarks with polarization $\uparrow$ or $\downarrow$ in A, $\hat{\sigma}_{\uparrow}$, $\hat{\sigma}_{\downarrow}$ are the lowest order cross-sections for the partonic reaction

$$\left(a^{\uparrow} \text{ or } a^{\downarrow}\right) + b \to c + d \tag{13.2.3}$$

and $\pm\mathcal{P}_c$ is the spin-polarization vector of parton c produced in the reaction (13.2.3) when the polarization vector of a is $\mathcal{P}_a = \pm\hat{\mathbf{e}}_{(y)}$. $D(\mathcal{P}_c)$ is the fragmentation function for

$$c(\mathcal{P}_c) \to C + X.$$

Now since $\hat{a}_N = 0$ in lowest order, we have that

$$\hat{\sigma}_\uparrow = \hat{\sigma}_\downarrow = \hat{\sigma}. \tag{13.2.4}$$

Therefore (13.2.1) can be written

$$\begin{aligned} d\sigma^\uparrow &\sim \tfrac{1}{2}\left(f_\uparrow^\uparrow + f_\downarrow^\uparrow\right)\hat{\sigma}\,[D(\mathcal{P}_c) + D(-\mathcal{P}_c)] \\ &\quad + \tfrac{1}{2}\left(f_\uparrow^\uparrow - f_\downarrow^\uparrow\right)\hat{\sigma}\,[D(\mathcal{P}_c) - D(-\mathcal{P}_c)] \\ &= f^\uparrow\hat{\sigma}D + \tfrac{1}{2}(\Delta_T f)\hat{\sigma}\tilde{\Delta}D(\mathcal{P}_c) \end{aligned} \tag{13.2.5}$$

where $f^\uparrow$ is simply the number density inside $A^\uparrow$, D is the unpolarized fragmentation function

$$D = \tfrac{1}{2}\,[D(\mathcal{P}_c) + D(-\mathcal{P}_c)]\,, \tag{13.2.6}$$

the difference $\Delta_T f$ is given by

$$\Delta_T f \equiv f_\uparrow^\uparrow - f_\downarrow^\uparrow \tag{13.2.7}$$

and

$$\tilde{\Delta}D(\mathcal{P}_c) \equiv D(\mathcal{P}_c) - D(-\mathcal{P}_c). \tag{13.2.8}$$

Similarly

$$d\sigma^\downarrow \sim f^\downarrow\hat{\sigma}D - \tfrac{1}{2}(\Delta_T)f\hat{\sigma}\tilde{\Delta}D(\mathcal{P}_c).$$

In relation to the spin asymmetry we then have

$$\Delta_N d\sigma \equiv d\sigma^\uparrow - d\sigma^\downarrow \approx (\tilde{\Delta}_N f)\hat{\sigma}D + (\Delta_T f)\hat{\sigma}\tilde{\Delta}D(\mathcal{P}_c) \tag{13.2.9}$$

where

$$\tilde{\Delta}_N f \equiv f^\uparrow - f^\downarrow \tag{13.2.10}$$

and the asymmetry is defined as

$$A_N = \frac{d\sigma^\uparrow - d\sigma^\downarrow}{d\sigma^\uparrow + d\sigma^\downarrow} = \frac{\Delta_N d\sigma}{2d\sigma}. \tag{13.2.11}$$

Now the problem is that the expression (13.2.9) vanishes in the usual parton model, where the momentum of a parton is taken as *collinear* with the momentum of its parent hadron! Thus the total number of partons with momentum fraction x cannot depend on the polarization of the parent hadron, so that $\tilde{\Delta}_N f = 0$, and the total number of hadrons with momentum $z\mathbf{p}_c$ cannot depend upon the polarization of c, so that $\tilde{\Delta}D(\mathcal{P}_c) = 0$.

It has been suggested, however, that with the inclusion of intrinsic parton transverse momentum these differences of number densities could be non-zero.

Thus Sivers (1990, 1991) proposed, for a hadron A that is transversely polarized,

$$f_a^A(\mathbf{p}_A, \mathcal{P}_A; x_a, \mathbf{k}_{aT}) = f_a^A(x_a, k_{aT}) + \tfrac{1}{2}\Delta_N f_a^A(x_a, k_{aT}) \frac{\mathcal{P}_A \cdot (\mathbf{p}_A \times \mathbf{k}_{aT})}{|\mathbf{p}_A \times \mathbf{k}_{aT}|} \tag{13.2.12}$$

where $\mathbf{k}_{aT}$ is transverse to $\mathbf{p}_A$.

However, *if* we are permitted to regard f_a^A as describing the independent physical reaction

$$A(\mathbf{p}_A, \mathcal{P}_A) \to a(x_a \mathbf{p}_A + \mathbf{k}_{aT}) + X \tag{13.2.13}$$

then the asymmetry in the decay distribution implied by (13.2.12) is impossible, as can be seen by looking at the reaction in the CM of a and X. Collins (1993) has given a more subtle argument against the Sivers effect. Using the field-theoretic formalism of Section 11.9, $\tilde{\Delta}_N f$ can be related to a nucleon matrix element of certain operators and is shown to vanish as a consequence of parity invariance and time-reversal invariance. But this argument relies on an absence of final state interactions and is thus analogous to treating (13.2.13) as an independent physical reaction, which is an essential element of the factorization of the reaction into universal soft and hard parts.

Despite these arguments, some authors have postulated a non-zero Sivers mechanism (Anselmino, Boglione, Murgia, 1995; Ratcliffe, 1998) on the grounds that the parton model totally ignores the question of the need to neutralize colour and to compensate for fractional charge and baryon number. So there must indeed be final state interactions, negating the anti-Sivers argument, but they must be fairly negligible otherwise the parton model would not work at all. In the papers quoted above there is no attempt to define the dynamics causing the final state interactions, so there is no reason to believe that a treatment relevant to one particular reaction is also relevant to any other process.

It turns out, in fact, that by extending the scope of the partonic reactions one can produce effective initial and final state interactions at the partonic level in a well-defined and factorizable dynamical way, as will be explained in Section 13.4. Thus we shall not discuss the Sivers mechanism any further.

Collins (1993) argued that, contrary to what happens in the Sivers case, the time-reversal argument does not forbid a non-zero $\tilde{\Delta}D(\mathcal{P}_c)$ when $\mathbf{k}_T$ is taken into account. He thus postulates, for the fragmentation process

$$c(\mathbf{p}_c, \mathcal{P}_c) \to C(z\mathbf{p}_c + \mathbf{k}_{CT}) + X \tag{13.2.14}$$

where $\mathbf{k}_{CT}$ is perpendicular to $\mathbf{p}_c$,

$$D(\mathbf{p}_c, \mathcal{P}_c; z\mathbf{p}_c + \mathbf{k}_{CT}) = D(z, k_{CT}) + \tfrac{1}{2}\Delta D(z, k_{CT})\mathcal{P}_c \cdot \frac{\mathbf{p}_c \times \mathbf{k}_{CT}}{|\mathbf{p}_c \times \mathbf{k}_{CT}|} \qquad (13.2.15)$$

so that $\tilde{\Delta} D(\mathcal{P}_c) \neq 0$.

Since the Collins mechanism relies on detection of the produced hadron C, it will not be operative in jet production. Suggestions on the use of various reactions to try to sort out what mechanisms are at work are given in Anselmino, Leader and Murgia (1997) and in Boros, Liang, Meng and Rittel (1998).

How large are the asymmetries expected to be? If p_T is the magnitude of the transverse component of $\mathbf{p}_C$ and $\langle k_T \rangle$ is a measure of the magnitude of the intrinsic $\mathbf{k}_T$, we would expect effects of the order of $\langle k_T \rangle / p_T$. This is, of course, a higher-twist effect at large p_T.

Let us now consider the detailed expression for the asymmetry due to the Collins effect. For concreteness we shall consider the reaction

$$p^{\uparrow} + p \rightarrow \pi + X \qquad (13.2.16)$$

at large p_T.

13.3 Collins mechanism for single-spin asymmetry

We consider the asymmetry arising from the second term in (13.2.9). The CM frame for the reaction $p^{\uparrow}p \rightarrow \pi X$ is chosen so that the reaction takes place in the XZ-plane, with the polarized proton, A, moving along OZ. We consider pions whose momentum lies in the positive XZ-quadrant. The pion momentum is specified by $p_{\pi T} (= p_{\pi x})$ and $x_F = 2p_{\pi z}/\sqrt{s}$.

The polarization direction $\uparrow$ is defined to be along OY. The partons from the polarized proton have momentum $x_a\mathbf{p}$ and those from the unpolarized proton have momentum, $-x_b\mathbf{p}$, where $\mathbf{p}$ is the CM momentum of A.

Since intrinsic partonic $\mathbf{k}_T$ effects are small we have taken them to be zero except where they are essential, i.e. in the fragmentation process.

Let quark c be produced at polar angles θ^*, ϕ^* in the *partonic* CM. The components of the polarization vector of c, with respect to the axes X_c, Y_c, Z_c in the helicity frame of c reached from the partonic CM (see Fig. 13.8), can be obtained from (5.6.20) together with (5.6.1), bearing in mind that the analysing power of the partonic reaction is zero. When the polarization vector of a is

$$\mathcal{P}_a = \mathbf{e}_{(y)} \qquad (13.3.1)$$

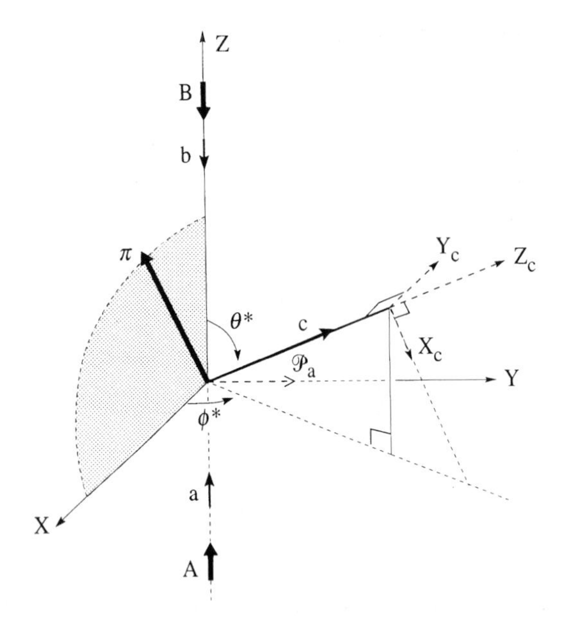

Fig. 13.8 Helicity frame for quark c reached from the parton CM in the reaction $ab \to cd$.

one has

$$\mathscr{P}^c_{x_c} \frac{d\hat{\sigma}(\boldsymbol{\mathcal{P}}_a)^{ab\to cd}}{d\hat{t}d\phi^*} = \left(\frac{1}{2\pi}\frac{d\hat{\sigma}}{d\hat{t}}\right)(X0|X0)\sin\phi^* \tag{13.3.2}$$

$$\mathscr{P}^c_{y_c} \frac{d\hat{\sigma}(\boldsymbol{\mathcal{P}}_a)^{ab\to cd}}{d\hat{t}d\phi^*} = \left(\frac{1}{2\pi}\frac{d\hat{\sigma}}{d\hat{t}}\right)(Y0|Y0)\cos\phi^*. \tag{13.3.3}$$

For all the relevant partonic processes one finds that

$$(X0|X0) = (Y0|Y0) \equiv \hat{d}_{NN}(\theta^*) \tag{13.3.4}$$

and also that $d\hat{\sigma}(\boldsymbol{\mathcal{P}}_a)/d\hat{t}d\phi^*$ is independent of ϕ^*. Hence (13.3.2) and (13.3.3) become

$$\mathscr{P}^c_{x_c} = \hat{d}_{NN}(\theta^*)\sin\phi^* \tag{13.3.5}$$

$$\mathscr{P}^c_{y_c} = \hat{d}_{NN}(\theta^*)\cos\phi^*. \tag{13.3.6}$$

From Fig. 13.8 one can read off the components of $\boldsymbol{\mathcal{P}}_c$ with respect to the partonic CM axes:

$$\mathscr{P}^c_x = \hat{d}_{NN}(\theta^*)\left[(\cos\theta^* - 1)\cos\phi^*\sin\phi^*\right] \tag{13.3.7}$$

$$\mathscr{P}^c_y = \hat{d}_{NN}(\theta^*)\left[\cos\theta^*\sin^2\phi^* + \cos^2\phi^*\right] \tag{13.3.8}$$

$$\mathscr{P}^c_z = -\hat{d}_{NN}(\theta^*)\sin\theta^*\sin\phi^*. \tag{13.3.9}$$

In fact one can see that $\mathcal{P}_c$ is just the vector

$$\mathcal{P}_c = \hat{d}_{NN}(\theta^*) \left[\mathscr{R}(\theta^*)\mathbf{e}_{(y)}\right], \tag{13.3.10}$$

where $\mathscr{R}(\theta^*)$ is the rotation about $\mathbf{p}_a \times \mathbf{p}_c$ which takes $\mathbf{p}_a$ into $\mathbf{p}_c$.

The expressions for $\hat{d}_{NN}(\theta^*)$ are given in Table 13.1.

We can write eqns (13.3.7)–(13.3.9) as

$$\mathcal{P}_c = \hat{d}_{NN}(\theta^*)\mathbf{e} \tag{13.3.11}$$

where the unit vector $\mathbf{e}$ is given by

$$\mathbf{e} = \Big((\cos\theta^* - 1)\cos\phi^* \sin\phi^*,\ \cos\theta^* \sin^2\phi^* + \cos^2\phi^*,\ -\sin\theta^* \sin\phi^*\Big) \tag{13.3.12}$$

Let

$$\mathbf{p}_\pi = (\mathbf{p}_{\pi T}, p_{\pi z}) \tag{13.3.13}$$

and E_π be the momentum and energy of the produced π in the CM of the reaction. With our choice of axes, $\mathbf{p}_{\pi T}$ lies along OX. Its momentum in the partonic CM is then

$$\mathbf{p}^*_\pi = (\mathbf{p}_{\pi T}, p^*_{\pi Z}) \tag{13.3.14}$$

where

$$p^*_{\pi Z} = \frac{1}{2\sqrt{x_a x_b}}\left[(x_a + x_b)p_{\pi z} - (x_a - x_b)E_\pi\right]. \tag{13.3.15}$$

Table 13.1. Partonic spin-transfer parameters

Reaction	$\hat{d}_{NN}$
$qq \to qq$, $\bar{q}\bar{q} \to \bar{q}\bar{q}$	$\dfrac{-2\hat{s}\hat{u}\left(1 - \frac{\hat{t}}{3\hat{u}}\right)}{\hat{s}^2 + \hat{u}^2 + (\hat{s}^2 + \hat{t}^2)\frac{\hat{t}^2}{\hat{u}^2} - 2\hat{s}^2\frac{\hat{t}}{3\hat{u}}}$
$q\bar{q} \to q\bar{q}$, $\bar{q}q \to \bar{q}q$	$\dfrac{-2\hat{s}\hat{u}\left(1 - \frac{\hat{t}}{3\hat{s}}\right)}{\hat{s}^2 + \hat{u}^2 + (\hat{u}^2 + \hat{t}^2)\frac{\hat{t}^2}{\hat{s}^2} - 2\hat{u}^2\frac{\hat{t}}{3\hat{s}}}$
$qq' \to qq'$, $q\bar{q} \to q'\bar{q}'$, $\bar{q}q \to \bar{q}'q'$, $qG \to qG$, $\bar{q}G \to \bar{q}G$	$\dfrac{-2\hat{s}\hat{u}}{\hat{s}^2 + \hat{u}^2}$

Now by the definition of $\mathbf{p}_{\pi T}$ we can write

$$\mathbf{p}_\pi^* = z\mathbf{p}_c + \mathbf{k}_{\pi T} \tag{13.3.16}$$

where

$$\mathbf{k}_{\pi T} \cdot \mathbf{p}_c = 0 \tag{13.3.17}$$

so that

$$z = \frac{\hat{\mathbf{p}}_c \cdot \mathbf{p}_\pi^*}{p_c} \tag{13.3.18}$$

and

$$\mathbf{k}_{\pi T} = \mathbf{p}_\pi^* - (\hat{\mathbf{p}}_c \cdot \mathbf{p}_\pi^*)\hat{\mathbf{p}}_c. \tag{13.3.19}$$

Further, for the vector product needed in (13.2.15),

$$\mathbf{p}_c \times \mathbf{k}_{\pi T} = \mathbf{p}_c \times \mathbf{p}_\pi^* \tag{13.3.20}$$

so that $\tilde{\Delta}D(\mathcal{P}_c)$ defined in (13.2.8) becomes, upon using (13.3.11) and (13.2.15),

$$\tilde{\Delta}D_\pi(\mathcal{P}_c) = \Delta D_\pi(z, k_{\pi T})\hat{d}_{NN}(\theta^*)\frac{\mathbf{e} \cdot (\mathbf{p}_c \times \mathbf{p}_\pi^*)}{|\mathbf{p}_c \times \mathbf{p}_\pi^*|}. \tag{13.3.21}$$

Finally the cross-section difference (13.2.9) becomes

$$\begin{aligned} E_\pi \left(\frac{d^3\sigma^\uparrow}{d\mathbf{p}_\pi^3} - \frac{d^3\sigma^\downarrow}{d\mathbf{p}_\pi^3} \right) = & \sum_{a,b,c,d} \int dx_b f_b(x_b) \int dx_a \Delta_T f_a(x_a)\, E_\pi^* \\ & \times \int d\cos\theta^* \left[\frac{d\hat{\sigma}^{ab\to cd}}{d\cos\theta^*} \right] \hat{d}_{NN}(\theta^*) \\ & \times \int \frac{d\phi^*}{2\pi} \Delta D_\pi(z, k_{\pi T}) \frac{\mathbf{e} \cdot (\mathbf{p}_c \times \mathbf{p}_\pi^*)}{p_c|\mathbf{p}_c \times \mathbf{p}_\pi^*|} \end{aligned} \tag{13.3.22}$$

where z and $k_{\pi T}$ are given via (13.3.18) and (13.3.19), $\mathbf{p}_\pi^*$ by (13.3.14) and (13.3.15) and $\mathbf{e}$ by (13.3.12).

As suggested by Artru, Czyzewski and Yabuki (1997), we can get some feeling for the size of the effect if we note that at large x_F the dominant contribution comes from u and d quarks, for π^+ and π^- respectively, and that the partonic scattering occurs predominantly at small θ^*. In that case $\hat{d}_{NN}(\theta^*) \approx 1$, $\mathbf{p}_c$ is approximately along OZ, $\mathbf{k}_{\pi T} \approx \mathbf{p}_{\pi T} = p_{\pi T}\mathbf{e}$, and $\mathbf{e} \approx \mathbf{e}_{(y)}$. The approximate asymmetry is then

$$A_{\pi^+ N} \approx \frac{\Delta_T u(\bar{x})}{u(\bar{x})} \Delta D_\pi(\bar{z}, p_{\pi T}) \tag{13.3.23}$$

$$A_{\pi^- N} \approx \frac{\Delta_T d(\bar{x})}{d(\bar{x})} \Delta D_\pi(\bar{z}, p_{\pi T}) \tag{13.3.24}$$

where we have used isospin invariance for the fragmentation. Here $\bar{x}$ and $\bar{z}$ are the most probable values of x and z subject to $\bar{x}\,\bar{z} \approx x_F$. Equations (13.3.23) and (13.3.24) are essentially upper bounds to the magnitude of the asymmetry, since the angular integrations will dilute the effect.

From Fig. 13.4 one sees that $A_{\pi^\pm N} \approx \pm 0.4$ for $x_F \approx 0.8$. Thus to produce the measured asymmetries entirely via the Collins mechanism requires firstly that

$$\frac{\Delta_T u(\bar{x})}{u(\bar{x})} \approx -\frac{\Delta_T d(\bar{x})}{d(\bar{x})}. \tag{13.3.25}$$

Given that for the longitudinal polarized-parton densities the measured values of $\Delta u(x)/u(x)$ and $\Delta d(x)/d(x)$ are in agreement with the sign, but not the magnitude, predicted using $SU(6)$ wave functions for the nucleon, it is not unreasonable to expect to find the negative sign in (13.3.25), which follows from $SU(6)$, while finding that the magnitudes violate the $SU(6)$ result

$$\frac{\Delta_T d(x)}{d(x)} = -\frac{1}{2}\frac{\Delta_T u(x)}{u(x)}. \tag{13.3.26}$$

Secondly, one requires

$$|\Delta D_\pi(\bar{z}, p_{\pi T})| \gtrsim \frac{0.4}{\min\{|\Delta_T u(\bar{x})/u(\bar{x})|,\ |\Delta_T d(\bar{x})/d(\bar{x})|\}} \tag{13.3.27}$$

Artru, Czyzewski and Yabuki (1997) parametrized ΔD_π, using a model based on the Lund string and the simple *anzatz*

$$\frac{\Delta_T u(x)}{u(x)} = \frac{-\Delta_T d(x)}{d(x)} = \mathscr{P}_{\max} x^n \tag{13.3.28}$$

and found a best fit to the x_F-dependence of the data with $\mathscr{P}_{\max} = 1, n = 2$. The factorized form $F(z)G(k_T)$ involved is not compatible, however, with LEP DELPHI collaboration data (Abreu *et al.*, 1995a, b). This treatment is, of course, very approximate and recently Anselmino, Boglione, Hansson and Murgia (2000) have adopted a more systematic approach, with ΔD_π parametrized as follows:

$$\Delta D_\pi(z, k_T) = N \frac{\langle k_T(z) \rangle}{M} z^\alpha (1-z)^\beta, \tag{13.3.29}$$

where the mean $\langle k_T(z) \rangle$ is z-dependent and taken from the data of Abreu *et al.* (1995a, b) and N, α, β are parameters to be fitted. For the transversely polarized quark densities it is assumed that $\Delta_T d(x)/d(x)$ and $\Delta_T u(x)/u(x)$ are both independent of x, with the $SU(6)$ value

$$\frac{\Delta_T u(x)}{u(x)} = \frac{2}{3}. \tag{13.3.30}$$

The excellent fit to the data produces the surprising result

$$\frac{\Delta_T d(x)}{d(x)} = -1.33 \frac{\Delta_T u(x)}{u(x)}. \tag{13.3.31}$$

Now recall that, in the region where it is measured, $\Delta d(x)$ is negative so that the very large value $\Delta_T d(x)/d(x) \approx -8/9$ implied by (13.3.31) and (13.3.32) will violate the Soffer bound (11.9.18) over a significant range of x. In addition the positivity condition $|\Delta D_\pi| \leq 2D_\pi$ is violated at large z. Thus the above treatment is inconsistent and must be disregarded.

An attempt at a consistent treatment by Boglione and Leader (2000) has led to some very surprising conclusions. The parametrizations of $\Delta_T u(x)$, $\Delta_T d(x)$ and $\Delta D_\pi(z, k_T)$ are constructed so that both the Soffer bound and the positivity bound are automatically respected. However, in almost all parametrizations of $\Delta d(x)$ obtained from fitting polarized DIS data, $\Delta d(x)$ is negative for all x. As a consequence the Soffer bound

$$|\Delta_T d(x)| \leq \tfrac{1}{2}\,[d(x) + \Delta d(x)] \tag{13.3.32}$$

is highly restrictive. This leads to a conflict with the demand that $|\Delta_T d(x)|$ be large in the region of large x, which is imposed by the fact that the $\pi^\pm$ asymmetries are big, and of roughly equal magnitude, for large x_F. As

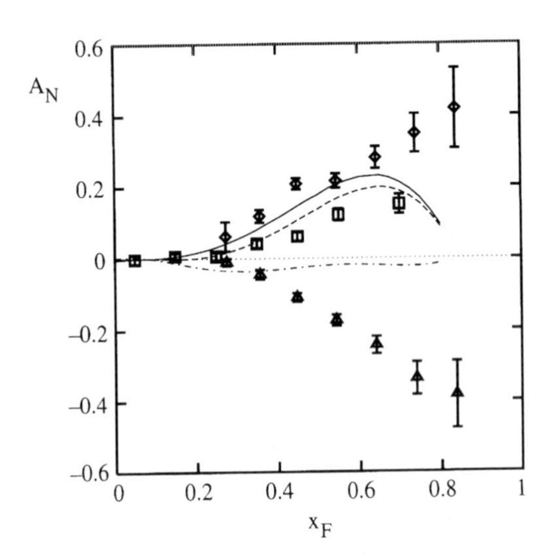

Fig. 13.9 The single-spin asymmetry A_N for pion production in $p^\uparrow p \to \pi X$ as a function of x_F when using the GS polarized parton densities. The failure of the theory to fit the data can be seen. Solid line, π^+; broken line π^0; broken and dotted line π^-. (Courtesy of M. Boglione.)

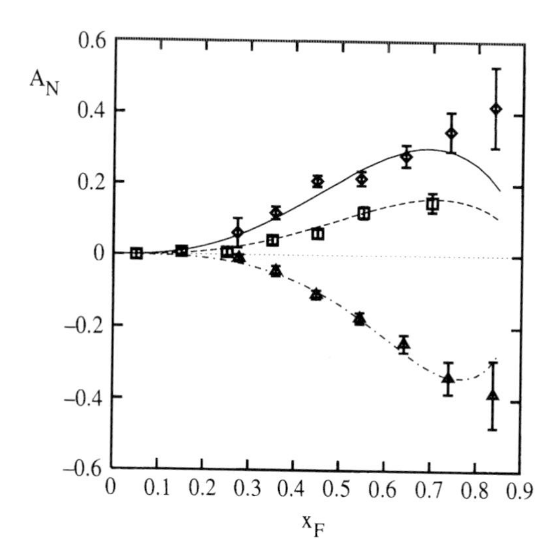

Fig. 13.10 The single-spin asymmetry A_N for pion production in the process $p^\uparrow p \to \pi X$ as a function of x_F, obtained by using the BBS polarized parton densities in the Soffer bound. Solid line, π^+; broken line, π^0; broken and dotted line, π^-. (Courtesy of M. Boglione.)

an example, in Fig. 13.9 we show the very poor fit to the data when the GS polarized densities, due to Gehrmann and Stirling (1996), are used: $\chi^2_{\rm DOF} = 25$! This raises an intriguing question. In (11.8.4) we pointed out that perturbative QCD arguments suggest that

$$\frac{\Delta q(x)}{q(x)} \to 1 \qquad \text{as } x \to 1 \tag{13.3.33}$$

For the d quark this would imply that $\Delta d(x)$ has to change sign and become positive at large x, thereby rendering the Soffer bound much less restrictive. In fact a more precise version of (13.3.33) is

$$\Delta q(x) = \left[1 - c(1-x)^2\right] q(x) \qquad \text{as } x \to 1 \tag{13.3.34}$$

where c is a positive constant. This is the origin of the fact that (13.3.33) is almost never imposed on $\Delta q(x)$ when fitting data on polarized DIS, the point being that (13.3.34) is inconsistent with the evolution equations. In truth, however, one should not use the evolution equations near the exclusive region $x = 1$, so there is not really a contradiction. There are two fits to the polarized DIS data in the literature that do respect (13.3.34). The first, the BBS, due to Brodsky, Burkhardt and Schmidt (1995), is somewhat incomplete since Q^2-evolution was not used. The second, the $(\rm LSS)_{BBS}$, due to Leader, Sidorov and Stamenov (1998), uses the BBS parametrization but includes evolution. There is a dramatic improvement to the fits to the π asymmetry data when these polarized densities are used, as seen in Figs. 13.10 and 13.11, which have $\chi^2_{\rm DOF}$-values 1.45 and 2.41 respectively.

Note, however, that it does not seem possible to fit the asymmetry data at the largest values of x_F, indicating that the Collins mechanism alone is probably unable to explain all the π asymmetry data.

13.4 Beyond the standard QCD parton model

Consider once again, for concreteness, the reaction $p_A^\uparrow p_B \to \pi X$. Recall that the asymmetries are largest at large x_F and that these $\pi^\pm$ are produced mainly from u and d quarks, respectively, having large values of x_a in the polarized proton and colliding with partons in the unpolarized proton with small values of x_b. We thus simplify by considering only valence quarks in $p_A^\uparrow$ and gluons and antiquarks in p_B. To explain the approach we shall limit ourselves to one flavour of quark in $p_A^\uparrow$ and of gluon in p_B.

The cross-section is proportional to the quantity W defined graphically in Fig. 13.12, where we do not show the fragmentation of the final quark $q(p)$ into the pion. All final state particles, including the gluon but excluding $q(p)$, are summed over. In this diagram the soft physics is in

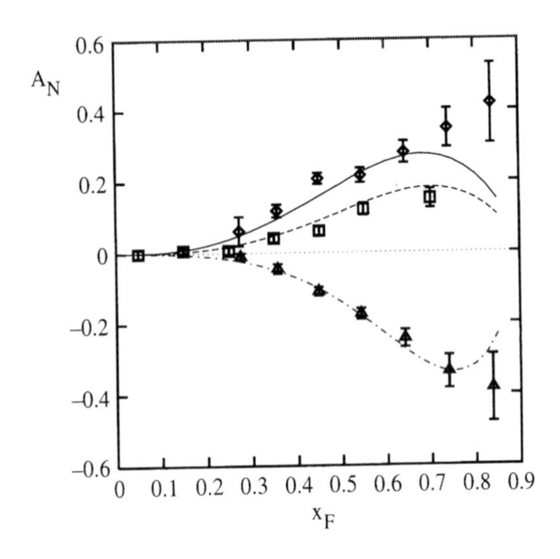

Fig. 13.11 The single-spin asymmetry A_N for pion production in the process $p^\uparrow p \to \pi X$ as a function of x_F, determined by the fit using the $({\rm LSS})_{\rm BBS}$ polarized parton densities in the Soffer bound. Solid line, π^+; broken line, π^0; broken and dotted line, π^-. (Courtesy of M. Boglione.)

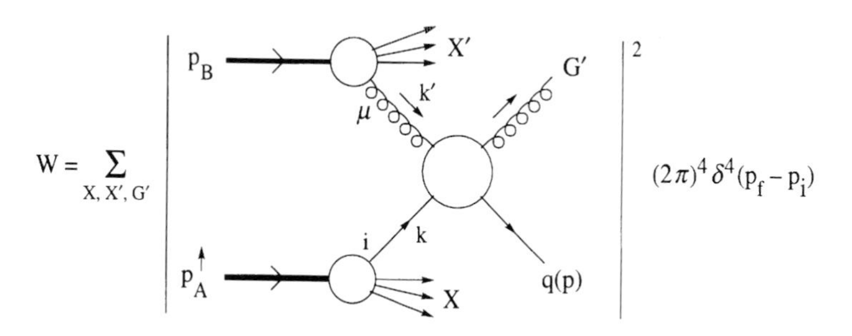

Fig. 13.12 Graphical definition of W for $p^{\uparrow}p \to \pi X$ in the standard QCD-parton model. μ is a Lorentz index, i a spinor index.

the amplitudes for $p_A^{\uparrow}$ and p_B to split into partons, and the $Gq \to G'q$ amplitude describes a hard process and is calculated in lowest-order perturbative QCD, i.e. using just the Born terms. Since the treatment of the *unpolarized* proton p_B is conventional, let us, to simplify the discussion, remove it and thus effectively consider

$$p_A^{\uparrow} + G \to \pi + X$$

and, to simplify even further, just consider one of the possible hard scattering Born terms, i.e. take

$$W = \sum_{X,G'} |M_q|^2 (2\pi)^4 \delta(p_{\mathrm{f}} - p_{\mathrm{i}}) \tag{13.4.1}$$

where M_q is shown in Fig. 13.13.

As in Section 11.5 the result for W can be written as a Feynman diagram with a cut propagator, in this case a gluon propagator, as shown in Fig. 13.14, with a similar structure to (11.5.14). Recall that the Collins mechanism for a transverse spin asymmetry came from the fragmentation

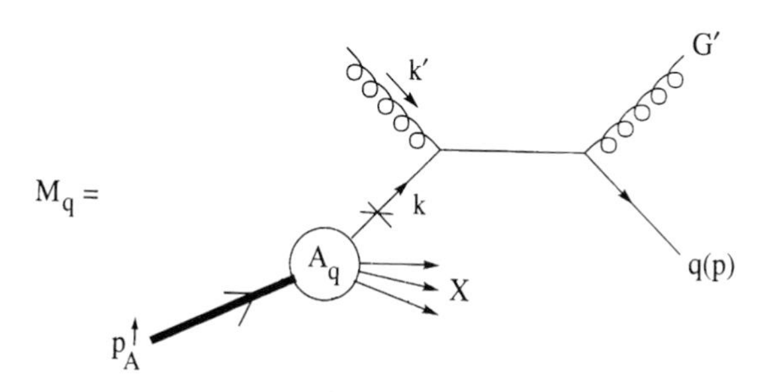

Fig. 13.13. Simplified version for M_q for the reaction $p^{\uparrow}G \to \pi X$. The cross on the fermion line indicates that there is no propagator for the quark of momentum **k**.

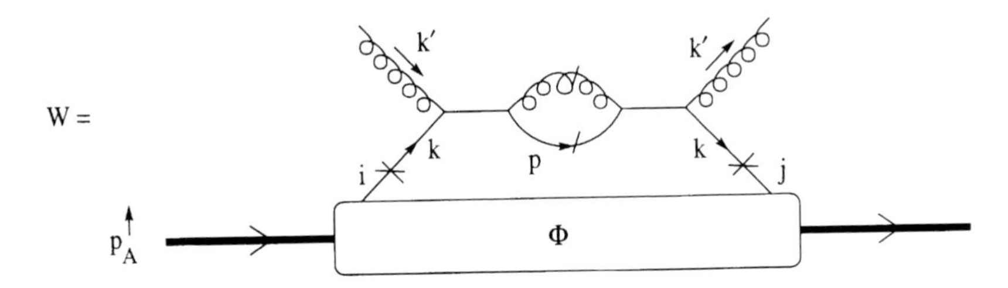

Fig. 13.14. Field-theoretic diagram corresponding to (13.4.1); i, j are spinor indices.

of $q(p)$ into a pion with non-zero transverse momentum. Our diagram really corresponds to

$$p^{\uparrow} + G \rightarrow \text{jet} + X \tag{13.4.2}$$

so the Collins mechanism is inoperative. Also, if time reversal is an exact symmetry then the Siver's mechanism that places the asymmetry in the spin-dependent quark density is absent and, as stressed earlier, we are unable to produce an asymmetry. To remedy this, and for several other reasons as well, Efremov and Teryaev (1984), following ideas of Ellis, Furmanski and Petronzio (1983), introduced a correlated quark–gluon density function, which, as we shall see, does yield an asymmetry.

Consider the soft amplitude for a proton to produce a quark, a gluon of colour a and index μ and some other set of particles X. To simplify the analysis pretend that X is fixed. The amplitude $A_{qG}^{\mu,a}$ is a Dirac spinor (see Section 11.5), and is shown graphically in Fig. 13.15, where as usual there are no propagators for partons. Conventionally, if the quark has momentum k_1 the gluon is given momentum $k_2 - k_1$. Combining the

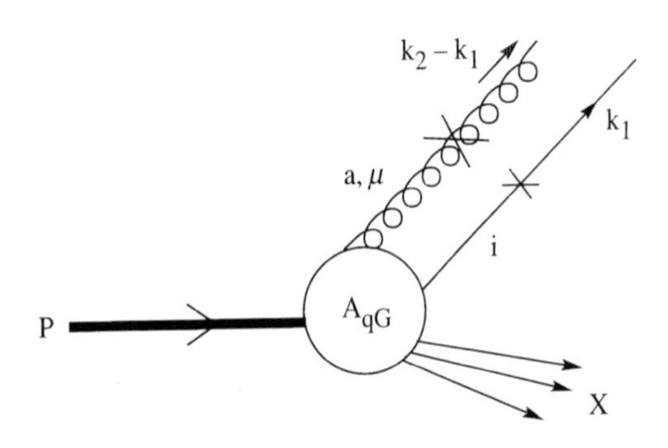

Fig. 13.15 The amplitude for a proton to produce a quark, a gluon and some set of particles X.

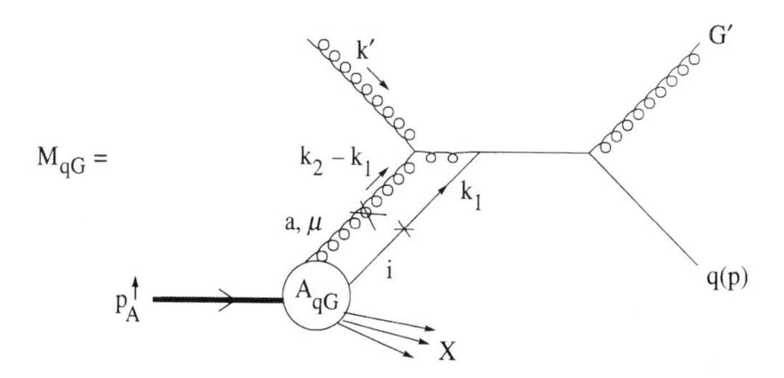

Fig. 13.16. A possible Feynman diagram for $p^{\uparrow}G \to \text{jet} + X$, utilizing A_{qG}.

amplitudes from Figs. 13.13 and 13.16 we now have

$$W = \sum_{X,G'} |M_q + M_{qG}|^2 (2\pi)^4 \delta(p_{\rm f} - p_{\rm i}). \tag{13.4.3}$$

Now, $|M_q|^2$ can be shown to correspond to twist 2 and $|M_{qG}|^2$ to twist 4, so for a large effect at moderately large p_T we must produce the asymmetry from the twist-3 interference term

$$\begin{aligned} I &\equiv (M_{qG}M_q^* + M_q M_{qG}^*)(2\pi)^4 \delta^4(p_{\rm f} - p_{\rm i}) \\ &= M_{qG}M_q^*(2\pi)^4 \delta^4(p_f - p_i) + \text{c.c.} \end{aligned} \tag{13.4.4}$$

Firstly, for there to be interference at all the states X in Figs. 13.13 and 13.16 must be identical, which will only be possible if the colour indices, which we have ignored thus far, are such that the quark–gluon pair transforms under a colour transformation like a quark. It should be

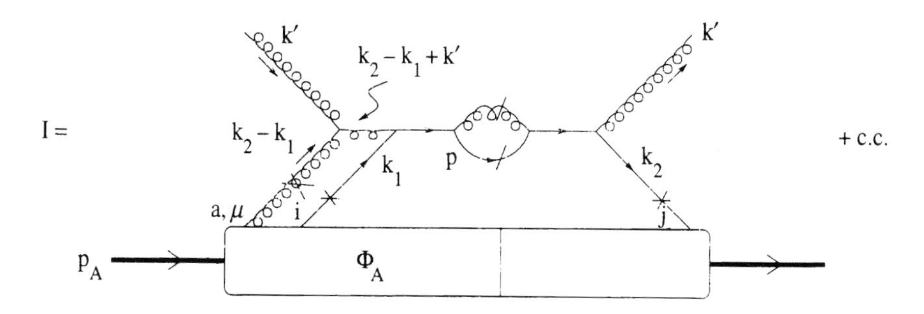

Fig. 13.17 Field-theoretic interpretation of the interference term in (13.4.4) (c.c. means complex conjugate; i, j are spinor indices).

clear that this will emerge automatically from the colour structure of the hard diagram.

Secondly, a non-zero interference term requires M_q and M_{qG} to be relatively real. Achieving this is the non-trivial step, as we shall see. As usual (13.4.3) and (13.4.4) can be given an interpretation as Feynman diagrams with a cut propagator. The interference term I then corresponds to the diagram shown in Fig. 13.17 plus its complex conjugate. Non-zero I then requires the Feynman amplitude to be real. The new soft function $\Phi_A^{\mu,a}$ is a 4×4 matrix in Dirac spinor space and can be shown, analogously to (11.5.17), to be given by

$$\Phi_{i\mu j}^{lam}(A;k_1,k_2;P,\mathcal{S}) = \int \frac{d^4y}{(2\pi)^4}\frac{d^4z}{(2\pi)^4} e^{ik_1\cdot z} e^{i(k_2-k_1)\cdot y} \times \langle P,\mathcal{S}|\overline{\Psi}_j^m(0)A_\mu^a(y)\Psi_i^l(z)|P,\mathcal{S}\rangle \quad (13.4.5)$$

where $A_\mu^a(y)$ is the usual gluon field operator of colour a and where we have now attached colour labels l,m to the quark fields, since at this point a careful treatment of the colour structure is essential.

To this end we redraw in Fig. 13.18 the hard part of the Feynman diagram with all colour labels displayed. The colour factor is

$$C_{asl} \equiv \sum_d t^b_{sr} t^d_{rn} t^d_{nm} t^c_{ml} f_{abc}. \quad (13.4.6)$$

Carrying out the sum over d produces a Kronecker delta δ_{rm}, so that

$$\begin{aligned} C_{asl} &\propto (t^b t^c)_{sl} f_{abc} \\ &= \frac{1}{2} f_{abc}[t^b, t^c]_{sl} \\ &= \frac{i}{2} f_{abc} f_{bcd} t^d_{sl} \\ &= \frac{3i}{2} t^a_{sl}. \end{aligned} \quad (13.4.7)$$

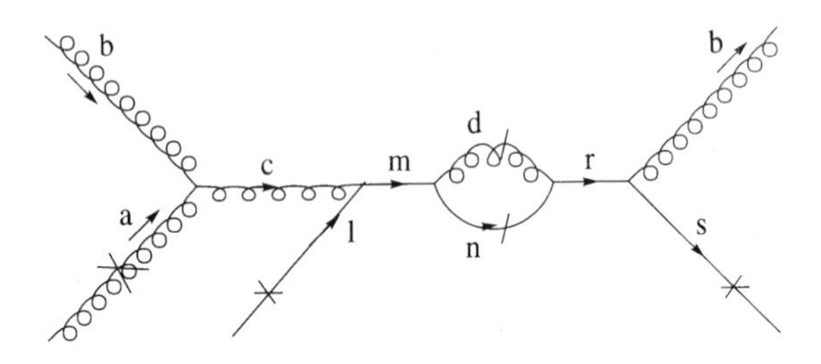

Fig. 13.18 The hard part of the Feynman diagram of Fig. 13.17 with all colour labels (a,b,c,d for gluons, l,m,n,r,s for quarks) shown.

Since this is the colour structure that will always occur it is convenient to absorb the t^a_{sl} into the soft amplitude, and, by convention, for reasons that will appear clear later, a factor of the strong coupling g. Thus the operator structure in (13.4.5) becomes

$$\sum_a g\overline{\Psi}_j t^a A^a_\mu \Psi^l_i = \overline{\Psi}_j g A_\mu \Psi_i \tag{13.4.8}$$

where, A_μ is the matrix

$$A_\mu = \sum_a t^a A^a_\mu \tag{13.4.9}$$

and Ψ_i is now a column vector in colour space.

Hence one utilizes the colour-singlet correlator

$$\Phi^\mu_{A_{ij}}(k_1, k_2; P, \mathcal{S}) = \int \frac{d^4y}{(2\pi)^4}\frac{d^4z}{(2\pi)^4} e^{ik_1\cdot z} e^{i(k_2-k_1)\cdot y} \times \langle P, \mathcal{S}|\overline{\Psi}_j(0) g A^\mu(y)\Psi_i(z)|P, \mathcal{S}\rangle. \tag{13.4.10}$$

Now the crucial point is that time-reversal invariance does not prohibit Φ^μ_A from having $\mathcal{S}_T$-dependent terms of the form

$$ib_\text{V}\epsilon^{\mu\alpha\beta\gamma}\mathcal{S}_\alpha P_\beta n_\gamma \not{P} + b_\text{A}\mathcal{S}^\mu \not{P}\gamma_5 \tag{13.4.11}$$

where b_V and b_A are *real* scalar functions and where n_μ is the null vector fixing the gauge $A^\mu n_\mu = 0$; see (11.5.24). Then, recalling that there is one γ-matrix at each vertex and in each fermion propagator (as usual, neglecting quark-mass terms), one sees that the hard part of Fig. 13.17 contains a product of seven γ-matrices. When the trace analogous to (11.5.14) is taken using Φ_A, the b_V term in (13.4.11) will involve a trace of eight γ-matrices, which will be real, whereas the b_A term involves eight γ-matrices and also γ_5 and will be imaginary. In consequence the traces over the terms in (13.4.11) produce a result proportional to i.

Next we count the factors of i coming from quark–gluon vertices and all non-cut propagators, both quark and gluon. There are seven of them, so that they yield a factor of i.

In total, then, we have a product of three factors i, from colour, from vertices and from the trace with the soft amplitude. Thus, contrary to our hope, the relevant spin-dependent part of the Feynman diagram in Fig. 13.17 appears to be imaginary.

However, in the loop integrations over k_1, k_2 (and k' when the upper gluon is attached to a hadron) we encounter the point where the gluon propagator on the left, carrying momentum $k_2 - k_1 + k'$, is on shell, i.e. where $(k_2 - k_1 + k')^2 = 0$. As can be understood from eqn (11.5.12), this will give a term $-i\pi\delta\left[(k_2 - k_1 + k')^2\right]$, which just provides the last i necessary to render the amplitude real!

Finally, then, we have a mechanism for producing a single-transverse-spin asymmetry that respects all the fundamental discrete symmetries of QCD. The asymmetry is calculated from the Feynman diagram of Fig. 13.17, in which, in the propagator for the gluon carrying momentum $k_2 - k_1 + k'$, one makes the replacement

$$\frac{i}{(k_2 - k_1 + k')^2 + i\epsilon} \quad \rightarrow \quad \pi\delta\left[(k_2 - k_1 + k')^2\right]. \tag{13.4.12}$$

The final result takes a form analogous to (11.5.14):

$$I = \int d^a k_1 d^4 k_2 \ \mathrm{Tr}\ \left[\Phi_A^\mu S_\mu(k_1, k_2)\right] + \mathrm{c.c.}, \tag{13.4.13}$$

where S_μ is the short-distance amplitude with the modification (13.4.12) and a factor gt^a_{sl} removed.

The detailed analysis is exceedingly complicated (Qiu and Sterman, 1999) and the above pedagogical treatment aims only at presenting the essential ideas.

In a more careful treatment the following points should be noted.

(1) The above discussion focussed on the pole in the gluon propagator and is referred to as the *gluonic pole mechanism*.
(2) There are other diagrams involving Φ_A^μ that contribute to the asymmetry. An example is shown in Fig 13.19. In this case the extra factor i is produced via the pole in the fermion propagator carrying momentum $k_1 + k'$, i.e. at $(k_1 + k')^2 = 0$. This is referred to as the *fermionic pole mechanism* and has been studied by Efremov, Korotkiyan and Teryaev (1995), by Teryaev (1995) and by Korotkiyan and Teryaev (1995).
(3) There is an unresolved dispute in the literature whether the gluon or fermion pole is expected to be the dominant mechanism. The kinematics are such that in the gluon pole case the gluon field in the proton corresponds to a static, constant, field. In the fermion pole case one has the somewhat strange concept of a static, constant, fermion

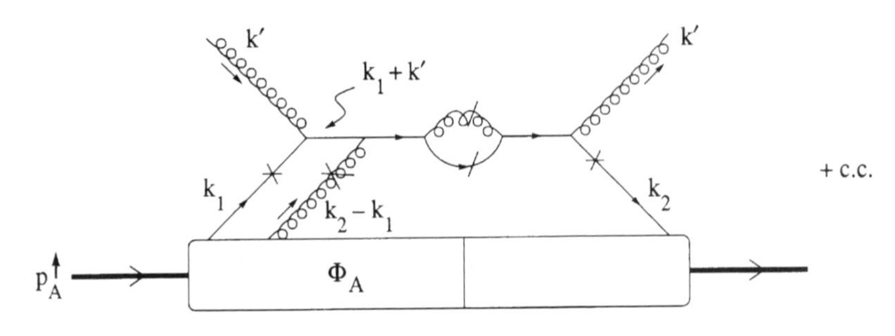

Fig. 13.19. Alternate type of interference term involving Φ_A.

field. A complete treatment, including both mechanisms, has not, to our knowledge, been carried out for the reaction $p^\uparrow p \to \pi X$, but we shall give below the result for the simpler process $p^\uparrow p \to \gamma X$.

(4) The above discussion involving 4-vectors $k^\mu_{1,2}$ is, as in the discussion of Section 11.5, too general and is not yet in parton-model form. One still has to make the leading collinear approximation $k_1 = x_1 P$, $k_2 = x_2 P$ in the hard amplitude and then carry out the integration $\int dk^- d^2\mathbf{k}_T$ for both k_1 and k_2 in the soft amplitude. Thereby one comes finally to the standard form of correlator

$$\Phi^\mu_A(x_1, x_2; P, \mathscr{S}) = \int \frac{d\lambda}{2\pi}\frac{d\zeta}{2\pi} e^{i\lambda x_1} e^{i\zeta(x_2 - x_1)} \times \langle P, \mathscr{S}|\overline{\Psi}_j(0) A^\mu(\zeta n)\Psi_i(\lambda n)|P, \mathscr{S}\rangle \quad (13.4.14)$$

involving operators on the light-cone. (recall that $n^2 = 0$).

(5) Since we are studying a twist-3 contribution we must, for consistency, include the non-leading twist-3 terms coming from the standard parton diagram Fig. 11.7. These arise when one goes beyond the collinear approximation $k = xP$ inside the hard amplitude S in the hadron–hadron analogue of (11.5.14). The inclusion of transverse momentum involves making a Taylor expansion of $S(k)$ about the point $k^\mu = xP^\mu$:

$$S(k^\mu) = S(k^\mu = xP^\mu) + (k^\mu - xP^\mu)\frac{\partial}{\partial k^\mu} S(k^\mu) + \cdots \quad (13.4.15)$$

and the term $(k^\mu - xP^\mu)\Phi_{ij}$ can be transformed, via partial integration, into a matrix element involving $\partial^\mu \Psi(z)$. The beautiful, and perhaps unsurprising, result is that this term can be combined with Φ^μ_A to produce a new correlator

$$\Phi^\mu_{D_{ij}}(k_1, k_2; P, \mathscr{S}) = \int \frac{d^4y}{(2\pi)^4}\frac{d^4z}{(2\pi)^4} e^{ik_1\cdot z} e^{i(k_2 - k_1)\cdot y} \times \langle P, \mathscr{S}|\overline{\Psi}_j(0)\vec{D}^\mu(y)\Psi_i(z)|P, \mathscr{S}\rangle \quad (13.4.16)$$

where

$$\vec{D}^\mu(y)\Psi_i(z) = \left[i\frac{\partial \Psi_i(z)}{\partial z_\mu} + gA^\mu(y)\Psi_i(z)\right] \quad (13.4.17)$$

(the arrows indicate that the operator acts only to the right) is similar to the standard covariant derivative that appears in the QCD equation of motion

$$\left[\vec{\not{D}}(z) - m_q \right] \Psi(z) = 0. \quad (13.4.18)$$

Finally, after making the collinear approximation in the twist-3 part involving A^μ, the correlator involved becomes

$$\Phi^\mu_{D_{ij}}(x_1, x_2; P, \mathscr{S}) = \int \frac{d\lambda}{2\pi}\frac{d\zeta}{2\pi} e^{i\lambda x_1} e^{i\zeta(x_2 - x_1)} \times \langle P, \mathscr{S}|\overline{\Psi}_j(0)\vec{D}^\mu(\zeta n)\Psi_i(\lambda n)|P, \mathscr{S}\rangle. \quad (13.4.19)$$

This is a very interesting approach and, although the basic idea is not new, it is only now that detailed calculations are beginning to be performed. It may well be that in combination with the Collins mechanism one can obtain a fit to all the data on the π asymmetry. However, it is clearly essential to study asymmetries in reactions where the Collins mechanism is inoperative, e.g. in hard γ or jet production, in order to learn more about the gluonic and fermionic pole mechanisms.

The reaction $p_A^\uparrow p_B \to \gamma X$ is the only case, to our knowledge, where the entire contribution of gluonic and fermionic poles has been taken into account (Qiu and Sterman, 1992) and the structure of their result is instructive.

Let the photon emerge with momentum $\mathbf{p}_\gamma$ and energy E_γ. Then

$$E_\gamma\left(\frac{d^3\sigma(\mathscr{S}_T)}{d^3\mathbf{p}_\gamma} - \frac{d^3\sigma(-\mathscr{S}_T)}{d^3\mathbf{p}_\gamma}\right) = \frac{\alpha\alpha_s}{s}\epsilon_{\mu\nu\rho\sigma}\mathscr{S}_T^\mu p_\gamma^\nu n^\rho p_A^\sigma \int\frac{dx_b}{x_b}G(x_b)\int\frac{dx_a}{x_a}\delta(\hat{s}+\hat{t}+\hat{u}) \times \sum_f e_f^2\left[G_f(x_a, x_b, p_\gamma) + F_f(x_a, x_b, p_\gamma)\right] \quad (13.4.20)$$

where G_f and F_f are the contributions of flavour f from gluonic (G) and fermionic (F) poles respectively, n^μ is given in (11.5.24) and $\hat{s}$, $\hat{t}$, $\hat{u}$ are the Mandelstam variables involved in the partonic process

$$\hat{s} = x_a x_b s \qquad \hat{t} = x_a t \qquad \hat{u} = x_b u. \quad (13.4.21)$$

In terms of quark–gluon correlators T and hard scattering terms H one has for the gluonic pole:

$$G_f = H_G^{(1)}(x_a, x_b, p_\gamma)T_G^{(V)}(x_a, x_a) + H_G^{(2)}(x_a, x_b, p_\gamma)\left[T_G^{(V)}(x_a, x_a) - x_a\frac{\partial}{\partial x_a}T_G^{(V)}(x_a, x_a)\right] \quad (13.4.22)$$

where

$$H_G^{(1)} = -\frac{3g}{8}\left(\frac{1}{\hat{t}}\right) \quad (13.4.23)$$

and

$$H_G^{(2)} = \frac{3g}{8}\left(\frac{1}{\hat{u}}\right)\left(\frac{\hat{s}}{\hat{t}} + \frac{\hat{t}}{\hat{s}}\right). \tag{13.4.24}$$

The soft quark–gluon correlator $T_G^{(\mathrm{V})}$ is real and is given by

$$T_G^{(\mathrm{V})} = n_\mu \epsilon_{\nu\alpha\beta\gamma} n^\alpha p_A^\beta \mathscr{S}_T^\gamma \int \frac{d\lambda d\xi}{4\pi} e^{i\lambda x} \langle p_A; \mathscr{S}_T | \overline{\Psi}(0) \not{n} G^{\mu\nu}(\xi n) \Psi(\lambda n) | p_A; \mathscr{S}_T \rangle \tag{13.4.25}$$

where $G^{\mu\nu}$ is the gluon field-strength tensor. For the fermionic pole one has

$$F_f = H_D^{(1)}(x_a, x_b, p_\gamma)\left[T_D^{(\mathrm{V})}(0, x_a) + iT_D^{(\mathrm{A})}(0, x_a)\right] \tag{13.4.26}$$

where $T_D^{(\mathrm{V})}$ is real, $T_D^{(\mathrm{A})}$ pure imaginary. Here

$$H_D^{(1)} = \frac{1}{24}\left(\frac{9}{\hat{s}} - \frac{1}{\hat{t}}\right) \tag{13.4.27}$$

and the correlators are

$$T_D^{(\mathrm{V})}(0, x) = \epsilon_{\mu\alpha\beta\gamma} n^\alpha p_A^\beta \mathscr{S}_T^\gamma \int \frac{d\lambda d\xi}{4\pi} e^{i\lambda x} \langle p_A; \mathscr{S}_T | \overline{\Psi}(0) \not{n} D^\mu(\xi n) \Psi(\lambda n) | p_A; \mathscr{S}_T \rangle \tag{13.4.28}$$

and

$$T_D^{(\mathrm{A})} = \mathscr{S}_T^\mu \int \frac{d\lambda d\xi}{4\pi} e^{i\xi x} \langle p_A; \mathscr{S}_T | \overline{\Psi}(0) \; \not{n} \gamma_5 D_\mu(\xi n) \Psi(\lambda n) | p_A; \mathscr{S}_T \rangle. \tag{13.4.29}$$

Qiu and Sterman (1999) argue that the correlators in the fermionic pole case are essentially the overlap of states in one of which the quark has momentum xp and in the other of which all this momentum is carried by the gluon, so that the overlap should be small. In the gluonic pole case, on the contrary, in both states the quark carries momentum xp, so that the overlap might be expected to be larger. For this reason Qiu and Sterman expect the gluonic pole mechanism to dominate. Further, they suggest that typically $T_G^{(V)}(x, x)$ will vanish like $(1 - x)^\beta$ as $x \to 1$ with $\beta > 0$, in which case the term $x(\partial/\partial x)T_G^V(x, x)$ in (13.4.22) will dominate at large x. (In their treatment of $p^\uparrow p \to \pi X$ mentioned earlier, Qiu and Sterman keep just this term.)

In conclusion to this section, we note that the theoretical developments are fascinating, but it will be a mammoth task to sort out the mechanisms and learn experimentally about the various correlators.

13.5 Phenomenological models

It is a historical fact that we have known ever since 1976 that hyperons, and in particular Λs, are produced in a highly polarized state in the high energy collision of unpolarized hadrons (Bunce *et al.*, 1976). A sample of the data was shown in Figs. 13.2, 13.3, 13.5 and 13.6. The general features were summarized at the beginning of Chapter 13.

In this section we shall briefly describe some of the phenomenological attempts to explain the hyperon data. None is really convincing, firstly because they are really semiclassical models, secondly because, while enjoying some success, they cannot account for all the main features of the data.

Concerning the semiclassical aspect there is an important point that should be noted. If, as is conventional, one works with helicity amplitudes, the asymmetry or polarization is always of the form

$$A \propto \text{Im}\left(\phi^*_{\text{flip}}\phi_{\text{non-flip}}\right) \tag{13.5.1}$$

where the ϕ are helicity amplitudes involving either helicity-flip or no helicity-flip. Thus one requires a model for the *amplitudes* and their phases, a concept beyond classical physics.

To evade this dilemma one can work in a basis where the spin states are transverse, see Section 11.9 and one then finds that (13.5.1) is replaced by

$$A \propto |f_{\text{non-flip}}|^2 - |f_{\text{flip}}|^2 \tag{13.5.2}$$

where f are amplitudes involving either flip or non-flip of the transverse spin. In this formalism one *can* make a probabilistic model for the moduli squared of the amplitudes; however, such a theory can never be totally satisfactory because there will be in general other spin-dependent variables that *do* involve interference between the transverse spin amplitudes, and these will then be outside the scope of the model.

At the time of writing there seems to be some hope of attacking the matter in a more fundamental way, using an analogue of the Collins mechanism, discussed in Section 13.4, in which an unpolarized quark can fragment into a polarized hadron if p_T is non-zero. However, it would be premature to comment on this approach, so we shall outline some of the phenomenological methods used over the past two-and-a-half decades. Our presentation owes much to the review of Soffer (1999).

In very broad terms the following features, specific to the hyperon polarization, require explanation:

$$\mathscr{P}_\Lambda \sim \mathscr{P}_{\Xi^-} \sim \mathscr{P}_{\Xi^0} \tag{13.5.3}$$

$$\mathscr{P}_{\Sigma^+} \sim \mathscr{P}_{\Sigma^-} \sim -\mathscr{P}_\Lambda \tag{13.5.4}$$

$$\mathscr{P}_{\bar{\Lambda}} \sim \mathscr{P}_{\bar{\Xi}^0} \sim 0 \tag{13.5.5}$$

yet

$$\mathscr{P}_{\bar{\Sigma}^-} \sim \mathscr{P}_{\Sigma^+} \qquad \mathscr{P}_{\bar{\Xi}^+} \sim \mathscr{P}_{\Xi^-}. \tag{13.5.6}$$

All our considerations will be directed at the beam fragmentation region where most of the data lie.

13.5.1 The Lund model

Consider $pp \to \Lambda X$, in which the Λ has transverse momentum p_T. The hadrons are assumed to have simple $SU(6)$ three-quark wave functions. Thus the Λ consists of an isosinglet (ud) diquark with spin $S = 0$ and a strange quark, which carries the spin of the Λ. The reaction is visualized as follows (Andersson, Gustafson and Ingelman, 1979; Andersson *et al.*, 1983). A suitable ud diquark from the proton moves forward, stretching the confining colour field, which ultimately 'snaps', producing an $s\bar{s}$ pair in a process that conserves angular momentum locally (see Fig. 13.20).

The momenta of the s and $\bar{s}$ are chosen to allow the s to combine with the essentially forward-going ud to produce a Λ with p_T as indicated. In this configuration the $s\bar{s}$ pair has orbital angular momentum along $\mathbf{p} \times \mathbf{p}_\Lambda$. To compensate for this, the spins s and $\bar{s}$ must be along $-(\mathbf{p} \times \mathbf{p}_\Lambda)$. Consequently the Λ emerges with polarization along $-(\mathbf{p} \times \mathbf{p}_\Lambda)$, as is found experimentally.

For the production of Σ^0, the $SU(6)$ wave function is built up from a ud diquark with $S = 1$. The two spin states of the Σ^0, referred to an axis along $\mathbf{p} \times \mathbf{p}_\Sigma$, are

$$\begin{aligned} |\tfrac{1}{2}\rangle &= \sqrt{\tfrac{2}{3}}\,|1; -\tfrac{1}{2}\rangle - \sqrt{\tfrac{1}{3}}\,|0; \tfrac{1}{2}\rangle \\ |-\tfrac{1}{2}\rangle &= \sqrt{\tfrac{1}{3}}\,|0; -\tfrac{1}{2}\rangle - \sqrt{\tfrac{2}{3}}\,|-1; \tfrac{1}{2}\rangle \end{aligned} \tag{13.5.7}$$

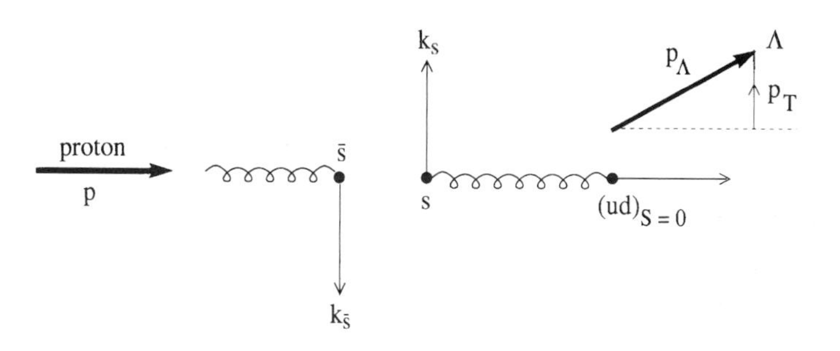

Fig. 13.20. Schematic diagram of the breaking of a Lund string to produce an $s\bar{s}$ pair.

and since the $s\bar{s}$ configuration that produces the required p_T has an s with spin projection 1/2 one sees easily that

$$\mathscr{P}_{\Sigma^0} = -(\tfrac{1}{3})\mathscr{P}_{\Lambda}. \tag{13.5.8}$$

A similar result holds for Σ^+.

There seems to be only one measurement of $\mathscr{P}_{\Sigma^0}$ and it is in agreement with the sign in (13.5.8). As seen in Fig. 13.6, the sign of $\mathscr{P}_{\Sigma^+}$ is also in agreement with (13.5.8), but not the magnitude.

However, the mechanism for Σ^- production must be quite different, since the string-breaking must provide a ds pair. Nonetheless, $\mathscr{P}_{\Sigma^-}$ is much like $\mathscr{P}_{\Sigma^+}$. It is equally unclear why Ξ^- and Ξ^0 have the same polarization as the Λ.

Finally, the vanishing of the polarizations for $\bar{\Lambda}$ and $\bar{\Xi}^0$ seems intuitive since the entire particle has to be created via the string-breaking. But then the significant polarizations of the $\bar{\Sigma}^-$ and the $\bar{\Xi}^+$ are a mystery.

In short, while the model has some success it in no way provides an adequate quantitative description of the data.

13.5.2 The Thomas precession model

This very clever semiclassical model, due to De Grand and Miettinen (1981), utilizes the Thomas precession to argue in favour of a higher probability for particular states of polarization.

Here it is assumed that a u quark from the beam proton of momentum $\mathbf{p}$ is wrenched off in the collision, leaving a fast forward-moving $S = 0$ ud diquark with momentum roughly $\frac{2}{3}\mathbf{p}$ and various low-momentum sea partons; one of these, an s, is then attracted towards and binds with the ud to form the Λ. The s quark is assumed to have transverse momentum $\mathbf{p}_T$. It is further assumed that the force that drags it towards the forward-moving ud diquark arises from a Lorentz scalar potential.

The process by which the s and the ud come together is viewed in a hybrid fashion. Firstly, one pictures the classical orbits involved to argue that the orbital angular momentum $\mathbf{L}$ of the s is opposite to $\mathbf{p} \times \mathbf{p}_{\Lambda}$. Next, one visualizes the interaction between the spinless ud and the spin-1/2 s in quantum mechanical terms involving a scalar attractive potential $V(r)$. As explained in subsection 2.2.8 the Thomas precession induces a rotation of the spin vector, given by (2.2.33), and this is equivalent to an $\mathbf{L} \cdot \mathbf{S}$ coupling, so that the effective attractive potential becomes

$$V_{\text{eff}} = V(r) - \frac{1}{2m_s^2 c^2}\left(\frac{1}{r}\frac{dV}{dr}\right)\mathbf{L}\cdot\mathbf{S} \tag{13.5.9}$$

where m_s is the strange-quark mass. Now, since $-(1/r)dV/dr$ is negative, $|V_{\text{eff}}|$ will be largest if $\mathbf{L} \cdot \mathbf{S}$ is positive. Hence the binding takes place preferentially when **S** is parallel to **L**, i.e. opposite to the normal to the scattering plane $\mathbf{n} = \mathbf{p} \times \mathbf{p}_\Lambda$. Thus the Λs are produced preferentially with spin opposite to **n**, i.e. are negatively polarized as required. The model also predicts that $|\mathscr{P}_\Lambda| \propto p_T$ as is seen experimentally.

An interesting prediction arises in the case of $K^- p \to \Lambda X$ for the Λ in the K^- fragmentation region. Now the s quark is initially in the K^- and moving too fast, so must decelerate to form the Λ. The Thomas precession is now reversed and **S** along **n** is favoured. Indeed the Λ polarization is found to be positive in this reaction, though its magnitude is twice as large as in $pp \to \Lambda X$ and this is not explained by the model.

Nor can the model explain the differing behaviours of the various antihyperons; see (13.5.5) and (13.5.6).

13.5.3 Concluding remarks

For access to the detailed literature and for a description of some other phenomenological models the reader is referred to Soffer (1999).

In summary, it has to be admitted that there is still, after 25 years of experiment and some decades of QCD, no coherent theory of the hyperon polarization data. Moreover the richness of the experimental data is continually growing, and none of the models can explain the beautiful discovery by the E704 experiment of Fermilab (Bravar *et al.*, 1995; 1997) that the analysing power A_N and the spin-transfer parameter

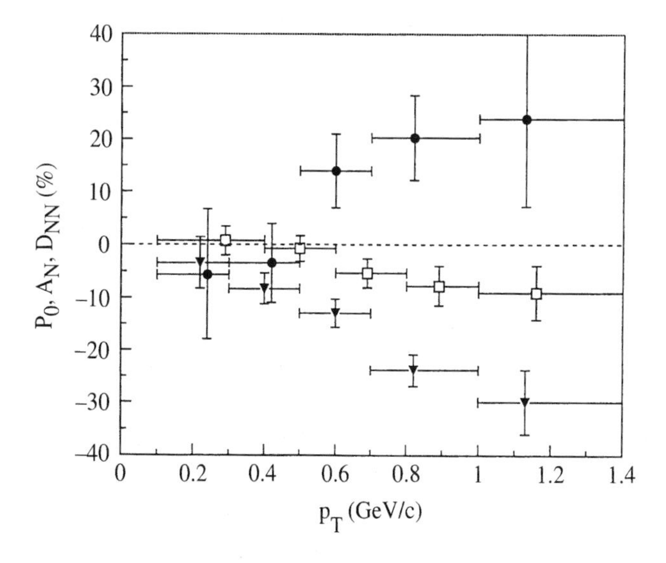

Fig. 13.21 Fermilab E704 data on Λ polarization P_0 (triangles), analysing power A_N (squares) and spin-transfer parameter D_{NN} (circles), each given as a percentage, for $pp \to \Lambda X$ at 200 GeV/c. (Courtesy of A. Penzo.)

D_{NN} are both large and growing with p_T in $pp \to \Lambda X$ at 200 GeV/c (see Fig. 13.21).

This whole area of high energy physics remains an open challenge to the theory of strong interactions.

14
Elastic scattering at high energies

Elastic scattering is in some sense the most fundamental type of reaction, but it is also the most difficult to understand theoretically. There is a huge amount of spin-dependent data at low to medium energies, but little understanding of the mechanisms at work. In several instances, however, spin-dependent data have played a crucial rôle in nailing the coffin of a current theoretical picture. Somehow, simple-minded ideas, which succeed in explaining gross features of cross-sections, angular distributions etc., run aground when faced with the more probing questions involved in spin-dependent reactions. Because of the lack of clear-cut theoretical ideas and because of the difficulty of the experiments there has generally been a lack of experimental effort in this field since the mid-1980s, but this situation is about to change with the commissioning of the RHIC collider at Brookhaven. There, besides a major programme of heavy-ion physics, it will be possible to study *pp* collisions, with both beams polarized and up to an energy of 250 GeV per beam. Consequently we shall concentrate in this chapter on nucleon–nucleon scattering.

Broadly speaking there are two kinematic regions of interest, small to medium values of momentum transfer and large momentum transfer. The first is, strictly speaking, in the domain of non-perturbative QCD, so there are no precise theoretical predictions, though there are very interesting suggestive hints. In the second region perturbative QCD ought to be applicable and, indeed, very powerful theoretical results have been derived. It is a well-known secret that there is a major disagreement between present data and these predictions. The usual argument for not therefore abandoning QCD, and it is a sound one, is that the data are not yet at large enough energy and momentum transfer to justify fully a perturbative treatment. With the increased kinematic range at RHIC we will thus be facing some very challenging questions: either the trend of the experimental results must begin to change or we must seriously begin to question the validity of QCD.

14.1 Small momentum transfer: general

Consider proton–proton elastic scattering with momenta as indicated:

$$p(p_1) + p(p_2) \to p(p_3) + p(p_4). \tag{14.1.1}$$

There are five independent helicity amplitudes corresponding to the following transitions:

$$\begin{aligned} \phi_1 &= \langle ++|T|++\rangle & \phi_2 &= \langle ++|T|--\rangle \\ \phi_3 &= \langle +-|T|+-\rangle & \phi_4 &= \langle +-|T|-+\rangle \\ & & \phi_5 = \langle ++|T|+-\rangle \end{aligned} \tag{14.1.2}$$

Relations between these and any other helicity amplitude can be determined via the symmetry relations given in Section 4.2. Each of the ϕ_j is a function of the Mandelstam variables

$$s = (p_1 + p_2)^2 \qquad t = (p_1 - p_3)^2 \qquad u = (p_1 - p_4)^2 \tag{14.1.3}$$

with

$$s + t + u = 4m^2. \tag{14.1.4}$$

In the CM of the reaction, where the protons all have a magnitude of momentum p, one has

$$\begin{aligned} s &= 4E^2 = 4(p^2 + m^2) \\ t &= -2p^2(1 - \cos\theta) \\ u &= -2p^2(1 + \cos\theta) \end{aligned} \tag{14.1.5}$$

where θ is the CM scattering angle.

There is a large number of spin-dependent observables that one can measure. A comprehensive list is given in Table A10.4, and expressions for the observables in terms of the ϕ_j are given in Tables A10.5 and A10.6.

The conservation of angular momentum imposes restrictions on the helicity-flip amplitudes in the forward direction (Section 4.3), namely

$$\phi_5 \propto \sqrt{-t} \qquad \phi_4 \propto t \tag{14.1.6}$$

as $t \to 0$.

In the region of very small t, as discussed in subsection 8.1.1, we have interference between the hadronic and electromagnetic amplitudes and we shall presently explain some new results in this field. Firstly, however, we shall summarize what is known about the hadronic amplitudes near the forward direction. It is convenient to analyse the high energy behaviour of the ϕ_j in terms of the quantum numbers of the system that can be exchanged between the protons, and the singularities at $J = \alpha(t)$ in the complex angular momentum plane associated with such a system. (For an introduction to the concept of complex angular momentum see Gasiorowicz, 1967, Chapter 28.) In this discussion it will be convenient to use helicity amplitudes normalized such that

$$\frac{d\sigma}{dt} = \frac{2\pi}{s^2}\left(|\phi_1|^2 + |\phi_2|^2 + |\phi_3|^2 + |\phi_4|^2 + 4|\phi_5|^2\right). \tag{14.1.7}$$

Then, via the optical theorem,

$$\sigma_{\text{tot}}(s) = \frac{4\pi}{s}\ \text{Im}\ [\phi_1(s,t) + \phi_3(s,t)]_{t=0}\,. \tag{14.1.8}$$

A singularity at $J = \alpha(t)$ then implies an asymptotic behaviour

$$|\phi(s,t)| \propto s^{\alpha(t)} \qquad \text{as } s \to \infty \tag{14.1.9}$$

up to possible logarithmic corrections. In this normalization, the rigorous Froissart–Martin bound (Froissart, 1961; Martin, 1966) reads

$$|\phi_1(s,t) + \phi_3(s,t)|_{t=0} \lesssim \text{constant} \times s \ln^2 s \qquad \text{as } s \to \infty \tag{14.1.10}$$

or

$$\sigma_{\text{tot}} \lesssim \text{constant} \times \ln^2 s \qquad \text{as } s \to \infty \tag{14.1.11}$$

so that the leading J-plane singularity cannot lie above $J = 1$ at $t = 0$.

A particular dynamical exchange mechanism is classified according to its quantum numbers: *parity* ($\mathscr{P}$), *charge conjugation* ($\mathscr{C}$) and *signature* (τ). An amplitude is called *even* or *odd* under *crossing*, i.e. under the analytic continuation

$$s \to e^{i\pi}s \tag{14.1.12}$$

for $\tau = \pm 1$, since

$$A_\tau(e^{i\pi}s, t) = \tau A_\tau^*(s,t). \tag{14.1.13}$$

For nucleon–nucleon scattering there are three classes of exchange (Leader and Slansky, 1966), as indicated in Table 14.1, which also shows to which

Table 14.1. Classification of exchanges and the amplitudes to which they contribute to in pp scattering

	Class 1 $\tau = \mathscr{P} = \mathscr{C}$	Class 2 $\tau = -\mathscr{P} = -\mathscr{C}$	Class 3 $\tau = -\mathscr{P} = \mathscr{C}$
amplitudes	$\phi_1 + \phi_3,\ \phi_5,\ \phi_2 - \phi_4$	$\phi_1 - \phi_3$	$\phi_2 + \phi_4$
particles or mechanism	$\mathsf{P}, \mathsf{O}, \rho, \omega, f, a_2$	a_1	π, η, b

amplitudes, or combination of amplitudes, each class contributes. Also shown in Table 14.1 are some particles whose quantum numbers coincide with each class. P, the *pomeron* and O, the *odderon* are not particles, but label dynamical systems with the quantum numbers of the vacuum, $\mathscr{P} = +1, \mathscr{C} = +1, \tau = +1$ (the pomeron), or $\mathscr{P} = +1, \mathscr{C} = -1, \tau = -1$ (the odderon).

The pomeron is important because if one single exchange mechanism dominates at asymptotic energies, it has to have the quantum numbers of the vacuum (Peierls and Trueman, 1964).

The singularities associated with the other particles in Table 14.1 all lie well below $J = 1$ and the pomeron is supposed to have a singularity at $J = 1$ when $t = 0$ in order to explain the fact that both σ_{pp} and $\sigma_{\bar{p}p}$ appear to be growing like $\ln^2 s$ at the highest energies measured.

The rôle of the odderon is interesting, because it is the quantum number $\mathscr{C}$ that determines the relative sign of the contribution of a given exchange system to $pp \to pp$ and $\bar{p}p \to \bar{p}p$:

$$A^{\bar{p}p}_{\tau,\mathscr{P},\mathscr{C}}(s,t) = \mathscr{C} A^{pp}_{\tau,\mathscr{P},\mathscr{C}}(s,t). \tag{14.1.14}$$

It was believed for decades that asymptotically one had to have

$$A^{\bar{p}p} - A^{pp} \to 0 \qquad \text{as } s \to \infty \tag{14.1.15}$$

but Łukaszuk and Nicolescu (1973) pointed out that in fact

$$A^{\bar{p}p} - A^{pp} \not\to 0 \qquad \text{as } s \to \infty \tag{14.1.16}$$

is compatible with all known general properties of field theory. The odderon is the name given to the putative mechanism responsible for this (Joynson, Leader and Nicolescu, 1975).

If for example one has

$$A^{pp} = A_{\mathsf{P}} + A_{\mathsf{O}} \tag{14.1.17}$$

then

$$A^{\bar{p}p} = A_{\mathsf{P}} - A_{\mathsf{O}} \tag{14.1.18}$$

and the analysis of Łukaszuk and Nicolescu showed that it was possible to have

$$\frac{|A_{\mathsf{O}}|}{|A_{\mathsf{P}}|} \not\to 0 \qquad \text{as } s \to \infty. \tag{14.1.19}$$

The pomeron and odderon mechanisms are believed to reflect two-gluon and three-gluon exchange in QCD. Although it is not possible to carry out a QCD calculation in the truly soft, non-perturbative, regime, powerful conformal field-theoretic methods have been utilized by Lipatov and co-workers (Lipatov, 1986; 1989; Braun, Gauron and Nicolescu, 1999) to study fully interacting two- and three-gluon exchange dynamics just outside the soft region. The two-gluon dynamics leads to a system with a singularity just above $J = 1$ (often called the QCD pomeron), and the three-gluon case to a singularity with $\mathscr{C} = -1$ just below but very close to $J = 1$, which is identified with the odderon.

There is no convincing experimental evidence for the odderon though there are hints of a difference between $d\sigma/dt$ for pp and $\bar{p}p$ at small t in the ISR data at $\sqrt{s} = 53$ GeV. But all in all the data on total cross-sections and on the ratio of real to imaginary parts of forward spin-averaged amplitudes suggest that the coupling of the odderon to $\phi_1 + \phi_3$ at $t = 0$ is much smaller that that of the pomeron:

$$\frac{|\phi_1 + \phi_3|_{\mathsf{O}}}{|\phi_1 + \phi_3|_{\mathsf{P}}} \lesssim 2\%. \tag{14.1.20}$$

On the one hand, almost nothing is known about the coupling of P or O to the other helicity amplitudes, though Hinotani, Neal, Predazzi and Walters (1979) claimed some evidence for a roughly energy-independent single-helicity-flip amplitude. For a more modern assesment see Buttimore *et al.* (1999).

On the other hand we do have some knowledge about the phases of the amplitudes. Because of the analytic properties of the scattering amplitude in the complex s-plane the phase of an amplitude, in the asymptotic regime, is governed by its energy dependence and its signature τ (Eden, 1971). If the asymptotic behaviour due to an exchange system with signature τ is

$$|A_\tau| \approx s^\alpha (\ln s)^p \qquad \text{as } s \to \infty \tag{14.1.21}$$

then one has for $\tau = +1$

$$A_+ \approx s^\alpha (\ln s)^p e^{-i\pi\alpha/2} \left(1 - \frac{i\pi p}{2 \ln s}\right), \tag{14.1.22}$$

whereas, for $\tau = -1$, the behaviour is

$$A_- \approx i s^\alpha (\ln s)^p e^{-i\pi\alpha/2} \left(1 - \frac{i\pi p}{2 \ln s}\right). \tag{14.1.23}$$

Measurements of spin-dependent observables often have helped and will continue to help to disentangle the dynamical effects.

A case in point is Regge pole theory. Many aspects of the behaviour of the small-t differential cross-sections, their shrinkage etc., in a wide range of reactions were well described by Regge pole exchange. There was even some success with polarizations since the Regge pole exchange amplitudes are not real and possess a natural phase needed to obtain non-zero polarization (see Table A10.5). But in $\pi^- p \to \pi^0 n$ only one Regge pole can be exchanged, the ρ, so that both helicity-flip and non-flip amplitudes have the same phase and the polarization vanishes (see Table A10.1). Nonetheless, significant polarizations were measured.

Another aspect of Regge pole theory that runs counter to spin-dependent results is the property of factorization (Fox and Leader, 1967). In $pp \to pp$, for example, one has

$$\langle \lambda_1', \lambda_2' | T | \lambda_1, \lambda_2 \rangle^{\text{R.pole}} \propto \beta_{\lambda_1' \lambda_1}(t) \beta_{\lambda_2' \lambda_2}(t) s^{\alpha(t)} \tag{14.1.24}$$

where the $\beta(t)$ are called *residue functions.*

This, via (14.1.2), leads to

$$\phi_2 = \phi_5^2 / \phi_1 \tag{14.1.25}$$

so that, from (14.1.6)

$$\phi_2^{\text{R.pole}} \propto t \qquad \text{as } t \to 0. \tag{14.1.26}$$

(A comprehensive account of the spin properties of Regge poles is given in Leader (1969).)

The vanishing of ϕ_2 at $t = 0$ would be a totally dynamic effect, but it and similar predictions do not seem to agree with the data, though it must be said that there is a real scarcity of data at really high energies.

For instance, the transverse cross-section difference $\Delta\sigma_T$ defined in (5.1.12) is proportional to $\text{Im}\,\phi_2$ at $t = 0$. It is certainly not zero in the low to medium energy region, but the rather limited data do suggest that

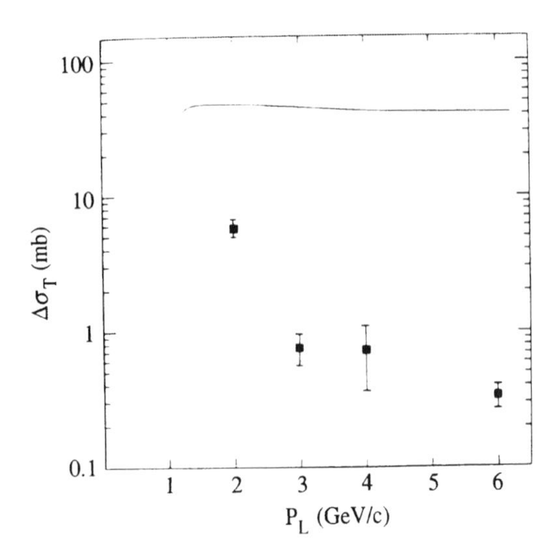

Fig. 14.1 The transverse cross-section difference $\Delta\sigma_T$ for $pp \to pp$. The line in the upper part of the graph gives σ_{tot} (spin ave.). (From de Boer *et al.*, 1975.)

it is decreasing rapidly with energy (see Fig. 14.1.) Another case is the longitudinal cross-section difference $\Delta\sigma_L$ in (5.1.11): it is proportional to Im $(\phi_1 - \phi_3)$ at $t = 0$. But factorization (14.1.24) together with a parity property of the contribution of a single Regge pole to the $pp \to pp$ helicity amplitudes,

$$\langle \lambda_1'; -\lambda_2'|T|\lambda_1; -\lambda_2\rangle = \tau\mathscr{P}\langle \lambda_1'; \lambda_2'|T|\lambda_1; \lambda_2\rangle, \tag{14.1.27}$$

leads to $\phi_1 = \phi_3$ for the dominant poles, which all have $\tau\mathscr{P} = +1$. Hence one would expect $\Delta\sigma_L$ to decrease with energy.

As seen in Fig. 14.2, $\Delta\sigma_L$ has a complicated structure at low-to-medium energies but is decreasing in magnitude fast with energy.

From our present-day perspective, we prefer to think of a Regge exchange contribution as something more complex than a pole, perhaps a so-called *cut* or a pole–cut combination, with a characteristic energy dependence and phase but without the factorization property of its couplings. The decrease of $\Delta\sigma_L$ with energy is then quite compatible with its being controlled by an exchange system with the quantum numbers of the a_1 (see Table 14.1), which is expected to have an effective $\alpha(0) \approx 1/2$.

The behaviour of $\Delta\sigma_T$ is more interesting, since in principle it could receive contributions from both the pomeron and odderon, leading to its growing at higher energies. This will be studied at the RHIC collider and will provide important information about the spin couplings of pomeron and odderon. (A detailed analysis can be found in Leader and Trueman, 2000.)

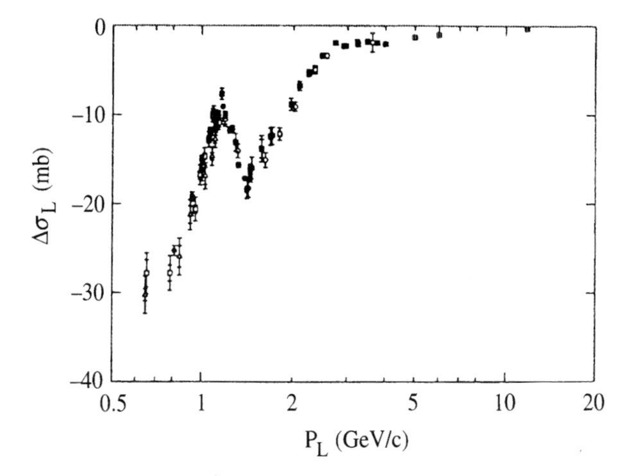

Fig. 14.2 Longitudinal cross-section difference $\Delta\sigma_L$ for $pp \to pp$. (From Grosnick *et al.*, 1997.)

It turns out that the study of spin dependence is greatly facilitated by studying the Coulomb interference region, where the interference between the hadronic amplitudes and the *known* electromagnetic amplitude helps in the process of identifying the details of the hadron dynamics.

14.2 Electromagnetic interference revisited

We shall now explain the newly discovered fact (Buttimore *et al.*, 1999) that pp elastic scattering is *self-calibrating*, in the sense that a sufficient number of measurements of spin-dependent observables at very small t, in the Coulomb interference region, allows one not only to determine most of the helicity amplitudes but also the polarization of the beam and target.

It will be seen that the method involves the taking of several ratios of possibly very small quantities, so that the precision needed may be difficult to achieve experimentally. However, so little is known about the amplitudes at high energy that we are unable to quantify this matter. We shall follow the treatment of Buttimore *et al.* (1999).

At the very small values of t in which we are interested, the interference between the strong and electromagnetic forces can be taken into account by writing

$$\phi_j = \phi_j^{\mathrm{N}} e^{-i\delta} + \phi_j^{\mathrm{EM}} \tag{14.2.1}$$

where the ϕ_j^{EM} are the one-photon exchange amplitudes given in (8.1.5), multiplied by $s/(2\sqrt{\pi})$. The Coulomb phase δ was shown by Buttimore, Gotsman and Leader (1978) to be the same for all helicity amplitudes. It is very small and we shall ignore it in our approximate treatment.

We assume the beam has polarization P and the target P'. In a pp collider it will be true to an extremely high degree of accuracy that $P = \pm P'$, depending on the machine setting. We consider the experimentally measured asymmetries, which are given by $PA d\sigma/dt$, $PP'A_{NN} d\sigma/dt$ etc. These contain singular terms at $t \to 0$ coming from interference between the one-photon and the hadronic amplitudes. To order α the asymmetries involving A_{NN}, A_{SS} and A_{LL} are singular like $1/t$ whereas A and A_{SL} go like $1/\sqrt{-t}$.[1] From the work of Buttimore, Gotsman and Leader (1978) we can write, for very small t,

$$-\frac{m\sqrt{-t}}{\sigma_{\mathrm{tot}}} PA \frac{d\sigma}{dt} = \alpha a_N + \frac{\sigma_{\mathrm{tot}}}{8\pi} b_N t + \cdots \tag{14.2.2}$$

$$\frac{t}{\sigma_{\mathrm{tot}}} PP' A_{LL} \frac{d\sigma}{dt} = \alpha a_{LL} + \frac{\sigma_{\mathrm{tot}}}{8\pi} b_{LL} t + \cdots \tag{14.2.3}$$

[1] The connection between these asymmetry parameters and the CM parameters is given in Table A10.7, and the relation to the helicity amplitudes then follows via Table 10.5. Note that here A stands for $A^{(A)}_{\mathrm{ARG}}$.

$$\frac{t}{\sigma_{\text{tot}}} PP' A_{NN} \frac{d\sigma}{dt} = \alpha a_{NN} + \frac{\sigma_{\text{tot}}}{8\pi} b_{NN} t + \cdots \tag{14.2.4}$$

$$\frac{t}{\sigma_{\text{tot}}} PP' A_{SS} \frac{d\sigma}{dt} = \alpha a_{SS} + \frac{\sigma_{\text{tot}}}{8\pi} b_{SS} t + \cdots \tag{14.2.5}$$

$$-\frac{m\sqrt{-t}}{\sigma_{\text{tot}}} PP' A_{SL} \frac{d\sigma}{dt} = \alpha a_{SL} + \frac{\sigma_{\text{tot}}}{8\pi} b_{SL} t + \cdots. \tag{14.2.6}$$

Expressions for the a_j and b_j are given in Table 14.2 in terms of the following rescaled amplitudes, which may be taken independent of t:

$$R_2 + iI_2 = \frac{\phi_2^N(s,t)}{2\ \text{Im}\ \phi_+^N(s)} \tag{14.2.7}$$

$$R_- + iI_- = \frac{\phi_-^N(s,t)}{\text{Im}\ \phi_+^N(s)} \tag{14.2.8}$$

$$R_5 + iI_5 = \left(\frac{m}{\sqrt{-t}}\right) \frac{\phi_5^N(s,t)}{\text{Im}\ \phi_+^N(s)} \tag{14.2.9}$$

where

$$\phi_\pm^N(s,t) = \tfrac{1}{2}\left[\phi_1^N(s,t) \pm \phi_3^N(s,t)\right] \tag{14.2.10}$$

and

$$\text{Im}\ \phi_+^N(s) \equiv \text{Im}\ \phi_+^N(s, t=0) = \frac{s}{8\pi}\sigma_{\text{tot}}, \tag{14.2.11}$$

the latter via the optical theorem. Also, as usual,

$$\rho = \frac{\text{Re}\ \phi_+^N(s,0)}{\text{Im}\ \phi_+^N(s,0)}. \tag{14.2.12}$$

In Table 14.2, terms of order αt are omitted. In this approximation $A_{NN} = A_{SS}$ and measurement of these quantities could be used as a check on the validity of the approximations. Indeed, the entire procedure can

Table 14.2. Expressions for the coefficients a_j and b_j in eqns (14.2.2) to (14.2.6). κ is the anomalous magnetic moment of the proton

Observable	a_j	b_j
A_{NN}	$PP'R_2$	$PP'[R_2(\rho + R_-) + I_2(1 + I_-)]$
A_{LL}	$PP'R_-$	$PP'[\rho R_- + I_- + R_2^2 + I_2^2]$
A_{SL}	$PP'\frac{\kappa}{2}(R_2 + R_-)$	$PP'[R_5(R_2 + R_-) + I_5(I_2 + I_-)]$
A	$P[I_5 - \frac{\kappa}{2}(1 + I_2)]$	$P[I_5(\rho + R_2) - R_5(1 + I_2)]$

be tested by checking whether the measured quantities in (14.2.2)–(14.2.6) are linear functions of t.

In addition to the above observables we need to know the cross-section differences $\Delta\sigma_L$ and $\Delta\sigma_T$ defined in (5.1.11) and (5.1.12). The measured observables we use are

$$\delta_T \equiv -\frac{1}{2} P P' \frac{\Delta\sigma_T}{\sigma_{\text{tot}}} = P P' I_2 \tag{14.2.13}$$

and

$$\delta_L \equiv \frac{1}{2} P P' \frac{\Delta\sigma_L}{\sigma_{\text{tot}}} = P P' I_-. \tag{14.2.14}$$

Having measured ρ, a_{NN}, a_{LL}, δ_L and δ_T we can substitute in the expression for b_{LL} to obtain

$$b_{LL} = \rho a_{LL} + \delta_L + \frac{a_{NN}^2 + \delta_T^2}{P P'} \tag{14.2.15}$$

from which one obtains an expression for the polarization,

$$P P' = \frac{a_{NN}^2 + \delta_T^2}{b_{LL} - \rho a_{LL} - \delta_L}, \tag{14.2.16}$$

whence, since it will be known whether $P' = P$ or $P' = -P$, one can obtain P. The sign ambiguity should be innocuous.

Knowing PP' one can now obtain the values of R_2, I_2, R_-, I_- from a_{NN}, δ_T, a_{LL} and δ_L respectively.

In practice, it may turn out that the errors on PP' obtained from (14.2.16) are unacceptably large. In that case there is an alternative procedure, which should be more accurate.

The analysing power of the reaction is given by

$$\begin{aligned}\frac{m\sqrt{-t}}{\sigma_{\text{tot}}} A \frac{d\sigma}{dt} &= \left(\frac{\kappa}{2} - I_5 + \frac{\kappa}{2} I_2\right) \\ &\quad + \frac{\sigma_{\text{tot}}}{8\pi} \left[R_5(1 + I_2) - I_5(\rho + R_2)\right] t \end{aligned} \tag{14.2.17}$$

and this is expected to be very largely dominated by the term $\kappa/2$. Thus the other terms in (14.2.17) are small corrections and large errors on them may be unimportant. We already have values for ρ, R_2 and I_2. There is a lengthy algebraic procedure for estimating R_5 and I_5 from the measurement of a_{SL}, b_{SL}, a_N and b_N and which uses (14.2.16). One finds

$$I_5 = \frac{\kappa}{2} \left(\frac{b_{SL}}{a_{SL}} - \frac{b_N}{a_N}\right) \bigg/ \left(\frac{\delta_T + \delta_L}{a_{NN} + a_{LL}} - \frac{b_N/a_N - \rho - a_{NN} O_1}{1 + \delta_T O_1}\right) \tag{14.2.18}$$

$$R_5 = \frac{\kappa}{2} \frac{b_{SL}}{a_{SL}} - \frac{\delta_T + \delta_L}{a_{NN} + a_{LL}} I_5 \tag{14.2.19}$$

where O_1 is the estimate for PP' given in (14.2.16).

Using these should provide a relatively accurate estimate of the analysing power, after which the reaction can be used directly to measure the proton polarization.

A more accurate treatment of the problem, including a discussion of the rôle of the Coulomb phase, can be found in Buttimore, Leader and Trueman (1999).

It turns out that the measurement of spin-dependent observables in the interference region could also be helpful in trying to understand the rôle of the odderon; in particular A_{NN} is sensitive to it. A detailed discussion of this issue is given in Leader and Trueman (2000).

14.3 Elastic scattering at large momentum transfer

Considered as a QCD reaction, elastic proton–proton scattering is manifestly a very complex process. Even in its simplest version, taking into account only the valence quarks, one has to deal with a six-quark → six-quark reaction. Examples of Feynman diagrams for such an interaction are shown in Fig. 14.3 (the Brodsky–Lepage hard-scattering mechanism)

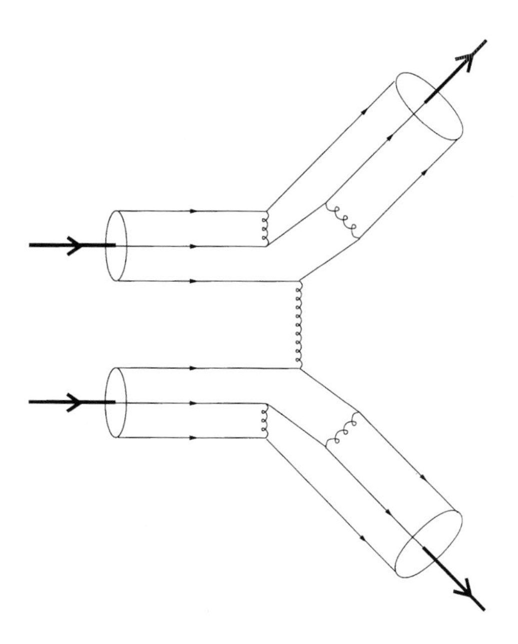

Fig. 14.3 A Brodsky–Lepage diagram for large-momentum-transfer $pp \to pp$ (Brodsky and Lepage, 1980).

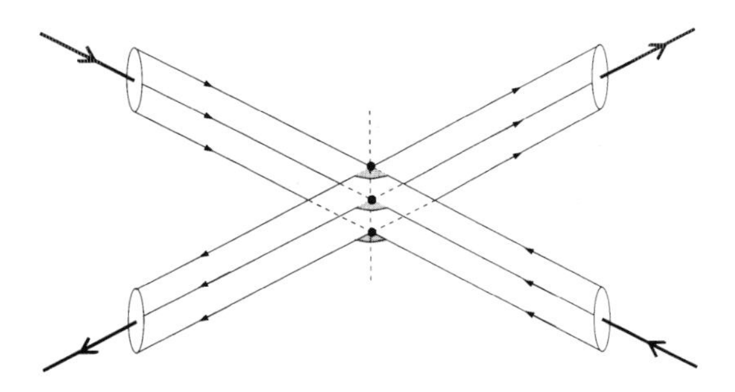

Fig. 14.4 A Landshoff diagram for large-momentum-transfer $pp \to pp$ (Landshoff, 1974.)

and in Fig. 14.4 (the Landshoff mechanism). Despite the complexity it turns out that one can deduce powerful results for the spin dependence in the asymptotic limit where $|s|$ and $|t|$ are $\gg m^2$. The problem, as will become clear, is precisely where one can expect the asymptotic behaviour to set in. It will be seen that the present experimental data badly contradict these asymptotic predictions but that there are theoretical arguments, indicating many subtle effects, which suggest that the present-day experiments are still far from the asymptotic regime. However, while these effects alter the momentum-transfer dependence, it is far from clear whether they affect the spin dependence significantly. Hopefully the RHIC collider, if it can probe large enough momentum transfer, will help to resolve the matter.

14.3.1 The asymptotic behaviour

For an exclusive reaction we need the actual wave function of the quarks that make up a hadron. That is, we require to know the amplitude, shown in Fig. 14.5, for the hadron, momentum $\mathbf{P}$, helicity λ to break up into quarks of momentum $\mathbf{q}_j$ and helicity λ_j. We are considering only reactions with large p_T.

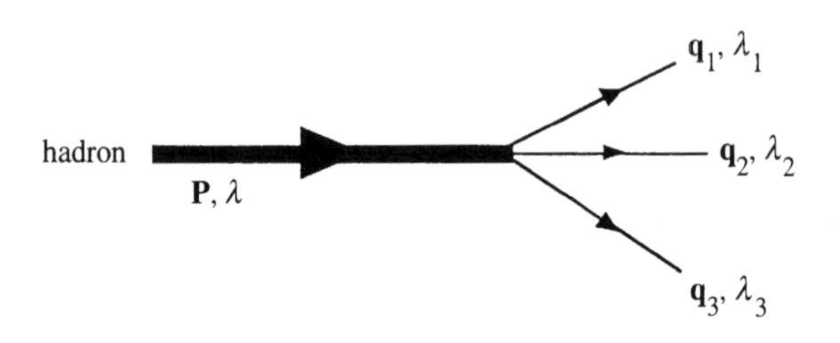

Fig. 14.5. Wave function for three quarks in a hadron.

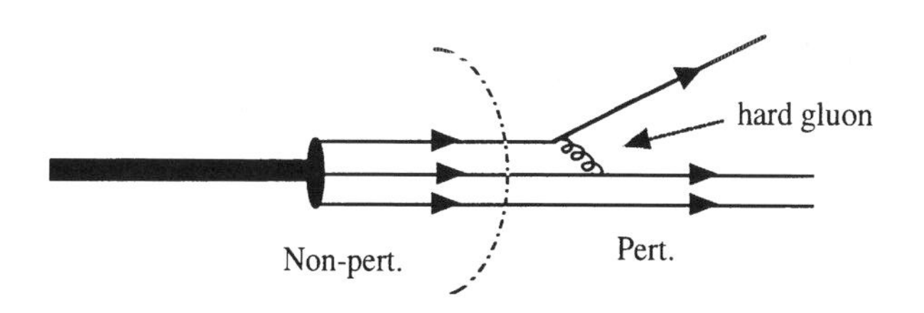

Fig. 14.6. Mechanism for generating a quark with large transverse momentum.

It is believed that the most efficient way to produce large p_T is for each hadron to produce a beam of essentially parallel quarks, which then get a high-p_T kick via a perturbative QCD interaction, shown in Fig. 14.6. So, roughly speaking, the only non-perturbative input is the *soft* amplitude, or wave function, where $\mathbf{P}$ and each $\mathbf{q}_j$ are essentially parallel, say along OZ as shown in Fig. 14.7. Although we cannot compute this soft amplitude we can deduce an important piece of information, as follows. Since all momenta are along OZ any orbital angular momentum must be perpendicular to OZ. Thus the only angular momentum along OZ is spin angular momentum. Conservation of J_z then implies

$$\lambda = \lambda_1 + \lambda_2 + \lambda_3 \tag{14.3.1}$$

for each hadron in the reaction.

Since each quark that interacts perturbatively conserves its helicity (see Section 10.4) we end up with a remarkable result, due to Brodsky and Lepage (1980): in any exclusive reaction

$$A + B \rightarrow C + D + E + \cdots$$

one has

$$\lambda_A + \lambda_B = \lambda_C + \lambda_D + \lambda_E + \cdots, \tag{14.3.2}$$

that is, *total* initial helicity equals *total* final helicity.

More precisely, what plays the rôle of the soft wave function is the *distribution amplitude*

$$\phi(x, Q^2) = \int_0^{Q^2} d^2\mathbf{k}_T \psi(x, \mathbf{k}_T), \tag{14.3.3}$$

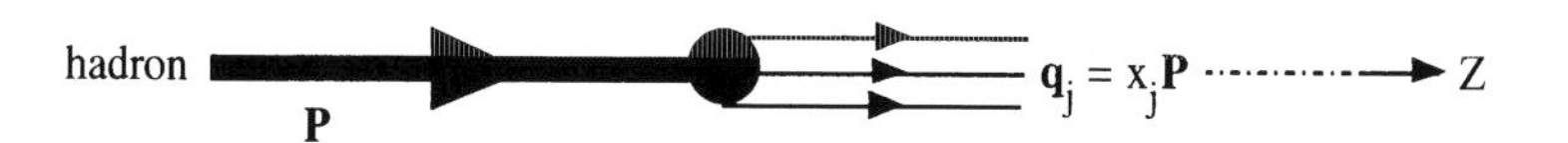

Fig. 14.7 Soft wave function with all quarks essentially parallel to the parent hadron.

i.e. the wave function integrated over a region of the transverse components of the quark momenta. Although the quarks are not strictly parallel to the hadron in this, the integral over $\mathbf{k}_T$ has the effect of eliminating any part of the wave function having $L_z \neq 0$. Thus (14.3.1) and (14.3.2) continue to hold.

Consequences abound! Perhaps the most dramatic example is that the analysing power A in $pp \to pp$ should vanish because it is proportional to the single-flip helicity amplitude ϕ_5:

$$A\frac{d\sigma}{dt} = -\ \mathrm{Im}\ \left[\phi_5^*(\phi_1 + \phi_3 + \phi_2 - \phi_4)\right]. \qquad (14.3.4)$$

The vanishing of A follows since ϕ_5 corresponds to the transition

$$|1/2, 1/2\rangle \to |1/2, -1/2\rangle$$

so that the initial total helicity (= 1) is not equal to the final total helicity (= 0).

Quite contrary to this prediction the analysing power in elastic pp scattering is large all the way out to $p_T^2 \approx 8\ (\mathrm{GeV}/c)^2$. The results of experiments at CERN (Antille *et al.*, 1981) and at the Brookhaven AGS (Crabb *et al.*, 1990) can be seen in Fig. 14.8. (Recall that for $pp \to pp$ the analysing power A is the same as the polarizing power P.)

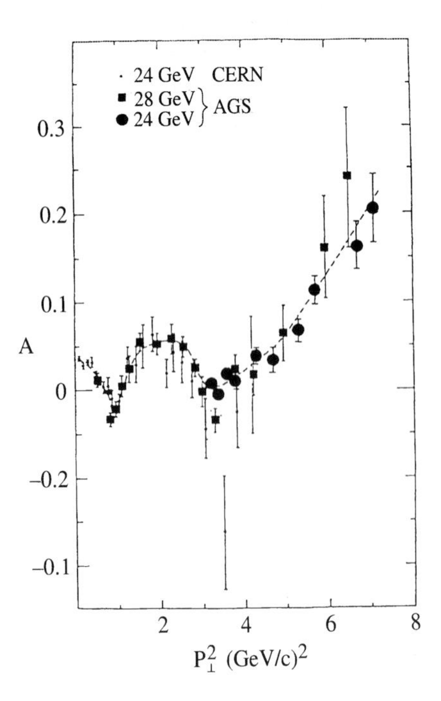

Fig. 14.8. Analysing power for $pp \to pp$. (From Crabb *et al.*, 1990.)

This contradiction between theory and experiment is usually glossed over by claiming that p_T is too small to expect the asymptotic predictions to hold. This may well be correct, though many papers have pointed out that other 'asymptotic' predictions, in inclusive and semi-inclusive reactions, seem to work at precociously low scales, 1–2 $(\text{GeV}/c)^2$. In fact, as we shall briefly explain, the asymptotic behaviour in exclusive reactions *should* be expected to be less precocious, but if the trend in A shown in Fig. 14.8 continues to much larger values of p_T^2 we will seriously have to question whether our QCD picture of the strong interactions is really correct.

14.3.2 Complications of exclusive reactions

The proton–proton amplitude is immensely complicated: there are some 100 000 Feynman diagrams of the Brodsky–Lepage type. Hence most analyses of the relevance of the asymptotic description have focussed on the much simpler question of electromagnetic form factors at large momentum transfer. A very clear discussion can be found in Kroll (1994) and Jakob and Kroll (1993), whose treatment we follow.

Consider, for simplicity, the pion electromagnetic form factor $F_\pi(Q^2)$. The asymptotic behaviour, as $Q^2 \to \infty$, is supposed to be controlled by the Feynman diagram in Fig. 14.9, where T_H is the hard scattering amplitude shown in Fig. 14.10 and the hadron $\to$ quark, antiquark vertices are soft wave functions analogous to those in Fig. 14.7. This leads, in its simplest form, to the remarkable result (Brodsky and Lepage, 1980)

$$Q^2 F_\pi(Q^2) \xrightarrow{Q^2\to\infty} 8\pi\alpha_s f_\pi^2 \tag{14.3.5}$$

where $f_\pi = 133$ MeV.

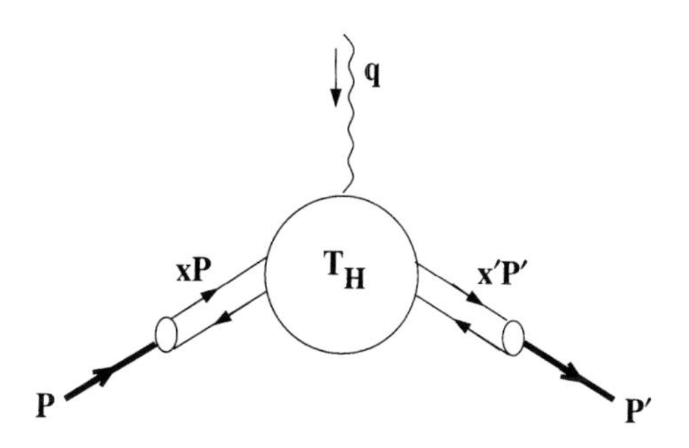

Fig. 14.9. Hard-scattering diagram for pion form factor.

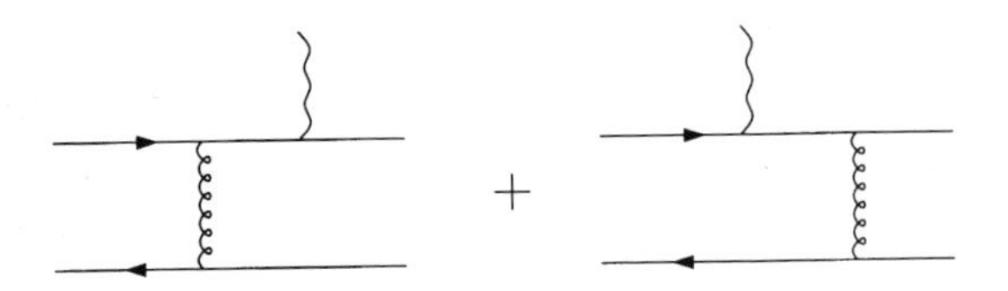

Fig. 14.10. Lowest-order Feynman diagrams for T_{H}, Fig. 14.9.

A major criticism of this approach was put forward by Isgur and Llewellyn-Smith (1989). Firstly, they showed that in the asymptotic calculation a very large fraction of the result was actually generated from a kinematic region that is not perturbative. The point is that the gluon virtuality in T_{H} is of order $xx'Q^2$, not Q^2, so that, for part of the range of integration x and x' are small, we are in a region of small virtuality and a perturbative treatment cannot be justified. *A priori* this is not surprising. What is a shock is the magnitude of the inconsistency. For example, for $F_\pi(Q^2)$ at $Q^2 = 4\,(\mathrm{GeV}/c)^2$ only 13% of the result comes from a region where the gluon virtuality is $> 1\,(\mathrm{GeV}/c)^2$.

Secondly, Isgur and Llewellyn-Smith pointed out that the contribution from the overlap of initial and final soft wave functions (see Fig. 14.11), given a reasonably gaussian k_T-dependence corresponding to a hadron radius of order 1 fm, is much larger that the asymptotic result (14.3.5). The reason is that even if the k_T-dependence of the wave function cuts off like a gaussian, the overlap only decreases as an inverse power of Q, the precise behaviour depending on the x-dependence of the wave function.

Their astounding conclusion was an estimate that for $\pi N \to \pi N$ the asymptotic behaviour would only set in for $p_T^2 \geq 10^8\,(\mathrm{GeV}/c)^2$!

This, however, is not the end of the story.

In a series of groundbreaking papers Sterman and collaborators (Botts and Sterman, 1989; Li and Sterman, 1992; Li, 1993) demonstrated that it is important to take into account the transverse momentum dependence in T_H (neglected in deriving (14.3.5)) and at the same time to include

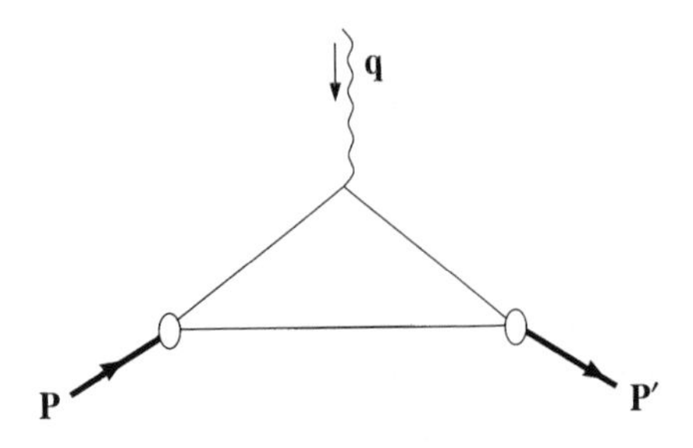

Fig. 14.11. Wave function overlap contribution to the pion form factor.

the effects of so-called Sudakov suppression (Sudakov, 1956), which we shall explain presently. This has a major effect, on the one hand largely negating the Isgur and Llewellyn-Smith criticisms, on the other hand making the treatment of elastic scattering vastly more complicated than in the asymptotic approach.

To understand the physics of Sudakov suppression, recall that in classical electrodynamics the scattering amplitude for non-forward $e+e \to e+e$ is exactly zero. The reason is that an accelerated electron always radiates photons, so the pure process $ee \to ee$ cannot occur.

The field-theoretic analogue is that $e+e \to e+e$ is highly suppressed at large momentum transfer by a 'Sudakov double logarithm'. For example, for the electromagnetic form factor of an electron at large Q^2 the suppression in the amplitude is (Sudakov, 1956)

$$\exp\left\{-\frac{e^2}{2\pi}\left[\ln\left(\frac{Q^2}{m_e^2}\right)\right]^2\right\}, \tag{14.3.6}$$

which goes to zero faster that any inverse power of Q.

Because of the running coupling in QCD, the analogous suppression for the elastic form factor of a quark gets softened to (Mueller, 1981; Sen, 1983)

$$\exp\left\{\frac{-4C_{\text{F}}}{11-(2/3)n_f}\ln\left(\frac{Q^2}{\lambda^2}\right)\ln\ln\left(\frac{Q^2}{\lambda^2}\right)\right\} \tag{14.3.7}$$

where λ is an infrared cut-off, and $C_F = 4/3$ for QCD.

In the electron case the probability of emission of a *finite* number of photons in total is also highly suppressed, but the probability to emit any number of photons within some specified energy range is less suppressed, and is the quantity that would be relevant, given the finite energy resolution of any electron detector. But, in the case of the elastic form factor of a hadron the scattered quarks cannot radiate gluons, since in the final state they have to combine to produce the lowest Fock state of the hadron. At first sight, therefore, it seems that the full suppression (14.3.7) should apply.

However, just as an electrically neutral point particle does not suffer the suppression (14.3.6) so a colour-neutral point-like object will not be suppressed by (14.3.7). Hadrons are, of course, colour neutral, but they are extended objects, not point-like. It is then intuitively clear what to expect. For small separations of the constituents, i.e. in the region where perturbative QCD is reliable, there will be little suppression, whereas in the non-perturbative region of large separations the Sudakov factor will drastically suppress the contribution.

The relevant separation turns out to be in a direction perpendicular to the hadron momentum, so one introduces the b-space transform of the wave function,

$$\hat{\psi}(x, \mathbf{b}) = \frac{1}{(2\pi)^2} \int d^2\mathbf{k}_T e^{-i\mathbf{b}\cdot\mathbf{k}_T} \psi(x, \mathbf{k}_T), \tag{14.3.8}$$

and the dominant term in the Sudakov suppression factor then takes the form

$$\exp\left\{-\frac{2}{3\beta_1} \ln\left(\frac{xQ}{\sqrt{2}\Lambda_{\mathrm{QCD}}}\right)\left[\ln\ln\left(\frac{xQ}{\sqrt{2}\Lambda_{\mathrm{QCD}}}\right) - \ln\ln\left(\frac{1}{b\Lambda_{\mathrm{QCD}}}\right)\right]\right\} \tag{14.3.9}$$

for $b \leq 1/\Lambda_{\mathrm{QCD}}$, where $\beta_1 = (33 - 2n_f)/12$.

The expression in (14.3.9) decreases to zero as b grows from zero to $1/\Lambda_{\mathrm{QCD}}$ and is taken as equal to zero for $b > 1/\Lambda_{\mathrm{QCD}}$. Thus the non-perturbative region of large b is damped out. Moreover it can be shown that the scale to use in $\alpha_s(\mu^2)$ in T_H is not $\mu^2 = xx'Q^2$ but, rather, $\max\left\{xx'Q^2, 1/b^2\right\}$, so that α_s remains perturbatively small in the calculation.

The net result is that the perturbative calculation, of course vastly more complicated now, should be trustworthy for $Q^2 \gtrsim 4\,(\mathrm{GeV}/c)^2$.

Analogous considerations apply to elastic scattering at large momentum transfer, usually expressed in terms of scattering at fixed angle θ in the CM.

The naive perturbative treatment of the Brodsky–Lepage diagrams leads to cross-sections that obey the *dimensional counting-rules* (ignoring logarithms)

$$\frac{d\sigma}{dt}(AB \to CD) = \left[\frac{\alpha_s(p_T^2)}{s}\right]^{n-2} f(\theta) \tag{14.3.10}$$

where n is the total number of partons in the lowest Fock states of all the particles. Thus for $\pi N \to \pi N$ we have $n = 10$, implying an s^{-8} behaviour, whereas for $NN \to NN$ we have $n = 12$, yielding s^{-10}.

There is not a great deal of data on large-momentum-transfer elastic scattering, but what does exist is in reasonable agreement with the counting rules.

However, the Landshoff-type diagrams, Fig. 14.4, lead to a slower decrease with s at fixed θ (Landshoff, 1974). For example, for $NN \to NN$ the behaviour is s^{-8}. It was argued, however, that the normalization of these contributions would be much smaller than that of the Brodsky–Lepage diagrams (Brodsky and Lepage, 1980). In fact Botts and Sterman (1989) demonstrated that Sudakov suppression is very important for the

Landshoff diagrams and they estimated that for $NN \to NN$ the naive behaviour is modified to $s^{-9.66}$, quite close to the dimensional counting-rule result s^{-10}.

So, as regards cross-sections, although it has not yet been possible to calculate the hundreds of thousands of Feynman diagrams involved, at least for the broad pattern of decrease with increasing momentum transfer there seems to be agreement between theory and experiment. Moreover, the inclusion of Sudakov effects negates much of the criticism against the premature use of the perturbative results. There has thus been considerable progress in understanding, at a deeper level, the large-p_T dependence.

14.3.3 Summary

Where does all this leave the problem of the analysing power in $pp \to pp$? At first sight we are no better off than before, since the sophisticated ingredients now included appear to have no effect upon the helicity rule (14.3.2). However, an interesting development was the discovery by Gousset, Pire and Ralston (1996), in the context of meson–meson scattering, that the Landshoff-type diagrams permit wave functions with non-zero L_z, which are not suppressed by $1/s$ as they are in the Brodsky–Lepage hard scattering diagrams. There is some suppression, but it is much milder, $\approx s^{-0.55}$. This would imply that the helicity rule (14.3.2) only becomes valid at extremely large momentum transfer. Unfortunately this discovery does not directly resolve the problem of the proton–proton analysing power, since it turns out that the permitted change in total helicity has to be an *even* number, at least for the meson–meson case studied.

It is hoped, though not yet demonstrated, that this kind of mechanism will lead to the possibility of single helicity-flip in the Landshoff diagrams for $pp \to pp$. There remains the question of generating a phase difference between the flip and non-flip amplitudes in (14.3.4). The Sudakov factors indeed possess a non-zero phase, but whether there is a significant difference between the phase of ϕ_5 and the non-flip amplitudes is unclear. Our own, perhaps simplistic, guess is that there will be no difference of phase.

It is also possible to generate a single helicity-flip if the nucleon is regarded as a quark–diquark system that includes a component in the wave function corresponding to a spin-1 vector diquark. In this way, Kroll and collaborators (see e.g. Jakob, Kroll, Schürmann and Schweiger, 1993) have been able to obtain, amongst other things, a reasonable description of the Pauli electromagnetic form factor $F_2(Q^2)$ of the proton, which involves a nucleon single helicity-flip matrix element.

Goloskokov and Kroll (1999) have attempted to estimate the analysing power in $pp \to pp$ using the quark–diquark picture. The helicity non-flip amplitude is modelled phenomenologically so that it corresponds to what one might expect from multiple pomeron exchange (Section 14.1). The helicity-flip amplitude ϕ_5 is calculated perturbatively using two-gluon exchange diagrams. The proton Fock state contains both scalar and vector diquarks, but to simplify the calculation only the scalar is used in estimating the non-flip vertex in ϕ_5, and only the vector in the flip vertex.

With all the approximations made, this model is not expected nor tuned to agree with the data, but numerical studies show that it does provide an acceptable, approximately energy-independent, analysing power, which, however, eventually decreases with increasing momentum transfer and finally merges into the Brodsky–Lepage hard scattering result.

The quark–diquark picture is best regarded as a model for higher-twist effects and as such would lead to an analysing power that ultimately decreases like $1/s$ at fixed angle. In the Gousset *et al.* picture, if it can really produce a non-zero analysing power, that too will eventually tend to zero, but probably more slowly than $1/s$.

In either case it seems unavoidable that ultimately QCD demands that $A \to 0$ as p_T increases. But we have no concrete predictions for A nor for the scale at which the decrease should begin to be seen. Given the exciting experimental possibilities about to open up at the RHIC collider, this is, alas, a most frustrating state of affairs and we can only hope for a major theoretical breakthrough.

Appendix 1

The irreducible representation matrices for the rotation group and the rotation functions $d^j_{\lambda\mu}(\theta)$

We present here some useful properties of the unitary matrices $\mathscr{D}^{(j)}_{m'm}(r)$, $j = 0, 1/2, 1, 3/2, \ldots, -j \leq m, m' \leq j$, which form an irreducible representation of the operation corresponding to an arbitrary rotation r. We also give a simple method for calculating them. Our conventions for describing r are explained in Section 1.1. We follow the notation of Jacob and Wick (1959). Detailed discussions of the rotation group can be found in the books of Rose (1957), whose notation is the same as ours, and of Edmonds (1957), whose notation differs from ours, as will be explained below.

The most general rotation, through Euler angles α, β, γ, is given in (1.2.17). The unitary operator $U(r)$ corresponding to this rotation can be expressed in terms of the angular momentum operators J_x, J_y, J_z, which are the generators of rotations. One has

$$U[r(\alpha, \beta, \gamma)] = e^{-i\alpha J_z} e^{-i\beta J_y} e^{-i\gamma J_z}. \tag{A1.1}$$

It follows from (1.1.20) that

$$\mathscr{D}^{(j)}_{\lambda\mu}(\alpha, \beta, \gamma) = e^{-i\lambda\alpha} d^j_{\lambda\mu}(\beta) e^{-i\mu\gamma} \tag{A1.2}$$

where[1]

$$d^j_{\lambda\mu}(\beta) = \left\langle j\lambda | e^{-i\beta J_y} | j\mu \right\rangle. \tag{A1.3}$$

Note that for clarity we are here using λ, μ instead of m' and m respectively.

The d-functions enjoy several symmetry properties:

$$d^j_{\lambda\mu}(\beta) = d^j_{-\mu-\lambda}(\beta) = (-1)^{\lambda-\mu} d^j_{\mu\lambda}(\beta) \tag{A1.4}$$

[1] Note that the functions $\mathscr{D}^{(j)}_{\lambda\mu}(\alpha, \beta, \gamma)$ and $d^j_{\lambda\mu}(\beta)$ in Edmond's book correspond to our $\mathscr{D}^{(j)}_{\lambda\mu}(-\alpha, -\beta, -\gamma)$ and $d^j_{\lambda\mu}(-\beta)$ respectively.

$$d^j_{\lambda\mu}(\beta) = (-1)^{j+\lambda} d^j_{\lambda,-\mu}(\pi - \beta) \tag{A1.5}$$

$$d^j_{\lambda\mu}(-\beta) = d^j_{\mu\lambda}(\beta). \tag{A1.6}$$

Procedures for the computation of the d-functions are described in Edmonds (1957) and in Jacob and Wick (1959). However, as we shall explain below, the disadvantage of these methods is that they produce expressions for the d-functions that do not show explicitly certain key features, for example that

$$d^j_{\lambda\mu}(\beta) = (\sin\beta/2)^{|\lambda-\mu|}(\cos\beta/2)^{|\lambda+\mu|} \times (\text{polynomial in } z = \cos\beta). \tag{A1.7}$$

The angular factors are of crucial importance for $\beta \to 0$ or π, as discussed in Chapter 4. We shall therefore modify the published methods so as to make explicit the structure (A1.7).

Our starting point is the relation given in Jacob and Wick (1959):

$$d^j_{\lambda,\mu\pm1}(\beta) = \frac{1}{\sqrt{(j\pm\mu+1)(j\mp\mu)}} \times \left(-\frac{\lambda}{\sin\beta} + \mu\cot\beta \mp \frac{d}{d\beta}\right) d^j_{\lambda,\mu}(\beta). \tag{A1.8}$$

Let us define

$$d^j_{\lambda,\mu}(\beta) \equiv (\sin\beta/2)^{|\lambda-\mu|}(\cos\beta/2)^{|\lambda+\mu|} \times \left[\frac{(j-\lambda)!(j-\mu)!}{(j+\lambda)!(j+\mu)!}\right]^{1/2} P^j_{\lambda,\mu}(\cos\beta) \tag{A1.9}$$

where $P^j_{\lambda,\mu}$ is a polynomial in $\cos\beta$. Then after some algebra (A1.8) can be written, for the case which will be of interest to us, as

$$d^j_{\lambda,\mu+1}(\beta) = \tfrac{1}{4}(\sin\beta/2)^{|\lambda-\mu|-1}(\cos\beta/2)^{|\lambda+\mu|-1} \times \left[\frac{(j-\lambda)!(j-\mu-1)!}{(j+\lambda)!(j+\mu+1)!}\right]^{1/2} \Big\{[(\mu-\lambda)-|\mu-\lambda|](1+\cos\beta) - [(\mu+\lambda)-|\mu+\lambda|](1-\cos\beta)2\sin^2\beta\frac{d}{d\cos\beta}\Big\} \times P^j_{\lambda,\mu}(\cos\beta). \tag{A1.10}$$

Now, the symmetry properties (A1.4) and (A1.5) imply that we only require expressions for $d^j_{\lambda\mu}$ for positive values of λ and μ. And (A1.6) implies that we can, in fact, choose to work with $\lambda \geq \mu \geq 0$. This will turn out to be very helpful since we have simple expressions for $d^j_{\lambda 0}$ when $j = l =$ integer and for $d^j_{\lambda,1/2}$ when $j = l + 1/2 =$ half-integer, and the other $d^l_{\lambda,\mu}$ for $\mu > 0$ or $\mu > 1/2$ can then be built up from these. In this case (A1.10) becomes a simple recursion formula for the polynomial $P^j_{\lambda,\mu}(\cos\beta)$.

Namely, for $\lambda \geq \mu + 1 \geq 1$ on the left-hand side one has:

$$P^j_{\lambda,\mu+1}(\cos\beta) = \left(\mu - \lambda + (1-\cos\beta)\frac{d}{d\cos\beta}\right) P^j_{\lambda,\mu}(\cos\beta). \qquad \text{(A1.11)}$$

The starting functions $P^j_{\lambda 0}$ or $P^j_{\lambda 1/2}$ are obtained as follows.

For $j = l =$ integer one has, for $\lambda \geq 0$,

$$d^l_{\lambda 0}(\beta) = (\sin\beta/2)^\lambda(\cos\beta/2)^\lambda(-2)^\lambda \times \sqrt{\frac{(l-\lambda)!}{(l+\lambda)!}}\frac{d^\lambda}{d\cos\beta^\lambda}P_l(\cos\beta) \qquad \text{(A1.12)}$$

where $P_l(\cos\beta)$ are the usual Legendre polynomials. Via (A1.9) we then have, for $\lambda \geq 0$

$$P^l_{\lambda 0}(\cos\beta) = (-2)^\lambda \frac{d^\lambda}{d\cos\beta^\lambda}P_l(\cos\beta). \qquad \text{(A1.13)}$$

For $j = l + 1/2 =$ half-integer, one can start with the relation given in Jacob and Wick (1959)

$$d^j_{\lambda,1/2}(\beta) = \frac{1}{\sqrt{j+1/2}}\left\{\sqrt{j+\lambda}\cos\beta/2\ d^l_{\lambda-1/2,0}(\beta) + \sqrt{j-\lambda}\sin\beta/2\ d^j_{\lambda+1/2,0}(\beta)\right\}, \qquad \text{(A1.14)}$$

which using (A1.9), can be rewritten, for $\lambda \geq 1/2$, $j = l + 1/2$, as

$$P^j_{\lambda,1/2}(\cos\beta) = (j+\lambda)P^l_{\lambda-1/2,0}(\cos\beta) + \frac{1-\cos\beta}{2}P^l_{\lambda+1/2,0}(\cos\beta). \qquad \text{(A1.15)}$$

Thus starting with (A1.13) or (A1.15) one can build up the required $P^j_{\lambda,\mu}(\cos\beta)$, from which the d-functions are obtained via (A1.9). In this approach all the d-functions appear with the correct angular factors explicit.

We list a few of the most often used d-functions.

- $j = 1/2$

$$d^{1/2}_{1/2,1/2}(\beta) = \cos\beta/2 \qquad \text{(A1.16)}$$

- $j = 1$

$$d^1_{11}(\beta) = \frac{1+\cos\beta}{2} \qquad d^1_{10}(\beta) = -\frac{\sin\beta}{\sqrt{2}} \qquad \text{(A1.17)}$$

$$d^1_{00}(\beta) = \cos\beta \qquad \text{(A1.18)}$$

- $j = 3/2$

$$d^{3/2}_{3/2,3/2}(\beta) = 6\cos^3\beta/2$$
$$d^{3/2}_{3/2,1/2}(\beta) = -\sqrt{3}(\sin\beta/2)(\cos^2\beta/2) \tag{A1.19}$$
$$d^{3/2}_{1/2,1/2}(\beta) = \tfrac{1}{2}\cos\beta/2(3\cos\beta - 1) \tag{A1.20}$$

- $j = 2$

$$d^2_{22}(\beta) = \frac{1}{4}(1+\cos\beta)^2 \qquad d^2_{21}(\beta) = -\frac{\sin\beta}{2}(1+\cos\beta) \tag{A1.21}$$
$$d^2_{20}(\beta) = \frac{1}{2}\sqrt{\frac{3}{2}}\sin^2\beta \tag{A1.22}$$
$$d^2_{11}(\beta) = \frac{1}{2}(1+\cos\beta)(2\cos\beta - 1) \tag{A1.23}$$
$$d^2_{10}(\beta) = -\sqrt{\frac{3}{2}}\sin\beta\cos\beta \tag{A1.24}$$
$$d^2_{00}(\beta) = \frac{1}{2}(3\cos^2\beta - 1). \tag{A1.25}$$

Appendix 2
Homogeneous Lorentz transformations and their representations

We present here a brief discussion of the homogeneous Lorentz transformations and some of their finite dimensional representation matrices.

A2.1 The finite-dimensional representations

The generators of rotations $\hat{J}_i$ and boosts $\hat{K}_i$, introduced in (1.2.1), can be shown (see, for example, Gasiorowicz, 1967) to satisfy the following commutation relations:

$$\begin{aligned} \left[\hat{J}_j, \hat{J}_k\right] &= i\epsilon_{jkl}\hat{J}_l \\ \left[\hat{J}_j, \hat{K}_k\right] &= i\epsilon_{jkl}\hat{K}_l \\ \left[\hat{K}_j, \hat{K}_l\right] &= -\, i\epsilon_{jkl}\hat{J}_l. \end{aligned} \tag{A2.1}$$

If we now define

$$\hat{\mathbf{A}} \equiv \tfrac{1}{2}\left(\hat{\mathbf{J}} + i\hat{\mathbf{K}}\right) \qquad \hat{\mathbf{B}} \equiv \tfrac{1}{2}\left(\hat{\mathbf{J}} - i\hat{\mathbf{K}}\right) \tag{A2.2}$$

then $\hat{\mathbf{A}}, \hat{\mathbf{B}}$ behave like angular momentum operators, and they commute with each other:

$$\begin{aligned} \left[\hat{A}_j, \hat{A}_k\right] &= i\epsilon_{jkl}\hat{A}_l \\ \left[\hat{B}_j, \hat{B}_k\right] &= i\epsilon_{jkl}\hat{B}_l \\ \left[\hat{A}_j, \hat{B}_k\right] &= 0. \end{aligned} \tag{A2.3}$$

The most general Lorentz transformation is of the form

$$U(\boldsymbol{\vartheta}, \boldsymbol{\alpha}) = \exp\left(-i\boldsymbol{\vartheta} \cdot \hat{\mathbf{J}} - i\boldsymbol{\alpha} \cdot \hat{\mathbf{K}}\right). \tag{A2.4}$$

The vector $\boldsymbol{\vartheta}$ specifies a positive rotation through angle θ about an axis along $\boldsymbol{\vartheta}$. The vector $\boldsymbol{\alpha}$ specifies a pure boost of speed $\beta = \tanh\alpha$ along the direction of $\boldsymbol{\alpha}$.

Equation (A2.4) can be rewritten as

$$U(\vartheta, \alpha) = \exp\left[-\hat{\mathbf{A}} \cdot (\alpha + i\vartheta) + \hat{\mathbf{B}} \cdot (\alpha - i\vartheta)\right]$$
$$= \exp\left[-i\hat{\mathbf{A}} \cdot (\vartheta - i\alpha)\right] \exp\left[-i\hat{\mathbf{B}} \cdot (\vartheta + i\alpha)\right] \qquad \text{(A2.5)}$$

the last step following because the $\hat{A}_j$ commute with the $\hat{B}_j$.

As discussed in Appendix 1, the $(2l + 1)$-dimensional representation matrices of the rotation group are the matrix elements of the rotation operator. Here we are not using the Euler angles to specify the rotation, but that is irrelevant. The matrix $\mathscr{D}^{(l)}_{m'm}(\vartheta)$ representing the rotation operator

$$U[(r(\vartheta)] \equiv e^{-i\hat{\mathbf{J}}\cdot\vartheta}$$

is given by

$$\mathscr{D}^{(l)}_{m'm}(\vartheta) = \left\langle j, m' | e^{-i\hat{\mathbf{J}}\cdot\vartheta} | j, m \right\rangle \qquad \text{(A2.6)}$$

with $-j \le m, m' \le j$ and $j =$ integer or half-integer.

From (A2.5) and (A2.6) we see that we can represent the Lorentz transformation $U(\vartheta, \alpha)$ by the $(2A + 1)(2B + 1)$-dimensional matrix

$$\mathscr{D}^{(A,B)}_{a'b',ab}(\vartheta, \alpha) \equiv \mathscr{D}^{(A)}_{a'a}(\vartheta - i\alpha)\mathscr{D}^{(B)}_{b'b}(\vartheta + i\alpha) \qquad \text{(A2.7)}$$

where A, B are integer or half-integer, $-A \le a, a' \le A$, $-B \le b, b' \le B$. Note that the operators $\hat{\mathbf{A}}$, $\hat{\mathbf{B}}$ are here represented by hermitian matrices $\mathscr{D}^{(A,B)}$, so these matrices are only unitary if $\boldsymbol{\beta} = 0$, i.e. for pure rotations. Generally they are not irreducible for pure rotations; they behave like the product of representations of spin $A \otimes$ spin B.

It is clear from the product structure of (A2.7) and from the theory of addition of angular momentum that if we take the direct product of two representations (A_1, B_1) and (A_2, B_2) then the Clebsch–Gordan decomposition will be of the general form

$$\begin{aligned}(A_1, B_1) \otimes (A_2, B_2) = {} & (A_1 + A_2, B_1 + B_2) \oplus (A_1 + A_2 - 1, B_1 + B_2) \oplus \\ & \cdots \oplus (|A_1 - A_2|, B_1 + B_2) \oplus (A_1 + A_2, B_1 + B_2 - 1) \\ & \cdots \oplus (A_1 + A_2, |B_1 - B_2|) \\ & \cdots \oplus (|A_1 - A_2|, |B_1 - B_2|). \end{aligned} \qquad \text{(A2.8)}$$

Perhaps the simplest representations are $(s, 0)$ and $(0, s)$, where

$$\mathscr{D}^{(s,0)}_{a'b',ab}(\vartheta, \alpha) = \delta_{b'b}\mathscr{D}^{(s)}_{a'a}(\vartheta - i\alpha). \qquad \text{(A2.9)}$$

Clearly the b, b' labels are irrelevant and we may use

$$\mathscr{D}^{(s,0)}_{a'a}(\vartheta, \alpha) = \mathscr{D}^{(s)}_{a'a}(\vartheta - i\alpha). \qquad \text{(A2.10)}$$

Similarly we may take

$$\mathscr{D}^{(0,s)}_{b'b}(\boldsymbol{\vartheta}, \boldsymbol{\alpha}) = \mathscr{D}^{(s)}_{b'b}(\boldsymbol{\vartheta} + i\boldsymbol{\alpha}). \tag{A2.11}$$

Note that from (A2.6) and (A2.7) that

$$\mathscr{D}^{(0,s)}(l) = \mathscr{D}^{(s,0)}(l^{-1})^{\dagger} = \left[\mathscr{D}^{(s,0)}(l)^{\dagger}\right]^{-1} \tag{A2.12}$$

for an arbitrary Lorentz transformation l.

Now, as mentioned in subsection 2.4.2, for a pure rotation the complex conjugate representation $\mathscr{D}^{(s)^*}$ is equivalent to $\mathscr{D}^{(s)}$ ($\mathscr{D}^{(s)^*} \approx \mathscr{D}^{(s)}$), i.e. there exists a unitary matrix C, which depends on s but not upon the parameters of the rotation, such that

$$\mathscr{D}^{(s)^*}(\boldsymbol{\vartheta}) = C\mathscr{D}^{(s)}(\boldsymbol{\vartheta})C^{-1} \tag{A2.13}$$

with $C^*C = (-1)^{2s}$ and $C^{\dagger}C = 1$. Conventionally one takes

$$C_{\lambda\lambda'} = (-1)^{s-\lambda}\delta_{\lambda,-\lambda'} \tag{A2.14}$$

Then from (A2.7) one can see that

$$\mathscr{D}^{(A,B)^*}(\boldsymbol{\vartheta}, \boldsymbol{\alpha}) = C\mathscr{D}^{(B,A)}(\boldsymbol{\vartheta}, \boldsymbol{\alpha})C^{-1} \tag{A2.15}$$

where here C is a direct product of the matrices in (A2.14):

$$C_{a'b',ab} = C_{a'a}C_{b'b}. \tag{A2.16}$$

In particular $\mathscr{D}^{(0,s)^*}$ is equivalent to $\mathscr{D}^{(s,0)}$.

A2.2 Spinors

The case of $s = 1/2$ is especially important, because of its relevance to the Dirac equation and the spinor calculus. There are four sets of 2×2 representation matrices of interest: $\mathscr{D}^{(1/2,0)}$; $\mathscr{D}^{(0,1/2)^*}$, which is equivalent to $\mathscr{D}^{(1/2,0)}$; $\mathscr{D}^{(0,1/2)}$; and $\mathscr{D}^{(1/2,0)^*}$ which is equivalent to $\mathscr{D}^{(0,1/2)}$. It is easy to check that (A2.13) and (A2.14) correspond to

$$(i\sigma_2)\mathscr{D}^{(1/2,0)}(i\sigma_2)^{-1} = \mathscr{D}^{(0,1/2)^*} \tag{A2.17}$$

Since we shall only discuss $s = 1/2$ it is conventional to define

$$\mathscr{D} \equiv \mathscr{D}^{(1/2,0)} \tag{A2.18}$$

and then to introduce

$$D_a{}^b \equiv \mathscr{D}_{ab} \equiv \mathscr{D}^{(1/2,0)}_{ab} \tag{A2.19}$$

$$D_{\dot{a}}{}^{\dot{b}} \equiv \mathscr{D}^{(1/2,0)^*}_{ab} = \mathscr{D}^*_{ab} \tag{A2.20}$$

i.e. a 'dot' on a row or column label signifies use of the complex conjugate representation.

We can then define two kinds of two-component spinors χ_a and $\chi_{\dot{a}}$ such that if the reference frame undergoes some Lorentz transformation, then the components of the spinors in the transformed frame are, analogously to (1.1.15),

$$\chi'_a = D_a{}^b \chi_b \tag{A2.21}$$

$$\chi'_{\dot{a}} = D_{\dot{a}}{}^{\dot{b}} \chi_{\dot{b}} \tag{A2.22}$$

where we have used the shorthand notation χ'_a for $(\chi_a)_{S'}$ used in Chapter 1.

One can introduce a kind of 'metric spinor'

$$\begin{aligned} \epsilon^{ab} &= \epsilon_{ab} = (i\sigma_2)_{ab} \\ &= \begin{pmatrix} 0 & 1 \\ -1 & 0 \end{pmatrix} && \text{(A2.23)} \\ &= \epsilon^{\dot{a}\dot{b}} = \epsilon_{\dot{a}\dot{b}} && \text{(A2.24)} \end{aligned}$$

and then define the 'contravariant' spinors

$$\chi^a = \epsilon^{ab} \chi_b \tag{A2.25}$$

and

$$\chi^{\dot{a}} = \epsilon^{\dot{a}\dot{b}} \chi_{\dot{b}}. \tag{A2.26}$$

Note that the inverse of (A2.25), for example, is

$$\chi_a = -\epsilon_{ab} \chi^b = \epsilon_{ba} \chi^b \tag{A2.27}$$

since

$$\epsilon_{ab} \epsilon^{bc} = -\delta^c_a. \tag{A2.28}$$

The minus sign in (A2.28) has the peculiar effect that if $\boldsymbol{\chi}$ and $\boldsymbol{\eta}$ are two spinors then

$$\chi^\alpha \eta_\alpha = -\chi_\alpha \eta^\alpha. \tag{A2.29}$$

Now using (A2.21) and (A2.25) one finds

$$\chi^{a'} = \mathscr{D}^{(0,1/2)^*}_{ab} \chi^b. \tag{A2.30}$$

Conventionally one defines

$$D^a{}_b \equiv \mathscr{D}^{(0,1/2)^*}_{ab} \tag{A2.31}$$

so that (A2.30) becomes

$$\chi^{a'} = D^a{}_b \chi^b \tag{A2.32}$$

and from (A2.12)

$$D^a{}_b = \left[\left(\mathscr{D}^{-1} \right)^T \right]_{ab}. \tag{A2.33}$$

Similarly, for (A2.26), under transformation of the reference frame

$$\chi^{\dot{a}'} = \mathscr{D}^{(0,1/2)}_{ac}\chi^{\dot{c}}. \qquad \text{(A2.34)}$$

One defines

$$D^{\dot{a}}{}_{\dot{b}} \equiv \mathscr{D}^{(0,1/2)}_{ab} \qquad \text{(A2.35)}$$

so that (A2.34) reads

$$\chi^{\dot{a}'} = D^{\dot{a}}{}_{\dot{b}}\chi^{\dot{b}} \qquad \text{(A2.36)}$$

and by (A2.12)

$$D^{\dot{a}}{}_{\dot{b}} = \left[\left(\mathscr{D}^{-1}\right)^{\dagger}\right]_{ab}. \qquad \text{(A2.37)}$$

Let us summarize the transformation laws for the various two-component spinors introduced:

$$\begin{aligned} \chi_a &: \quad \mathscr{D}^{(1/2,0)} \\ \chi_{\dot{a}} &: \quad \mathscr{D}^{(1/2,0)^*} \approx \mathscr{D}^{(0,1/2)} \\ \chi^a &: \quad \mathscr{D}^{(0,1/2)^*} \approx \mathscr{D}^{(1/2,0)} \\ \chi^{\dot{a}} &: \quad \mathscr{D}^{(0,1/2)}. \end{aligned} \qquad \text{(A2.38)}$$

An important question is how to form invariants from these. The Clebsch–Gordan decomposition (A2.8) tells us that both $(1/2, 0) \otimes (1/2, 0)$ and $(0, 1/2) \otimes (0, 1/2)$ will contain the invariant representation $(0, 0)$.

Hence if χ_a and η_b are spinors of type $(1/2, 0)$ then we expect some linear combination $f^{ab}\chi_a\eta_b$ to be invariant. In fact the combination is just

$$\epsilon^{ab}\chi_a\eta_b = \chi_a\eta^a \qquad \text{(A2.39)}$$

since

$$\begin{aligned} \chi'_a\eta^{a'} &= D_a{}^b D^a{}_c \chi_b\eta^c \\ &= \mathscr{D}_{ab}\left[\left(\mathscr{D}^{-1}\right)^T\right]_{ac}\chi_b\eta^c = \chi_a\eta^a \end{aligned} \qquad \text{(A2.40)}$$

i.e. it is indeed invariant.

Similarly

$$\epsilon^{\dot{a}\dot{b}}\chi_{\dot{a}}\eta_{\dot{b}} = \chi_{\dot{a}}\eta^{\dot{a}} \qquad \text{(A2.41)}$$

is invariant.

Finally, by using complex conjugation, we can build up an invariant out of spinors ξ_a of type $(1/2, 0)$ and $\zeta^{\dot{a}}$ of type $(0, 1/2)$. Namely, under transformation of the reference frame, writing (A2.21), (A2.36) and (A2.37) in matrix form, the spinors transform as

$$\xi' = \mathscr{D}\xi \qquad \text{and} \qquad \zeta' = (\mathscr{D}^{-1})^{\dagger}\zeta$$

so that

$$\zeta'^{\dagger}\xi' = \zeta^{\dagger}\xi \tag{A2.42}$$

i.e. is invariant.

A2.3 Connection between spinor and vector representations

Let A^{μ} be a 4-vector. Under a Lorentz transformation l applied to the reference frame, the components of A^{μ} in the transformed frame are (see (1.2.14))

$$A^{\mu\prime} = \Lambda^{\mu}{}_{\nu}(l^{-1})A^{\nu} \tag{A2.43}$$

where $A^{\mu\prime}$ is short for $(A^{\mu})_{S_l}$.

The $\Lambda^{\mu}{}_{\nu}$ are the transformation matrices for the *vector representation* and are the basic blocks for building up tensor representations, the latter being generally reducible.

We shall now demonstrate that the representation $\mathscr{D}^{(1/2,1/2)}$ is equivalent to the vector representation. This is a result of great importance since it gives a fundamental connection between spinors and 4-vectors.

Firstly, from the form of Clebsch–Gordan decomposition (A2.8) we have that

$$\mathscr{D}^{(1/2,0)} \otimes \mathscr{D}^{(0,1/2)} = \mathscr{D}^{(1/2,1/2)}. \tag{A2.44}$$

But from (A2.15), $\mathscr{D}^{(0,1/2)}$ is equivalent to $\mathscr{D}^{(1/2,0)^*}$. Hence

$$\mathscr{D}^{(1/2,0)^*} \otimes \mathscr{D}^{(1/2,0)} \approx \mathscr{D}^{(1/2,1/2)}. \tag{A2.45}$$

We thus need to show that transformation under the left-hand side of (A2.45) is equivalent to the vector transformation. Hence if ξ is a two-component spinor of type $(1/2, 0)$ we need to prove the existence of a set of coefficients $C^{\mu ab}$ such that $C^{\mu ab}\xi_a^*\xi_b$ transforms like a vector. But it is well known that if one adds the two-dimensional unit matrix to a set of Pauli matrices to form

$$\sigma^{\mu} = (I, \boldsymbol{\sigma}) \tag{A2.46}$$

then

$$V^{\mu} \equiv \xi^{\dagger}\sigma^{\mu}\xi \tag{A2.47}$$

transforms as a 4-vector, i.e.

$$V^{\mu\prime} = \xi'^{\dagger}\sigma^{\mu}\xi' = \Lambda^{\mu}{}_{\nu}V^{\nu}. \tag{A2.48}$$

This is easily shown for rotations or pure boosts upon using

$$e^{i\boldsymbol{\vartheta}\cdot\boldsymbol{\sigma}/2} = \cos\theta/2 + i\hat{\boldsymbol{\vartheta}}\cdot\boldsymbol{\sigma}\sin\theta/2 \tag{A2.49}$$

and

$$e^{\boldsymbol{\alpha}\cdot\boldsymbol{\sigma}/2} = \cosh\alpha/2 + \hat{\boldsymbol{\alpha}}\cdot\boldsymbol{\sigma}\sinh\alpha/2. \tag{A2.50}$$

Of course we can write (A2.47) in the form

$$V^{\mu} = (\sigma^{\mu})_{ab}\,\xi^{*}_{a}\xi_{b}, \tag{A2.51}$$

which casts an interesting new light on the matrices σ^{μ}. The elements $(\sigma^{\mu})_{ab}$ are the elements of the transformation matrix from the $(1/2,0)^{*}\otimes(1/2,0) \approx (1/2,1/2)$ representation to the equivalent usual 4-vector representation.

Note that we have been a little cavalier with the group-theoretical aspects. Strictly speaking, the representations (A, B) with which we have been dealing are representations of the group $SL(2,c)$, whereas the 4-vector representation is the vector representation of the group $O(1,3)$.

For a detailed discussion of the spinor calculus and its use in constructing relativistic wave equations the reader is referred to Carruthers (1971), where there is also a treatment of the *unitary* (hence, infinite-dimensional) representations of the homogeneous Lorentz group. For applications to supersymmetry see Sohnius (1985).

Appendix 3
Spin properties of fields and wave equations

This is not a book on field theory, so we do not wish to get involved in a comprehensive discussion of field equations. But the transformation laws for particle states examined in Section 2.4 shed an interesting light upon the problem of constructing fields for arbitrary spin-particles and upon the wave equations they satisfy.

In particular, concerning the Dirac equation, many readers will have followed the beautiful derivation by Dirac of his famous equation for spin-1/2 particles (See Dirac, 1947). Here we shall look at the Dirac equation from a different point of view which provides an alternative insight into the origin and meaning of the equation.

A3.1 Relativistic quantum fields

The essence of the physical states that were discussed in Chapter 1 is that for a particle at rest they transform *irreducibly* under rotations. It would be possible to deal with quantum field operators that also had this property (Weinberg, 1964a), i.e. spin-s fields, which have only $2s+1$ components. This, as we shall see, is not very convenient for constructing Lagrangians and building-in symmetry properties so that, for example, we normally use a four-component field for spin-1/2 Dirac particles and a 4-vector A_μ to describe spin-1 mesons or photons etc. Thus we usually carry redundant components, and the free-field equations, other than the Klein–Gordon equation, do nothing other than place Lorentz-invariant constraints on the redundant components. It is instructive to compare the approach via irreducible fields with the conventional approach, especially in the case of the Dirac equation.

A local field is constructed by taking a linear combination of creation and annihilation operators in the form of a Fourier transform. Under an arbitrary homogeneous Lorentz transformation l and space–time translation a^μ an N-component field is required to transform as

$$U(l,a)\Psi_n(x)U^{-1}(l,a) = \sum_m D_{nm}(l^{-1})\Psi_m(lx+a) \tag{A3.1}$$

where D_{nm} is an N-dimensional representation of the homogeneous Lorentz group. These properties make it relatively simple to write down Lorentz-invariant lagrangians and interactions.

In the following we shall briefly survey the relationship between the physical states introduced earlier and the local fields related to them. We shall see that quanta that have spin s can be embedded in many ways in a field with $N \geq 2s+1$ components. For a more detailed discussion the reader is referred to Weinberg's seminal paper.

We shall present the analysis in terms of the helicity states defined in (1.2.26). With obvious modifications one can base the discussion on the canonical states.

Let $a^\dagger(\mathbf{p},\lambda)$ be the creation operator of the state $|\mathbf{p};\lambda\rangle$ when acting on the bare vacuum. With the invariant normalization

$$\langle\mathbf{p}';\lambda'|\mathbf{p};\lambda\rangle = 2p^0\delta(\mathbf{p}'-\mathbf{p})\delta_{\lambda'\lambda} \tag{A3.2}$$

we take

$$|\mathbf{p};\lambda\rangle = a^\dagger(\mathbf{p},\lambda)|0\rangle \tag{A3.3}$$

so that a and $a^\dagger$ are annihilation and creation operators satisfying commutation or anticommutation relations

$$\left[a(\mathbf{p},\lambda), a^\dagger(\mathbf{p}',\lambda')\right]_{\mp} = 2p^0\delta(\mathbf{p}'-\mathbf{p})\delta_{\lambda'\lambda} \tag{A3.4}$$

according as $(-1)^{2s} = \pm 1$.

From the transformation properties of the state vector and the invariance of the vacuum, one sees via (2.1.9) that under a Lorentz transformation l

$$U(l)a^\dagger(\mathbf{p},\lambda)U^{-1}(l) = \mathscr{D}^{(s)}_{\lambda'\lambda}\left[h^{-1}(l\mathbf{p})lh(\mathbf{p})\right]a^\dagger(l\mathbf{p},\lambda') \tag{A3.5}$$

where $h(\mathbf{p})$ is given in (1.2.22).

Taking the adjoint and using the unitarity of the representations of the rotation group we get

$$U(l)a(\mathbf{p},\lambda)U^{-1}(l) = \mathscr{D}^{(s)}_{\lambda\lambda'}\left[h^{-1}(\mathbf{p})l^{-1}h(l\mathbf{p})\right]a(l\mathbf{p},\lambda') \tag{A3.6}$$

where the argument of $\mathscr{D}$ is just a Wick helicity rotation.

Because of this complicated $\mathbf{p}$-dependence, a local field built from the $a(\mathbf{p},\lambda)$ via a Fourier transform will not transform in a simple covariant fashion.

A3.2 Irreducible relativistic quantum fields

To construct a field transforming according to (A3.1) it is necessary to split off the **p**-dependent factors appearing in (A3.5) and to absorb them into new creation operators. The problem is that $\mathscr{D}^{(s)}$ is a representation matrix of the *rotation* group, not the Lorentz group, so that we cannot simply use the property $\mathscr{D}_{ij}(l_1 l_2) = \mathscr{D}_{ik}(l_1)\mathscr{D}_{kj}(l_2)$. To proceed we require certain properties of the representations of the Lorentz group that were discussed in Appendix 2.

As explained there the finite-dimensional representations are labelled (A, B), where A is either integer $(0, 1, 2, \ldots)$ or half integer $(1/2, 3/2, \ldots)$. The simplest representations are the $(s, 0)$ and $(0, s)$ representations, of dimension $2s + 1$.

Consider now fields based upon the use of the $(s, 0)$ representation given in (A2.10). In (A3.6) we can now put

$$\begin{aligned}\mathscr{D}^{(s)}(r_{\text{Wick}}) &= \mathscr{D}^{(s,0)}(r_{\text{Wick}}) \\ &= \mathscr{D}^{(s,0)}\left[h^{-1}(\mathbf{p})\right]\mathscr{D}^{(s,0)}\left(l^{-1}\right)\mathscr{D}^{(s,0)}\left[h(l\mathbf{p})\right].\end{aligned} \tag{A3.7}$$

If we then define

$$\mathscr{A}(\mathbf{p}, \lambda) \equiv \mathscr{D}^{(s,0)}_{\lambda\lambda'}\left[h(\mathbf{p})\right] a(\mathbf{p}, \lambda') \tag{A3.8}$$

then from (A3.6) and (A3.7) we get the simple result

$$U(l)\mathscr{A}(\mathbf{p}, \lambda)U^{-1}(l) = \mathscr{D}^{(s,0)}_{\lambda\lambda'}\left(l^{-1}\right)\mathscr{A}(l\mathbf{p}, \lambda') \tag{A3.9}$$

so that the transformation matrix is no longer a function of **p**.

In order that the field include both particles and antiparticles we must now consider the operator $b^{\dagger}(\mathbf{p}, \lambda)$ that creates the antiparticle of the particle which $a(\mathbf{p}, \lambda)$ annihilates. It must transform just like $a^{\dagger}(\mathbf{p}, \lambda)$, as given in (A3.5). However, the ordering of summation indices in (A3.6) and (A3.5) is different, so we first rewrite (A3.5) in a form analogous to (A3.6) using (A2.13).

Because $\mathscr{D}^{(s)}(r)$ is unitary, we can write

$$\mathscr{D}^{(s)}(r) = \mathscr{D}^{(s)}(r^{-1})^{\dagger} = \left[C\mathscr{D}^{(s)}(r^{-1})C^{-1}\right]^{T} \tag{A3.10}$$

where the last step follows from (A2.13), so that (A3.5) becomes

$$U(l)a^{\dagger}(\mathbf{p}, \lambda)U^{-1}(l) = \left\{C\mathscr{D}^{(s)}\left[h^{-1}(\mathbf{p})l^{-1}h(l\mathbf{p})\right]C^{-1}\right\}_{\lambda\lambda'} a^{\dagger}(l\mathbf{p}, \lambda') \tag{A3.11}$$

and the same result will hold for $b^{\dagger}(l\mathbf{p}, \lambda)$. We now define

$$\mathscr{B}^{\dagger}(l\mathbf{p}, \lambda) = \left\{\mathscr{D}^{(s,0)}\left[h(\mathbf{p})\right]C^{-1}\right\}_{\lambda\lambda'} b^{\dagger}(l\mathbf{p}, \lambda') \tag{A3.12}$$

from which follows, just as in (A3.9),

$$U(l)\mathscr{B}^\dagger(\mathbf{p},\lambda)U^{-1}(l) = \mathscr{D}^{(s,0)}_{\lambda\lambda'}\left(l^{-1}\right)\mathscr{B}^\dagger(l\mathbf{p},\lambda'). \tag{A3.13}$$

The local spin-s field of type $(s,0)$,

$$\phi^{(s,0)}_\lambda(x) = \int \frac{d^3\mathbf{p}}{(2\pi)^{3/2}2p^0}\left[\mathscr{A}(\mathbf{p},\lambda)e^{-ip\cdot x} + \mathscr{B}^\dagger(\mathbf{p},\lambda)e^{ip\cdot x}\right], \tag{A3.14}$$

transforms according to (A3.1) with $D_{nm} \to \mathscr{D}^{(s,0)}_{nm}$ and can be shown to satisfy causal commutation or anticommutation relations, according as $(-1)^{2s} = \pm 1$. Note that we could introduce a phase factor ξ, $|\xi| = 1$, in front of the $\mathscr{B}^\dagger$ term without altering any of the relevant properties of the local field.

By rewriting ϕ in terms of the original a and b operators, i.e.

$$\phi^{(s,0)}_\lambda(x) = \int \frac{d^3\mathbf{p}}{(2\pi)^{3/2}2p^0}\left\{\mathscr{D}^{(s,0)}_{\lambda\lambda'}[h(\mathbf{p})]a(\mathbf{p},\lambda')e^{-ip\cdot x} + \left[\mathscr{D}^{(s,0)}[h(\mathbf{p})]C^{-1}\right]_{\lambda\lambda'} b^\dagger(\mathbf{p},\lambda')e^{ip\cdot x}\right\} \tag{A3.15}$$

one can see, for example, that $\phi^{(s,0)}_\lambda(x)$ creates particles of momentum $\mathbf{p}$ and helicity λ' with wave function

$$\frac{1}{(2\pi)^{3/2}2p^0}\mathscr{D}^{(s,0)*}_{\lambda\lambda'}[h(\mathbf{p})]e^{ip\cdot x}.$$

The field (A3.15) obeys only the Klein–Gordon equation.

Clearly we can introduce a field $\phi^{(0,s)}_\lambda(x)$ in an analogous fashion, and it will transform according to (A3.1) with $D_{nm} \to \mathscr{D}^{(0,s)}_{nm}$. It turns out to be most useful to define $\phi^{(0,s)}_\lambda(x)$ with a phase factor $(-1)^{2s}$ in front of the creation operators:

$$\phi^{(0,s)}_\lambda(x) = \int \frac{d^3\mathbf{p}}{(2\pi)^{3/2}2p^0}\left\{\mathscr{D}^{(0,s)}_{\lambda\lambda'}[h(\mathbf{p})]a(\mathbf{p},\lambda')e^{-ip\cdot x} + (-1)^{2s}\left[\mathscr{D}^{(0,s)}[h(\mathbf{p})]C^{-1}\right]_{\lambda\lambda'} b^\dagger(\mathbf{p},\lambda')e^{ip\cdot x}\right\}. \tag{A3.16}$$

A3.3 Parity and field equations

We shall now see that these fields, by themselves, are not suited to a parity-conserving theory. From (A3.3) and eqn (2.3.7) of Chapter 2, we deduce that

$$\mathscr{P}a(p,\theta,\varphi;\lambda)\mathscr{P}^{-1} = \eta_{\mathscr{P}}e^{i\pi s}a(p,\pi-\theta,\varphi+2\pi;-\lambda). \tag{A3.17}$$

Then after some labour, one finds that

$$\mathscr{P}\phi_\lambda^{(s,0)}(t,\mathbf{x})\mathscr{P}^{-1} = \eta_{\mathscr{P}}\phi_\lambda^{(0,s)}(t,-\mathbf{x}) \tag{A3.18}$$

provided that the intrinsic parity $\bar{\eta}_{\mathscr{P}}$ of the antiparticle is chosen in such a way that

$$\bar{\eta}_{\mathscr{P}} = (-1)^{2s}\eta_{\mathscr{P}}. \tag{A3.19}$$

Thus parity transforms the $(s,0)$ field into the $(0,s)$ field and we are forced to use both to set up a parity-conserving theory.

It is then helpful to combine the $(s,0)$ and $(0,s)$ fields into one $2(2s+1)$-component field

$$\psi_\alpha(x) = \begin{pmatrix} \phi^{(s,0)}(x) \\ \phi^{(0,s)}(x) \end{pmatrix}, \tag{A3.20}$$

which then transforms according to

$$U(l)\psi_\alpha(x)U^{-1}(l) = D_{\alpha\beta}^{(s)}(l^{-1})\psi_\beta(lx) \tag{A3.21}$$

where

$$D_{\alpha\beta}^{(s)}(l) = \begin{pmatrix} \mathscr{D}^{(s,0)}(l) & 0 \\ 0 & \mathscr{D}^{(0,s)}(l) \end{pmatrix} \tag{A3.22}$$

i.e. ψ transforms according to the $(s,0)\oplus(0,s)$ representation.

It can be shown that the fields $\psi_\alpha(x)$ satisfy causal commutation or anticommutation relations. The factor $(-1)^{2s}$ in (A3.16) is crucial for this.

Each field $\psi_\alpha(x)$ will clearly satisfy a Klein–Gordon equation. But there will be other equations of constraint. To see where these come from consider the matrices

$$\Pi^{(s)}(p) \equiv m^{2s}\mathscr{D}^{(s,0)}[h(\mathbf{p})]\mathscr{D}^{(0,s)}[h^{-1}(\mathbf{p})] \tag{A3.23}$$

and

$$\overline{\Pi}^{(s)}(p) \equiv m^{2s}\mathscr{D}^{(0,s)}[h(\mathbf{p})]\mathscr{D}^{(s,0)}[h^{-1}(\mathbf{p})], \tag{A3.24}$$

which will convert $\mathscr{D}^{(0,s)}[h(\mathbf{p})]$ to $\mathscr{D}^{(s,0)}[h(\mathbf{p})]$ and vice versa respectively.

Using (A2.10), (A2.11) and (1.2.22), Weinberg has shown that $\Pi^{(s)}$ and $\overline{\Pi}^{(s)}$ are homogeneous polynomials of order $2s$ in the components p^μ of the 4-vector $(E,\mathbf{p})$. Hence we can define a matrix differential operator $\overline{\Pi}^{(s)}(i\partial)$ and consider its action on $\phi_\lambda^{(s,0)}(x)$. When taken under the integral sign in (A3.15) and acting on $e^{-ip\cdot x}$, $\overline{\Pi}^{(s)}(i\partial)$ becomes $\overline{\Pi}^{(s)}(p)$ and thus converts the first $\mathscr{D}^{(s,0)}$ to $\mathscr{D}^{(0,s)}$. Acting on $e^{ip\cdot x}$, it converts the second $\mathscr{D}^{(s,0)}$ to $(-1)^{2s}\mathscr{D}^{(0,s)}$. In other words, using (A3.16),

$$\overline{\Pi}_{\nu\lambda}^{(s)}(i\partial)\phi_\lambda^{(s,0)}(x) = m^{2s}\phi_\nu^{(0,s)}(x). \tag{A3.25}$$

Similarly

$$\Pi^{(s)}_{\nu\lambda}(i\partial)\phi^{(0,s)}_{\lambda}(x) = m^{2s}\phi^{(s,0)}_{\nu}(x). \tag{A3.26}$$

Thus $\psi(x)$ satisfies the equation

$$\begin{pmatrix} 0 & \Pi^{(s)}(i\partial) \\ \overline{\Pi}^{(s)}(i\partial) & 0 \end{pmatrix} \psi(x) = m^{2s}\psi(x). \tag{A3.27}$$

A3.4 The Dirac equation

The classic example of the above construction is the Dirac equation for spin 1/2. For the (1/2, 0) representation (see eqn (A2.2)) $\hat{\mathbf{A}} \to (1/2)\boldsymbol{\sigma}$ so that

$$\mathscr{D}^{(1/2,0)}(\boldsymbol{\vartheta}, \boldsymbol{\alpha}) = e^{-i\boldsymbol{\sigma}\cdot(\boldsymbol{\vartheta}-i\boldsymbol{\alpha})/2}. \tag{A3.28}$$

For (0, 1/2), $\hat{\mathbf{B}} \to (1/2)\boldsymbol{\sigma}$ so that

$$\mathscr{D}^{(0,1/2)}(\boldsymbol{\vartheta}, \boldsymbol{\alpha}) = e^{-i\boldsymbol{\sigma}\cdot(\boldsymbol{\vartheta}+i\boldsymbol{\alpha})/2} \tag{A3.29}$$

and the physical meanings of $\boldsymbol{\vartheta}$ and $\boldsymbol{\alpha}$ are given after equation (A2.4).

Then, using (1.2.22), (A3.23) and (A3.24), one finds

$$\Pi^{(1/2)}(p) = me^{-\boldsymbol{\sigma}\cdot\boldsymbol{\alpha}} = p^0 - \mathbf{p}\cdot\boldsymbol{\sigma} \tag{A3.30}$$

and

$$\overline{\Pi}^{(1/2)}(p) = me^{\boldsymbol{\sigma}\cdot\boldsymbol{\alpha}} = p^0 + \mathbf{p}\cdot\boldsymbol{\sigma} \tag{A3.31}$$

where we have used the fact that for a boost from rest to 4-momentum $(E, \mathbf{p})$, $\tanh\alpha = \beta = |\mathbf{p}|/E$.

Finally, (A3.27) for $s = 1/2$ can be recognized as the Dirac equation

$$(i\gamma^\mu\partial_\mu - m)\,\psi(x) = 0 \tag{A3.32}$$

in the representation where

$$\gamma^0 = \begin{pmatrix} 0 & I \\ I & 0 \end{pmatrix} \qquad \boldsymbol{\gamma} = \begin{pmatrix} 0 & \boldsymbol{\sigma} \\ -\boldsymbol{\sigma} & 0 \end{pmatrix}. \tag{A3.33}$$

This is just a representation in which γ_5 is diagonal. Weinberg (1964a) showed how the above generalizes to a Dirac-like equation for arbitrary spin, i.e where the fields transform like $(s, 0) \oplus (0, s)$.

As mentioned earlier, these minimal fields, with the exception of the spin-1/2 case, are not those normally used in constructing lagrangians for particle interactions. For developments concerning more general fields consult the fundamental papers of Weinberg (1964a,b).

Appendix 4

Transversity amplitudes

We briefly introduce the concept of transversity amplitudes and mention some of their key properties.

A4.1 Definition of transversity amplitudes

It has been known for a long time that certain simplifications occur if the spin quantization axis for each particle in the reaction

$$A + B \to C + D$$

is taken along the normal to the reaction plane (Dalitz, 1966). The usefulness of *transversity states* and *transversity amplitudes* in a modern context was emphasized by Kotanski (1970).

The transversity amplitudes $T_{cd;ab}(\theta)$ are defined by

$$\begin{aligned} T_{cd;ab}(\theta) = & \sum_{\text{all } \lambda} \mathscr{D}^{(s_C)^*}_{c\lambda_C} \mathscr{D}^{(s_D)^*}_{d\lambda_D} e^{i\pi(\lambda_D - \lambda_B)} \\ & \times H_{\lambda_C\lambda_D;\lambda_A\lambda_B}(\theta) \mathscr{D}^{(s_A)}_{\lambda_A a} \mathscr{D}^{(s_B)}_{\lambda_B b} \end{aligned} \qquad \text{(A4.1)}$$

where the argument of each $\mathscr{D}$-function is

$$r_x(-\pi/2) = r(\pi/2, \pi/2, -\pi/2)$$

so that

$$\mathscr{D}^{(s)}_{\lambda\mu}\left(r_x(-\pi/2)\right) = \exp\left[i\pi(\mu - \lambda)/2\right] d^s_{\lambda\mu}(\pi/2). \qquad \text{(A4.2)}$$

The transversity amplitudes measure the probability amplitudes for transitions amongst states of the type, $|\mathbf{p}_A; a\rangle_T$, which corresponds to particle A having spin component $s_z = a$ in the transversity rest frame S^T_A of A. S^T_A is obtained from the helicity rest frame S_A of A by a rotation

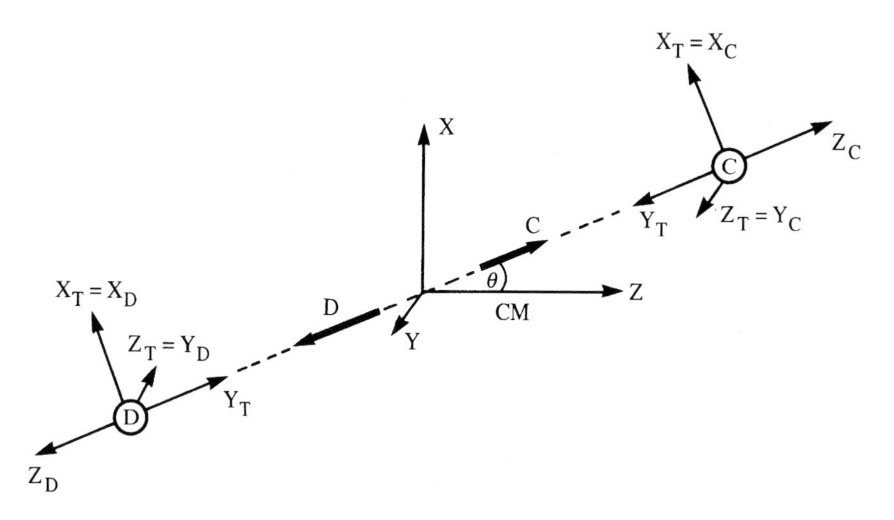

Fig. A4.1. Transversity rest frames for final particles in $A + B \to C + D$.

through $-\pi/2$ about the X axis of S_A. This is illustrated in Fig. A4.1[1] for particles C and D.

A4.2 Symmetry of transversity amplitudes

The symmetry properties of helicity amplitudes give rise to analogous properties for the transversity amplitudes as follows.

(*a*) *Parity.* With the intrinsic parities η_i one finds

$$T_{cd;ab}(\theta) = \frac{\eta_C \eta_D}{\eta_A \eta_B} (-1)^{a+b+c+d} T_{cd;ab}(\theta). \qquad \text{(A4.3)}$$

Thus invariance under space inversion makes

$$T_{cd;ab}(\theta) = 0 \qquad \text{if} \qquad \frac{\eta_C \eta_D}{\eta_A \eta_B} (-1)^{a+b+c+d} = -1. \qquad \text{(A4.4)}$$

This simplifies the appearance of the density matrix in the transversity basis giving it a 'chequer board' pattern, as discussed in subsection 5.4.1.

(*b*) *Time reversal.* In general

$$T_{cd;ab}(AB \to CD) = (-1)^{b-a+c-d} T_{ab;cd}(CD \to AB). \qquad \text{(A4.5)}$$

and for elastic reactions $A + B \to A + B$

$$T_{a'b';ab} = (-1)^{b-a+a'-b'} T_{ab;a'b'}. \qquad \text{(A4.6)}$$

---

[1] Note that some authors use a different convention. We have followed the original paper of Kotanski cited above.

(*c*) *Identical particles.* For the correctly symmetrized amplitudes one finds the following.

For $A + B \to C + C$,

$$T^{\mathscr{S}}_{cc';ab}(\theta) = (-1)^{s_B - s_A + a + b + c + c'} T^{\mathscr{S}}_{-c'-c;-a-b}(\pi - \theta). \tag{A4.7}$$

For $A + A \to C + D$,

$$T^{\mathscr{S}}_{cd;aa'}(\theta) = (-1)^{s_D - s_C + a + a' + c + d} T^{\mathscr{S}}_{-c-d;-a'-a}(\pi - \theta). \tag{A4.8}$$

For $A + A \to C + C$, both the above, as well as

$$T^{\mathscr{S}}_{cc';aa'}(\theta) = T^{\mathscr{S}}_{c'c;a'a}(\theta). \tag{A4.9}$$

For states of definite isospin the right-hand side of (A4.7) and (A4.8) should contain an extra factor $(-1)^{I+1}$.

A4.3 Some analytic properties of transversity amplitudes

As remarked in Section 4.3 the analytic properties of the transversity amplitudes are only simple at thresholds and pseudothresholds. Their behaviour at $\theta = 0, \pi$ is just given by using (4.3.1) in (A4.1) and does not simplify.

In high energy models based on *t-channel* amplitudes the behaviour at the thresholds and pseudothresholds is important (Kotanski, 1970):

$$T^{(t)}_{cd;ab} \sim \varphi_{ab}^{\epsilon(a+b)} \varphi_{cd}^{\epsilon(c+d)} \psi_{ab}^{\epsilon\epsilon_{AB}(a-b)} \varphi_{cd}^{\epsilon\epsilon_{CD}(c-d)} \tag{A4.10}$$

where

$$\begin{aligned} \varphi_{ij} &= [t - (m_i + m_j)^2]^{1/2} \\ \psi_{ij} &= [t - (m_i - m_j)^2]^{1/2} \\ \epsilon &= \ \text{sign}\left\{t(s-u) + (m_A^2 - m_B^2)(m_C^2 - m_D^2)\right\} \\ \epsilon_{ij} &= \ \text{sign}\left\{m_i - m_j\right\}. \end{aligned} \tag{A4.11}$$

If any of these thresholds or pseudothresholds is close to the physical region then the correct behaviour (A4.10) must be built into the models of $T^{(t)}_{cd;ab}$.

Appendix 5

Common notations for helicity amplitudes

We list here some of the conventional notations for the helicity amplitudes in specific reactions.[1]

(1) Meson–baryon scattering (Halzen and Michael, 1971):

$$H_{++} = H_{0\,1/2;0\,1/2} \qquad H_{+-} = H_{0\,1/2;0-1/2}$$

(2) Nucleon-nucleon scattering (Goldberger *et al.*, 1960):

$$\phi_1 = H_{1/2\,1/2;1/2\,1/2} \qquad \phi_3 = H_{1/2-1/2;1/2-1/2}$$
$$\phi_2 = H_{1/2\,1/2;-1/2-1/2} \qquad \phi_4 = H_{1/2-1/2;-1/2\,1/2}$$
$$\phi_5 = H_{1/2\,1/2;1/2-1/2}.$$

(3) Baryon–baryon scattering with non-identical particles, e.g. $\Lambda p \to \Lambda p$ (Buttimore *et al.*, 1978): In addition to the five ϕ_i listed above for $NN \to NN$ one has also

$$\phi_6 = H_{1/2\,1/2;-1/2\,1/2}.$$

For identical particles one has $\phi_6 = -\phi_5$.

(4) Photoproduction of a pseudoscalar meson (Storrow, 1978):

$$N = H_{0-1/2;1\,1/2} \qquad S_2 = H_{0\,1/2;1\,1/2}$$
$$S_1 = H_{0-1/2;1-1/2} \qquad D = H_{0\,1/2;1-1/2}.$$

(5) Vector meson production amplitudes in $0^-(1/2)^+ \to 1^-(1/2)^+$:

$$P^0_{\lambda\mu} = H_{0\mu;0\lambda} \qquad P^{\pm}_{\lambda\mu} = \frac{1}{\sqrt{2}}\left(H_{1\mu;0\lambda} \pm H_{-1\mu;0\lambda}\right).$$

[1] We have not included normalization factors.

(6) Baryon resonance production amplitudes in $0^-(1/2)^+ \to 0^-(3/2)^+$: there are four s-channel helicity amplitudes, two single-flip, one non-flip and one double-flip:

$$M_0 = H_{0\,1/2;0\,1/2} = H_{0-1/2;0-1/2} \qquad M_1' = H_{0\,1/2;0-1/2} = -H_{0-1/2;0\,1/2}$$
$$M_1 = H_{0\frac{3}{2};0\,1/2} = -H_{0-\frac{3}{2};0-1/2} \qquad M_2 = H_{0\frac{3}{2};0-1/2} = H_{0-\frac{3}{2};0\,1/2}.$$

Appendix 6
The coefficients $\mathscr{A}_{l'm'}(lm)$

The coefficients $\mathscr{A}_{l'm'}(lm)$ involved in the parity-invariance relations amongst the dynamical reaction parameters (subsection 5.3.1(v)) are given, for a spin-s particle, in terms of vector addition coefficients as follows.

If both $(-1)^{m \pm m' + 2s} = 1$ and $|m \pm m'| \leq 2s$ then

$$\mathscr{A}_{l'm'}(lm) = e^{i\pi(m'-m)/2} \sqrt{\frac{2l'+1}{2l+1}} \left\langle l, m \middle| s, \frac{m+m'}{2}; s, \frac{m-m'}{2} \right\rangle \times \left\langle l', m' \middle| s, \frac{m+m'}{2}; s, \frac{m'-m}{2} \right\rangle .$$

Otherwise $\mathscr{A}_{l'm'}(lm) = 0$.

The following symmetry properties reduce drastically the number of computations required:

$$\begin{aligned} \mathscr{A}_{l'm'}(lm) &= (-1)^{2s} \frac{2l'+1}{2l+1} \mathscr{A}_{lm}(l'm') \\ \mathscr{A}_{l'-m'}(lm) &= (-1)^{l+m} \mathscr{A}_{l'm'}(lm) \\ \mathscr{A}_{l'm'}(l-m) &= (-1)^{l'+m'} \mathscr{A}_{l'm'}(lm). \end{aligned}$$

We list the independent, non-zero, coefficients for spins $1/2$, 1 and $3/2$.

Spin 1/2:

$$\mathscr{A}_{10}(11) = -\frac{i}{\sqrt{2}} \qquad \mathscr{A}_{00}(11) = -\frac{i}{\sqrt{6}}.$$

Spin 1:

$$\mathscr{A}_{20}(22) = -\frac{1}{\sqrt{6}} \qquad \mathscr{A}_{10}(22) = -\sqrt{\frac{3}{10}} \qquad \mathscr{A}_{00}(22) = -\frac{1}{\sqrt{15}}$$

$$\mathscr{A}_{21}(21) = \frac{1}{2} \qquad \mathscr{A}_{11}(21) = \frac{1}{2}\sqrt{\frac{3}{5}} \qquad \mathscr{A}_{00}(20) = -\frac{1}{3}\sqrt{\frac{2}{3}}$$

$$\mathscr{A}_{20}(20) = \frac{2}{15} \qquad \mathscr{A}_{11}(11) = \frac{1}{2} \qquad \mathscr{A}_{00}(00) = \frac{1}{3}.$$

Spin 3/2:

$$\mathscr{A}_{30}(33) = \frac{i}{2\sqrt{5}} \qquad \mathscr{A}_{20}(33) = \frac{i\sqrt{5}}{2\sqrt{7}} \qquad \mathscr{A}_{10}(33) = \frac{i3\sqrt{3}}{2\sqrt{35}}$$

$$\mathscr{A}_{31}(32) = -\frac{i}{\sqrt{10}} \qquad \mathscr{A}_{21}(32) = -\frac{i\sqrt{5}}{2\sqrt{7}} \qquad \mathscr{A}_{11}(32) = -\frac{i3}{2\sqrt{35}}$$

$$\mathscr{A}_{30}(31) = -\frac{i3\sqrt{3}}{10} \qquad \mathscr{A}_{22}(31) = \frac{i}{\sqrt{14}} \qquad \mathscr{A}_{20}(31) = -\frac{i\sqrt{3}}{2\sqrt{7}}$$

$$\mathscr{A}_{10}(31) = \frac{i3}{10\sqrt{7}}$$

$$\mathscr{A}_{11}(30) = -\frac{i3\sqrt{3}}{5\sqrt{14}}$$

$$\mathscr{A}_{21}(22) = -\frac{i}{2}$$

$$\mathscr{A}_{10}(11) = -\frac{i}{5\sqrt{2}}.$$

Appendix 7

The coefficients $\mathscr{C}^{lm;l'm'}_{l_1m_1;l'_1m'_1}$

The coefficients involved in the additional invariance constraints on the dynamical reaction parameters for a spin-s particle (subsection 5.3.1(v)) are real and are given in terms of vector addition coefficients as follows.

(i) $\mathscr{C}^{lm;l'm'}_{l_1m_1;l'_1m'_1} = 0$ unless *all* the following conditions are satisfied:

$$\begin{gathered} l + l' + l_1 + l'_1 \quad \text{is even} \\ m_1 + m'_1 = m + m' \\ |m - m_1| \leq 2s \qquad |m - m'_1| \leq 2s \\ |m' - m'_1| \leq 2s \qquad |m' - m_1| \leq 2s. \end{gathered}$$

(ii) If the above are satisfied, then

$$\begin{aligned} \mathscr{C}^{lm;l'm'}_{l_1m_1;l'_1m'_1} = (-1)^{m_1-m} & \sqrt{\frac{(2l_1+1)(2l'_1+1)}{(2l+1)(2l'+1)}} \\ & \times \sum_{\mu} \langle l, m | s, -\mu; s, m+\mu \rangle \langle l', m' | s, m' - m'_1 - \mu; s, \mu + m'_1 \rangle \\ & \times \langle l_1, m_1 | s, m' - m'_1 - \mu; s, \mu + m \rangle \langle l'_1, m'_1 | s, -\mu; s, \mu + m'_1 \rangle . \end{aligned}$$

The coefficients satisfy the following symmetry properties:

$$\begin{aligned} \mathscr{C}^{lm;l'm'}_{l_1m_1;l'_1m'_1} &= \mathscr{C}^{l'm';lm}_{l'_1m'_1;l_1m_1} \\ (2l+1)(2l'+1)\mathscr{C}^{lm;l'm'}_{l_1m_1;l'_1m'_1} &= (2l_1+1)(2l'_1+1)\mathscr{C}^{l_1m_1;l'_1m'_1}_{lm;l'm'} \\ \mathscr{C}^{l-m;l'-m'}_{l_1-m_1;l'_1-m'_1} &= \mathscr{C}^{lm;l'm'}_{l_1m_1;l'_1m'_1} . \end{aligned}$$

We list the non-zero independent coefficients for a spin 1/2 particle:

$$\mathscr{C}^{11;11}_{11;11} = 1$$

$$\mathscr{C}^{11;10}_{11;10} = \tfrac{1}{2} \qquad \mathscr{C}^{11;10}_{11;00} = \tfrac{1}{6}$$

$$\mathscr{C}^{11;1-1}_{00;00} = \tfrac{1}{6} \qquad \mathscr{C}^{11;1-1}_{10;10} = -\tfrac{1}{2} \qquad \mathscr{C}^{11;1-1}_{10;00} = -\frac{1}{2\sqrt{3}}$$

$$\mathscr{C}^{11;00}_{11;00} = \tfrac{1}{2}$$

$$\mathscr{C}^{10;10}_{10;10} = \tfrac{1}{2} \qquad \mathscr{C}^{10;10}_{00;00} = \tfrac{1}{6}$$

$$\mathscr{C}^{10;00}_{00;10} = \tfrac{1}{2}$$

$$\mathscr{C}^{00;00}_{00;00} = \tfrac{1}{2}.$$

Appendix 8

Symmetry properties of the Cartesian reaction parameters

We consider $A + B \to A + B$, where all particles have spin 1/2 but A and B need not be identical. The additional symmetries when $A \equiv B$ are given separately. Many of the results given were derived by Thomas (Thomas, 1969). Results for $0 + 1/2 \to 0 + 1/2$ are obtained by simply suppressing the α, α' labels everywhere. For $1/2 + 1/2 \to 0 + 0$ the labels α' and β' are suppressed everywhere.

A8.1 The CM reaction parameters

To begin with there are 256 parameters.

(*a*) *Parity.* Use of parity in both H amplitudes gives (see (5.6.4))

$$(\alpha\beta|\alpha'\beta') = \xi_\alpha^{\mathscr{P}} \xi_\beta^{\mathscr{P}} \xi_{\alpha'}^{\mathscr{P}} \xi_{\beta'}^{\mathscr{P}} (\alpha\beta|\alpha'\beta') \tag{A8.1}$$

where

$$\xi_0^{\mathscr{P}} = \xi_Y^{\mathscr{P}} = 1 \qquad \xi_X^{\mathscr{P}} = \xi_Z^{\mathscr{P}} = -1.$$

This implies that the parameter is zero when the number of X labels plus the number of Z labels in it is an odd number. This eliminates one half of the coefficients, leaving 128.

Use of parity in just one amplitude leads to

$$(\alpha\beta|\alpha'\beta') = \xi_\alpha \xi_\beta \xi_{\alpha'}^* \xi_{\beta'}^* (\alpha_{\mathscr{P}}\beta_{\mathscr{P}}|\alpha'_{\mathscr{P}}\beta'_{\mathscr{P}}) \tag{A8.2}$$

where

$$\xi_0 = \xi_Y = 1 \qquad \xi_Z = -\xi_X = i$$

and for any label α, $\alpha_{\mathscr{P}}$ means

$$0 \longleftrightarrow Y \qquad X \longleftrightarrow Z.$$

An example: under $\mathscr{P}$, $(00|Y0) = (YY|0Y)$.

This eliminates 64 coefficients, leaving 64.

(*b*) *Time reversal.*[1] Use of time reversal in both H amplitudes gives

$$(\alpha\beta|\alpha'\beta') = \xi_\alpha^{\mathcal{T}}\,\xi_\beta^{\mathcal{T}}\,\xi_{\alpha'}^{\mathcal{T}}\,\xi_{\beta'}^{\mathcal{T}}\,(\alpha'\beta'|\alpha\beta) \tag{A8.3}$$

where

$$\xi_0^{\mathcal{T}} = \xi_Y^{\mathcal{T}} = \xi_Z^{\mathcal{T}} = 1 \qquad \xi_X^{\mathcal{T}} = -1.$$

This eliminates 12 coefficients, and the use of parity combined with time-reversal a further 12, leaving 40. Use of time reversal in one H only gives

$$(\alpha\beta|\alpha'\beta') = \eta_\alpha\eta_\beta\eta_{\alpha'}^*\eta_{\beta'}^* \sum_{\nu,\nu',\gamma,\gamma'} C_{\gamma'\gamma}^{\alpha'_{\mathcal{T}}\alpha_{\mathcal{T}}}\,C_{\nu'\nu}^{\beta'_{\mathcal{T}}\beta_{\mathcal{T}}}\,(\gamma\nu|\gamma'\nu') \tag{A8.4}$$

where, for any label α, $\alpha_{\mathcal{T}}$ means

$$X \longleftrightarrow Y \qquad 0 \longleftrightarrow Z.$$

Also,

$$\eta_0 = \eta_Z = 1 \qquad \eta_X = -\eta_Y = i$$

and

$$C_{\gamma'\gamma}^{\alpha'\alpha} = -\tfrac{1}{16}\zeta_\alpha\zeta_\gamma\ \mathrm{Tr}\,(\sigma_{\gamma'}\sigma_{\alpha'}\sigma_\gamma\sigma_\alpha)$$

with

$$\zeta_0 = \zeta_X = \zeta_Z = 1 \qquad \zeta_Y = -1.$$

This leads to four new conditions:

$$\begin{aligned}(XX|ZZ) &= (XX|XX) - (YY|00) - 1\\ (XZ|XZ) &= (XX|XX) + (0Y|0Y) - 1\\ (ZX|XZ) &= -(XX|XX) + (Y0|Y0) + 1\\ (Z0|Z0) &= (0Z|0Z) + (X0|X0) - (0X|0X).\end{aligned} \tag{A8.5}$$

The first three of these were given by Thomas for NN scattering.[2]

We are now left with 36 linearly independent reaction parameters, just what is expected since there are six independent helicity amplitudes in the reaction. The expressions for the 36 observables parameters in terms of the helicity amplitudes are given in Appendix 10.

[1] Clearly the results of this section do not hold for $1/2 + 1/2 \to 0 + 0$.

[2] Thomas found these by 'brute force' from studying the relations between observable parameters and helicity amplitudes – he knew that three extra conditions had to exist.

(*c*) *Identical particles.* When $A \equiv B$ as in nucleon–nucleon scattering we get

$$(\alpha\beta|\alpha'\beta') = \xi_\alpha^{\mathscr{S}} \xi_\beta^{\mathscr{S}} \xi_{\alpha'}^{\mathscr{S}} \xi_{\beta'}^{\mathscr{S}} (\beta\alpha|\beta'\alpha') \tag{A8.6}$$

where

$$\xi_X^{\mathscr{S}} = \xi_Y^{\mathscr{S}} = -1 \qquad \xi_O^{\mathscr{S}} = \xi_Z^{\mathscr{S}} = 1.$$

This eliminates 11 of the parameters, leaving the customary 25.

A8.2 The Argonne Lab reaction parameters

The label 'ARG' is only appended where confusion is possible.

(*a*) *Parity.* An Argonne Lab parameter vanishes if the number of S labels plus the number of L labels in it is an odd number.

Also one has

$$(\alpha\beta|\alpha'\beta')_{\text{Lab}}^{\text{ARG}} = \xi_\alpha \xi_\beta \xi_{\alpha'}^* \xi_{\beta'}^* (\alpha_{\mathscr{P}}\beta_{\mathscr{P}}|\alpha'_{\mathscr{P}}\beta'_{\mathscr{P}})_{\text{Lab}}^{\text{ARG}} \tag{A8.7}$$

where ξ_α and the parity operation $\mathscr{P}$ are defined in (A8.2). Of course the $\mathscr{P}$ operation now reads

$$0 \longleftrightarrow N \qquad S \longleftrightarrow L$$

and

$$\xi_0 = \xi_N = 1 \qquad \xi_L = -\xi_S = i.$$

Some examples are the following

$$C_{NN} = A_{NN} \qquad D_{NN}^{(A)} = D_{NN}^{(B)} \qquad K_{NN}^{(A)} = K_{NN}^{(B)}$$
$$(LS|NN)_{\text{Lab}}^{\text{ARG}} = (SL|00)_{\text{Lab}}^{\text{ARG}} = A_{SL}.$$

(*b*) *Time reversal.* One finds

$$(\alpha, \beta|\alpha', \beta')_{\text{Lab}}^{\text{ARG}} = (\alpha'_{\mathscr{T}}, \beta'_{\mathscr{T}}|\alpha_{\mathscr{T}}, \beta_{\mathscr{T}})_{\text{Lab}}^{\text{ARG}} \tag{A8.8}$$

where for particle A (i.e. for α, α')

$$0_{\mathscr{T}} = 0 \qquad N_{\mathscr{T}} = N$$
$$S_{\mathscr{T}} = -\cos\alpha_C\, S + \sin\alpha_C\, L \qquad L_{\mathscr{T}} = \sin\alpha_C\, S + \cos\alpha_C\, L$$

while for particle B (i.e. for β, β')

$$0_{\mathscr{T}} = 0 \qquad N_{\mathscr{T}} = N$$
$$S_{\mathscr{T}} = \cos\theta_{\text{R}}\, S + \sin\theta_{\text{R}}\, L \qquad L_{\mathscr{T}} = \sin\theta_{\text{R}}\, S - \cos\theta_{\text{R}}\, L.$$

Here α_C is of course the Wick helicity rotation angle for C = final particle A; $\alpha_C = \theta_{\text{L}}$, the Lab scattering angle, for $NN \to NN$, see subsection 2.2.4 and θ_{R} is the Lab recoil angle.

Some examples are the following:

$$A^{(A)}_{\text{ARG}} \equiv (N0|00)^{\text{ARG}}_{\text{Lab}} = (00|N0)^{\text{ARG}}_{\text{Lab}} \equiv \mathscr{P}^{(A)}_{\text{ARG}}$$

that is, the analysing power for particle A = the polarizing power for particle A.

Also

$$\begin{aligned} A_{SS} &\equiv (SS|00)^{\text{ARG}}_{\text{Lab}} \\ &= (00| -\cos\alpha_C\, S + \sin\alpha_C\, L, \cos\theta_{\text{R}}\, S + \sin\theta_{\text{R}}\, L)^{\text{ARG}}_{\text{Lab}} \\ &= -\cos\alpha_C \cos\theta_{\text{R}}\, C_{SS} - \cos\alpha_C \sin\theta_{\text{R}}\, C_{SL} \\ &\quad + \sin\alpha_C \cos\theta_{\text{R}}\, C_{LS} + \sin\alpha_C \sin\theta_{\text{R}}\, C_{LL} \end{aligned}$$

and

$$\frac{D^{(A)}_{LS} + D^{(A)}_{SL}}{D^{(A)}_{LL} - D^{(A)}_{SS}} = \tan\alpha_C \qquad \frac{D^{(B)}_{LS} + D^{(B)}_{SL}}{D^{(B)}_{SS} - D^{(B)}_{LL}} = \tan\theta_{\text{R}} \qquad \text{etc.}$$

Appendix 9

'Shorthand' notation and nomenclature for the Argonne Lab reaction parameters

We consider $A + B \to A + B$, all particles having spin 1/2. The order is

$$(\text{beam, target}|\text{scattered, recoil}).$$

Since *all* the parameters listed are the Argonne Lab ones we shall not keep repeating those labels.

Argonne Lab parameters $(\alpha\beta|\alpha'\beta')^{\text{ARG}}_{\text{Lab}}$	Shorthand notation	Name
$(N0|00)$	$A^{(A)}$	Analyzing power for particle A
$(0N|00)$	$A^{(B)}$	Analyzing power for particle B
$(00|N0)$	$P^{(A)}$	Polarizing power for particle A
$(00|0N)$	$P^{(B)}$	Polarizing power for particle B
$(00|\alpha'\beta')$	$C_{\alpha'\beta'}$	Final state correlation parameters
$(\alpha\beta|00)$	$A_{\alpha\beta}$	Initial state correlation parameters
$(\alpha0|\alpha'0)$	$D^{(A)}_{\alpha\alpha'}$	Depolarization parameters for A
$(0\beta|0\beta')$	$D^{(B)}_{\beta\beta'}$	Depolarization parameters for B
$(\alpha0|0\beta')$	$K^{(A)}_{\alpha\beta'}$	Polarization transfer parameters for A
$(0\beta|\alpha'0)$	$K^{(B)}_{\beta\alpha'}$	Polarization transfer parameters for B

It has been agreed (Ann Arbor Convention 1977; see Krisch, 1978) that no special names shall be given to the three-and four-spin parameters.

It should be noted that in the days when very few spin measurements seemed feasible, certain of the above parameters were given specific, but not very systematic, symbols. These are no longer appropriate, but to facilitate comparison with the older literature we list the most important:

$$D = D_{NN}; \qquad R = D_{SS}^{(A)}; \qquad \bar{R} = D_{SS}^{(B)};$$

$$A = D_{LS}^{(A)}; \qquad \bar{A} = D_{LS}^{(B)}; \qquad R' = D_{SL}^{(A)}; \qquad \bar{R}' = -D_{SL}^{(B)};$$

$$A' = D_{LL}^{(A)}; \qquad \bar{A}' = -D_{LL}^{(B)}.$$

Appendix 10

The linearly independent reaction parameters for various reactions and their relation to the helicity amplitudes

We give here the expressions for the fundamental CM reaction parameters (Section 5.6) in terms of the helicity amplitudes for various reactions. We also list the relation between the CM reaction parameters and those used in the Argonne convention.

A10.1 $0 + 1/2 \to 0 + 1/2$

An example is $\pi p \to \pi p$. We have

$$d\sigma/dt = |H_{++}|^2 + |H_{+-}|^2.$$

In Table A10.1 the expressions in the right-hand column correspond to the CM reaction parameters multiplied by $d\sigma/dt$. The order of the labels is (target|recoil).

The relation to Argonne Lab parameters is given in Table A10.2.

Table A10.1.

CM parameter	Shorthand notation	Expression in terms of helicity amplitudes
$(0Y\|00) = (00\|0Y)$	P	$2\ \mathrm{Im}\,(H^*_{+-}H_{++})$
$(0X\|0X)$	D_{XX}	$\|H_{++}\|^2 - \|H_{+-}\|^2$
$(0Z\|0X)$	D_{ZX}	$-2\ \mathrm{Re}\,(H_{++}H^*_{+-})$

Table A10.2.

CM parameters	Argonne Lab parameters
P	$-P_{\mathrm{ARG}}$
D_{XX}	$-\cos\theta_{\mathrm{R}}\ D_{SS}+\sin\theta_{\mathrm{R}}\ D_{LS}$
D_{ZX}	$\sin\theta_{\mathrm{R}}\ D_{SS}+\cos\theta_{\mathrm{R}}\ D_{LS}$

In the old literature one finds $R = D_{SS}$, $A = D_{LS}$.

A10.2 $A(1/2) + B(1/2) \to 0 + 0$

An example is $\bar{p}p \to \pi\pi$. We have

$$\frac{d\sigma}{dt} = \frac{1}{2}\left(|H_{++}|^2 + |H_{+-}|^2\right).$$

In Table A10.3 the expressions in the right-hand column are the CM reaction parameters multiplied by $d\sigma/dt$.

A10.3 $A + B \to A + B$ all with spin 1/2

Because of their complexity we list in the second column of Table A10.4 a convenient set of 36 linearly independent Argonne Lab parameters. The cases $A \neq B$ and $A = B$ are both included.

Next, in Tables A10.5 and A10.6 we give expressions for the CM reaction parameters in terms of the helicity amplitudes. The results hold whether A is different from B or is identical to it. If $A = B$ then one should put $\phi_6 = -\phi_5$. Also, those parameters marked † are then no longer independent.

We have

$$\frac{d\sigma}{dt} = \frac{1}{2}\left(|\phi_1|^2 + |\phi_2|^2 + |\phi_3|^2 + |\phi_4|^2 + 2|\phi_5|^2 + 2|\phi_6|^2\right). \tag{A10.1}$$

In Tables A10.5 and A10.6 the entries in the right hand column correspond to the CM reaction parameters multiplied by $d\sigma/dt$.

Table A10.3.

CM parameters	Shorthand notation	Expression in terms of helicity amplitudes
$(OY\|00)$	$A^{(B)}$	$2\ \mathrm{Im}\ (H_{++}H^*_{+-})$
$(Y0\|00)$	$A^{(A)} = -A^{(B)}$	$2\ \mathrm{Im}\ (H_{++}H^*_{-+})$
$(XX\|00)$	A_{XX}	$\|H_{++}\|^2 - \|H_{+-}\|^2$
$(XZ\|00)$	A_{XZ}	$-2\ \mathrm{Re}\ (H_{++}H^*_{+-})$

Table A10.4. 36 linearly independent Argonne Lab parameters. As usual θ_L is the Lab scattering angle of A and θ_R is the recoil angle of B

Type of measurement	$A = B$	Additional parameters if $A \neq B$	Relation when $A = B$
No spin	$d\sigma/dt$		
One spin	$A_{ARG}^{(A)}$	$A_{ARG}^{(B)}$	$A_{ARG}^{(A)} = A_{ARG}^{(B)}$
Two spins	A_{SS}, A_{LL}, A_{NN}		
	A_{SL}	A_{LS}	$A_{LS} = A_{SL}$
	$D_{LL}^{(B)}, D_{NN}^{(B)}$		
	$D_{SS}^{(B)}$	$D_{SS}^{(A)}$	$D_{SS}^{(A)} = -\sin(\theta_R + \theta_L)D_{SL}^{(B)} - \cos(\theta_R + \theta_L)D_{SS}^{(B)}$
	$D_{SL}^{(B)}$	$D_{SL}^{(A)}$	$D_{SL}^{(A)} = -\cos(\theta_R + \theta_L)D_{SL}^{(B)} + \sin(\theta_R + \theta_L)D_{SS}^{(B)}$
	$K_{SS}^{(A)}, K_{LL}^{(A)}, K_{NN}^{(A)}$		
	$K_{SL}^{(A)}$	$K_{LS}^{(A)}$	$K_{LS}^{(A)} = -K_{SL}^{(A)} + \tan\theta_R \left[K_{SS}^{(A)} - K_{LL}^{(A)}\right]$
Three spins	$(SN\|0S)$	$(SN\|S0)$	$-(SN\|S0) = \cos(\theta_R + \theta_L)(NS\|0S) + \sin(\theta_R + \theta_L)(NS\|0L)$
	$(NS\|0S)$		
	$(SN\|0L)$	$(SN\|L0)$	$(SN\|L0) = \sin(\theta_R + \theta_L)(NS\|0S) - \cos(\theta_R + \theta_L)(NS\|0L)$
	$(LS\|0N)$		
	$(SL\|0N)$		
	$(LN\|0S)$	$(LN\|S0)$	$(LN\|S0) = (\cos\theta_L/\cos\theta_R)(NL\|0S) - \sin(\theta_L + \theta_R) [(NS\|0S) + \tan\theta_R(NS\|0L)]$
	$(NL\|0S)$		
	$(SS\|0N)$	$(SS\|N0)$	$(SS\|N0) = (SS\|0N)$
		$(LN\|0L)$	$(LN\|0L) = (SN\|0S)\tan\theta_R [(LN\|0S) + (SN\|0L)]$
Four spins	$(SS\|SS)$		
	$(SS\|LS)$	$(SS\|SL)$	$(SS\|SL) = (SS\|LS) + \sin(\theta_R + \theta_L)(A_{NN} - 1)$

Table A10.5. Relation between CM reaction parameters and helicity amplitudes for one- and two-spin measurements

Shorthand	CM parameter	Formula
$A^{(B)}$	$(0Y\|00)$	$\mathrm{Im}\,[\phi_5^*(\phi_1+\phi_3)-\phi_6^*(\phi_2-\phi_4)]$
$A^{(A)}$	$\dagger(Y0\|00)$	$\mathrm{Im}\,[\phi_6^*(\phi_1+\phi_3)-\phi_5^*(\phi_2-\phi_4)]$
$D_{XX}^{(B)}$	$(0X\|0X)$	$\mathrm{Re}\,(\phi_1\phi_3^*+\phi_2\phi_4^*)-\|\phi_5\|^2+\|\phi_6\|^2$
$D_{ZZ}^{(B)}$	$(0Z\|0Z)$	$(1/2)(\|\phi_1\|^2-\|\phi_2\|^2+\|\phi_3\|^2 -\|\phi_4\|^2-2\|\phi_5\|^2+2\|\phi_6\|^2)$
$D_{ZX}^{(B)}$	$(0Z\|0X)$	$-\,\mathrm{Re}\,[(\phi_1+\phi_3)\phi_5^*+(\phi_2-\phi_4)\phi_6^*]$
$D_{YY}^{(B)}$	$(0Y\|0Y)$	$\mathrm{Re}\,[\phi_1\phi_3^*-\phi_2\phi_4^*]+\|\phi_5\|^2+\|\phi_6\|^2$
$D_{XX}^{(A)}$	$\dagger(X0\|X0)$	$\mathrm{Re}\,(\phi_1\phi_3^*+\phi_2\phi_4^*)+\|\phi_5\|^2-\|\phi_6\|^2$
$D_{XZ}^{(A)}$	$\dagger(X0\|Z0)$	$\mathrm{Re}\,[(\phi_1+\phi_3)\phi_6^*+(\phi_2-\phi_4)\phi_5^*]$
A_{XX}	$(XX\|00)$	$\mathrm{Re}\,(\phi_1\phi_2^*+\phi_3\phi_4^*)$
A_{YY}	$(YY\|00)$	$\mathrm{Re}\,(\phi_3\phi_4^*-\phi_1\phi_2^*+2\phi_5\phi_6^*)$
A_{ZZ}	$(ZZ\|00)$	$(1/2)(\|\phi_1\|^2+\|\phi_2\|^2-\|\phi_3\|^2-\|\phi_4\|^2)$
A_{XZ}	$(XZ\|00)$	$\mathrm{Re}\,[(\phi_1-\phi_3)\phi_6^*-(\phi_2+\phi_4)\phi_5^*]$
A_{ZX}	$\dagger(ZX\|00)$	$\mathrm{Re}\,[(\phi_1-\phi_3)\phi_5^*-(\phi_2+\phi_4)\phi_6^*]$
$K_{XX}^{(A)}$	$(X0\|0X)$	$\mathrm{Re}\,\left(\phi_1\phi_4^*+\phi_3\phi_2^*\right)$
$K_{XZ}^{(A)}$	$(X0\|0Z)$	$\mathrm{Re}\,\left[(\phi_1-\phi_3)\phi_6^*+(\phi_2+\phi_4)\phi_5^*\right]$
$K_{ZZ}^{(A)}$	$(Z0\|0Z)$	$(1/2)(\|\phi_1\|^2-\|\phi_2\|^2-\|\phi_3\|^2+\|\phi_4\|^2)$
$K_{ZX}^{(A)}$	$\dagger(Z0\|0X)$	$-\,\mathrm{Re}\,\left[(\phi_1-\phi_3)\phi_5^*+(\phi_2+\phi_4)\phi_6^*\right]$
$K_{YY}^{(A)}$	$(Y0\|0Y)$	$\mathrm{Re}\,\left(\phi_1\phi_4^*-\phi_3\phi_2^*+2\phi_5\phi_6^*\right)$

In Tables A10.7 and A10.8 we show the relationship between the CM reaction parameters and the Argonne Lab parameters. (Many of these are due to N.H. Buttimore, unpublished.) The tables are arranged so that pairs of relations can be used to solve for the Argonne Lab parameters in terms of the CM reaction parameters and then, via Tables A10.5 and A10.6, in terms of the helicity amplitudes. A few entries, marked $*$, involve parameters not listed in Table A10.4. α_C is, as usual, the Wick helicity rotation angle for particle $C = A$. The formulae (A10.2) enable one to eliminate them in favour of the Table A10.4 parameters, if so desired (for the meaning of the dagger see the text above (A10.1)):

Table A10.6. Relation between CM reaction parameters and helicity amplitudes for the three- and four-spin measurements

CM parameter	Formula
$(XX\|0Y)$	$-\,\text{Im}\,\left[(\phi_1-\phi_3)\phi_6^*-(\phi_2+\phi_4)\phi_5^*\right]$
$\dagger(ZZ\|0Y)$	$\text{Im}\,\left[(\phi_1-\phi_3)\phi_5^*-(\phi_2+\phi_4)\phi_6^*\right]$
$(XZ\|0Y)$	$\text{Im}\,(\phi_2\phi_3^*-\phi_1\phi_4^*)$
$(ZX\|0Y)$	$\text{Im}\,(\phi_2\phi_4^*-\phi_1\phi_3^*)$
$(ZY\|0X)$	$\text{Im}\,(\phi_1\phi_3^*+\phi_2\phi_4^*)$
$\dagger(ZY\|0Z)$	$\text{Im}\,\left[(\phi_1-\phi_3)\phi_5^*+(\phi_2+\phi_4)\phi_6^*\right]$
$(XY\|0X)$	$\text{Im}\,\left[(\phi_1-\phi_3)\phi_6^*+(\phi_2+\phi_4)\phi_5^*\right]$
$(XY\|0Z)$	$\text{Im}\,(\phi_1\phi_2^*+\phi_3\phi_4^*)$
$(YX\|0X)$	$\text{Im}\,[(\phi_1+\phi_3)\phi_6^*+(\phi_2-\phi_4)\phi_5^*]$
$(YX\|0Z)$	$\text{Im}\,\left(\phi_1\phi_2^*-\phi_3\phi_4^*+2\phi_5\phi_6^*\right)$
$\dagger(XY\|X0)$	$\text{Im}\,\left[(\phi_1+\phi_3)\phi_5^*+(\phi_2-\phi_4)\phi_6^*\right]$
$\dagger(XY\|Z0)$	$\text{Im}\,\left(\phi_1\phi_2^*-\phi_3\phi_4^*-2\phi_5\phi_6^*\right)$
$(YZ\|0X)$	$\text{Im}\,\left(\phi_1\phi_4^*+\phi_2\phi_3^*-2\phi_5\phi_6^*\right)$
$\dagger(ZY\|X0)$	$\text{Im}\,\left(\phi_1\phi_4^*+\phi_2\phi_3^*+2\phi_5\phi_6^*\right)$
$(XX\|XX)$	$(1/2)(\|\phi_1\|^2+\|\phi_2\|^2+\|\phi_3\|^2+\|\phi_4\|^2-2\|\phi_5\|^2-2\|\phi_6\|^2)$
$(XX\|XZ)$	$\text{Re}\,\left[(\phi_1+\phi_3)\phi_5^*-(\phi_2-\phi_4)\phi_6^*\right]$
$\dagger(XX\|ZX)$	$\text{Re}\,\left[(\phi_1+\phi_3)\phi_6^*-(\phi_2-\phi_4)\phi_5^*\right]$

$$\begin{aligned} D_{LS}^{(B)} &= -D_{SL}^{(B)} + \tan\theta_{\text{R}}\left(D_{SS}^{(B)} - D_{LL}^{(B)}\right) \\ (NL|0L) &= (NS|0S) + \tan\theta_{\text{R}}\left[(NS|0L) + (NL|0S)\right] \\ (LN|L0) &= (SN|S0) - \tan\alpha_C\left[(LN|S0) + (SN|L0)\right] \\ (SS|LL) &= -(SS|SS) + \tan(\alpha_C+\theta_{\text{R}})\left[(SS|LS) - (SS|SL)\right] \\ &\quad + \sec(\alpha_C+\theta_{\text{R}})[A_{NN}-1]. \end{aligned} \tag{A10.2}$$

A10.4 Photoproduction of pseudoscalar mesons

The simplest observables to measure are the *photon beam asymmetry* Σ, the *target asymmetry* T and the *polarizing power* P.

The asymmetry Σ is measured by comparing the differential cross-sections for the photon beam linearly polarized parallel to and perpendicular to the reaction plane, using an unpolarized proton target. One has

$$\Sigma = \frac{d\sigma_\perp - d\sigma_\parallel}{d\sigma_\perp + d\sigma_\parallel}.$$

Table A10.7. Relation between CM and Argonne Lab parameters for one- and two-spin measurements. See the text for explanation

CM	Argonne
$(0Y\|00)$	$-A^{(B)}_{\mathrm{ARG}}$
†$(Y0\|00)$	$A^{(A)}_{\mathrm{ARG}}$
$(0X\|0X)$	$-\cos\theta_{\mathrm{R}}\, D^{(B)}_{SS} - \sin\theta_{\mathrm{R}}\, D^{(B)}_{SL}$
*$(0Z\|0Z)$	$\sin\theta_{\mathrm{R}}\, D^{(B)}_{LS} - \cos\theta_{\mathrm{R}}\, D^{(B)}_{LL}$
*$(0Z\|0X)$	$\cos\theta_{\mathrm{R}}\, D^{(B)}_{LS} + \sin\theta_{\mathrm{R}}\, D^{(B)}_{LL}$
$(0Y\|0Y)$	$D^{(B)}_{NN}$
†$(X0\|X0)$	$\cos\alpha_C\, D^{(A)}_{SS} - \sin\alpha_C\, D^{(A)}_{SL}$
†$(X0\|Z0)$	$\sin\alpha_C\, D^{(A)}_{SS} + \cos\alpha_C\, D^{(A)}_{SL}$
$(XX\|00)$	A_{SS}
$(YY\|00)$	$-A_{NN}$
$(ZZ\|00)$	$-A_{LL}$
$(XZ\|00)$	$-A_{SL}$
†$(ZX\|00)$	A_{LS}
$(X0\|0X)$	$-\cos\theta_{\mathrm{R}}\, K^{(A)}_{SS} - \sin\theta_{\mathrm{R}}\, K^{(A)}_{SL}$
$(X0\|0Z)$	$-\sin\theta_{\mathrm{R}}\, K^{(A)}_{SS} + \cos\theta_{\mathrm{R}}\, K^{(A)}_{SL}$
$(Z0\|0Z)$	$-\sin\theta_{\mathrm{R}}\, K^{(A)}_{LS} + \cos\theta_{\mathrm{R}}\, K^{(A)}_{LL}$
†$(Z0\|0X)$	$-\cos\theta_{\mathrm{R}}\, K^{(A)}_{LS} - \sin\theta_{\mathrm{R}}\, K^{(A)}_{LL}$
$(Y0\|0Y)$	$-K^{(A)}_{NN}$

T is measured using an unpolarized photon beam incident on a transversely polarized proton target with spin polarization P_T along ($\uparrow$) or opposite ($\downarrow$) to the normal to the reaction plane. One has

$$T = \frac{1}{P_T}\frac{d\sigma_\uparrow - d\sigma_\downarrow}{d\sigma_\uparrow + d\sigma_\downarrow}.$$

Both Σ and T function as analysing powers of the reaction.

The polarizing power P is just the degree of polarization along the normal to the reaction plane of the recoil proton beam when an unpolarized photon beam is incident on an unpolarized proton target.

Table A10.8. Relation between CM and Argonne Lab parameters for the three- and four-spin measurements

CM	Argonne
$(XX\|0Y)$	$-(SS\|0N)$
$\dagger(ZZ\|0Y)$	$-(SS\|N0)$
$(XZ\|0Y)$	$(SL\|0N)$
$(ZX\|0Y)$	$-(LS\|0N)$
$(ZY\|0X)$	$\cos\theta_R\,(LN\|0S) + \sin\theta_R\,(LN\|0L)$
$\dagger(ZY\|0Z)$	$\sin\theta_R\,(LN\|0S) - \cos\theta_R\,(LN\|0L)$
$(XY\|0X)$	$\cos\theta_R\,(SN\|0S) + \sin\theta_R\,(SN\|0L)$
$(XY\|0Z)$	$\sin\theta_R\,(SN\|0S) - \cos\theta_R\,(SN\|0L)$
$(YX\|0X)$	$-\cos\theta_R\,(NS\|0S) - \sin\theta_R\,(NS\|0L)$
$(YX\|0Z)$	$-\sin\theta_R\,(NS\|0S) + \cos\theta_R\,(NS\|0L)$
$\dagger(XY\|X0)$	$\sin\alpha_C\,(SN\|L0) - \cos\alpha_C\,(SN\|S0)$
$\dagger(XY\|Z0)$	$-\cos\alpha_C\,(SN\|L0) - \sin\alpha_C\,(SN\|S0)$
$*(YZ\|0X)$	$\cos\theta_R\,(NL\|0S) + \sin\theta_R\,(NL\|0L)$
$\dagger * (ZY\|X0)$	$\sin\alpha_C\,(LN\|L0) - \cos\alpha_C\,(LN\|S0)$
$*(XX\|XX)$	$\sin\alpha_C\,[\cos\theta_R\,(SS\|LS) + \sin\theta_R\,(SS\|LL)]$
	$-\cos\alpha_C\,[\cos\theta_R\,(SS\|SS) + \sin\theta_R\,(SS\|SL)]$
$*(XX\|XZ)$	$\cos\alpha_C\,[\cos\theta_R\,(SS\|SL) + \sin\theta_R\,(SS\|SS)]$
	$+\sin\alpha_C\,[\sin\theta_R\,(SS\|LS) - \cos\theta_R\,(SS\|LL)]$
$\dagger * (XX\|ZX)$	$-\cos\alpha_C\,[\sin\theta_R\,(SS\|LL) + \cos\theta_R\,(SS\|LS)]$
	$-\sin\alpha_C\,[\cos\theta_R\,(SS\|SS) + \sin\theta_R\,(SS\|SL)]$

These observables are given in terms of the amplitudes defined in Appendix 5 by the following expressions:

$$\Sigma\frac{d\sigma}{dt} = 2\,\mathrm{Re}\,(S_1S_2^* - ND^*)$$
$$T\frac{d\sigma}{dt} = 2\,\mathrm{Im}\,(S_1N^* - S_2D^*)$$
$$P\frac{d\sigma}{dt} = 2\,\mathrm{Im}\,(S_2N^* - S_1D^*)$$
$$\frac{d\sigma}{dt} = |N|^2 + |S_1|^2 + |S_2|^2 + |D|^2.$$

For a very general discussion of the observables and possible measurements, see the review paper of Storrow (Storrow, 1978). Care should be taken regarding sign conventions for the various axes.

A10.5 Vector meson production in $0^-(1/2)^+ \to 1^-(1/2)^+$

The vector density matrix elements $\rho_{mm'}$ and transversely polarized target asymmetries T_0, T_+, T_-, which are commonly used observables, are given in terms of the vector meson production amplitudes P, defined above (Appendix 5); see also Field and Sidhu (1974) and Irving and Worden (1977):

$$\sigma_0 \equiv \rho_{00}\sigma = |P^0_{++}|^2 + |P^0_{+-}|^2 \equiv |P_0|^2$$
$$\sigma_\pm \equiv (\rho_{11} \pm \rho_{1-1})\sigma = |P^\pm_{++}|^2 + |P^\pm_{+-}|^2 \equiv |P_\pm|^2$$
$$\sqrt{2}\ \mathrm{Re}\ (\rho_{10}\sigma) = \ \mathrm{Re}\ (P^0_{++}P^{-*}_{++} + P^0_{+-}P^{-*}_{+-}$$
$$T_0\sigma_0 = -2\ \mathrm{Im}\ (P^0_{++}P^{0*}_{+-})$$
$$T_+\sigma_+ = -2\ \mathrm{Im}\ (P^+_{++}P^{+*}_{+-})$$
$$T_-\sigma_- = -2\ \mathrm{Im}\ (P^-_{++}P^{-*}_{+-})$$

with

$$\sigma = d\sigma/dt = \sigma_0 + \sigma_+ + \sigma_-.$$

A10.6 Baryon resonance production in $0^-(1/2)^+ \to 0^-(3/2)^+$

The density matrix elements $\rho_{mm'}$ are given in terms of the M amplitudes defined above (Appendix 5):

$$\rho_{33}\sigma = \tfrac{1}{2}\left(|M_1|^2 + |M_2|^2\right) \qquad \rho_{11} + \rho_{33} = \tfrac{1}{2}$$
$$\left\{\begin{matrix}\mathrm{Re}\\ \mathrm{Im}\end{matrix}\right\}\rho_{31}\sigma = \frac{1}{2}\left\{\begin{matrix}\mathrm{Re}\\ \mathrm{Im}\end{matrix}\right\}(M_0M_1^* + M_1'M_2^*)$$
$$\left\{\begin{matrix}\mathrm{Re}\\ \mathrm{Im}\end{matrix}\right\}\rho_{3-1}\sigma = \frac{1}{2}\left\{\begin{matrix}\mathrm{Re}\\ \mathrm{Im}\end{matrix}\right\}(M_0M_2^* - M_1'M_1^*)$$
$$\mathrm{Im}\ \rho_{3-3}\sigma = \ \mathrm{Im}\ (M_1M_2^*) \qquad \mathrm{Im}\ \rho_{1-1}\sigma = \ \mathrm{Im}\ (M_0M_1^*)$$

with

$$\sigma = d\sigma/dt = |M_0|^2 + |M_1|^2 + |M_1'|^2 + |M_2|^2.$$

Appendix 11
The Feynman rules for QCD

We present here a list of the Feynman rules for QCD that are valid in two classes of gauge:

- the *covariant* gauges labelled by a parameter 'a' ($a = 1$ is the Feynman gauge; $a = 0$ the Landau gauge) in which the subsidiary condition, at least at the classical level, is $\partial^\mu A^c_\mu = 0$ for all values of the colour label c, and the gauge-fixing term in the lagrangian is $\frac{-1}{2a}\sum_c(\partial^\mu A^c_\mu)^2$;
- an *axial* gauge, one of a family again labelled by 'a', in which the subsidiary condition is $n^\mu A^c_\mu = 0$ for all c, where n^μ is a fixed space-like or null 4-vector, and where the gauge-fixing term in the lagrangian is $\frac{-1}{2a}\sum_c(n^\mu A^c_\mu)^2$.

We allow the quarks to have a mass parameter m, which should be put to zero when working with massless quarks.

(*a*) *The propagators*

lepton ——→—— p $\qquad \dfrac{i(\not{p}+m)}{p^2 - m^2 + i\epsilon}$

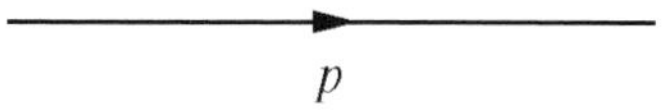

quark j ——→—— l, p $\qquad \delta_{jl}\dfrac{i(\not{p}+m)}{p^2 - m^2 + i\epsilon}$

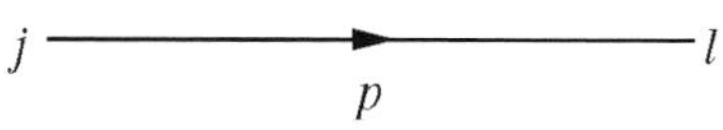

In the above the arrow indicates the flow of fermion number and p is the 4-momentum in that direction. (Note: j, l are quark colour labels, b, c

gluon and ghost colour labels.)

gluon: 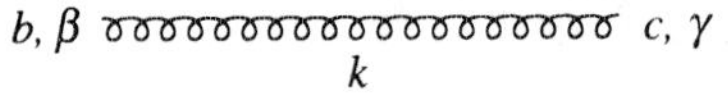

$$\delta_{bc}\frac{i}{k^2+i\epsilon}\times\begin{cases}\text{covariant gauges:}\\ \left[-g_{\beta\gamma}+(1-a)\dfrac{k_\beta k_\gamma}{k^2+i\epsilon}\right]\\ \text{axial gauges with } a=0\text{:}\\ \left[-g_{\beta\gamma}+\dfrac{n_\beta k_\gamma+n_\gamma k_\beta}{n\cdot k}-\dfrac{n^2 k_\beta k_\gamma}{(n\cdot k)^2}\right]\end{cases}$$

Note that in the above axial gauges the propagator is orthogonal to n^β, and it is orthogonal to k^β when $k^2=0$.

ghost:

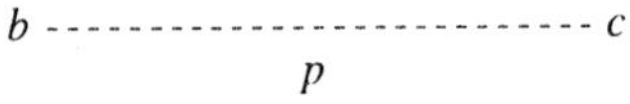

$$\delta_{bc}\frac{i}{p^2+i\epsilon}$$ (covariant gauges only).

(*b*) *The vertices*

quark–gluon vertex:

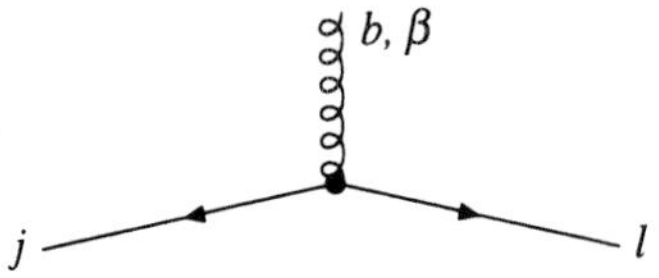

$$ig(\mathbf{t}^b)_{lj}\gamma^\beta$$

triple-gluon vertex:

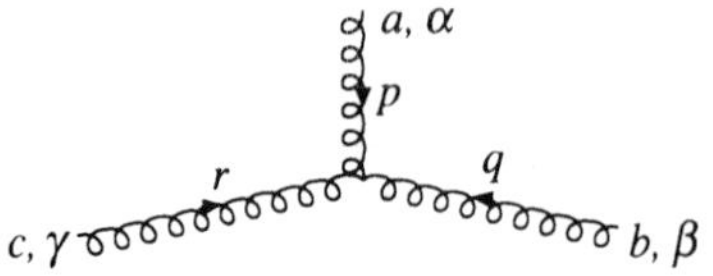

$$gf_{abc}\left[g^{\alpha\beta}(p-q)^\gamma + g^{\beta\gamma}(q-r)^\alpha + g^{\gamma\alpha}(q-r)^\beta\right]$$

where p, q, r are momenta, with $p+q+r=0$.

quartic gluon vertex:

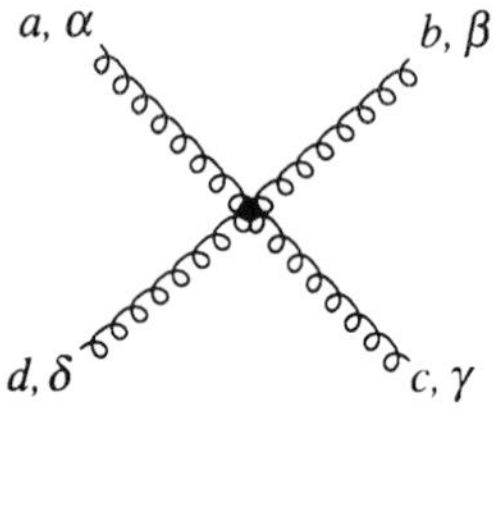

$$-ig^2\left[f_{eac}f_{ebd}(g^{\alpha\beta}g^{\gamma\delta}-g^{\alpha\delta}g^{\beta\gamma}) + f_{ead}f_{ebc}(g^{\alpha\beta}g^{\gamma\delta}-g^{\alpha\gamma}g^{\beta\delta}) + f_{eab}f_{ecd}(g^{\alpha\gamma}g^{\beta\delta}-g^{\alpha\delta}g^{\beta\gamma})\right]$$

gluon–ghost vertex:

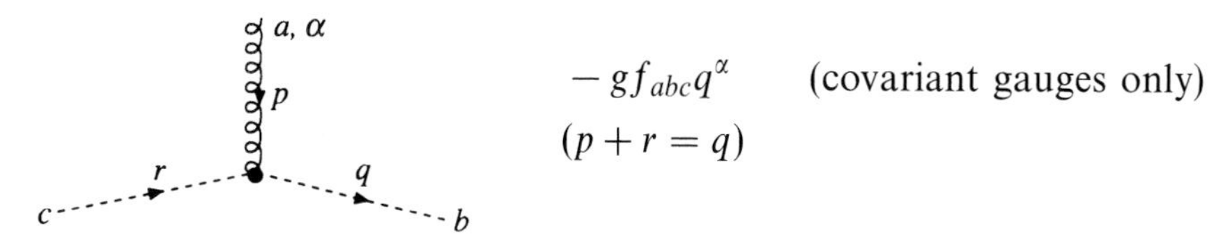

Note that the ghosts are scalar fields, but a factor -1 must be included for each closed loop, as in the case for fermions.

Appendix 12

Dirac spinors and matrix elements

A12.1 General properties

We discuss here some properties of four-component spinors and the Dirac matrices, which are particularly useful in the computation of helicity amplitudes. We shall not touch on the usual elementary considerations of the Dirac equation and the finding of its free-particle solutions. For that, the reader should consult Bjorken and Drell (1964).

The γ-matrices satisfy

$$\gamma^\mu \gamma^\nu + \gamma^\nu \gamma^\mu = 2g^{\mu\nu}, \tag{A12.1}$$

and we define

$$\gamma_5 \equiv \gamma^5 = i\gamma^0\gamma^1\gamma^2\gamma^3 \tag{A12.2}$$

and

$$\sigma^{\mu\nu} \equiv \frac{i}{2}\,[\gamma^\mu, \gamma^\nu]\,. \tag{A12.3}$$

For any 4-vector A^μ we use

$$A\!\!\!/ \equiv A_\mu \gamma^\mu = A^0\gamma^0 - A^1\gamma^1 - A^2\gamma^2 - A^3\gamma^3. \tag{A12.4}$$

The particle spinors u and the antiparticle spinors v satisfy the Dirac equations

$$(p\!\!\!/ - m)u(p) = 0 \tag{A12.5}$$

$$(p\!\!\!/ + m)v(p) = 0. \tag{A12.6}$$

Our normalization is

$$u^\dagger u = 2E \qquad v^\dagger v = 2E \tag{A12.7}$$

which implies

$$\bar{u}u = 2m \qquad \bar{v}v = -2m, \tag{A12.8}$$

the above holding also if $m = 0$.

With this normalization the cross-section formula (B.1) of Appendix B of Bjorken and Drell (1964) holds for both mesons and fermions, massive or massless.

Of very great importance are the properties of the traces of products of the γ-matrices. The most useful ones are:

$$\mathrm{Tr}\,(\gamma^{\mu_1}\gamma^{\mu_2}\ldots\gamma^{\mu_N}) = 0 \qquad \text{if } N \text{ is odd} \tag{A12.9}$$

$$\mathrm{Tr}\,(\gamma^{\mu}\gamma^{\nu}) = 4g^{\mu\nu} \tag{A12.10}$$

$$\mathrm{Tr}\left(\gamma^{\alpha}\gamma^{\beta}\gamma^{\mu}\gamma^{\nu}\right) = 4\left(g^{\alpha\beta}g^{\mu\nu} - g^{\alpha\mu}g^{\beta\nu} + g^{\alpha\nu}g^{\beta\mu}\right) \tag{A12.11}$$

$$\mathrm{Tr}\,(\gamma^{\mu_1}\gamma^{\mu_2}\ldots\gamma^{\mu_{N-1}}\gamma^{\mu_N}) = \mathrm{Tr}\,(\gamma^{\mu_N}\gamma^{\mu_{N-1}}\ldots\gamma^{\mu_2}\gamma^{\mu_1}) \tag{A12.12}$$

$$\mathrm{Tr}\,\gamma_5 = 0 \tag{A12.13}$$

$$\mathrm{Tr}\,(\gamma_5\gamma^{\alpha}) = \mathrm{Tr}\left(\gamma_5\gamma^{\alpha}\gamma^{\beta}\right) = \mathrm{Tr}\left(\gamma_5\gamma^{\alpha}\gamma^{\beta}\gamma^{\mu}\right) = 0 \tag{A12.14}$$

$$\mathrm{Tr}\left(\gamma_5\gamma^{\alpha}\gamma^{\beta}\gamma^{\mu}\gamma^{\nu}\right) = -4i\epsilon^{\alpha\beta\mu\nu}, \tag{A12.15}$$

where $\epsilon_{\alpha\beta\mu\nu}$, the totally antisymmetric tensor in Minkowski space, is defined with $\epsilon_{0123} = +1$.

From the γ-matrices one can construct a set of 16 linearly independent matrices,

$$\begin{array}{ccc} \Gamma_S \equiv I & \Gamma_V^{\mu} \equiv \gamma^{\mu} & \Gamma_T^{\mu\nu} \equiv \sigma^{\mu\nu} \\ \Gamma_A^{\mu} = \gamma^{\mu}\gamma_5 & & \Gamma_P = i\gamma_5, \end{array} \tag{A12.16}$$

chosen so that for each of them

$$\gamma^0\Gamma^{\dagger}\gamma^0 = \Gamma. \tag{A12.17}$$

Moreover, from the above,

$$\mathrm{Tr}\,\Gamma_j = 0 \qquad \text{for all } j \neq S \tag{A12.18}$$

and

$$\mathrm{Tr}\,\Gamma_S = 4.$$

An arbitrary 4×4 matrix can thus be written

$$M = SI + V_{\mu}\gamma^{\mu} + \tfrac{1}{2}T_{\mu\nu}\sigma^{\mu\nu} + A_{\mu}\gamma^{\mu}\gamma_5 + Pi\gamma_5 \tag{A12.19}$$

where $T_{\mu\nu} = -T_{\nu\mu}$.

It is *not* implied that the coefficients are Lorentz vectors or tensors etc. for a general M.

The expansion coefficients can be found from M simply by taking appropriate traces:

$$\begin{array}{ccc} S = \tfrac{1}{4}\,\mathrm{Tr}\,M & V_{\mu} = \tfrac{1}{4}\,\mathrm{Tr}\,\gamma_{\mu}M & T_{\mu\nu} = \tfrac{1}{4}\,\mathrm{Tr}\,\sigma_{\mu\nu}M \\ A_{\mu} = \tfrac{1}{4}\,\mathrm{Tr}\,\gamma_5\gamma_{\mu}M & & iP = \tfrac{1}{4}\,\mathrm{Tr}\,\gamma_5 M \end{array} \tag{A12.20}$$

A12.2 Helicity spinors and Lorentz transformations

Corresponding to the definition of helicity states given in subsection 1.2.1, the particle (or antiparticle) spinors for momentum p^μ are related to those at rest (see eqn (2.4.14)) by

$$u_n(\mathbf{p},\lambda) = D_{nm}[h(\mathbf{p})]u_m(\overset{\circ}{p},\lambda). \tag{A12.21}$$

For Dirac spinors, the representation matrices are given by
(i)

$$D[r_j(\theta)] = e^{-i\theta\Sigma_j/2} \tag{A12.22}$$

where $r_j(\theta)$ is a rotation through angle θ about the j-axis and

$$\Sigma_j \equiv \tfrac{1}{2}\epsilon_{jkl}\sigma^{kl}; \tag{A12.23}$$

(ii)

$$D\left[l_j(v)\right] = e^{-\omega\gamma^j\gamma^0/2}$$

where $l_j(v)$ is a boost along the j-axis and $\tanh\omega = v/c$.
Thus for $\mathbf{p} = (p,\theta,\phi)$

$$D\left[h(\mathbf{p})\right] = e^{-i\phi\Sigma_3/2}e^{-i\theta\Sigma_2/2}e^{-\omega\gamma^3\gamma^0/2} \tag{A12.24}$$

where now

$$\cosh\omega/2 = \sqrt{(E+mc^2)/(2mc^2)}$$

and

$$\tanh\omega/2 = pc/(E+mc^2).$$

The actual form of the spinors depends upon the representation used for the γ-matrices. Two useful choices will be discussed later.

The helicity spinors of course represent eigenstates of helicity. They are thus eigenspinors of the matrix that represents the helicity operator. The most convenient form is in terms of the covariant helicity spin vectors $\mathscr{S}^\mu(\mathbf{p},\lambda)$ introduced in Section 3.4. One finds that

$$\begin{aligned}\gamma_5\,\mathscr{S}\!\!\!/(\mathbf{p},\lambda)u(\mathbf{p},\lambda) &= mu(\mathbf{p},\lambda)\\ \gamma_5\,\mathscr{S}\!\!\!/(\mathbf{p},\lambda)v(\mathbf{p},\lambda) &= mv(\mathbf{p},\lambda)\end{aligned} \tag{A12.25}$$

where

$$\mathscr{S}^\mu(\mathbf{p},\lambda) = 2\lambda(p, E\hat{\mathbf{p}}) \tag{A12.26}$$

and

$$p_\mu\mathscr{S}^\mu(\mathbf{p},\lambda) = 0. \tag{A12.27}$$

Sometimes it is convenient to label the spinors by p^μ and $\mathscr{S}^\mu$, but it should be remembered that the explicit form for the spinors depends upon the representation used for the γ-matrices and so cannot be written as a covariant combination of p^μ and $\mathscr{S}^\mu$.

The following results for the 4×4 matrices formed from the spinors are important in calculating physical cross-sections:

$$\sum_\lambda \left[u_\alpha(\mathbf{p}, \lambda)\bar{u}_\beta(\mathbf{p}, \lambda) - v_\alpha(\mathbf{p}, \lambda)\bar{v}_\beta(\mathbf{p}, \lambda)\right] = \delta_{\alpha\beta} 2m \qquad \text{(A12.28)}$$

$$\begin{aligned} \sum_\lambda u(\mathbf{p}, \lambda)\bar{u}(\mathbf{p}, \lambda) &= \not{p} + m \\ \sum_\lambda v(\mathbf{p}, \lambda)\bar{v}(\mathbf{p}, \lambda) &= \not{p} - m \end{aligned} \qquad \text{(A12.29)}$$

and, using (A12.25),

$$\begin{aligned} u(\mathbf{p}, \lambda)\bar{u}(\mathbf{p}, \lambda) &= \frac{\not{p} + m}{2}\left[1 + \frac{\gamma_5 \not{\mathscr{S}}(\mathbf{p}, \lambda)}{m}\right] \\ v(\mathbf{p}, \lambda)\bar{v}(\mathbf{p}, \lambda) &= \frac{\not{p} - m}{2}\left[1 + \frac{\gamma_5 \not{\mathscr{S}}(\mathbf{p}, \lambda)}{m}\right]. \end{aligned} \qquad \text{(A12.30)}$$

Now note that, irrespective of whether we have particle or antiparticle spinors, *any* matrix element of the form $\bar{u}(1)\Gamma u(2)$ can be written as a trace:

$$\begin{aligned} \bar{u}(1)\Gamma u(2) &= \bar{u}_\alpha(1)\Gamma_{\alpha\beta}u_\beta(2) = \Gamma_{\alpha\beta}u_\beta(2)\bar{u}_\alpha(1) \\ &= \text{Tr}\ [\Gamma u(2)\bar{u}(1)]. \end{aligned} \qquad \text{(A12.31)}$$

Generally this trick is not useful because of the complexity of $u(2)\bar{u}(1)$ when $m \neq 0$. However, when $p_1^\mu = p_2^\mu$ we can use (A12.30) to derive the following helpful results:

$$\begin{aligned} \bar{u}(\mathbf{p}, \mathscr{S})\gamma^\mu u(\mathbf{p}, \mathscr{S}) &= 2p^\mu \\ \bar{u}(\mathbf{p}, \mathscr{S})\sigma^{\mu\nu} u(\mathbf{p}, \mathscr{S}) &= \frac{2}{m}\epsilon^{\alpha\beta\mu\nu} p_\alpha \mathscr{S}_\beta \\ \bar{u}(\mathbf{p}, \mathscr{S})\gamma^\mu\gamma_5 u(\mathbf{p}, \mathscr{S}) &= 2\mathscr{S}^\mu \\ \bar{u}(\mathbf{p}, \mathscr{S})\gamma_5 u(\mathbf{p}, \mathscr{S}) &= 0 \end{aligned} \qquad \text{(A12.32)}$$

We consider now the specific form of the spinors in particular representations of the γ-matrices.

A12.3 The Dirac–Pauli representation

One takes

$$\gamma^0 = \begin{pmatrix} I & 0 \\ 0 & -I \end{pmatrix} \qquad \gamma^k = \begin{pmatrix} 0 & \sigma_k \\ -\sigma_k & 0 \end{pmatrix} \quad k = 1,2,3$$

$$\gamma^5 = \begin{pmatrix} 0 & I \\ I & 0 \end{pmatrix} \tag{A12.33}$$

$$\Sigma_k = \begin{pmatrix} \sigma_k & 0 \\ 0 & \sigma_k \end{pmatrix} \qquad \gamma^j\gamma^0 = -\begin{pmatrix} 0 & \sigma_j \\ \sigma_j & 0 \end{pmatrix} \quad j = 1,2,3$$

where, as usual,

$$\sigma_1 = \begin{pmatrix} 0 & 1 \\ 1 & 0 \end{pmatrix} \qquad \sigma_2 = \begin{pmatrix} 0 & -i \\ i & 0 \end{pmatrix} \qquad \sigma_3 = \begin{pmatrix} 1 & 0 \\ 0 & -1 \end{pmatrix}. \tag{A12.34}$$

For the transpose (T) of the matrices one has

$$\begin{aligned} {\gamma^j}^T &= \gamma^j \qquad j = 0,2,5 \\ {\gamma^j}^T &= -\gamma^j \qquad j = 1,3 \end{aligned} \tag{A12.35}$$

and for the hermitian conjugate $(\dagger)$

$$\gamma^{0\dagger} = \gamma^0 \qquad {\gamma_5}^\dagger = \gamma_5 \qquad \gamma^{j\dagger} = -\gamma^j \qquad j = 1,2,3 \tag{A12.36}$$

For the rest-frame spinors one usually takes

$$u(\overset{\circ}{p},\lambda) = \sqrt{2m}\begin{pmatrix} \chi_\lambda \\ 0 \end{pmatrix} \qquad v(\overset{\circ}{p},\lambda) = \sqrt{2m}\begin{pmatrix} 0 \\ \chi_{-\lambda} \end{pmatrix} \tag{A12.37}$$

where χ_λ is a two-component spinor and

$$\chi_+ = \begin{pmatrix} 1 \\ 0 \end{pmatrix} \qquad \chi_- = \begin{pmatrix} 0 \\ 1 \end{pmatrix}. \tag{A12.38}$$

The explicit form of the helicity spinors is, then, from (A12.24),

$$\begin{aligned} u(\mathbf{p},\lambda) &= \frac{1}{\sqrt{E+m}}\begin{pmatrix} E+m \\ 2p\lambda \end{pmatrix}\chi_\lambda(\hat{\mathbf{p}}) \\ v(\mathbf{p},\lambda) &= \frac{1}{\sqrt{E+m}}\begin{pmatrix} -2p\lambda \\ E+m \end{pmatrix}\chi_{-\lambda}(\hat{\mathbf{p}}) \end{aligned} \tag{A12.39}$$

where

$$\chi_\lambda(\hat{\mathbf{p}}) \equiv e^{-i\phi\sigma_3/2}e^{-i\theta\sigma_2/2}\chi_\lambda. \tag{A12.40}$$

One finds

$$\begin{aligned} \chi_+(\mathbf{p}) &= \begin{pmatrix} e^{-i\phi/2}\cos\theta/2 \\ e^{i\phi/2}\sin\theta/2 \end{pmatrix} \\ \chi_-(\mathbf{p}) &= \begin{pmatrix} -e^{-i\phi/2}\sin\theta/2 \\ e^{i\phi/2}\cos\theta/2 \end{pmatrix} \end{aligned} \tag{A12.41}$$

Note that in this representation we have

$$v(\mathbf{p},\lambda) = i\gamma^2 u^*(\mathbf{p},\lambda) \tag{A12.42}$$

A12.4 The Weyl representation

This is particularly useful in the relativistic limit and in the massless case. One takes

$$\begin{gathered}
\gamma^0 = \begin{pmatrix} 0 & I \\ I & 0 \end{pmatrix} \qquad \gamma^j = \begin{pmatrix} 0 & -\sigma_j \\ \sigma_j & 0 \end{pmatrix} \quad j = 1,2,3 \\
\gamma^5 = \begin{pmatrix} I & 0 \\ 0 & -I \end{pmatrix} \\
\Sigma_k = \begin{pmatrix} \sigma_k & 0 \\ 0 & \sigma_k \end{pmatrix} \qquad \gamma^j\gamma^0 = \begin{pmatrix} -\sigma_j & 0 \\ 0 & \sigma_j \end{pmatrix}
\end{gathered} \tag{A12.43}$$

Equations (A12.35) and (A12.36) for transpose and hermitian conjugate continue to hold.

The explicit form of the helicity spinors can be taken, via (A12.24), as

$$\begin{aligned}
u(\mathbf{p},\lambda) &= \frac{1}{\sqrt{2(E+m)}}\begin{pmatrix} E+m+2p\lambda \\ E+m-2p\lambda \end{pmatrix}\chi_\lambda(\hat{\mathbf{p}}) \\
v(\mathbf{p},\lambda) &= \frac{1}{\sqrt{2(E+m)}}\begin{pmatrix} p-2\lambda(E+m) \\ p+2\lambda(E+m) \end{pmatrix}\chi_{-\lambda}(\hat{\mathbf{p}})
\end{aligned} \tag{A12.44}$$

in the Weyl representation. Note that if $m \neq 0$ this corresponds to the choice

$$\begin{aligned}
u(\overset{\circ}{p},\lambda) &= \sqrt{m}\begin{pmatrix} \chi_\lambda \\ \chi_\lambda \end{pmatrix} \\
v(\overset{\circ}{p},\lambda) &= 2\lambda\sqrt{m}\begin{pmatrix} -\chi_{-\lambda} \\ \chi_{-\lambda} \end{pmatrix}
\end{aligned} \tag{A12.45}$$

and in this representation

$$v(\mathbf{p},\lambda) = -i\gamma^2 u^*(\mathbf{p},\lambda). \tag{A12.46}$$

A12.5 Massless fermions

When $m = 0$, remarkable simplifications occur in the Weyl representation. For example, if $m \ll E$ or $m = 0$ then (A12.44) becomes

$$\begin{aligned}
u(\mathbf{p},+) &= \sqrt{2E}\begin{pmatrix} \chi_+(\mathbf{p}) \\ 0 \end{pmatrix} = v(\mathbf{p},-) \\
u(\mathbf{p},-) &= \sqrt{2E}\begin{pmatrix} 0 \\ \chi_-(\mathbf{p}) \end{pmatrix} = v(\mathbf{p},+)
\end{aligned} \tag{A12.47}$$

with

$$\not{p}u(\mathbf{p}) = \not{p}v(\mathbf{p}) = 0$$
$$\bar{u}_\lambda(\mathbf{p})u_\lambda(\mathbf{p}') = \bar{v}_\lambda(\mathbf{p})v_\lambda(\mathbf{p}') = 0. \tag{A12.48}$$

If in (A12.26) we write

$$\begin{aligned}\mathscr{S}^\mu(\mathbf{p},\lambda) &= 2\lambda(E,\mathbf{p}) + 2\lambda(p - E, (E-p)\hat{\mathbf{p}}) \\ &= 2\lambda p^\mu + 2\lambda(E-p)(-1,\hat{\mathbf{p}})\end{aligned} \tag{A12.49}$$

where $\hat{\mathbf{p}}$ is the unit vector along $\mathbf{p}$ then, using (A12.5) and (A12.25), we see that for $m \ll E$ or $m = 0$

$$\left.\begin{aligned}\gamma_5 u(\mathbf{p},\lambda) &= 2\lambda u(\mathbf{p},\lambda) \\ \gamma_5 v(\mathbf{p},\lambda) &= -2\lambda v(\mathbf{p},\lambda)\end{aligned}\right\} \quad (m=0) \tag{A12.50}$$

from which follow

$$\begin{aligned}\bar{u}(\mathbf{p},\lambda)\gamma_5 &= -2\lambda\bar{u}(\mathbf{p},\lambda) \\ \bar{v}(\mathbf{p},\lambda)\gamma_5 &= 2\lambda\bar{v}(\mathbf{p},\lambda).\end{aligned} \tag{A12.51}$$

The consequences of these and the connection with chirality are discussed in subsection 4.6.3.

The relations (A12.29) become

$$u_+(\mathbf{p})\bar{u}_+(\mathbf{p}) + u_-(\mathbf{p})\bar{u}_-(\mathbf{p}) = \not{p} = v_+(\mathbf{p})\bar{v}_+(\mathbf{p}) + v_-(\mathbf{p})\bar{v}_-(\mathbf{p}). \tag{A12.52}$$

Multiplying by $(1 \pm \gamma_5)/2$ we obtain

$$\begin{aligned}u_\pm(\mathbf{p})\bar{u}_\pm(\mathbf{p}) &= \tfrac{1}{2}(1 \pm \gamma_5)\not{p} \\ v_\pm(\mathbf{p})\bar{v}_\pm(\mathbf{p}) &= \tfrac{1}{2}(1 \mp \gamma_5)\not{p}\end{aligned} \tag{A12.53}$$

Consider now the very important matrix $M = u_+(\mathbf{p}_1)\bar{u}_+(\mathbf{p}_2)$. It can be expanded as in (A12.19). Using the fact that

$$\gamma_5 M = M = -M\gamma_5 \tag{A12.54}$$

the coefficients in the expansion must satisfy

$$\mathrm{Tr}\,(\Gamma M) = -\,\mathrm{Tr}\,(\Gamma\gamma_5 M\gamma_5) = -\,\mathrm{Tr}\,(\gamma_5\Gamma\gamma_5 M).$$

But $\gamma_5\{I, \sigma^{\mu\nu}, \gamma_5\}\gamma_5 = \{I, \sigma^{\mu\nu}, \gamma_5\}$, so that $S = T_{\mu\nu} = P = 0$. Also $A_\mu = -V_\mu$.

Thus for the massless case the only relevant coefficients are

$$V_\mu = \tfrac{1}{4}\,\mathrm{Tr}\,(\gamma_\mu u_+(\mathbf{p}_1)\bar{u}_+(\mathbf{p}_2)) \tag{A12.55}$$

However, using (A12.31) we can rewrite (A12.55) as

$$V_\mu = \tfrac{1}{4}\bar{u}_+(\mathbf{p}_2)\gamma_\mu u_+(\mathbf{p}_1)$$

so that finally

$$u_+(\mathbf{p}_1)\bar{u}_+(\mathbf{p}_2) = \tfrac{1}{4}\left[\bar{u}_+(\mathbf{p}_2)\gamma_\mu u_+(\mathbf{p}_1)\right]\gamma^\mu(1-\gamma_5). \tag{A12.56}$$

In similar fashion one finds

$$u_-(\mathbf{p}_1)\bar{u}_-(\mathbf{p}_2) = \tfrac{1}{4}\left[\bar{u}_-(\mathbf{p}_2)\gamma_\mu u_-(\mathbf{p}_1)\right]\gamma^\mu(1+\gamma_5). \tag{A12.57}$$

One can check explicitly that

$$\bar{u}_-(\mathbf{p}_2)\gamma_\mu u_-(\mathbf{p}_1) = \bar{u}_+(\mathbf{p}_1)\gamma_\mu u_+(\mathbf{p}_2) \tag{A12.58}$$

so that (A12.57) becomes

$$u_-(\mathbf{p}_1)\bar{u}_-(\mathbf{p}_2) = \tfrac{1}{4}\left[\bar{u}_+(\mathbf{p}_1)\gamma_\mu u_+(\mathbf{p}_2)\right]\gamma^\mu(1+\gamma_5). \tag{A12.59}$$

Using similar techniques one can show that

$$u_+(\mathbf{p}_1)\bar{u}_-(\mathbf{p}_2) - u_+(\mathbf{p}_2)\bar{u}_-(\mathbf{p}_1) = \left[\bar{u}_-(\mathbf{p}_2)u_+(\mathbf{p}_1)\right]\tfrac{1}{2}(1+\gamma_5). \tag{A12.60}$$

Equations (A12.56), (A12.59) and (A12.60) are very useful in deriving the rules for Feynman diagrams with massless particles (Chapter 10).

A12.6 The Fierz rearrangement theorem

It sometimes happens, when dealing with the matrix element corresponding to a Feynman diagram involving spin-1/2 particles, that it is convenient to rearrange the order of the spinors in relation to the order they acquire directly from the Feynman diagram.

In general, let $\tilde{\Gamma}^i (i = 1, \ldots, 16)$ stand for any of the independent combinations of unit matrix and γ-matrices $I, \gamma^\mu, \sigma^{\mu\nu}, i\gamma^\mu\gamma_5, \gamma_5$.

Let $\tilde{\Gamma}_i$ stand for the above set of matrices with their Lorentz indices lowered when relevant, i.e. $\tilde{\Gamma}_i$ contains for example γ_μ, whereas $\tilde{\Gamma}^i$ contains γ^μ etc.

As a result of the algebraic properties of the set $\tilde{\Gamma}^i$ it can be shown that

$$\tfrac{1}{4}\sum_i \left(\tilde{\Gamma}_i\right)_{\alpha\beta}\left(\tilde{\Gamma}^i\right)_{\gamma\delta} = \delta_{\alpha\delta}\delta_{\beta\gamma}. \tag{A12.61}$$

If now A and B are any 4×4 matrices, then on multiplying (A12.61) by $A_{\rho\alpha}B_{\nu\gamma}$ we obtain

$$\begin{aligned}\tfrac{1}{4}\sum_i A_{\rho\alpha}\left(\tilde{\Gamma}_i\right)_{\alpha\beta}B_{\nu\gamma}\left(\tilde{\Gamma}^i\right)_{\gamma\delta} &= A_{\rho\delta}B_{\nu\beta}\\ A_{\rho\delta}B_{\nu\beta} &= \tfrac{1}{4}\sum_i \left(A\tilde{\Gamma}_i\right)_{\rho\beta}\left(B\tilde{\Gamma}^i\right)_{\nu\delta}.\end{aligned} \tag{A12.62}$$

Since the 16 $\tilde{\Gamma}^i$ are a complete set of 4×4 matrices, each product $A\tilde{\Gamma}_i$ etc. will reduce to a sum of $\tilde{\Gamma}_i$.

After some labour one can obtain the following relation:

$$[\gamma^\mu(1-\gamma_5)]_{\rho\delta}\,[\gamma_\mu(1-\gamma_5)]_{\nu\beta} = -[\gamma^\mu(1-\gamma_5)]_{\rho\beta}\,[\gamma_\mu(1-\gamma_5)]_{\nu\delta}. \quad \text{(A12.63)}$$

Clearly, analogous relations can be worked out for any product of the $\tilde{\Gamma}$-matrices. Results may be found in Section 2.2B of Marshak, Riazuddin and Ryan (1969).

References

Abe, K. *et al.* (1997a). *Phys. Lett.*, **B404**, 377
Abe, K. *et al.* (1997b). *Phys. Lett.*, **B405**, 180
Abe, K. *et al.* (1997c). *Phys. Rev. Lett.*, **79**, 26
Abreu, P. *et al.* (1995a). *Z. Phys.*, **C65**, 11
Abreu, P. *et al.* (1995b). *Z. Phys.*, **C67**, 183
Abreu, P. *et al.* (1997). *Phys. Lett.*, **B404**, 194
Acciarri, P. *et al.* (1994). *Phys. Lett.*, **B341**, 245
Ackerstaff, K. *et al.* (1997). *Phys. Lett.*, **B404**, 783
Adams, D.L. *et al.* (1991a). *Phys. Lett.*, **B261**, 197
Adams, D.L. *et al.* (1991b). *Phys. Lett.*, **B264**, 462
Adams, D.L. *et al.* (1997). *Phys. Lett.*, **B396**, 338
Adeva, B. *et al.* (1994). *Nucl. Instrum. Meth.*, **A343**, 363
Adeva, B. *et al.* (1997). *Phys. Lett.*, **B412**, 414
Adler, S.I. (1969). *Phys. Rev.*, **177**, 2426
Adloff, C. *et al.* (1997). *Nucl. Phys.*, **B497**, 3
Akchurin, N. *et al.* (1993). *Phys. Rev.*, **D48**, 3026
Akchurin, N. *et al.* (1996). In *Proc. 12th Intl. Symp. on High Energy Spin Physics*, Amsterdam 1996 (World Scientific, Singapore), p. 263
Alexander, G. *et al.* (1996). *Z. Phys.*, **C72**, 365
Alley, R.I. *et al.* (1995). *Nucl. Instrum. and Meth. in Phys. Res.*, **A365**, 1
Altarelli, G. (1982). *Phys. Rep.*, **81**, 1
Altarelli, G. and Parisi, G. (1977). *Nucl. Phys.*, **B126**, 298
Altarelli, G. and Ross, G.G. (1988). *Phys. Lett.*, **B212**, 391
Altarelli, G. *et al.* (1997). *Nucl. Phys.*, **B496**, 337
Andersson, B., Gustafson, G. and Ingelman, G. (1979). *Phys. Lett.*, **B85**, 417
Andersson, B. *et al.* (1983). *Phys. Rep.*, **97**, 31
Anselmino, M. (1979). *Phys. Rev.*, **D19**, 2803
Anselmino, M. and Leader, E. (1988). Santa Barbara Prep. NSF-88-142, unpublished

Anselmino, M. and Murgia, F. (1998). *Phys. Lett.*, **B442**, 470

Anselmino, M., Boglione, M. and Murgia, F. (1995). *Phys. Lett.*, **B362**, 164

Anselmino, M., Boglione, M. and Murgia, F. (1999). *Phys. Rev.*, **D60**, 054027

Anselmino, M., Efremov, A. and Leader, E. (1995). *Phys. Rep.*, **261**, 1

Anselmino, M., Leader, E. and Murgia, F. (1997). *Phys. Rev.*, **D56**, 6021

Anselmino, M., Boglione, M., Hansson, J. and Murgia, F. (2000). *Eur. Phys. J.* **C13**, 519

Anthony, P. L. *et al.* (2000). *Phys. Lett.*, **B458**, 529

Antille, J. *et al.* (1981). *Nucl. Phys.*, **B185**, 1

Appel, H. (1968). In *Landalt-Bornstein*, Group 1, Vol. 3 (Springer-Verlag)

Arneodo, M. *et al.* (1987). *Z. Phys.*, **C34**, 277

Artru, X. and Mekhfi, M. (1990). *Z. Phys.*, **C45**, 669

Artru, X., Czyzewski, J. and Yabuki, H. (1997). *Z. Phys.*, **C73**, 527

Ashman, J. *et al.* (1988). *Phys. Lett.*, **B206**, 364

Ashman, J. *et al.* (1989). *Nucl. Phys.*, **B328**, 1

Assmann, R. *et al.* (1995). In *Proc. 1995 Particle Accelerator Conf.*, Dallas 1995 (American Physical Society), 219

Babcock, J., Monsay, E. and Sivers, D. (1979). *Phys. Rev.*, **D19**, 1483

Bailin, D. (1982). *Weak Interactions*, 2nd edn. (Sussex University Press)

Ball, R.D., Forte, S. and Ridolfi, G. (1995). *Nucl. Phys.*, **B444**, 287

Ball, R.D., Forte, S. and Ridolfi, G. (1996). *Phys. Lett.*, **B378**, 255

Band, H.R. (1997). In *Proc. 12th Intl. Symp. on High Energy Spin Physics*, Amsterdam 1996 (World Scientific, Singapore), 765

Barber, D.P. (1994). *Nucl. Instrum. Meth.*, **A338**, 166

Barber, D.P. (1996). In *Proc. 12th Intl. Symp. on High Energy Spin Physics*, Amsterdam 1996 (World Scientific, Singapore), 98

Bargmann, V., Michel, L. and Telegedi, V.A. (1959). *Phys. Rev. Lett.*, **2**, 435

Barnett, R.M. *et al.* (1996). *Phys. Rev.*, **D54**, 1

Bartels, J., Ermolatev, B.I. and Ryskin, M.G. (1996a). *Z. Phys.*, **C70**, 273

Bartels, J., Ermolatev, B.I. and Ryskin, M.G. (1996b). *Z. Phys.*, **C72**, 627

Bell, J.S. and Jackiw, R. (1969). *Nuovo Cimento*, **51A**, 47

Benn, Z. *et al.* (1997). *Nucl. Phys.*, **B489**, 3

Berends, F.A. and Giele, W. (1987). *Nucl. Phys.*, **B294**, 700

Berrera, A. (1992). *Phys. Lett.* **B293**, 445

Berman, S.M. and Jacob, M. (1965). *Phys. Rev.*, **B139**, 1023

Bialkowski, G. (1970). *Acta Physica Polonica*, **B1**, 77

Bjorken, J.D. and Drell, S.D. (1964). *Relativistic Quantum Mechanics* (McGraw-Hill, New York)

Blondel, A. (1998). *Phys. Lett.*, **202B**, 145

Blümlein, J. and Kochelev, N. (1997). *Nucl. Phys.* **B498**, 285

Blümlein, J. *et al.* (1997). In *Proc. Workshop on Deep Inelastic Scattering off Polarized Targets: Theory Meets Experiments*, DESY-Zeuthen, September 1997 (DESY-97-200)

Boer, D. (1998). Azimuthal Asymmetries in Hard Scattering Processes, Ph.D. thesis, NIKHEF, Amsterdam

Boglione, M. and Leader, E. (2000). *Phys. Rev.*, **D61**, 114001

Boros, C., Liang, Z.T., Meng, T.C. and Rittel,R. (1998). *J. Phys*, **G24**, 75

Botts, J. and Sterman, G. (1989). *Nucl. Phys.*, **B325**, 62

Bouchiat, M.A. *et al.* (1960). *Phys. Rev. Lett.*, **5**, 373

Bourrely, C., Leader, E. and Soffer, J. (1980). *Phys. Rep.*, **59**, 95

Bourrely, C. and Soffer, J. (1975). *Phys. Rev.*, **D12**, 2932

Bourrely, C. and Soffer, J. (1993). *Phys. Lett.*, **B314**, 132

Bourrely, C., Renard, F.M., Soffer, J. and Taxil, P. (1989). *Phys. Rep.*, **177**, 319

Bourrely, C., Soffer, J. and Teryaev, O.V. (1998). *Phys. Lett.*, **B420**, 375

Braun, M.A., Gauron, P. and Nicolescu, B. (1999). *Nucl. Phys.*, **B542**, 329

Bravar, A. *et al.* (1995). *Phys. Rev. Lett.*, **75**, 3073

Bravar, A. *et al.* (1996). *Phys. Rev. Lett.*, **77**, 2626

Bravar, A. *et al.* (1997). *Phys. Rev. Lett.*, **78**, 4003

Breitenlohner, P. and Maison, D. (1977). *Comm. Math. Phys.*, **52**, 11

Brodsky, S.J. and Lepage, G.P. (1980). *Phys. Rev.*, **D22**, 2157

Brodsky, S.J., Burkhardt, M. and Schmidt, I. (1995). *Nucl. Phys.*, **B441**, 197

Bunce, G. *et al.* (1976). *Phys. Rev. Lett.*, **36**, 1113

Bunce, G. *et al.* (1992). *Particle World*, **3**, (1), 1

Buon, J. and Steffen, K. (1986). *Nucl. Instr. Meth.*, **A245**, 248

Buskulic, D. *et al.* (1996). *Z. Phys.*, **C69**, 183

Buttimore, N.H., Gotsman, E. and Leader, E. (1978). *Phys. Rev.*, **D18**, 694 (See also *Phys. Rev.* **D35** (1987), 407)

Buttimore, N.H., *et al.* (1978). *Phys. Rev.*, **D18**, 694

Buttimore, N.H., *et al.* (1999). *Phys. Rev.*, **D59**, 114010

Button, J. and Mermod, R. (1960). *Phys. Rev.*, **118**, 1333

Cahn, R.M. (1975). *Phys. Lett.*, **B78**, 269

Carey, D.C. *et al.* (1990). *Phys. Rev. Lett.*, **64**, 357

Carlitz, R.D., Collins, J.C. and Mueller, A.H. (1988). *Phys. Lett.*, **B214**, 229

Carruthers, P.A. (1971). *Spin and Isospin in Particle Physics* (Gordon and Breach, New York)

Chupp, T.E. *et al.* (1992). *Phys. Rev.*, **C45**, 915

Close, F.E. (1979). *An Introduction to Quarks and Partons* (Academic, London)

Cohen-Tannoudji, G., Morel, A. and Navelet, H. (1968a). *Ann. Phys.*, **46**, 239

Cohen-Tannoudji, G., Salin, P. and Morel, A. (1968b). *Nuovo Cimento*, **55A**, 412

Colegrove, F.D., Schearer, L.D. and Walters, G.K. (1963). *Phys. Rev.*, **132**, 2561

Collins, J. (1993). *Nucl. Phys.* **B396**, 161

Collins, P.D.B. and Martin, A. (1984). *Hadron Interactions* (Adam Hilger, Bristol)

Combley, F. and Picasso, E. (1974). *Phys. Rep.*, **C14**, 20

Commins, E.D. and Bucksbaum, P.H. (1983). *Weak Interactions of Leptons and Quarks* (Cambridge University Press, Cambridge)

Consoli, M. and Hollik, A. (1989). In *Z Physics at Lep I*, CERN Yellow Report, CERN 89-08. The report contains several other articles of relevance

Conte, M., Penzo, A. and Pusterla, M. (1995). *Nuovo Cimento*, **108A**, 127
Contogouris, A.P. and Papadopoulos, S. (1991). *Phys. Lett.* **B260**, 204
Cortes, J.L. and Pire, B. (1988). *Phys. Rev.*, **D38**, 3586
Cortes, J.L., Pire, B. and Ralston, J.P. (1992). *Z. Phys.*, **C55**, 409
Crabb, D.G. *et al.* (1990). *Phys. Rev. Lett.*, **65**, 3241
Crabb, D.G. and Day, D.B. (1995). *Nucl. Instrum. Meth. Phys*, **A356**, 9
Crabb, D.G., Higley, C.B., Krisch, A.R., Raymond, E.S., Roser, T., Stewart, J.A. and Court, G.R. (1990). *Phys. Rev. Lett.*, **64**, 2627
Crabb, D.G. and Meyer, W. (1997). *Annu. Rev. Nucl. Part. Sci.*, **47**, 67
Craigie, N.S., Hidaka, K., Jacob, M. and Renard, F.M. (1983). *Phys. Rep.*, **99**, 69
Cvitanović, P., Lauwers, P.G. and Schabarch, P.N. (1981). *Nucl. Phys.*, **B186**, 165
Dalitz, R., (1966). In *Proc. Intl. School of Physics 'Enrico Fermi'*, Vol. 33 (Academic Press)
Davier, M., Duflot, L., Le Diberder, F. and Rougé, A. (1993). *Phys. Lett.*, **B306**, 411
de Boer, W. *et al.* (1975). *Phys. Rev. Lett.*, **34**, 558
De Causmaecker, P. *et al.* (1981). *Phys. Lett.*, **105B**, 215
De Grand, I.A. and Miettinen, H.I (1981). *Phys. Rev.*, **D24**, 2419
Derbenev, Ya.S. (1990). In *Proc. Polarized Collider Workshop,* eds. J. Collins, S.F. Heppleman and R.W. Robinett, p. 327 (AIP Conf. Proc. No. 223, New York)
Derbenev, Ya.s. and Kondratenko, A.M. (1976). *Sov. Phys. Doklady*, **20**, 562
Dick, L., Jeanneret, J.B., Kubišchta, W. and Antille, J. (1980). In *Proc. 4th Int. Symp. on High Energy Spin Physics*, Lausanne, p. 212
Dicus, D.A. (1972). *Phys. Rev.*, **D5**, 1367
Dirac, P.A.M. (1927). *Proc. Roy. Soc. London*, **A117**, 610
Dirac, P.A.M. (1947). *The Principles of Quantum Mechanics* (Clarendon Press, Oxford)
Doncel, M.G., Michel, L. and Minnaert, P. (1970). In *Ecole d'Eté de Gif-sur-Yuette*, ed. R. Salmeron (Ecole Polytechnique)
Doncel, M.G., Michel, L. and Minnaert, P. (1972). *Nucl. Phys.*, **B38**, 477
Doncheski, M.A. and Robinett, R.W. (1994). *Z. Phys.*, **C63**, 611
Eden, R.J. (1971). *Rev. Mod. Phys.*, **43**, 15
Edmonds, A. (1957). *Angular Momentum in Quantum Mechanics* (Princeton University Press)
Efremov, A.V. and Korotkiyan, V.M. (1984). *Nucl. Phys.*, **39**, 962 (*Yad. Fiz.*, **39**, 1517)
Efremov, A.V., Korotkiyan, V.M. and Teryaev, O.V. (1995). *Phys. Lett.*, **B348**, 577
Efremov, A.V., Leader, E. and Teryaev, O.V. (1997). *Phys. Rev.*, **D55**, 4307
Efremov, A.V. and Teryaev, O.V. (1984). *Sov. J. Nucl. Phys*, **39**, 962 (*Yad. Fiz.*, **39**, 1517)
Efremov, A.V. and Teryaev, O.V. (1988). JINR Report E2-88-287, unpublished
Ellis, R.K., Furmanski, W. and Petronzio, R. (1983). *Nucl. Phys.*, **B212**, 29
Fano, U. (1957). *Rev. Mod. Phys.*, **29**, 74
Farrar, G.R. and Jackson, D.R. (1975). *Phys. Rev. Lett.*, **35**, 1416

Farrar, G.R. and Neri, F. (1983). *Phys. Lett.*, **B130**, 109
Feltham, A. and Steiner, H. (1997). In *Proc. 12th Intl. Symp. on High Energy Spin Physics*, Amsterdam 1996 (World Scientific, Singapore), p. 782
Fernow, R.C. and Krisch, A.D. (1981). *Ann. Rev. Nucl. Part. Sci.*, **31**, 107
Field, R.D. and Sidhu, D. (1974). *Phys. Rev.*, **D10**, 89
Fox, G.C.. and Leader, E. (1967). *Phys. Rev. Lett.*, **18**, 628
Froissart, M. (1961). *Phys. Rev.*, **123**, 1053
Froissart, M. and Stora, R. (1959). *Nucl. Instrum. Meth.*, **7**, 297
Gasiorowicz, S. (1967). *Elementary Particle Physics* (John Wiley, New York)
Gasser, J. and Leutwyler, H. (1982). *Phys. Rep.*, **87**, 77
Gehrmann, T. (1997). In *Proc. 2nd Topical Workshop on Deep Inelastic Scattering off Polarized Targets*, Zeuthen, September 1997 (DESY, Hamburg)
Gehrmann, T. and Stirling, W.J. (1996). *Phys. Rev.*, **D53**, 6100
Glück, M. and Reya, E. (1988). *Z. Phys.*, **C39**, 569
Glück, M., Reya, E., Stratmann, M. and Vogelsang, W. (1996). *Phys. Rev.*, **D53**, 4775
Glück, M., Reya, E. and Vogelsang, W. (1992). *Phys. Rev.*, **D45**, 2552
Goldberger, M.L. *et al.* (1960). *Phys. Rev.*, **120**, 2250
Goldberger, M.L., Grisaru, M.T., MacDowell, S.W. and Wang, D.Y. (1960). *Phys. Rev.*, **120**, 2250
Goldstein, G. and Owens, J. (1976). *Nucl. Phys.*, **B103**, 145
Goloskokov, S.V. and Kroll, P. (1999). *Phys. Rev.*, **D60**, 014019
Gordon, L.E. and Vogelsang, W. (1993). *Phys. Rev.*, **D48**, 3136
Gordon, L.E. and Vogelsang, W. (1994). *Phys. Rev.*, **D49**, 170
Gousset, T., Pire, B. and Ralston, J.P. (1996). *Phys. Rev.*, **D53**, 1202
Gribov, V.N. and Lipatov, L.N. (1972). *Sov. J. Nucl. Phys.*, **15**, 438 and 675
Grosnick, D.P. *et al.* (1990). *Nucl. Instrum. Meth.*, **A290**, 269
Grosnick, D.P. *et al.* (1997). *Phys. Rev.*, **D55**, 1159
Halzen, F. and Michael, C. (1971). *Phys. Lett.*, **36B**, 367
Hamermesh, M. (1964). *Group Theory and its Application to Physical Problems* (Addison-Wesley), Ch. 9.2
Heimann, R.L. (1973). *Nucl. Phys.*, **B64**, 429
Heller, K. (1981). In *Proc. Int. Conf. on High Energy Phys.*, Madison, 1981 (AIP Conf. Proc. No. 68, New York), p. 61
Heller, K. ed. (1988). In *Proc. 8th Intl. Symp. on High Energy Spin Physics* (AIP Conf. Proc. No. 187, New York)
Hidaka, K., Monsay, E. and Sivers, D. (1979). *Phys. Rev.*, **D19**, 1503
Hinotani, K., Neal, H.A., Predazzi, E. and Walters, G. (1979). *Nuovo Cimento* **52A**, 363
Hollick, W. (1990). *Fortsch. Phys.*, **38**, 165
Huang, H. and Roser, T. (1994). *Phys. Rev. Lett.*, **73**, 2982
Ioffe, B.L. and Khodjamirian, A. (1995). *Phys. Rev.*, **D51**, 3373
Irving, A. and Worden, R. (1977). *Phys. Rep.*, **34C**, 117

Isgur, N. and Llewellyn-Smith, C.H. (1989). *Nucl. Phys.*, **B317**, 526

ISHESP (1996). *Proc. 12th Intl. Symp. on High Energy Spin Physics,* Amsterdam 1996 (World Scientific, Singapore)

Jacob, M. (1958). *Nuovo Cimento*, **9**, 826

Jacob, M. and Wick, G.C. (1959). *Ann. Phys.*, **7**, 404

Jaffe, R.L and Ji, X. (1991). *Phys. Rev. Lett.*, **67**, 552

Jaffe, R.L and Saito, N. (1996). *Phys. Lett.*, **B382**, 165

Jackson, J.D., (1965). In *High Energy Physics, Proc. Les Houches Summer School*, eds. C. Dewitt and M. Jacob (Gordon and Breach)

Jackson, J.D. (1976). *Rev. Mod. Phys.*, **48 (**3**)**, 417

Jakob, R. and Kroll, P. (1992). *Z. Phys.*, **A344**, 87

Jakob, R. and Kroll, P. (1993). *Phys. Lett.*, **B315**, 463

Jakob, R. *et al.* (1993). *Z. Phys.*, **A347**, 109

Jones, L.M. and Wyld, H.W. (1978). *Phys. Rev.*, **D17**, 759

Johnson, J.R. *et al.* (1995). *Nucl. Instrum. Meth.*, **A356**, 148

Joynson, D., Leader, E. and Nicolescu, B. (1975). *Lett. Nuovo Cimento*, **30**, 345

Kleiss, R. (1986). *Z. Phys.*, **C33**, 433

Kleiss, R. and Stirling, W.J. (1985). *Nucl. Phys.*, **B262**, 235

Klepikov, N., Kogan, Z. and Shamanin, S. (1967). *Sov. J. Nucl. Phys.*, **5**, 928 (*Yad. Phz.*, **5**, 1299)

Kleppner, D. and Greytak, T.J. (1982). In *Proc. 5th Intl. Symp. on High Energy Spin Physics* (AIP Conf. Proc. No. 95), p. 546

Knudsen, L. *et al.* (1991). *Phys. Lett.*, **B270**, 97

Korotkiyan, V.M. and Teryaev, O.V. (1995). *Phys. Rev.*, **D52**, R4775

Koutchouk, J.P. and Limberg, P. (1988). In *Polarization at LEP*, CERN Yellow Report 88-06, Vol. 2, 204

Kotanski, A. (1970). *Acta Phys. Polonica*, **B1**, 45

Krisch, A.D. (1978). In *Proc. AIP Conf. on High Energy Polarized Proton Beams*, Ann Arbor 1977, eds. A. Krisch and A. Salthouse, No. 42 (American Institute of Physics, New York), p. 142

Krisch, A.D. *et al.* (1989). *Phys. Rev. Lett.*, **63**, 1137

Kroll, P. (1994). In *Proc. Conf. Physics on GeV Particle Beams*, July 1994, hep-ph/9409262

Kunszt, Z. (1989). *Phys. Lett.*, **B218**, 243

Kuroda, K. (1982). In AIP Conf. Proc. No. 95 (American Institute of Physics, New York)

Lakin, W. (1955). *Phys. Rev.*, **98**, 139

Lam, C.S. and Li, B. (1982). *Phys. Rev.*, **D28**, 683

Landau, L. (1948). *Doklady Akad. Nauk, SSSR*, **60**, 207

Landshoff, P.V. (1974). *Phys. Rev.*, **D10**, 1024

Leader, E. (1968). *Phys. Rev.*, **166**, 1599

Leader, E. (1969). *Boulder Lectures in Theoretical Physics.*, Vol XII-B (Gordon and Breach, New York), p. 1

Leader, E. (1977). *Phys. Lett.*, **71B**, 353

Leader, E. (1986). *Phys. Rev. Lett.*, **56**, 1542

Leader, E. (1997). In *Proc. Workshop on Deep Inelastic Scattering off Polarized Targets*, DESY-Zeuthen, Germany

Leader, E. and Anselmino, M. (1988). *Z. Phys.*, **C41**, 239

Leader, E. and Predazzi, E. (1996). *An Introduction to Gauge Theories and Modern Particle Physics,* Vols. 1 and 2 (Cambridge University Press, Cambridge)

Leader, E. and Slansky, R. (1966). *Phys. Rev.*, **148**, 1491

Leader, E. and Sridhar, K. (1994). *Nucl. Phys.*, **B419**, 3

Leader, E. and Trueman, L. (1997). In *Proc. RIHEN-BNL Research Center Workshop*, Vol. 2

Leader, E., Sidorov, A.V. and Stamenov, D.B. (1998a). *Phys. Rev.*, **D58**, 114028

Leader, E., Sidorov, A.V. and Stamenov, D.B. (1998b). *Phys. Lett.*, **D445**, 232

Leader, E., Sidorov, A.V. and Stamenov, D.B. (1998c). *Int. J. Mod. Phys.*, **A13**, 5573

Leader, E., Sidorov, A.V. and Stamenov, D.B. (1999). *Phys. Lett.*, **B462**, 189

Leader, E. and Trueman, T.L. (1997). In *Proc. RIKEN-BNL Research Center Workshop*, Vol. 2 (Brookhaven Natl. Laboratory, Upton LI)

Leader, E. and Trueman, T.L. (2000). *Phys. Rev.*, **D61**, 077504

Lee, K. *et al.* (1993). *Nucl. Instrum. Meth.*, **A333**, 294

Li, H. (1993). *Phys. Rev.*, **D48**, 4243

Li, H. and Sterman, G. (1992). *Nucl. Phys.*, **B381**, 129

Lipatov, L.N. (1986). *Sov. Phys. JETP*, **63**, 904

Lipatov, L.N. (1989). In *Perturbative Quantum Chromodynamics*, ed. A.H. Muller (World Scientific, Singapore) p. 4.

Lipps, F.W. and Tolhoek, H.A. (1954). *Physica*, **XX**, 85 and 395

Longo, M.J. *et al.* (1989). In *Proc. 8th Intl. Symp. on High Energy Spin Physics*, Minneapolis (AIP Conf. Proc. No. 187), p. 80

Łukaszuk, L. and Nicolescu, B. (1973). *Lett. Nuovo Cimento*, **8**, 405

Lundberg, B. *et al.* (1989). *Phys. Rev.*, **D40**, 3557

Luppov, V.G. *et al.* (1996). In *Proc. 12th Intl. Symp on High Energy Spin Physics*, Amsterdam

Mahlon, G. and Parke, S.J. (1997). *Phys. Rev.*, **D55**, 7249

Manayenkov, S. and Ryskin, M.G. (1998). *European Phys. J.*, **C1**, 271

Mangano, M.L., Parke, S.J. and Xu, Z. (1987). *Nucl. Phys.*, **B298**, 653

Mangano, M.L. and Parke, S.J. (1991). *Phys. Rep.*, **200**, 301

Margolis, B. and Thomas, G.H. (1978). In AIP Conf. Proc. No. 42 (American Institute of Physics, New York), p. 173

Marshak, R.E. Riazuddin and Ryan, C.P. (1969). *Theory of Weak Interactions in Particle Physics* (Wiley Interscience, New York)

Martin, A. (1966). *Nuovo Cimento*, **42**, 930 (and *ibid*, **44**, 1219)

Martin, O., Schäfer, A., Stratmann, M. and Vogelsang, W. (1997). *Phys. Rev.*, **D57**, 3084

Mele, B. (1994). *Mod. Phys. Lett.*, **A9**, 1239

Mehra, J. and Rechenberg, H. (1982). *The Historical Development of Quantum Theory* (Springer Verlag)

Mertig, R. and van Neerven, W.L. (1996). *Z. Phys.*, **C70**, 637

Messiah, A. (1958). *Quantum Mechanics* (North Holland)

Montague, B.W. (1984). *Phys. Rep.*, **113**(1), 1

Mueller, A.H. (1970). *Phys. Rev.*, **D2**, 2963

Mueller, A.H. (1981). *Phys. Rep.*, **73**, 237

Mulders, P.J. (1997). *Nucl. Phys.*, **A622**, 239c

Niinikoski, T.O. (1980). In *Proc. 4th Intl. Symp. on High Energy Spin Physics*, Lausanne, p. 91

Niinikoski, T.O. and Rossmanith, R. (1985). In *Proc. Symp on Polarized Antiprotons*, Bodega Bay (AIP Conf. Proc. No. 145), p. 244

Onel, Y., Penzo, A. and Rossmanith, R. (1986). In *Proc. Workshop on Antimatter Physics at Low Energies*, Fermilab, April 1986

Particle Data Group (1978). *Phys. Lett.*, **75B**, 1

Particle Data Group (1982). *Phys. Lett.*, **B111**, 1

Peierls, R.F. and Trueman, T.L. (1964). *Phys. Rev.*, **134**, B1365

Placidi, M. and Rossmanith, R. (1985). LEP Note, 545 (CERN, Geneva)

Prepost, R. (1996). In *Proc. 12th Intl. Symp. on High Energy Spin Physics*, Amsterdam 1996 (World Scientific, Singapore), p. 127

Qiu, J. and Sterman, G. (1992). *Nucl. Phys.*, **B378**, 52

Qiu, J. and Sterman, G. (1999). *Phys. Rev.*, **D59**, 014004

Ralston, J.P. and Soper, D.E. (1979). *Nucl. Phys.*, **B152**, 109

Ramberg, E.J. *et al.* (1994). *Phys. Lett.*, **B338**, 403

Ramsey, G.P. and Sivers, D. (1991). *Phys. Rev.*, **D43**, 2861

Ranft, J. and Ranft, G. (1978). *Phys. Lett.*, **77B**, 309

Ratcliffe, P.G. (1983). *Nucl. Phys.*, **B223**, 45

Ratcliffe, P.G. (1998). In *Proc. 8th Intl. Symp. on High Energy Spin Physics*, Protvino 1998, No. 218 (Protvino, Russia), p. 210

Rose, M. (1957). *Elementary Theory of Angular Momentum* (Wiley)

Roser, T. *et al.* (1991). *Nucl. Instrum. Meth.*, **A301**, 42

Rossmanith, R. and Schmidt, R. (1985). *Nucl. Instrum. Meth.*, **A236**, 231

Sen, A. (1983). *Phys. Rev.*, **D28**, 860

Sinclair, C.K. (1988). In *Proc. 8th Intl. Symp. on High Energy Spin Physics*, Minneapolis (AIP Conf. Proc. No. 187), p. 1412

Sivers, D. (1990). *Phys. Rev.*, **D41**, 83

Sivers, D. (1991). *Phys. Rev.*, **D43**, 261

Smith, A., *et al.* (1987). *Phys. Lett.*, **B185**, 209

Soffer, J. (1995). *Phys. Rev. Lett.*, **74**, 1292

Soffer, J. (1999). In *Proc. Hyperon 99 Phys. Symp., Fermilab, September 99*

Sohnius, M.F. (1985). *Phys. Rep.*, **128**, 41

Sokolov, A.A. and Ternov, I.M. (1963). *Dokl. Akad. Nauk. USSR*, **153**, 1052 (*Sov. Phys. Dokl.*, **8**, 1203 (1964))

Sridhar, K. and Leader, E. (1992). *Phys. Lett.*, **B295**, 283
Stapp, H.P. (1962). *Phys. Rev.*, **125**, 2139
Steenberg, N. (1953). Ph.D. thesis, Oxford University
Storrow, J. (1978). In *Electromagnetic Interactions of Hadrons*, Vol. 1, eds. A. Donnachie and G. Shaw (Plenum Press, New York), p. 263
Sudakov, V.V. (1956). *Sov. Phys. JETP*, **3**, 65
Teryaev, O.V. (1995). In *Proc. Zeuthen Workshop on Prospects of Spin Physics at HERA* (DESY 95-200), p. 132
Teryaev, O.V. and Müller, D. (1997). *Phys. Rev.*, **D56**, 2607
Thomas, G. (1969). Thesis, University of California, Los Angeles
Thomas, L.H. (1926). *Nature*, **117**, 514
Thomas, L.H. (1927). *Phil. Mag.*, **3**, 1
't Hooft, G. and Veltman, M., (1972). *Nucl. Phys.*, **B44**, 189
Timmermans, J. (1998). In *Proc. XVIII Intl. Symp. on Lepton–Photon Ints.*, Hamburg 1997 (World Scientific, Singapore)
Trueman, T.L. and Wick, G. (1964). *Ann. Phys.*, **26**, 322
Trueman, T.L. (1996). RHIC Detector Note (Brookhaven National Laboratory, January 1996)
Uhlenbeck, G.E. and Goudsmit, S. (1925). *Nature*, **117**, 264
Underwood, D.G. (1977). Argonne Preprint ANL-HEP-PR-77-56, unpublished
Vogelsang, W. (1993). Ph.D. thesis, University of Dortmund: DO-TH 93/28
Vogelsang, W. (1996). *Phys. Rev.*, **D54**, 2023
Vogelsang, W. (1998). *Phys. Rev.*, **D57**, 1886
Vogelsang, W. and Weber, A. (1992). *Phys. Rev.*, **D45**, 4069
Vogelsang, W. and Weber, A. (1993). *Phys. Rev.*, **D48**, 2073
Wang, L. (1966). *Phys. Rev.*, **142**, 1187
Watson, A.D. (1982). *Z. Phys.*, **C12**, 123
Weinberg, S. (1964a). *Phys. Rev.*, **133**, B1318
Weinberg, S. (1964b). *Phys. Rev.*, **134**, B882
Werle, J. (1963). *Nucl. Phys.*, **44**, 579
Werle, J. (1966). *Relativistic Theory of Reactions* (North Holland, Amsterdam)
Wick, C.G. (1962). *Ann. Phys.*, **18**, 65
Xu, Z. *et al.* (1987). *Nucl. Phys.*, **B291**, 392
Yang, C.N. (1950). *Phys. Rev.*, **77**, 242

Index